IMAGE PROCESSING:
THEORY
AND APPLICATIONS

IMAGE PROCESSING: THEORY AND APPLICATIONS

Proceedings of IPTA '93
San Remo, Italy, 14-16 June, 1993

edited by

GIANNI VERNAZZA
DIBE-University of Genoa
Genoa, Italy

ANASTASIOS N. VENETSANOPOULOS
DEE-University of Toronto
Toronto, Canada

CARLO BRACCINI
DIST-University of Genoa
Genoa, Italy

1993
ELSEVIER
AMSTERDAM • LONDON • NEW YORK • TOKYO

ELSEVIER SCIENCE PUBLISHERS B.V.
Sara Burgerhartstraat 25
P.O. Box 211, 1000 AE Amsterdam, The Netherlands

ISBN: 0 444 89969 3

© 1993 Elsevier Science Publishers B.V. All rights reserved.

No part of this publication may be reproduced, stored in a retrieval system or transmitted in any form or by any means, electronic, mechanical, photocopying, recording or otherwise, without the prior written permission of the publisher, Elsevier Science Publishers B.V., Copyright & Permissions Department, P.O. Box 521, 1000 AM Amsterdam, The Netherlands.

Special regulations for readers in the U.S.A. – This publication has been registered with the Copyright Clearance Center Inc. (CCC), Salem, Massachusetts. Information can be obtained from the CCC about conditions under which photocopies of parts of this publication may be made in the U.S.A. All other copyright questions, including photocopying outside of the U.S.A., should be referred to the copyright owner, Elsevier Science Publishers B.V., unless otherwise specified.

No responsibility is assumed by the publisher for any injury and/or damage to persons or property as a matter of products liability, negligence or otherwise, or from any use or operation of any methods, products, instructions or ideas contained in the material herein.

pp. 177-180, 193-196, 263-266, 321-324, 385-388, 447-450: Copyright not transferred

This book is printed on acid-free paper.

Printed in The Netherlands.

CHAIRMEN's MESSAGE

It is our pleasure to invite you to participate in IPTA '93 to be held in the beautiful town of San Remo, on the Italian Riviera. The Conference venue will be the Casino Theatre (courtesy of San Remo's Municipality), located in the center of San Remo. The Welcome Reception will be held at the Villa Nobel (courtesy of Imperia's Province), where you will be the guests of San Remo's Authorities.

The response to IPTA call-for-papers has been remarkable; therefore, we could not accept all contributions. At the end of the review process, about one hundred papers were selected, with a rejection rate of about 30%.

Contributions have been allocated for oral or poster presentations, not on the basis of quality but rather according to their characteristics with respect to presentation. In particular, poster presentations will engage authors for the whole duration of the session, but will have the advantage of more direct discussions with interested attendees.

Seven Oral Sessions and four Poster Sessions have been organized. In addition, the conference will be highlighted by seven Invited Lectures given by outstanding experts.

Accepted papers cover almost all aspects of image processing. For organization purposes, they have been grouped into sessions according to the following categories:
- Image Coding
- Image Segmentation and Feature Extraction
- Image Filtering
- Image Classification
- Neural Networks for Image Processing
- Applications of Image Processing

Two Special Sessions will also be held, one on Hybrid Systems and the other on Industrial Applications of Image Processing. Hybrid Systems might represent a key to the solution of complex image processing and classification problems, which often require the integration of different approaches to data representation and processing. A fundamental task in the '90s will be the transfer of Image Processing theories and methodologies to the industrial environment: this will be the subject of the second Special Session.

We wish to thank all the contributors to IPTA, including, but not limited to, all authors, reviewers, sponsors, Elsevier Science Publishers, and the personnel of the University of Genoa. Special thanks are due to all participants, who will make IPTA a successful Conference.

We're looking forward to warmly welcoming you at IPTA.

Prof. G.Vernazza, General Chair
Prof. A.N.Venetsanopoulos, Program Chair
Prof. C.Braccini, Technical Committee Chair

SPONSOR LIST:

AEI Italian Association of Electrical and Electronic Engineers
(Liguria Chapter)

CEC Commission of the European Community - Directorate-General XII
Brite EURAM Program

EURASIP European Association for Signal Processing

GTTI Italian Group of Telecommunications and Information Theory
(National Research Council - CNR)

IAPR Italian Association for Pattern Recognition
(Italian Chapter)

IEEE The Institute of Electrical and Electronics Engineering
(North-Italy Chapter)

CONFERENCE COMMITTEE

General Chair

Gianni Vernazza　　　　　　　　　DIBE, University of Genoa
　　　　　　　　　　　　　　　　　Via Opera Pia 11A, Genoa, I-16145 Italy
　　　　　　　　　　　　　　　　　Ph:　+39-10-353-2755
　　　　　　　　　　　　　　　　　Fax:　+39-10-353-2777
　　　　　　　　　　　　　　　　　e-mail: vernazza@dibe.unige.it

Program Chair

Anastasios N. Venetsanopoulos　　　DEE, University of Toronto
　　　　　　　　　　　　　　　　　35 St. George Str., Toronto, Canada
　　　　　　　　　　　　　　　　　Ph:　+1-416-978-8670
　　　　　　　　　　　　　　　　　Fax:　+1-416-978-7423
　　　　　　　　　　　　　　　　　e-mail　anv@dsp.com.toronto.edu

Technical Committee Chairman

Carlo Braccini　　　　　　　　　　DIST, University of Genoa
　　　　　　　　　　　　　　　　　Via Opera Pia 11A, Genoa, I-16145 Italy
　　　　　　　　　　　　　　　　　Ph:　+39-10-353-2708
　　　　　　　　　　　　　　　　　Fax:　+39-10-353-2948
　　　　　　　　　　　　　　　　　e-mail: carlo@dist.unige.it

Technical Committee

M.Barlaud, France　　　　　　　　　J.Biemond, The Netherlands
H.Bunke, Switzerland　　　　　　　 V.Cappellini, Italy
L.P.Clarke, U.S.A.　　　　　　　　 T.S.Durrani, U.K.
T.Hinamoto, Japan　　　　　　　　　G.Jacovitti, Italy
J.Kittler, U.K.　　　　　　　　　　S.Levialdi, Italy
T.Matsuyama, Japan　　　　　　　　 S.K.Mitra, U.S.A.
A.Murat Tekalp, U.S.A.　　　　　　 Y.Neuvo, Finland
H.Niemann, Germany　　　　　　　　 I.Pitas, Greece
F.Rocca, Italy　　　　　　　　　　 H.Samet, U.S.A.
G.L.Sicuranza, Itlay　　　　　　　 I.Sobel, U.S.A.
P.Zamperoni, Germany

IPTA '93
San Remo, Italy, June 14-16, 1993

TABLE OF CONTENTS

Chairmen's message v

INVITED LECTURES

M.Kunt
Recent HDTV Systems 1

T.S.Huang
Polynomial Methods in Motion Analysis 13

J.W.Woods, J.Kim
Motion Compensated Spatiotemporal Kalman Filter 19

C.H.Chen
Artificial Neural Networks in Pattern Recognition and Signal Classification 25

B.Chandrasekaran, J.Josephson
Form and Content Issues in the Abductive Framework for Recognition 29

M.R.Coble, D.C.Munson
Inverse SAR Planetary Imaging 31

P.E.Trahanias, A.N.Venetsanopoulos
Multispectral Image Processing 41

IMAGE CODING I (Chair: S.K.Mitra)

D.M.Monro
Fractal Transforms: Complexity versus Fidelity 45

R.Baseri, J.W.Modestino
Region-Based Coding of Images Using a Spline Model 49

X.Song, Y.Neuvo
Compatible HDTV Coding Using MDPCM and VQ 53

IMAGE CODING II (Chairs: M.Barlaud and J.W.Modestino)

W.Li, M.Kunt
Coding of Motion Compensated Prediction Error Images by Morphological Segmentation 59

E.Takemasa, T.Komatsu, T.Saito
Two Dimensional Delayed-Decision Lossy Image Coding via Copying and the Application to Intraframe Low Rate Image Compression 63

D.W.Redmill, N.G.Kingsbury
Improving the Error Resilience of Entropy Coded Video Signals 67

G.Poggi
Low Bit-Rate Image VQ with Progressive Transmission 71

T.H.Yu, S.K.Mitra
Perceptually Based Contrast Enhancement Algorithm 75

IMAGE SEGMENTATION AND FEATURE EXTRACTION I (Chairs: H.Bunke and M.Petrou)

S.Urago, M.Berthod, J.Zerubia
Restoration of Incomplete Contour Images Using Markov Random Fields 79

G.J. Brelstaff
A Coarse-to-Fine Scheme for Region Growing Using Statistical Tests 83

H.Kalviainen
Computational Considerations on Randomized Hough Transform and Motion Analysis 87

A.J.Abrantes, J.S.Marques
A New Algorithm for Active Contours 91

M.Salotti, C.Garbay
Cooperation Between Edge Detection and Region Growing: The Problem of Control 95

IMAGE CODING III (poster)

S.Yao, R.J.Clarke
Motion-Compensated Wavelet Coding of Color Images Using Adaptive Vector Quantisation 99

M.Corvi, E.Ottaviani
High-Rate Still-Image Compression by DWT Coding 103

F.Bellifemine, A.Chimienti, R.Picco
Motion Compensation for Compatible Multiresolution Coding
Schemes 107

C.F.Harris, J.W.Modestino
Entropy-Constrained Design of an Image Coder with a Composite
Source Model 111

S.Simon, I.Bruyland
Layered and Packetized Subband Coding for Compatible Digital
HDTV 115

B.Aiazzi, L.Alparone, S.Baronti, A.Casini, F.Lotti
Pyramid Structures for Hierarchical Content-Driven Image Coding 119

G.Poggi
Fast VQ Codebook Design with the Kohonen Algorithm 123

L.Boroczky, K.Fazekas, P.Csillag
Advanced Motion Estimation for Subband Coding of Image Sequences 127

C.Braccini, S.Curinga, A.Grattarola, F.Lavagetto
3D Modeling of Human Heads from Multiple Views 131

F.G.B.DeNatale, G.S.Desoli, S.Fioravanti, D.D.Giusto
A Novel Approach to Data Interpolation for Image-Data Compression 137

IMAGE SEGMENTATION AND FEATURE EXTRACTION II (poster)

R.O.H.Thami, J.P.Asselin de Beauville
Contour Detection Using the Valleys of a Similarity Histogram 143

A.Broggi, G.Destri
Expectation-Driven Segmentation: A Pyramidal Approach 147

L.Najman, M.Schmitt
Definition and Some Properties of the Watershed of a Continuous
Function 151

J.Hornegger, D.W.R.Paulus
Detecting Elliptic Objects Using Inverse Hough-Transform 155

D.Chetverikov
Multiresolution Detection of Texture Defects for Surface Inspection . 159

S.Denasi, G.Quaglia
The Contribution of Region Segmentation Maps to the Localization
of Geometrical Structures in Real Scene Images 163

V.Murino, C.S.Regazzoni, G.L.Foresti
A Markov Random Field Approach to Grouping of Descriptive
Primitives 167

Q.Huang, G.C.Stockman, A.J.M.Smucker
A New Perspective on Segmentation: Token Grouping at Multiple
Abstract Levels 173

I.Pitas, C.Lialios
Seismic Image Segmentation Techniques 177

G.Venturi
Edge and Region Integration for Image Segmentation 181

J.Santos-Victor, J.Sentieiro
Image Matching for Underwater 3D Vision 185

IMAGE FILTERING I (Chairs: G.Jacovitti and I.Pitas)

Y.Zhang, D.Hatzinakos, A.N.Venetsanopoulos
Identification of Multichannel and Multidimensional Systems Using
Cumulants: Application to Colour Images 189

I.Pitas, P.Kilindris
Application of Multichannel Two-Dimensional AR Modelling to
Color Image Processing 193

M.Petrou, S.Yusof
Separable Optimal Filters for Edge Detection 197

P.Carrai, G.M.Cortelazzo, G.A.Mian
On the Design of FIR Filters for Scanning Rate Conversion 201

J.M.Jolion
Contrast Enhancement Through Multiresolution Decomposition 205

NEURAL NETWORKS FOR IMAGE PROCESSING (Chairs: O.K.Ersoy and G.L.Sicuranza)

D.Mingyue, F.M.Wahl
3-D Reconstruction of Line-Like Objects in Trinocular Vision 209

C.H.Chen, G.H.You, P.-S.King
Pattern Wafer Image Segmentation Using Neural Networks 213

B.G.Mertzios, D.Mitzias
Fast Shape Discrimination with a System of Neural Networks Based
on Scaled Normalized Central Moments 219

O.K.Ersoy, R.Sundaram
A Multistage Neural Network Approach to Image Restoration 223

E.Costamagna, A.Podda, A.Turno, A.Vargiu
Image Compression by Hadamard Transform and Neural Network
Spectrum Extrapolation 227

IMAGE FILTERING II (poster)

T.Hinamoto, M.Muneyasu
A Simplified 2D IIR Digital Filter and its Application to Image Processing 231

B.Bundschuh
Nonlinear Methods for the Restoration and Reconstruction of Images with Quasilinear Solutions 235

R.D.Morris, W.J.Fitzgerald
Detection and Correction of Speckle Degradation in Image Sequences Using a 3D Markov Random Field 239

M.Hamdi, R.W.Hall
Embedding Wavelet Transforms Onto Parallel Architectures 243

J.De Vriendt
Analysis of the Influence of Sampling Quantization and Noise on the Performance of the Second Directional Derivative Edge Detector 247

K.Tsirikolias, B.G.Mertzios
Edge Extraction and Enhancement Using Coordinate Logic Filters 251

M.E.Zervakis
A Structured Regularized Approach in Multichannel Image Restoration 255

Y.Iwata, M.Kawamata, T.Higuchi
Design of VLSI Array Processor for Very High Speed 2-D Digital Filtering of High Definition Images 259

L.Bedini, M.G.Pepe, E.Salerno, A.Tonazzini
Non-Convex Optimization for Image Reconstruction with Implicitly Referred Discontinuities 263

S.Bataouche, J.M.Jolion
Asyncronous Pyramids 267

L.Alparone, S.Baronti, R.Carlà
A Nonlinear Adaptive Filter for Edge-Preserving Reduction of Additive or Multiplicative Image Noise 271

N.Pendock
Prior Distributions for Tomographic Reconstruction 275

K.Tang, J.Astola, Y.Neuvo
Multivariate Estimation Based on Gradient Search 279

R.Yang, L.Yin, M.Gabbouj, Y.Neuvo
Optimal Weighted Median Filtering for Image Restoration 283

A.M.Guertz, R.Leonardi, M.Kunt
Hierarchical Fitting of Constrained Composite Curves 287

R.Suoranta, K.-P.Estola
2D Subset Averaged Median Filters 291

IMAGE CLASSIFICATION (Chairs: H.Niemann and S.Vitulano)

F.Samaria, F.Fallside
Face Identification and Feature Extraction Using Hidden Markov Models 295

L.Najman, R.Vaillant, E.Pernot
From Face Sideview to Identification 299

H.Bunke, B.T.Messmer
A New Algorithm for Efficient Subgraph Matching 303

C.J.Pye, J.A.Bangham, S.J.Impey, R.W.Aldridge
2D Pattern Recognition Using Multiscale Median and Morphological Filters 309

G.Jacovitti, A.Minniti, A.Neri, T.Tasselli
Imaging of Vehicles with Linear Array 313

APPLICATIONS OF IMAGE PROCESSING (Chairs: V.Cappellini and L.Fiore Donati)

Y.B.Karasik
Optical Image Processing Technique Speeds Up Algorithms of Computer Graphics 317

V.Cappellini, A.Chiuderi, S.Fini
Neural networks for remote sensing data fusion 321

S.Vinitski, S.Seshagiri, F.B.Mohamed, G.Frazer, A.E.Flanders, C.F.Gonzalez, D.G.Mitchell
Tissue Characterization by MRI: Data Segmentation Using 3-D Feature Map 325

L.P.Clarke, W.Qian, H.Li, R.Clark, E.Saff, B.Lucier
Mammographic Image Analysis with Tree Structured Order Statistic and Wavelet Filters 329

J.Bertrand, P.Bertrand, M.Roisin
Bistatic Radar Imaging in Laboratory 333

IMAGE CLASSIFICATION AND APPLICATIONS (poster)

M.Barni, V.Cappellini, M.Mattioli, A.Mecocci, S.Zipoli
Knowledge-Map Based Automatic Inspection and Change Detection in Complex Heterogeneous Environments 337

C.Di Ruberto, N.Di Ruocco, S.Vitulano
A Structural Approach in Stereo Vision 341

J.Flusser, T.Suk
Classification of Objects and Patterns by Affine Moment Invariants . 345

T.Saito, T.Komatsu, K.Aizawa
On Substantial Improvements in an Image-Processing-Based High-Resolution Imaging Scheme Using Multiple Imagers 349

L.Raffo, D.D.Caviglia, L.DeVena
Neural Defect Detection in Magnetic Particle Inspection Test's Images 353

S.Carrato
Deinterlacing of HDTV Images Using a Two-Layer Perceptron 357

P.Cosquer, M.Cattoen
System of Active Coded Targets for Motion Analysis 361

G.K.Matsopoulos, S.Marshall, J.C.H.Mackie, S.E.Solomonidis
3-D Imaging of the Residual Limb of an Above-Knee Amputee Using Morphological Edge Detection Algorithms 365

A.Kaup, T.Aach
A Spatial-Frequency Selective Approach for Modelling the Phenomenon of Perceptual 'Filling in' in Human Vision 369

R.Schettini
Object Identification Using Shape and Color Information 373

N.Iwama, M.Teranishi
GCV-Aided Linear Image Regularizations in Sparse-Data Computed Tomography and Their Applications to Plasma Imaging 377

J.Heikkonen
Recovering 3-D Motion from Optical Flow Field 381

P.A.Brivio, I.Doria, E.Zilioli
Analysis of Semivariograms in Satellite Images for Territorial Applications 385

G.Benelli, A.Garzelli, A.Mecocci
Radar Image Processing for Automatic Traffic Control of Sea-Ports .. 389

E.Binaghi, A.Mazzetti, A.Rampini
Integration of Neural Network Techniques in Fuzzy-Logic-Based Classification of Multi-Source Remote Sensing Data 393

F.Pedersini, S.Tubaro
Accurate Calibration of a Binocular Stereoscopic System and Its Application to 3D Object Measurements 397

M.Atiquzzaman, M.G.Hartley
Extraction of Road Traffic Data from Image Sequences 401

R.P.Velthuizen, L.P.Clarke, M.L.Silbiger
Measuring Tumor Volume Using Pattern Recognition in Magnetic Resonance Images 405

L.P.Clarke, W.Qian, H.Li
Nonlinear Filter and Neural Network for Nuclear Medicine Image Maximum Entropy Restoration 409

I.Erenyi, T.Holka, Z.Fazekas, A.Dekany
Automatic Honey Qualification via Computerizd Microscopy 413

S.B.Serpico
Feature Selection in Neural Networks for Image Classification 417

S.Dellepiane, F.Fontana
Isocontour Detection from Context-Dependent Region Growing 423

A.Chella, R.Pirrone, F.Sorbello
Neural Implementation of Shape from Shading Algorithms 429

A.Jozwik, F.Roli, C.Dambra
A Multistage Synthesis of Modified NN Rule and Its Application for Remote-Sensing Images 435

A.Blumenkrans, G.Viano
Application of Moiré Fringes Techniques to the Inspection of Aluminum Bars 439

M.A.T.Figueiredo, J.M.N.Leitao
Image Restoration Using a Standard Hopfield Neural Network 447

D.Anguita, G.C.Parodi, R.Zunino
Associative Reduction of Dimensionality in Image Classification 451

Author Index 455

Recent HDTV Systems

Murat Kunt

Signal Processing Laboratory, Swiss Federal Institute of Technology, CH-1015 Lausanne, Switzerland

Invited Paper

1. Introduction

This paper will duscuss a few main issues related to recent developments on high definition television (HDTV) systems, including a brief description of selected systems from Europe, USA and Japan. It has been previously presented at various conferences published a few times following the corresponding invitations [1] - [4]. Before discussing high definition, let us have a look to our good old regulat TV system. Doubtlessly, the best tool man ever made for mass education is television. It is also the most succesful commercial product: there are more than 1 billion TV sets on the earth (more than telephones!) using a 50 years ols design. The main issue in TV is that of programs. Instead of putting all the necessary efforts into high-quality programs for better education, look the mess we have done with it. We broadcast greed, violence, hate, injustice, religious crookery, Hollywood bad taste and, if there is time left, some superficial and biased information. Our good old TV system is undergoing important changes to improve its technical quality, and what may happen is to end up with much better images showing poorer quality programs. It may well be another example of improving technicality, but decreasing global quality. Some other well-known cases are fast-foods, 8 mm films which become super 8 and then become video with increased practicality but poorer quality, without forgetting publishing and book writing. Today everybody writes, nobody reads, and the same is true for scientific papers as well. 'Cut & paste' helps proliferation and kills quality. Nowaday, films are still edited by mechanical cut and paste, which is rather time consuming. If ever this becomes electronic and more efficient, imagine how many cheap and bad-taste films one can produce per week to pollute all the TV channels around the world.

2. Short history of TV

The first attempt to try to convert a visual scene into a electrical signal is done by Nipkow in 1884. From improvement to perfection, we move to the iconoscope of Zworkin in 1923, and then to the black and white TV tubes in 1941, and finally to colour in 1950. Let us summarize in a few words how a TV system works. It is a typical communication system including a transmitter, a channel and a receiver. To be more specific, consider figure 1, where such a system is shown. The first fact is that the 4-D space we are living in (three space variables and one time variable) is compressed in the system to a 1-D signal which is transmitted in the channel, and then expended to a 3-D signal which is displayed in front of the viewer. One dimension is lost by projecting the real world into an image plane, and the remaining 3-D space is sampled in time (thirty images per second in the US or 25 images per second in Europe). 2-D image space is scanned line by line producing a signal called video signal. If the line by line scanning and the time sampling is repeated fast enough, one creates the illusion of a continuity to fool the eye.

3. Video signal

Since we are transmitting the video signal over the channel, it is interesting to determine its bandwidth. Assuming a square screen to display the image of size N x N points refreshed every F second, the highest frequency we can have is $N(N/2)F = N^2F/2$. For example, for 25 frames per second and 625 lines per screen, the base bandwidth of the video signal is around 5 MHz. Notice that so far there is no NTSC, PAL or SECAM. The previous bandwidth represents a black and white image. Colour was introduced through 3 primaries. There are either additive primaries like red, green and blue (RGB), or substractive primaries, like yellow, magenta and cyan. A combination of either group of primaries may reproduce any desired colour. Furthermore, there is no necessity to work necessarily with these primaries. Any reversible three by three transformation is acceptable. For example, from the three additive primaries, R, G, B, we can obtain three other composants, called Y, I, Q, (used in the NTSC system) or another set D,H,S corresponding to the intensity, hue and saturation, respectively. We can also have another transformation leading to one luminance and to chrominance signals. Accordingly, we can have a colour TV by using three different video signals, one for each basic colour primary. Why one should change the basis of colour representation from basic primaries to any other transformation ? The main reason was that of compatibility. To make the new colour television, compatible with the existing black and white TV system, it was preferable to convert the RGB primaries into one luminance and two chrominance signals. Thus, the black and white receiver can receive only the luminance component, and allowed viewer to watch a colour program in black and white or a black and white program can be broadcasted and received with both black and white and colour TV sets. If the base-band of a black and white video signal is 5MHz, one may think that for colour signal requiring three such signals, the bandwidth should be 15 MHz. In fact, because of the line repetition or frame repetition, video signal is a quasi periodical signal having a line spectrum. Thus, it is possible to include in the original base-band two more colour components without increasing the bandwidth. Different TV systems, such as NTSC, PAL and SECAM, are in fact various ways of modulating the basic signal for transmission without increasing its bandwidth. In modulation, a parameter of a transmissible signal is varied as a function of the instantaneous value of the signal we want to transmit. There are three such parameters : amplitude, frequency or phase. If amplitude modulation is used, then the bandwidth of the transmitted signal is increased by factor of 2 with respect to the base-band of the video signal. It is possible to use single-side-band type modulation, but because of the very important DC component in the video signal, the prefered modulation scheme is the so-called vestigion side band modulation. Figure 2 shows the spectrum of the luminance signal modulated for transmission with the main carrier and audio carrier corresponding to the European system. In NTSC, the two chrominance signals I and Q modulate in amplitude and phase a seconder carrier situated in the upper part of the band at a frequence which enterleaves various frequency components. The PAL system tries to remedy the NTSC incovenience of being sensitive to phase distorsions, so that the phase of one of the chrominance signals is alternated every two lines. In 625 lines systems, the frequency of the chrominance carrier is higher by 4.4 MHertz to that of the main carrier. The SECAM system assumes that the chrominance signal does not need the same spacial resolution as the luminance signal. Chrominance resolution is divided by a factor of two. Thus, at each line of the luminance video signal, one transmits only one of the two chrominance signal which alternates with the other chrominance signal at the next line. At the receiver, one of them needs to be delayed (64 msec.) to recombine them. In contrast with PAL and NTSC, SECAM uses a frequency modulation for chrominance signals with two carriers situated around 4.2 MHz.

4. What is wrong with the existing TV system ?

The first point is the synchrony. Because of the non-availabilty of image memories in the early years of television, the present days system, all the way from the camera to the display in front of the viewer, is a synchronous system. It requires high rates for motion rendition, but waste bandwidth for transmission. The second point is unbalanced resolutions. For a given bandwidth, the gross video signal trades-off time resolution with space resolution. Today's systems for motion rendition at 50 or 60 fields per second are rather poor on vertical resolution. Endless experience at the movie theater shows that higher space resolution is preferable at the price of lower temporal resolution (24 frames per second). Yet another point is related to the frequency multiplex. Frequency multiplex was well mastered at least on theory at the turn of the century. Although selective filters can be built, frequency multiplex led to the so-called composite signal, which with cheap filters in the receivers are at the origin of luminance used as chrominance and vice versa. Another point of criticism could be made on interlace. Introduced for better motion rendition, at the expense of poorer vertical resolutions, interlace creates more problem than what is solves. It does not perform as well as expected, emphasizes interline flicker, has poorer vertical motion definition, and complicates transcoding. With image memory in receivers, interlace should not survive in the future. Another point is related to the vestigial side band modulation, an obsolete technique leading to a larger RF bandwidth than the base bandwidth. Vestigial side band modulation should be replaced either by quadrature modulation or frequency division multiplex. Gamma correction is a measure of the deviation from an ideal input/output characteristics at the grey level of the entry signal. This correction is introduced prior to encoding. It violates the constant luminance principle mixing luminance and chrominance. Today's TV with the aspect ratio of 4/3 gives to the viewer a quite different viewing experience than of the movie theater. A more rectangular screen (aspect ratior 16/9) will close this gap. Surprisingly, this point seems to be the only one on which there is a planetary consensus. There are many more wrong points one could list. However, since there are other consequences of the same basic principles, we shall not pursue the list. There are many improvements we can think of that could be incorporated into a new system. For example, receivers may have storage, avoiding interlace and relating frame rate to the motion rendition. We can do better signal processing (VLSI). One may introduce and expand programmability, which facilitates transcoding and upgrading. Frequency multiplex could be replaced by time multiplex avoiding cross talk. More contemporary modulation techniques could be used to avoid increase of bandwidth between the base band and radio frequency. Modern vision knowledge may be introduced in the design of a new system. We may pay attention to respect the constant luminance principle and many more small improvements could be added to such a list. Anyway, a TV system should not be viewed as the one going from the camera to the display. It obviously includes the real world before the camera and the viewer before the display. Accordingly, such a system should be investigated end to end including the real world and the viewer. Due to the foreseeble future of multimedia services, attention should be payed for them, emphasizing programmability and thus flexibility. Furthermore, technological updates should not affect the functions and should keep an extremely user-friendly receiver.

5. High definition television

Let's move now to high definition television. It became such a hot political and industrial topic that one can find it in almost everyday's newspapers. Three continents, Japan, Europe, and United States, are competing to establish a world's standard for the next generation of TV system. Standard is necessary whenever there are more than one solution competing for the same problem. In the case of television, there will be very likely standards on production, transmission, recording, copying, and so fort. Figure 3 shows the size of the actual TV image and the new high definition TV image with equal space resolution. In figure 4, the two systems are shown for equal surfaces, but with different aspect ratios. The new TV screen will be more rectangular with an aspect ratio of 16 to 9. The essential parameters of a new system are given as a succession of 3 numbers. The first one is the number of line per screen height. The second one is the number of images per second, and the third one is the number of frames per second. For example, the European proposition is characterized by 1250-50-50.

6. The MUSE system

The Japanese HDTV system is the result of long-range efforts started in late seventies within a program called "high vision", especially under the leadership of the state owned broadcast company NHK (Nippon Hoso Kyokai). First prototype demonstration was available around 1986. It was almost a world standard as proposed at the CCIR meeting in Dubrovnik. Sample-minded compression scheme gives the name of the system MUSE standing for multiple subnyquist sampling encoding. This system has been designed for satellite broadcasting in the bands of 12 GHz. The transmission system adopted is analog frequency modulation requiring a transmission bandwidth of 27 MHz for a base band within 8 MHz. The video signal of the MUSE system has 1125 lines, 30 frames and 60 fields per second. It is an interlaced scan. The video signal is processed component by component, chrominance signals being compressed through the so-called time compressed integration. Compression by a factor of 4 is used to fit chrominance signal in the line blanking interval. The digital sound is inserted in the field blanking interval. Figure 5 shows the block diagram of the MUSE transmitter. The time compressed integrated signal is sampled at 64.8 MHz and processed to determine fix areas and mobile areas. Each type of area is a sub-sample accordingly to fit the global signal within the 8 MHz bandwidth. Fix areas are sub-sampled by a factor of 4 so that 4 fields are needed to reconstruct the full resolution of scene. For mobile areas, this type of sub-sampling cannot be used because during 4 fields the motion can be very important and may lead to blur. For mobile areas, the motion is transmitted field by field, but with much less spatial resolution. Figure 6 shows the spatial resolution of this system, as well as the overlaps induced by sub-sampling. Because of the poor motion rendition of this basic system, it has been improved by reducing the spatial resolution, so that the time compressed integration is not sampled at 64.8 MHz, but for the 8.6, reducing the sub-sampling to a ratio of 3 instead of 4. Figure 7 summarizes various processing steps of this system for mobile and fix areas. The transmissible luminance information corresponding to this new scheme and its spectrum before and after sub-sampling are shown at figure 8. Probably, the greatest merit of the MUSE system is its existence. In 1991, they already have daily broadcast for about one or two hours a day, which has been increased to 8 hours since January 1992. Its second merit is that it driggered an international competition between Europe, Japan and the US, for the development of a new television system.

7. The HD-MAC system

European answer to the MUSE system was the set up of a European project called EUREKA 95, which came up with an equivalent system called HD-MAC. In contrast with the MUSE system, which is not compatible with the existing TV systems, like NTSC, PAL or SECAM, HD-MAC has been introduced in 1988 with the argument of compatibility. It refers to compatibility with a non existing D2-MAC system, which is even today not yet fuly operational. The D2-MAC system is a component system, in which luminance and chrominance signals are sampled and time-compressed to fit within 64 μsec. of one video scan-line, including audio and data. Because of time compression, the initial bandwidths of 5.6 MHz and 1.6 MHz for luminance and chrominance, respectively, increase to 8.4 and 4.8, requiring a global sampling frequency of 20.25 MHz. The D2-MAC system is the result of an evolution lasted for about ten years, starting with A-MAC, B-MAC, etc. The HD-MAC system is so-to-speak an upscaled version of D2-MAC. Accordingly by under-sampling high-resolution video signal one can bring it down to D2-MAC format. The input to HD-MAC system is a video signal with 1250 pictures lines, interlaced, scanned with 50 fields and 25 frames. This signal is sampled in a quinconce manner as shown in figure 9a. Then the resulting picture point set is decimated into 4 subsets using odd and even numbered lines and columns. After some special arrangement, this subset became the fields of the D2-MAC system, which could be recombined to make the interlaced HD-MAC frames. This scheme is for the still picture case, where there is no motion. The initial signal is processed to determine three categories of areas in each picture : fix areas, slowly moving areas and moving areas. Each area is analyzed according to these three hypothesis and processed for compression using a trade-off between time resolution and space resolution. The results are then compared to the original area and a decision is made to put the appropriate level where the similarity is the highest. Figure 10 shows the space and time resolution of the HD-MAC system. Compared to MUSE, HD-MAC system reveals similarities as striking as those between the Tupolev and Concorde supersonic airplanes! 1992 winter and summer Olympic Games are broadcasted to some view points through HD-MAC system. Although there are not yet regular daily broadcast, one may consider HD-MAC as close to existence. Both MUSE and HD-MAC systems have very simple-minded data compression schemes and obsolete analog transmission modes.

8. American systems

At the first stages of the competition between Japan and Europe, America was a careful observer. The Federal Communication Commission, better known as FCC, fix the rules of the game. It is required that American high definition TV, HDTV, should be based on terrestrial transmission and should fit within the 6 MHz bandwidth as presently used for regular NTSC, with no interference to the existing NTSC system. In early 1989, there were many systems being developed in the US. We can mention the so-called Advanced Compatible Television or AC TV developped by David Sarnoff Center and Thomson Consumer Electronics. The North American Philips developped another system based on NTSC, as well as the GLEN system. Both of these sytems where using NTSC in one channel and an augmentation signal in the next one. The Zenith system was yet another candidate. Professor Schreiber designed two systems at MIT. Finally let us mention two versions of the MUSE system, one being incompatible and the other one compatible with NTSC receivers. These systems were mainly hybrid combining some analog and some digital signal processing and modulation. Probably, the most up-to-date system was the early MIT CC system, as proposed by Schreiber, based on subband coding. The FFC opened a competition which will be closed sometime in 1993, to find the best HDTV system for American needs. While the entire world was hesitating around various hybrid schemes, courageously engineers from the General Instrument Corporation came up with the first all-digital HDTV system. Then, three more opponents followed. Out of 6 systems competing today for FCC, four of them are all-digital.

The space in this paper is not enough to describe in detail all these four systems. Luckily, they share many common principles that we can summarize, mentioning briefly their specificities, which make them unique. Important points on these systems are data compression and channel coding. Data compression is important to reduce the number of bits necessary to transmit, to represent faithfully the original data, i.e. the original scenes. Because compressed data is much more sensitive to noise existing in the transmission channel, compressed data must be protected against such disturbances. This is done with the channel coding. The use of digital techniques in both areas allows flexibility and efficiency. The data compression can be subdivided into spatial data compression and time data compression. The first refers to the removal or the attenuation of the spatial correlation existing between picture points of the same frame, whereas the second one refers to the redundancy existing from one frame to the next. Although we know in theory that these two types of redundancy should not be processed independently for the best efficiency, as a good engineering practice, they will be processed independently. There are some old techniques to remove spatial correlation within a single image. Introduced some 20 years ago, the so-called linear transform coding techniques allow to compact the energy of the image signal in two specific areas of a transform domain and then assigning bits to each transform coefficient in a very specific manner compressions around 10 ot 1 or 15 to 1 are easily obtained with good quality reconstructed pictures. Since its early years, this technique has been continuously improved by intensive fine tuning, so that today recommendation from international standardization bodies exist on the label of JPEG. The JPEG coder is shown in Figure 11. Basically, the input image is subdivided into blocks of 8 by 8 pixels. Each block is transformed by the so-called discrete cosine transform (DCT). The transform coefficients are quantized according to some specifications and then coded for transmission. The decoder shown in Figure 12 implements the inverse operations to reconstruct the picture. The JPEG scheme is information lossy because of the quantization of transformed coefficients. If good specifications are used to take into account the properties of the human visual system, distorsions introduced may be very small. The use of transform on small picture blocks is justified by the statistical nature of image data.

To reduce temporal redundancy, it is easy to realize that from one frame to the next, there are great similarities since they are only 16 or 20 msec. apart. In such a case, there is no need to code each frame independently from the preceeding one. The common principle used is to fully encode the first frame and then code only the changes appearing in the forthcoming frames with respect to the first one. These changes are best described if the objects moving in the scene are identified. Such an identification, however, is a very difficult task even for today's most developped computer vision systems. Again, a good engineering approximation of this problem is to trace the motion of small image blocks, for example 8 by 8 from one frame to the next. Figure 13 shows the subdivision of a given frame with numbered blocks and the previous frame showing the corresponding position of each block. It is possible to go one step further by taking into account the previous situation, to predict where the given block will be in a third frame. If the prediction is good, there is no need to transmit additional information. If there is a slight prediction error, then it is sufficient to transmit this small error with a very reduced number of bits, to recover fully the position of this block in the third frame. This technique is known as motion compensation. A better description would be the coding of the prediction error. Accordingly, using the previous techniques, one can produce a sequence of error images containing the prediction error of block motions from one frame to the next. This is called the displaced frame difference sequence. A common procedure is to view each frame of this new error sequence as an image and apply the spatial decorrelation technique previously discussed, using for example a discrete cosine transform. A recommendation called MPEG is issued by the International Standardization Organisation for video coding at about 1.5 Mbit per second. The comment one can make is that the statistics of the displaced frame differences do not behave like those of a real image. So applying a technique which is suitable for one does not mean necessarily that it is also suitable for the other. As far as the transmission is concerned, digital modulation allows the protection of the messages so that transmission errors can be detected and even be corrected at the receiver.

Among the competing American systems, the Zenith AT & T consortium designed a system which uses the previous principles. It has a very elaborate motion detection scheme, which uses hierarchy so that a coarse motion analysis is made in all cases and then depending on the bit available for transmission, more refined motion analysis is implemented where such a precision is needed. The spatial redundancy is attenuated using the discrete cosine transform. A rather important number of different quantizers are designed to quantize and code the transformed coefficients. The selected quantizer is indicated to the receiver through vector quantization. The coded data is modulated in the system, using 2 types of vestigial side band modulation with 4 or 2 levels. The main motivation for such a double scheme is to introduce graceful degradation of the transmitted data as a function of the distance from the emitter. The so-called Digicypher system proposed by General Instruments Corporation uses also similar principles. One level straightforward motion compensation reduces the time redundancy followed by a discrete cosine transform for spatial redundancy reduction. Quantized coefficients are then encoded prior to quadrature amplitude modulation. The advanced television research consortium (ATLC) brings together Philips, Thomson, NBC and the David Research Center of Stanford Research International. This system uses the MPEG recommendation adapted to the high definition picture format. They call it MPEG++. An interesting originality of this system is the introduction of priorities to the data to be transmitted. Very important data receives high priority and is coded with high security as opposed to low priority data which is less protected for transmission, against transmission error. In this scheme, the modulation used is again the quadrature amplitude modulation. Among all digital HDTV systems, let us mention finally the new system proposed by MIT. This system uses also the same general principles. It is mentionned that in a newer version of the system, a subband analysis will be introduced to reduce the spatial correlation of the data.

9. Conclusion

Today's situation to replace our existing TV system gives a rather unique opportunity to implement our best developments in information and signal processing applied to a vital area of communication between human beings. Because we cannot change an existing TV system with the same ease with which we change our shirts, our bicycles, our photocameras, we have to think deeply to design a system which has the highest chances of survival for the longest possible period ahead of us. It is not difficult today to foresee important progress in multi media communications and needs, requiring not only conventional image communication with various formats and sizes but also the introduction of the 3-D displays reproducing the illusion of space information on the TV screen, as well as user friendly interaction. If in the 90's, a closed HDTV system is introduced, there will be no chance to offer the additional flexibilities of multi media and 3-D system to the consumer before very long time. It is the author's opinion that priority should be given to the HDTV system which has the most open architecture and design, so that future developments can be easily incorporated into the existing system.

References

[1] M. Kunt (invited paper), "HDTV", IEEE Information Theory Society Newsletter, N. Mehravari (Ed), vol. 42, No. 3, pp. 1-15, September 1992.

[2] M. Kunt (invited paper), "Recent HDTV Systems", Signal Processing VI, Theories and Applications, J. Vandewalle and al. Edts, North Holland, Amsterdam, pp. 83-89, 1992.

[3] M. Kunt (invited paper), "An overview on current HDTV efforts", Proc. of Canadian Conference on Electrical and Communications Engineering, 13-16 September, Toronto, Canada, pp. TA5.16.1-TA5.16.7, 1992.

[4] M. Kunt (invited paper), "Recent HDTV Systems", in D. Thalmann (Edt.), "Virtual Worlds and Multimedia", John Wiley & Sons, New York, 1993.

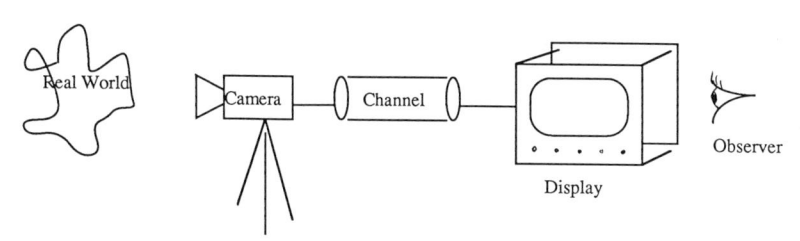

Fig. 1 TV as a communication system

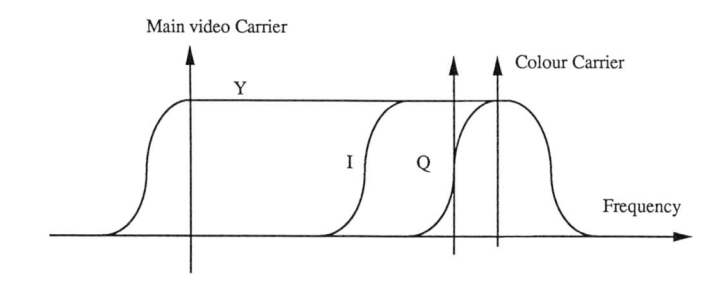

Fig. 2 The NTSC Spectrum

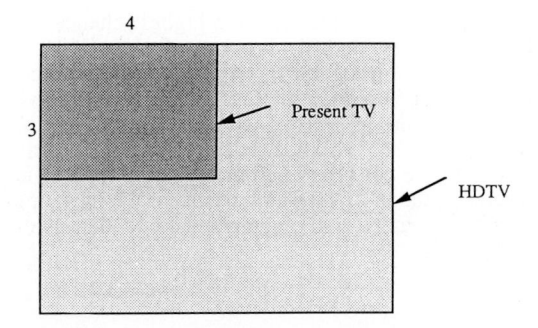

Fig. 3 Image sizes for TV and HDTV with equal resolution

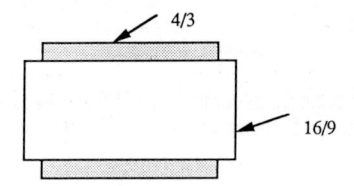

Fig. 4 Image shapes for Tv and HDTV with equal surface

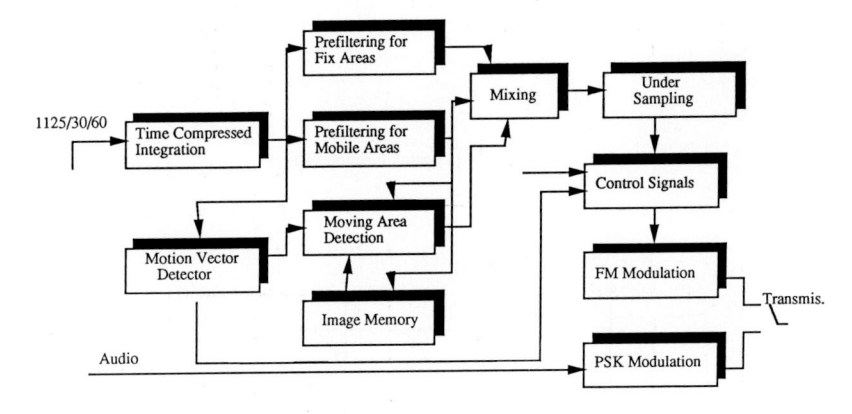

Fig. 5 The MUSE transmitter

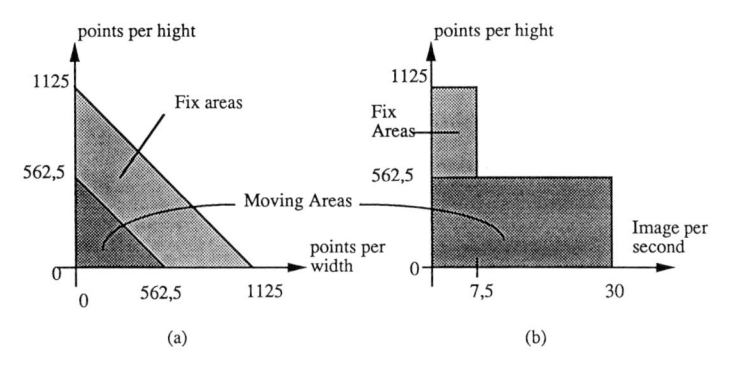

Fig. 6 Space (a) and time (b) resolutions of the MUSE system

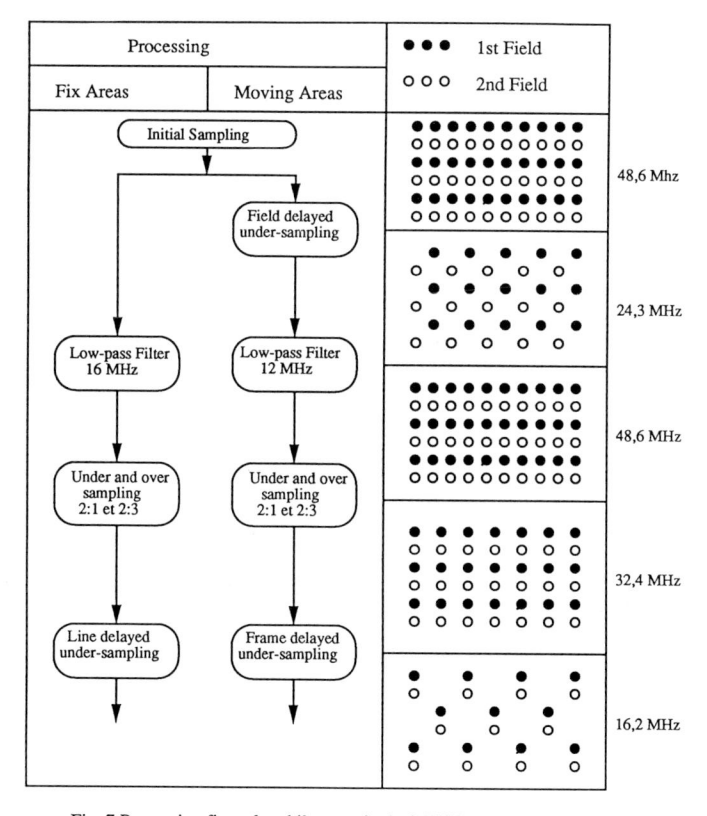

Fig. 7 Processing fix and mobile areas in the MUSE system

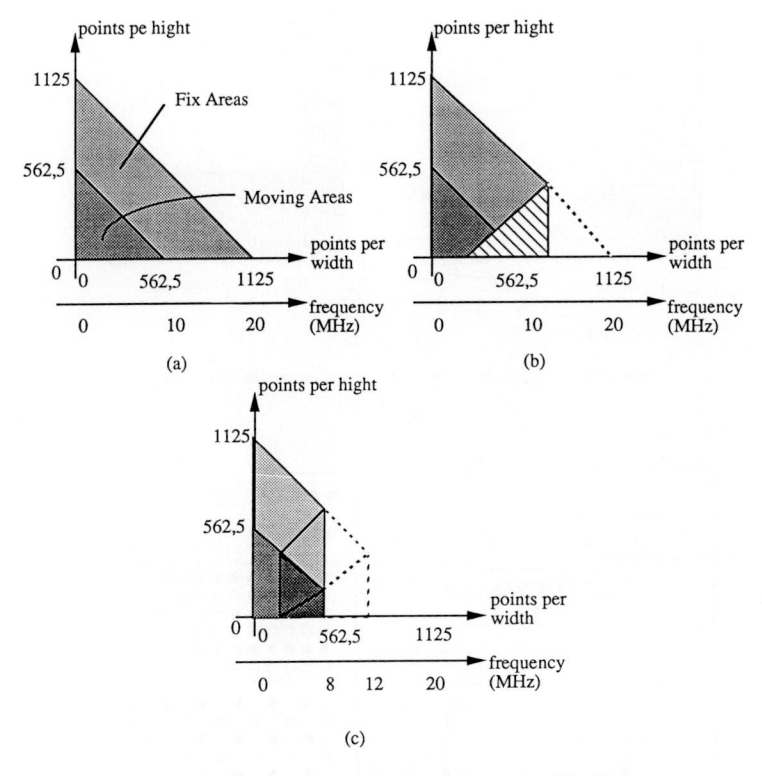

Fig. 8 Transmissible luminance spectrum and its aliasing after successive sub-samplings

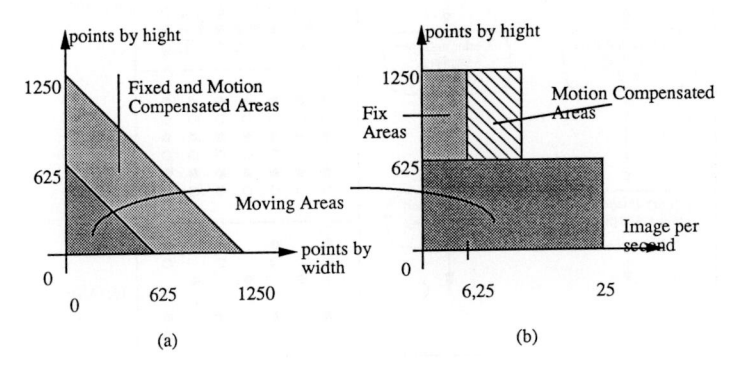

Fig. 10 Space (a) and time (b) resolutions of the HD-MAC System

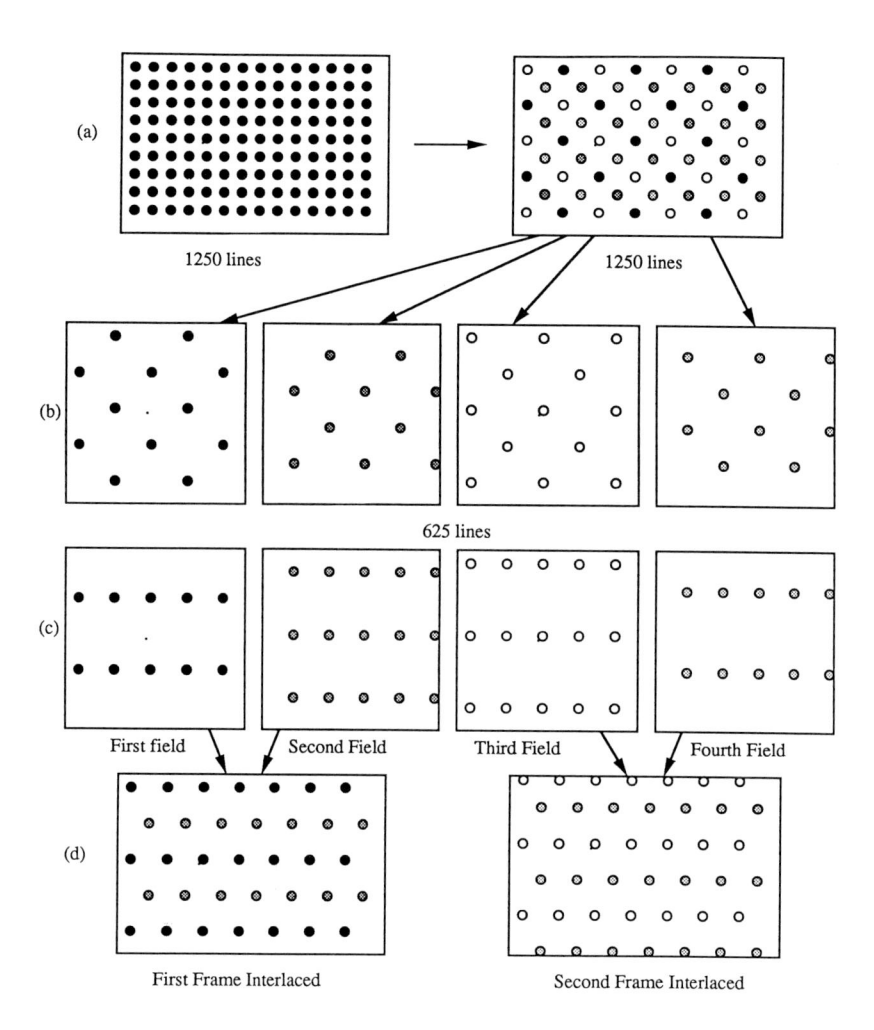

Fig. 9 Frame and field processing in the HD-MAC System

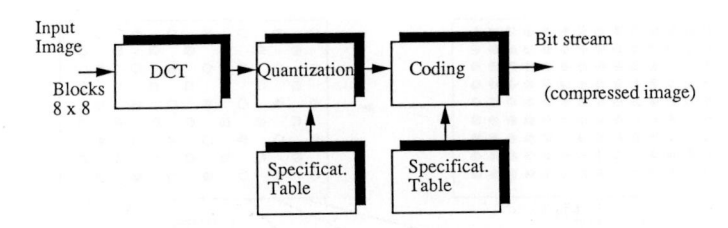

Fig. 11 The JPEG Coder

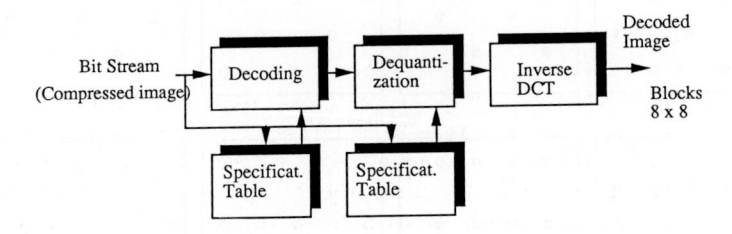

Fig. 12 The JPEG Decoder

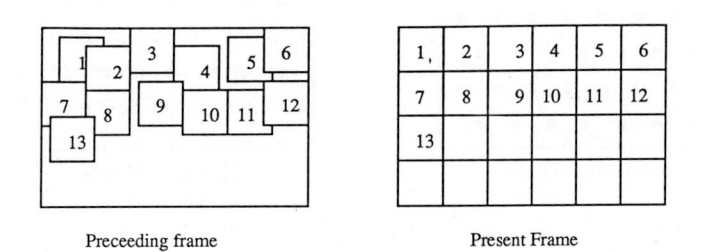

Preceeding framePresent Frame

Fig. 13 Block matching motion estimation

POLYNOMIAL METHODS IN MOTION ANALYSIS

THOMAS S. HUANG

University of Illinois at Urbana-Champaign, USA

Abstract

Almost all motion estimation problems can be formulated in terms of the solution of a set of simultaneous polynomial equations. With respect to these polynomial formulations, many open questions remain. These include: (i) How to find minimal formulations - the smallest total degree? (ii) How to find all the real solutions of a polynomial system? (iii) How to find robust algorithms in the presence of data noise? The main goal of this paper is to challenge the image processing research community to answer some of these questions.

1. INTRODUCTION AND PROBLEM STATEMENT

A straightforward formulation of motion estimation problems leads to the solution of simultaneous transcendental equations. The solution is typically carried out by iterative methods, which may diverge or may converge to a wrong local minimum unless one has a very good initial guess solution. Fortunately, in almost all cases, it is possible to formulate the problem in such a way that polynomial equations result. In some cases, it is even possible to obtain linear equations. However, many open mathematical questions remain with regard to these linear and polynomial methods. The goal of the present paper is to raise some of the open questions in the context of a specific motion estimation problem, that of two-view motion determination from point correspondences. The hope is the the paper will stimulate interest of the image processing community so that they may contribute to answering some of these open questions. For a more general review paper on motion analysis, see Ref. 1.

We first state the two-view motion determination problem. The basic imaging geometry is shown in Fig. 1. A pinhole camera model is used. Two images are taken of a moving rigid object at time instants t_1 and t_2, respectively. By processing these two images, one aims to determine the motion (in a sense specified below) of the object from t_1 to t_2. The solution approach consists of two steps: (i) Extract and match feature points (e.g., corners of man-made objects) over the two images. (ii) From the image coordinates of the corresponding points, solve equations to determine the motion parameters. In this paper, we are concerned with step (ii).

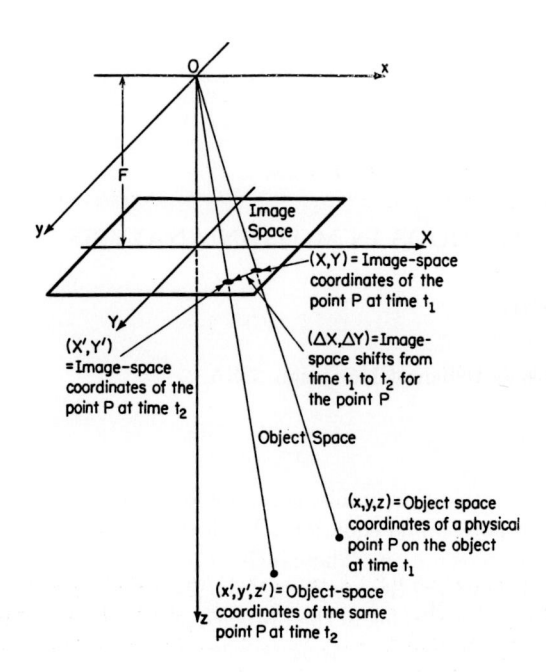

Figure 1: Basic Imaging Geometry

Referring to Figure 1, we shall use the following notation:

$p = (x, y, z)$ = coordinates of a 3-D point at time instant t_1

$p' = (x', y', z')$ = coordinates of the same 3-D point at t_2

$P = (X, Y, 1)$ = coordinates of the image point at t_1

$P' = (X', Y', 1)$ = coordinates of the image point at t_2

Without lost of generality, we have set the focal length F to 1. And we shall use a subscript i to denote the ith point.

From kinematics,

$$p'_i = Rp_i + t \qquad (1)$$

where R is a 3 x 3 rotation matrix (orthonormal and det $(R) = +1$)

$$R = \begin{bmatrix} r_{11} & r_{12} & r_{13} \\ r_{21} & r_{22} & r_{23} \\ r_{31} & r_{32} & r_{33} \end{bmatrix} \qquad (2)$$

t is a translation vector (or 3 x 1 column matrix)

$$t = \begin{bmatrix} t_x \\ t_y \\ t_z \end{bmatrix} \qquad (3)$$

and p'_i and p_i are considered as 3 x 1 column matrices.
From simple geometry, the object point and image point coordinates are related by:

$$X_i = x_i/z_i, Y_i = y_i/z_i \\ X'_i = x'_i/z'_i, Y'_i = y'_i/z'_i \qquad (4)$$

The mathematical problem can now be stated precisely:

Given $P_i \longleftrightarrow P'_i, i = 1, 2, ..., n$

Find $R, \frac{t}{\|t\|} ; \frac{z_i}{\|t\|}$.

Note that with a single camera t and z_i can be determined only to within a scale factor. The reason is that: If the object is moved d times farther away from the camera and enlarged d times (i.e., multiply all z_i by d), also the translation t is multiplied by d, then one will get exactly the same two images independent of d.

2. POLYNOMIAL METHODS

In the following sections (with the exception of A), results are described briefly and without derivations. The readers are referred to the references for details.

A. Using Rigidity Constraints [2]

From rigidity constraints, z_i and $z'_i (i = 1, 2, ..., n)$ are determined. Then R, t are obtained by solving Eq. (1).
The points p_i are on a rigid object, if and only if the distance between any pair of them is the same at t_1 and t_2 :

$$\|p_i - p_j\|^2 = \|p'_i - p'_j\|^2, for\ all\ i, j \qquad (5)$$

For n point correspondences ($n \geq 2$) there are 6+3 (N-4)= (3N-6) independent such equations. Eq. (5) can be written as

$$(x_i - x_j)^2 + (y_i - y_j)^2 + (z_i - z_j)^2 = (x'_i - x'_j)^2 + (y'_i - y'_j)^2 + (z'_i - z'_j)^2$$

or

$$(X_i z_i - X_j z_j)^2 + (Y_i z_i - Y_j z_j)^2 + (z_i - z_j)^2 \qquad (6)$$

$$= (X'_i z'_i - X'_j z'_u)^2 + (Y'_i z'_i - Y'_j z'_j)^2 + (z'_i - z'_j)^2$$

The unknowns are z_i, z_j, z'_i, z'_j. The equation is 2nd-order and homogeneous in the unknowns.

With five point correspondences, we have ten unknowns and nine homogeneous equations and can attempt to solve the equations to find the unknowns to within a scale factor. Note that the total degree of the set of nine 2nd-order polynomial equations is $2^9 = 512$.

B. Using Quaternions [3]

Rotation can be represented by a unit quaternion. Equivalently, one can scale the quaternion to make the scalar component 1:

$$q = [1, q_x, q_y, q_z] \tag{7}$$

If we can find q, then we can renormalize it to a unit quaternion.

Given five point correspondences, it is possible, by eliminating t and z_i, to obtain three 4th-order polynomial equations in the three unknowns $q_x, q_y,$ and q_z. The total degree is $4^3 = 64$.

C. Using the E-Matrix [4]

With five point correspondences, it turns out one can write five linear homogeneous equations in the nine unknowns e_i which are the components of the so-called "essential Matrix" E. The fact that the matrix E is decomposable into a skew symmetrical matrix (containing the translation components) post-multiplied by a right-handed orthonormal matrix R can be expressed in terms of three homogeneous polynomial equations in e_i. These equations are of degrees 3, 4, and 4, respectively. Taken together, we have eight homogeneous equations in the nine unknowns e_i. The total degree of the system is 3 x 4 x 4 = 48.

D. Using Projective Geometry [5]

By using results and techniques of algebraic projective geometry, it is possible to formulate a single 10th-order polynomial equation in one unknown. The derivation of the equation was tedious, and a symbolic mathematics package, MAPLE, was used.

E. Open Questions

From Sections A-D, we see that the number of equations equals to the number of unknowns when we have five point correspondences. Since the equations are polynomial, the number of solutions is finite. From Bezout's theorem [6] the number of solutions (generally complex) is equal to the total degree of the polynomial system, if one includes solutions at infinity. We are only interested in real solutions. The total degree of a polynomial system from any formulation is an upperbound to the number of real solutions. For our problem (two-view motion/range determination from point correspondences), the smallest upperbound to date is 10 (from Section D). A major question is: For a specific motion estimation problem, how does one derive a set of polynomial equations with the *minimum* total degree? A polynomial system of a smaller total degree not only gives a tighter upperbound to the number of real solutions, but also is obviously easier to solve.

With respect to finding solutions of a set of simultaneous polynomial equations, many questions remain unresolved. First, given a set of M polynomial equations with K unknowns (where $K \leq M$), how can one find *all* the real solutions? For $K = 1 = M$, Sturm's method can be used. However, its extension to the general case appears very difficult. One approach is the U-resultants [6], which unfortunately are computationally infeasible except for very simple polynomial systems. Second, a related and simpler question: How can one determine the number of *real* solutions of a polynomial system (without solving it)?

Numerically, the best hope to date of solving a set of polynomial equations (with $K = M$) to find all solutions (generally complex) appears to be the use of homotopy methods [7]. We have used this approach to solve the equation set of Section C with good results. However, this approach is still computationally intensive, and may become infeasible if the total degree of the polynomial system is larger than 50 or so. Furthermore, with numerical methods, because of round off and other quantization errors, one can never be completely sure that a specific solution is purely real.

For practical applications, an important third question arises: Can we find computationally efficient methods of estimating all the real solutions in the presence of data noise (which introduces errors into the coefficients of the polynomial equations)? In this connection, a careful noise sensitivity analysis needs to be made for polynomial systems.

Although in the above we have asked the three questions mainly from a numerical point of view, we are indeed interested in some of the deeper questions: (a) How does the number of solutions depend on 3-D point configurations and motion parameters? (b) How does the noise sensitivity depend on 3-D point configurations and motion parameters?

Acknowledgment. This work was supported by the Joint Services Electronics program under Grant N00014-90-5-1270.

3. REFERENCES

[1] T.S. Huang, Motion Analysis, in *Encyclopedia of Artificial Intelligence*, 2nd ed. by S.C. Shapiro, Wiley (1991).

[2] A. Mitichi and J.K. Aggarwal, A Computational Analysis of Time-Varying Images,

in *Handbook of Pattern Recognition and Image Processing*, ed. by T.Y. Young and K.S. Fu, Academic Press (1986).

[3] C. Jerian and R. Jain, Polynomial Methods for Structure from Motion, *Proc. 2nd International Conference on Computer Vision*, December 19988, Tarpon Springs, FL.

[4] T.S. Huang and Y.S. Shim, Linear Algorithm for Motion Estimation: How to Handle Degenerate Cases, *Proc. British Pattern Recognition Association Conference*, April 1987, Cambridge, England.

[5] D.D. Faugeras and S.J. Maybank, Multiplicity of Solutions for Motion Problems, Technical Report, INRIA, France, 1988.

[6] W.V.D. Hodge and D. Pedoe, *Methods of Algebraic Geometry*, 1, Cambridge University Press, 1953.

[7] A. Morgan, *Solving Polynomial Systems Using Continuation for Engineering and Science Problems*, Prentice-Hall, 1987.

Motion Compensated Spatiotemporal Kalman Filter

J. W. Woods and J. Kim

Center for Image Processing Research
Rensselaer Polytechnic Institute, Troy NY 12180-3590, USA

Abstract

This paper presents various 3-D or spatiotemporal Kalman based filters as an extension of the 2-D reduced update Kalman filter (RUKF) approach for still images. We start out with a basic steady state shift invariant filter or 3-D RUKF. Then, in order to make the most use of existing temporal correlation, we introduce a motion compensated 3-D Kalman filter, herein termed MC-RUKF. Since motion compensation has variable quality or success, with resulting variations in temporal correlation of the motion compensated frames, we conclude with a multi-model extension of the MC-RUKF, which achieves the best performance.

1 INTRODUCTION

In a noise-free video, each pair of consecutive frames has high correlation in the temporal direction. In contrast, any observation noise that is present will be mostly uncorrelated. When there is little or no movement between the frames, this temporal correlation can be used directly for noise reduction, without losing high spatial frequencies. A moving object still has strong temporal correlation, but along the trajectory of the motion. So motion compensation will thus be required for spatiotemporal filtering to be successful in preserving much of the high spatial frequencies.

Video has strong spatial correlation as well as temporal correlation. In order to use both correlations, a 3-D Wiener filter and motion-compensated 3-D Wiener filter have been presented in [1]. Under certain 3-D Markovian assumptions, the spatiotemporal Kalman filter has been presented in [2]. In two dimensions, the reduced update Kalman filter (RUKF) had already been shown to effectively solve the computational load problem of spatial recursive estimation [3]. In this paper we first present the 3-D or spatiotemporal reduced update Kalman filter (3-D RUKF), the natural extension of the 2-D RUKF for still images. In this spatiotemporal Kalman filter, the image sequence is filtered line-by-line and frame-by-frame on a non-interlaced or progressive raster, using a space-invariant approach. When motion in the video is large, adaptive filtering and motion-compensated filtering have been suggested [4] to improve performance. Thus in this paper, we also introduce a new motion-compensated Kalman filter, herein termed MC-RUKF.

In this motion-compensated (MC) filter, the 3-D RUKF is applied to the MC video, thus effectively filtering along the trajectory of motion.

A solution of the motion compensation problem does not always exist, so the generally high temporal correlation of the MC video is expected to be variable. Therefore, the video signal model should adapt to these different correlations. Thus we finally present an extension of MC-RUKF to a multi-model MC-RUKF.

The paper is organized as follows: The 3-D spatiotemporal Kalman filtering is presented first. This is followed in Section 3 by the introduction of the motion compensated MC-RUKF. Then in Section 4, the MC-RUKF is extended to a three-model, space variant Kalman filter, which accounts for expected variability in temporal correlation of MC video. Finally some experimental results are presented and discussed in Section 5.

2 SPATIOTEMPORAL KALMAN FILTER

When there is little motion in a video, it can be modeled as temporally stationary and Markov. Since the previous frame contains the information from all the previous frames in the stationary areas, the temporal Markov hypothesis is then natural for slowly moving areas. On the basis of such a spatial and temporal stationarity assumption, the spatiotemporal invariant or steady-state Kalman filter can then be used on such a video. Thus, in this section we summarize the spatiotemporal reduced update Kalman filter (3-D RUKF). More details on various related issues are presented in [5].

2.1 Image model

For a 3-D recursive filter, we must separate the *past* from *future* of the 3-D random field sequence. One way is to assume that the random field sequence is scanned in line-by-line and frame-by-frame mode. In this way, we can divide the random field sequence into two halves, the past and future. In the 2-D case, the NSHP (nonsymmetric half-plane) has been widely used. A simple extension of the NSHP concept leads to the nonsymmetric half-space or NSHS. Using this notion of the *past*, we can arrive at the Markov random field sequence model,

$$s(m,n,t) = \sum_{S_c} c(k,l,\tau) s(m-k, n-l, t-\tau) + w(m,n,t) \qquad (1)$$

where $S_c = \{k \geq 0, l \geq 0, \tau = 0\} \cup \{k < 0, l > 0, \tau = 0\} \cup \{\tau > 0\}$, and where $w(m,n,t)$ is a white noise field sequence.

In recursive filtering, only a finite subset of the NSHS is updated at each step. This updated region is called the *local state region* and is only slightly enlarged from the model support region.

Our observation model is given by:

$$r(m,n,t) = \sum_{k,l,\tau \in S_h} h(k,l,\tau) s(m-k, n-l, t-\tau) + v(m,n,t). \qquad (2)$$

The S_h is the support of the point-spread function (psf) $h(k,l,\tau)$. The additive observation noise $v(m,n,t)$ is assumed to be a zero-mean, homogeneous Gaussian field sequence.

2.2 Reduced update Kalman filter (RUKF)

Kalman filtering is well established for the 1-D and 2-D cases [3]. In the 2-D case many methods have been introduced to address the computation time problem. The reduced update Kalman filter (RUKF) is widely known as a good suboptimal filter. Here we sketch our extension of RUKF to the three dimensions of space and time.

When the original video is modeled with an $(1 \times 1 \times 1)$th order model for example, the local state vector is given by:

$$\begin{aligned}\mathbf{s}(m,n,t) = &[s(m,n,t),...,s(m-2,n,t);\\ &s(m+2,n-1,t),...,s(m-2,n-1,t);\\ &s(m+2,n-2,t),...,s(m-2,n-2,t);\\ &s(m+2,n+2,t-1),...,s(m-2,n+2,t-1);...;\\ &s(m+2,n-2,t-1),...,s(m-2,n-2,t-1)].\end{aligned} \quad (3)$$

If we assume that the received image sequence $r(m,n,t)$ is scanned line-by-line and frame-by-frame, then the approximate steady state Kalman filter equations are:

$$\hat{s}_b(m,n,t) = \sum_{k,l,u \in S_c} c_{klu} \hat{s}_a(m-k, n-l, t-u)$$

$$\hat{s}_a(i,j,v) = \hat{s}_b(i,j,v) + k(m-i, n-j, t-u)$$
$$\cdot [r(m,n,t) - \sum_{k,l,u \in S_h} h_{klu}\hat{s}_b(m-k, n-l, t-u)] \quad (i,j,t) \in \mathcal{R}_\oplus^{(m,n,t)} \quad (4)$$

where $\mathcal{R}_\oplus^{(m,n,t)}$ is the update region. The approximate equations for the FIR Kalman gain array are shown in [5]. Since we assume stationary random field sequences, we can expect a steady-state RUKF to develop as we move into the data, i.e. after the initial frames and away from the boundaries. As in the 2-D case, the needed steady-state gain can be determined off-line on a fictitious and small image sequence, say $20 \times 20 \times 20$ in size.

3 MOTION COMPENSATED SPATIOTEMPORAL RUKF

When the amount of motion is large, the simple steady-state RUKF has a limitation on its performance due to blurring. To reduce object blurring and sometimes even double images, we shift the temporal axis of the filter to be aligned with motion trajectories. When a moderate moving object is so aligned, we can then apply the spatiotemporal invariant filter along the object's trajectory of motion, by filtering the MC video. Since the MC video has a strong temporal correlation, its image sequence model will have a small prediction error variance. This suggests that high spatial frequencies can be retained even at low input SNR's via motion compensated filtering.

The overall block diagram of our motion compensated 3-D Kalman filter, or MC-RUKF, is shown in Figure 1. This motion compensated spatiotemporal filter consists of three major parts: the motion estimator, the motion compensator, and the 3-D reduced update Kalman filter.

While filtering a video, two different previous frames could be used for motion estimation; one could use either the previous smoothed frame $E\{s(t-1)|r(t), r(t-1),...\}$, or the previous noisy frame $r(t-1)$. In our work we have generally found it best to use the smoothed previous frame, as the best estimate currently available. For motion estimation, we used a hierarchical block matching method. The details of this motion estimation method are presented in [5]. The motion estimate is used to align a set of local frames along the motion trajectory. This MC video is then filtered by a 3-D RUKF. The current frame is the noisy image, $r(t)$, and the previous frame is the spatially smoothed previous frame $E\{s(t-1)|r(t-1),r(t-2),...\}$. To effect this local alignment, we displaced the previous frames with respect to the current frame.

In an iterative method extension, two smoothed frames are used to improve upon the initial motion estimates. These smoothed frames retain spatial high frequencies and have reduced noise, so that these frames can now be used for motion estimation with a smaller size block. A motion vector field is then estimated from these two frames, i.e. $E\{s(t-1)|r(t),r(t-1),...\}$ and $E\{s(t)|r(t+1),r(t),...\}$, followed by a second application of the steady-state MC-RUKF.

4   MULTI-MODEL MC-RUKF

The local correlation between two MC frames depends on the accuracy of the motion estimation. When the motion estimation is very accurate, the correlation is very high, but when the motion estimation is not accurate, for example due to the uncovering of a new area, the temporal correlation can be quite low. Since the SNR of the observed video will be low, the motion estimation has a further limitation on its accuracy, due to statistical variations. In uncovered regions, the use of a homogeneous 3-D AR model causes some unwanted extra distortion. So we divided the motion compensated frames into three regions: *still region, predictable region,* and *unpredictable region*. These three regions will then have much different temporal correlation. In the still region, the correlation coefficient is almost 1.

As mentioned earlier, we employ a block matching method for motion estimation. Even when there is no correspondence between two motion compensated frames, this method chooses a pixel in the search area which minimizes the displaced frame difference measure. This then leads to a low temporal correlation in the unpredictable region.

To detect the still region, the frame difference was filtered to decrease the effect of observation noise using a simple 7×7 box filter. A 3×3 box filter was used, to detect the unpredictable region. The detection criteria used was local variance. A carefully chosen threshold value will decrease the adverse effects of the unpredictable region. The results were sensitive to this value. Isolated pixels were rejected. When a pixel at motion boundaries was also detected as a pixel in the still region, the visual error in the filtered image sequence was noticeable. Hence, we detected the still region again at the filtering step. Three spatiotemporal AR models were obtained from the residual video of the original sequence for our simulation. This training step thus used noise-free video data.

5   EXPERIMENTAL RESULTS

The experimental video was selected as a monochrome version of the well-known *salesman* sequence. We used only the even numbered frames, because there is significant difference between odd and even numbered frames. This difference had to do with NTSC chrominance data that was present in an original color version of *salesman*. The resulting frame rate was 15 frames per second. For our experiment, we pre-shifted the mean values of the salesman sequence to obtain zero mean as required for the filtering. We used this modified sequence as our original sequence. The noisy image was obtained by adding a white Gaussian noise to this sequence. The input SNR (variance-to-variance) was set to 10 dB. The size of video model support was $(1 \times 1 \times 1)$ - order. The gain support region was $(2 \times 2 \times 1)$ size. The 3-D autoregressive (AR) model obtained from the original (modified) video is used. This model could have been also obtained from the noisy video or from a prototype noise-free, with some loss of performance. Based on existing work in the identification of 2-D image models, it is our feeling that the additional loss would not be great.

We used a steady-state gain array, calculated off-line on a small fictitious image sequence. The SNR improvements of the spatiotemporal RUKF and MC-RUKF with both single model and multiple models are shown in Figure 2. The global SNR improvements of these filters at steady state, about frame 20, range from 7 to 9 dB, with the best result obtained by the multi-model MC-RUKF.

6 CONCLUSION

We presented the spatiotemporal reduced update Kalman filter (3-D RUKF) and introduced its motion compensated extension (MC-RUKF). Then, the approach was extended to a multi-model MC-RUKF, to deal with the uneven quality of the motion compensation process. These new methods have been seen to be effective, both objectively (MSE) and visually, for the suppression of noise in the several image sequences tested.

References

[1] M. K. Ozkan, I. Sezan, A. T. Erdem, and A. M. Tekalp, "Motion compensated multi-frame Wiener restoration of blurred and noisy image sequences," in *Proc. IEEE Int. Conf. Acoust., Speech, Signal Processing*, San Francisco, Califonia, Mar. 1992.

[2] D. Cano and M. Benard, "3D Kalman filtering of image sequences," in *Image Sequence Processing and Dynamic Scene Analysis*, pp. 563–579, Berlin Germany: Springer-Verlag, 1983.

[3] J. W. Woods and V. K. Ingle, "Kalman filtering in two dimensions: Futher results," *IEEE Trans. Acoust., Speech, Signal Process.*, vol. 29, pp. 188–97, 1981.

[4] A. K. Katsaggelos, R. P. Kleihorst, S. N. Efstratiadis, and R. L. Lagendijk, "Adaptive image sequence noise filtering methods," *Proc. SPIE Conf. Visual Communications and Image Process.*, vol. 1606, pp. 716–727, Nov. 1991.

[5] J. W. Woods and J. Kim, "Motion Compensated Spatiotemporal Kalman Filter," in *Motion Analysis and Image Sequence Processing*, Kluwer Academic, To be published.

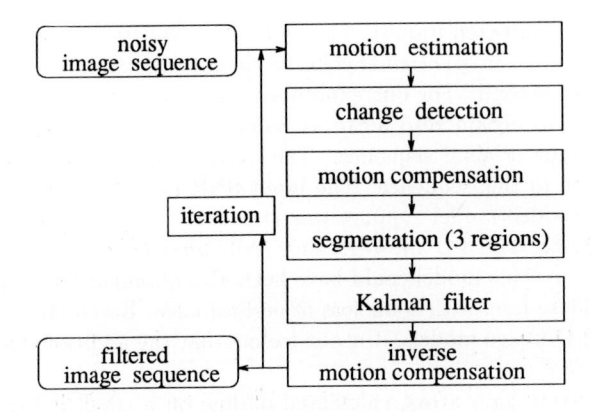

Figure 1: The Overall Block Diagram

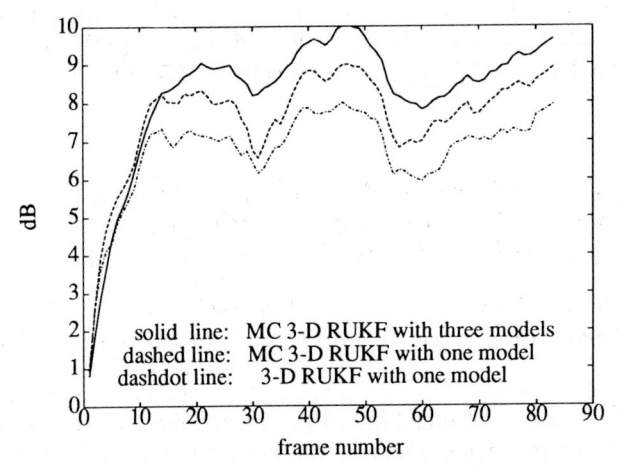

Figure 2: the SNR improvement of the salesman sequence

solid line: MC 3-D RUKF with three models
dashed line: MC 3-D RUKF with one model
dashdot line: 3-D RUKF with one model

Artificial Neural Networks in Pattern Recognition and Signal Classification

C. H. Chen

Signal/Image Processing and AI Laboratory, Electrical and Computer Engineering Dept.
University of Massachusetts Dartmouth, N. Dartmouth, MA 02747 USA

Abstract

The revitalization of artificial neural networks (ANN) in the last few years has already had a great impact on research and development in pattern recognition, signal classification and, in general, problems involving decision, estimation, prediction and control. A large number of applications using ANN have also emerged. In this tutorial paper, the emphasis is on pattern recognition and signal classification using ANN. The issues in ANN pattern recognition are examined. Relative merits of several feedforward networks are considered. A new class-sensitive neural network is presented. Signal classification applications are reviewed with specific examples given on classification with underwater acoustic signals and ultrasonic nondestructive testing signals for material evaluation. It is concluded that the effective use of ANN as pattern classifiers remains to be a challenging problem.

1. INTRODUCTION

In the late fifties, Rosenblatt's perception and Widrow's Adaptive Linear Networks were regarded as major nonstatistical pattern recognition systems. Both are now called artificial neural networks (ANN). Through the revitalization of ANN in the last few years, many new neural networks have been developed for pattern recognition. Also there is now a much better understanding of the neural network classification capabilities. Actually there is a close relationship between feedforward ANN and statistical pattern recognition [1], [2]. The issues in neural network pattern recognition include the training algorithms and training time, the sample sizes, feature effectiveness, and the classification performance. These issues will be addressed in the paper. Examples of signal classification using ANN are then presented for applications in underwater acoustics, ultrasonic nondestructive evaluation (NDE) of materials, and others. The results all demonstrate that neural networks can be a lot more effective in classification than the traditional classifiers.

2. NEURAL NETWORK PATTERN RECOGNITION

The artificial neural networks can be trained to perform certain functions like classification or decision making, estimation, prediction, filtering, and control, et. al. For an outstanding presentation of the general computing capabilities of neural networks, see e.g. [3-

6]. Though there are different opinions about objectives of pattern recognition [7], the main objective of minimum classification error remains to be the same in all pattern recognition problems. The minimum error criterion may be replaced by the Neyman-Pearson criterion that minimizes the probability of detection in problems like radar and sonar detection. However, most network configurations and training algorithms are developed for fast learning but not for best classification. Even the popular backpropagation training algorithm is based on minimizing the mean square error between the actual and desired classification outputs. The asymptotic performance of the backpropagation network may approach that of Bayes [8], but such asymptotic performance may never be obtained in practice. However since the true Bayes error probability is not known, it is not unusual for the networks to achieve superior classification performance with proper network configuration, training algorithms, and adequate training samples.

Among all neural networks considered, the feedforward ANN appear to be most suitable for classification. Our experience with the popular multilayer backpropagation trained network is that it performs almost the same as the traditional nearest neighbor decision rule, while the latter requires no training time. The probabilistic neural network [9,10] which is an implementation of the nonparametric density estimate requires no iterative computation and is thus very fast in training. Again its classification performance is fairly similar to that of the nearest neighbor decision rule. With an objective to minimize the classification error, we have developed a Class-Sensitive Neural Network (CSNN) [11] which decouples all classes by using one subnet for each class. All subnets share the same input but each is trying to minimize the error between the actual and desired outputs. Fig. 1 is a CSNN with 2 inputs, 1 hidden layer, 3 hidden nodes in each subnet and 3 outputs or classes.

It is noted that for multiple classes, some classes may be learned more easily. We can give more training passes to the hard-to-learn classes. Each subnet is trained for best performance and thus the whole network has better chance to get better performance. The training strategy is basically the same as that of backpropagation, but the learning coefficient and weights are updated more frequently according to the correlation between the actual and desired outputs. There is no doubt that CSNN will need more training time than the

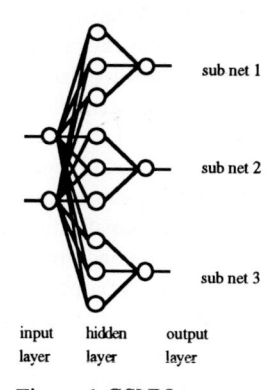

Figure 1 CSNN structure

backpropagation network, but the improvement in correct classification, typically 6 to 8 percent over the backpropagation network, may well justify the extra computation time. It is our view that the change of network configuration as stated above is necessary to improve the classification performance and that the training algorithm improvement alone is not enough to do the job.

Among other issues, we may note that it still takes good features for the network to perform well in classification. Although the networks may be used to extract some features, the advantage of using neural network is not evident. The backpropagation trained network, however has been found to be much less sensitive to the training sample size than the traditional classifiers. This advantage can be quite significant in practice when we have to work with a limited number of training samples. For a finite training sample size, the network classification performance will still reach a peak at certain number of features, beyond which the performance may start to decrease slightly. Though not extensively verified, neural networks can perform better at low signal-to-noise ratio conditions as well as when the data are degraded by multipaths or distortions. The reason is that the network has better adaptive capability to new environments, which again is a significant advantage with the use of neural networks for classification. Other advantage with neural networks is the ease with sensor fusion or integration of information from different sources.

3. APPLICATIONS TO SIGNAL CLASSIFICATION

Neural networks have now been considered for application in almost all signal classification problems. In this paper we shall consider mainly the applications in underwater acoustics and the ultrasonic NDE. The early work of Gorman and Sejnowski [12] has generated much interest in using neural network for sonar target classification. For the passive transient data, a number of papers presented in [13] have made an extensive comparison of classifications by different neural networks. For the active sonar data typically represented by high-dimensional vectors [14], a typical comparison, based on four pattern classes, 13-dimensional vectors, is as follows.

	percentage correct in testing
NNDR (nearest-neighbor decision rule)	57.96
BPN (backpropagation network)	62.23
PNN (probabilistic neural network)	62.19
CSNN	68.50

For the same data in two classes, the above numbers become 70.69, 70.63, 72.23, and 78.26 respectively. The improvement of CSNN is evident in both cases.

For a simulated set of multipath data, we have also determined that the backpropagation network outperforms the nearest neighbor decision rule by several percentage points.

For the ultrasonic NDE data for testing different categories of hidden geometrical defects in aluminum blocks, we have been able to show [15] that based on a very limited set of waveforms, backpropagation trained networks can classify all defects correctly while the nearest neighbor decision rule can achieve only up to 83.3% correct classification.

For speech recognition, neural networks are shown to perform very comparatively with or better than the nearest neighbor decision rule [16]. For the character recognition, landsat image classification, seismic signal classification [17], and many other applications, the superiority of neural networks is again demonstrated not only in classification performance, but also in the practical implementation, as well as in meeting the real-time application needs. As a concluding remark, it may be stated that it is still a challenge to use neural networks as effective pattern classifiers which can perform well in different situations.

4. REFERENCES

1. C.H. Chen, On the relationships between statistical pattern recognition and artificial neural networks, International Journal of Pattern Recognition and Artificial Intelligence, vol. 5, no. 3, pp. 655-661, 1991.
2. R.P. Lippmann, An introduction to computing with neural nets, IEEE ASSP Magazine, pp. 4-22, 1989.
3. P.G. Morasso, Neural network computing, in C.H. Chen (ed.), Computer Engineering Handbook, Chapter 18, McGraw-Hill, New York 1992.
4. Y.H. Pao, Neural net computing for pattern recognition, in C.H. Chen, L.F. Pao and P.S.P. Wang (eds.), Handbook of Pattern Recognition and Computer Vision, World Scientific Publishing, Singapore and New Jersey, 1993.
5. J.M. Zurada, Introduction to Artificial Neural Systems, West Publishing Co., St. Paul, MN 1992.
6. N.B. Karayiannis and A.N. Venetsanopoulos, Artificial Neural Networks, Kluwer Academic Publishers, Boston, MA 1993.
7. W. Chang, et. al., unpublished internal report, NUWC, New London, CT, September 1992.
8. F.Kanaya and S.Miyake, Bayes statistical behavior and valid generalization of pattern classifying neural networks, IEEE Trans. on Neural Networks, vol. 2, no. 4, pp. 471-475, July 1991.
9. D.F. Specht, Probabilistic neural network, Neural Networks, vol. 3, pp. 109-118, 1990.
10. D.F. Specht, Enhancement to probabilistic neural networks, Proc. of the International Joint Conference on Neural Networks, Baltimore, MD, 1993.
11. C.H. Chen and G.H. You, Class-sensitive neural network, Neural, Parallel and Scientific Computing, vol. 1, 1993.
12. R.P. Gorman and T.J. Sejnowski, Analysis of hidden units in a layered network trained to classify sonar targets, Neural Networks, vol. 1, no. 1, pp. 75-89, 1988.
13. Proc. of the IEEE International Conference on Neural Networks in Ocean Environments, Washington, D.C., Aug. 1991.
14. C.H. Chen, Issues on neural networks for active sonar classification, IEEE OCEANS '92, Newport, RI, Oct. 1992.
15. C.H. Chen and G.G. Lee, Neural networks for ultrasonic NDE signal classification using time-frequency analysis, IEEE ICASSP '93
16. R.P. Lippmann, Review of neural networks for speech recognition, Neural Computation, vol, pp. 1-38, 1989.
17. H.J. Shyu, J. M. Libert and S.D. Mann, Classifying seismic signals via RCE Proc. of the International Conference on Neural Networks, 1990.

Form and Content Issues in the Abductive Framework for Recognition

B. Chandrasekaran and J. Josephson

Laboratory for AI Research, Dept of Computer & Information Science,
The Ohio State University, Columbus, OH 43210-1277 USA.
Email: {chandra, jj}@cis.ohio-state.edu

SUMMARY

In an earlier paper, Chandrasekaran and Goel [1] considered different approaches to the task of classification. The classificartion problem is one of mapping from observations (data) to pre-enumerated classes. As the classification problem grows complex, the solutions evolve from simple one step numerical mapping to use of intermediate abstractions (symbols) to rules that use relations between abstractions to complex reasoning with knowledge, summarized in the progression:

numbers --> abstractions (symbols) ---> relations ---> knowledge structures.

We compared pattern classification approaches, connectionist networks, syntactic methods and finally knowledge-based classification. The task of classification particularly was deemed important since it is so ubiquitous, playing a role in visual and speech recognition, diagnosis and numerous other tasks of importance.

In this talk, we will review those arguments, correcting, updating and strengthening them where necessary. One particular idea that didn't seem to have come through very clearly is the issue of form versus content. Often much of the discussion in the field seems to revolve around whether this mechanism or that (connectionist nets, or logic or symbolic approaches) are the right way to go about building a system (say, for visual recognition). We have argued elsewhere (see [2]) that, for many purposes, mechanism questions are not the right kind of questions to ask, at least not in the beginning. These questions should be deferred until an understanding of the content of a task is obtained. When the task structure is properly understood, then decisions about what kinds of mechanisms are appropriate for what subtasks can be made.

Marr of course has also made similar arguments in his book on vision and Newell is also credited with similar remarks in his discussion on the Knowledge Level/Symbol Level distinction.

In this talk, we will emphasize the form-content distinctions. More importantly, we will move from consideration of the task of classification to the task of abductive problem solving. Abduction is a more general framework for problems of recognition. Abduction is the problem of generating an explanation, specifically a "best" explanation, of a set of observations. Visual and speech recognition, natural language understanding and diagnosis are all examples of abductive tasks.

We will end with an outline of a layered abduction architecture that is especially suited for problems of recognition. The architecture is a high-level architecture and is potentially compatible with a number of different implementation architectures, such as neural nets or traditional symbolic programming.

REFERENCES

1 B. Chandrasekaran and Ashok Goel, "From numbers to symbols to knowledge structures: Pattern Recognition and Artificial Intelligence perspectives on the classification task," IEEE Trans. Systems, Man & Cybernetics, Vol. 18, No. 3, May/June 1988, pp 415- 424.

2 B. Chandrasekaran, A. Goel and D. Allemang, "Connectionism and information processing abstractions: the message still counts more than the medium," AI Magazine, 9:4, pp. 24-34, 1989.

3 John and Susan Josephson, editors, "Abductive Inference: Computation, Philosophy, Technology," Cambridge University Press, 1993 (forthcoming).

Inverse SAR Planetary Imaging

M. R. Coble[a] and D. C. Munson, Jr.[b]

[a]Phillips Lab/LIMI, Kirtland Air Force Base, 3550 Aberdeen Ave. S.E., Albuquerque, New Mexico 87117-5776, USA

[b]Coordinated Science Laboratory and Department of Electrical & Computer Engineering, University of Illinois, 1308 W. Main St., Urbana, Illinois 61801, USA

Abstract

Microwave radars have been used to produce high-resolution images of planets and other heavenly bodies since the 1960's. The traditional image formation algorithm used in radar astronomy is based on ideas from range-Doppler imaging. In this paper we first review range-Doppler imaging and then explore the possibility of using an inverse Synthetic Aperture Radar (SAR) model instead. This latter model is based on theorems from computer tomography and leads to an improved method of processing for some scenarios.

1. INTRODUCTION

In the early 1960's, Paul Green [1] presented a complete theory for using radar to image planets and other rotating heavenly bodies. A range-Doppler imaging principle was developed, where points on the planet's surface are mapped according to their iso-range contours (concentric rings on the planet's surface) and iso-Doppler contours (lines parallel to the axis of rotation). Mathematically, the processing involves a bank of matched filters that are matched to frequency-shifted (i.e., Doppler-shifted) versions of the transmitted signal. Other researchers [2] have made this the standard processing model for radar astronomy and have imaged numerous astronomical objects, including Venus and the Moon.

The range-Doppler processing model assumes that points on the rotating planet surface do not move through resolution cells during the period of microwave illumination. While this assumption is reasonable for slowly rotating bodies, such as Venus, it is a poor assumption for imaging other objects such as asteroids or the Moon. In this paper we consider an alternative processing model based on Inverse Synthetic Aperture Radar (ISAR), which removes the restriction on motion through resolution cells. The paper is fairly tutorial in nature, and begins with a review of range-Doppler planetary imaging. We then discuss 3-D spotlight-mode SAR and ISAR from a tomographic perspective and show how it applies to the planetary imaging scenario. The paper ends with some conclusions and ideas for further work.

2. REVIEW OF RANGE-DOPPLER PLANETARY IMAGING

Since the surface of a planet is composed of point reflectors located at varying ranges from the radar, different portions of the surface will return energy at different delays. The loci of all points about the transmitter-receiver location having constant delay τ are concentric spheres of radius $c\tau/2$. The intersections of these spheres with the planetary surface are circles (solid lines in Fig. 1), which are the loci of constant range as the target is viewed head-on from the radar.

The rotation of the planet will cause the various point reflectors on the planetary surface to have different line-of-sight velocities v_l with respect to the radar. Therefore, a Doppler shift

$$\nu = \frac{2f_0}{c} v_l. \tag{1}$$

will be imposed on the incident signal by each reflector, and the amount of shift will depend on the location of the reflector on the planet's surface. Here, f_0 is the transmitter frequency and c is the speed of propagation. Using the coordinate system of Fig. 1, with the origin at the center of the planet, loci of points having constant Doppler shift are planes parallel to the xz-plane. Along any curve of intersection between such a plane and the planetary surface, the Doppler shift is constant.

If one looks head-on at the target, as in Fig. 1, contours of equal range (solid lines) intersect contours of equal Doppler shift (dashed lines) in at most two points as shown by the shaded regions. To isolate and study such a pair of point reflectors, it simply remains to isolate that part of the returned signal having a given time delay τ and simultaneously a given frequency shift ν. For unambiguous mapping it is necessary to discriminate between these two points [2].

2.1 The ambiguity function

The simplest form of target is one whose reflectivity density function in delay-Doppler coordinates is an impulse located at the origin of the illuminated scene, i.e., $\delta(\tau,\nu)$. The form of receiver that will develop maximum signal-to-noise ratio from such a target is a matched filter [3]. Letting $s(t)$ be the transmitted complex baseband signal, then the returned echo from a point reflector at the origin is simply $q(t) = s(t)$ and the matched filter output is

$$y(t) = \int_{-\infty}^{\infty} q(\beta) s^*(\beta - t) d\beta. \tag{2}$$

The matched filter, or correlation detection operation, described above is optimum only if the reflectivity density function is an impulse located at the origin. Thus, if the impulse-like reflectivity density is not at the origin of the scene, but instead at the point (τ_0, ν_0) in delay-Doppler coordinates, then $q(t) = s(t - \tau_0) e^{j2\pi\nu_0\tau_0}$, and the output given by Eq. (2) is diminished. As a result, the filter impulse response must be readjusted by the frequency offset ν_0, and its output observed τ_0 seconds later to restore the optimal matched filter condition.

Suppose the input waveform $q(t)$ in Eq. (2) is $s(t + \tau) e^{-j2\pi\nu t}$. Then the output at $t = 0$ of this matched filter, which has been set for a point reflector (impulse) at $\tau = 0$ and $\nu = 0$, is

$$\begin{aligned} y(0) &= \int_{-\infty}^{\infty} s(\beta + \tau) e^{-j2\pi\nu\beta} s^*(\beta) d\beta \\ &= e^{j2\pi\tau\nu} \int_{-\infty}^{\infty} s(\beta) s^*(\beta - \tau) e^{-j2\pi\nu\beta} d\beta. \end{aligned}$$

If this matched-filter output is plotted as a function of τ and ν, the result is the quantity $\chi(\tau,\nu)$ called the ambiguity function. A slightly different form which is asymmetric but essentially equivalent is defined as follows:

$$\chi(\tau,\nu) = \int_{-\infty}^{\infty} s(\beta) s^*(\beta - \tau) e^{-j2\pi\nu\beta} d\beta. \tag{3}$$

This second form of the ambiguity function will be used throughout this paper.

2.2 Imaging

A range-Doppler radar can be modeled as a device that forms a two-dimensional convolution of the reflectivity density function of an illuminated target with the ambiguity function of the radar waveform [4]. This unifying description, which will be reviewed here, is extremely powerful because it models the radar processor as a simple two-dimensional filter whose point spread function is the radar ambiguity function. Hence, the quality and resolution of the reconstructed target image are related directly to the shape of the radar ambiguity function.

For a rotating planet, the return from a point reflector with reflectivity ρ on the planetary surface is

$$q(t) = \rho \cdot s(t - 2R(t)/c)$$

where $R(t)$ is the time-dependent range of the point reflector given by

$$R(t) = R_o + \dot{R}_o t + \ddot{R}_o \frac{t^2}{2} + \cdots .$$

Usually $R(t)$ can be adequately approximated by the first several terms [4]. If the origin of the imaging coordinate system is located at the radar, then in the simplest case, which will be used here, $R(t)$ can be approximated by

$$R(t) = R_o - v_l t$$

where R_o is the distance from the radar to the point reflector at $t = 0$. The received complex baseband signal from a single point reflector is now

$$q(t) = \rho \cdot s\left(\left(1 + \frac{2v_l}{c}\right)t - 2R_o/c\right).$$

The velocity v_l is very small compared to c, so that the dilation term $(1 + \frac{2v_l}{c})$ in the above equation can be ignored. Thus, rotation of the planet has negligible effect on the properties of the returned echo when the incident signal is baseband. On the other hand, a high-frequency passband signal is changed quite significantly by the planet's motion; its frequency is changed. That is, after a single range delay, the transmitted complex passband signal $s(t)e^{j2\pi f_o t}$ is received as

$$q(t) = \rho \cdot s\left(\left(1 + \frac{2v_l}{c}\right)t - 2R_o/c\right) e^{j2\pi f_o(1 + \frac{2v_l}{c})t} e^{-j2\pi f_o \frac{2R_o}{c}}.$$

Again, the dilation term in the baseband signal $s(t)$ is ignored, and the received signal can be expressed compactly as the complex passband signal

$$q(t) = \rho \cdot s(t - \tau_o) e^{j2\pi \nu_o t} e^{-j2\pi\gamma} e^{j2\pi f_o t}$$

or the complex baseband signal

$$q(t) = \rho \cdot s(t - \tau_o) e^{j2\pi \nu_o t} e^{-j2\pi\gamma}$$

where $\tau_o = 2R_o/c$ is the time delay at $t = 0$, $\nu_o = \frac{2f_o}{c} v_l$ is a Doppler shift, and $\gamma = f_o \tau_o$ is a phase offset.

More generally, if $s(t)e^{j2\pi f_o t}$ is the transmitted signal, then the complex baseband representation of the returned echo having delay τ and Doppler shift ν is

$$q(t) = \rho(\tau, \nu) \cdot s(t - \tau) e^{j2\pi\nu t} e^{-j2\pi\gamma'}$$

where $\gamma' = f_o\tau$ is a phase delay due to the time of propagation [4]. It is important to note that here $p(\tau, \nu)$ is the sum of reflectivity densities of all point reflectors at delay τ and Doppler shift ν. For example, as Fig. 1 indicates, $\rho(\tau, \nu)$ for a spherical target will have contributions from two points on the target surface for each (τ, ν). For a planetary surface having a complex reflectivity density of $\rho(\tau, \nu)$ in delay-Doppler coordinates, the returned complex baseband signal is by superposition

$$q(t) = \int_{-\infty}^{\infty} \int_{-\infty}^{\infty} e^{-j2\pi\gamma'} \rho(\tau', \nu') s(t - \tau') e^{j2\pi\nu' t} d\tau' d\nu' . \tag{4}$$

If $q(t)$ given by the above equation is the input to a filter matched to $q_o(t) = s(t - \tau)e^{j2\pi\nu t}$, then by Eq. (2) the output at $t = 0$ will be

$$y(t=0) = M(\tau, \nu) = \int_{-\infty}^{\infty} q(\beta) \left(s(\beta - \tau) e^{j2\pi\nu\beta} \right)^* d\beta \tag{5}$$

$$= \int_{-\infty}^{\infty} \left[\int_{-\infty}^{\infty} \int_{-\infty}^{\infty} e^{-j2\pi\gamma'} \rho(\tau', \nu') s(\beta - \tau') e^{j2\pi\nu'\beta} d\tau' d\nu' \right]$$
$$\cdot s^*(\beta - \tau) e^{-j2\pi\nu\beta} d\beta$$

$$= \int_{-\infty}^{\infty} \int_{-\infty}^{\infty} e^{-j2\pi\gamma'} \rho(\tau', \nu') \left[\int_{-\infty}^{\infty} s(\beta) s^*(\beta - (\tau - \tau')) \right.$$
$$\left. \cdot e^{-j2\pi(\nu - \nu')\beta} d\beta \right] e^{-j2\pi(\nu - \nu')\tau'} d\tau' d\nu' . \tag{6}$$

In most applications $(\nu - \nu')\tau'$ is either much less than one or fairly constant over the range of (τ', ν') for which $\chi(\tau - \tau', \nu - \nu')$ has significant magnitude. In either case, $e^{-j2\pi(\nu - \nu')\tau}$ can be treated as approximately constant and can therefore be ignored in Eq. (6). Consequently,

$$M(\tau, \nu) = \int_{-\infty}^{\infty} \int_{-\infty}^{\infty} e^{-j2\pi\gamma'} \rho(\tau', \nu') \chi(\tau - \tau', \nu - \nu') d\tau' d\nu' . \tag{7}$$

Hence, the output of the matched filter as a function of time delay and frequency offset is proportional to the two-dimensional convolution of the reflectivity density function with the radar ambiguity function. Therefore, the resolution achieved with range-Doppler imaging is entirely dependent upon the shape of $\chi(\tau, \nu)$. In fact, if $\chi(\tau, \nu) = \delta(\tau, \nu)$, then $\rho(\tau, \nu)$ could be recovered exactly. Yet, $\chi(\tau, \nu) = \delta(\tau, \nu)$ is impossible, and range-Doppler imaging systems can only strive for impulse-like ambiguity functions. Since $\rho(\tau, \nu)$ is actually a function of time for a rotating target, the main approximations needed for the validity of Eq. (7) are that the $\ddot{R}_o$ and other higher-order range terms can be neglected in the range function and that $\rho(\tau, \nu)$ remains essentially constant during target illumination. As we shall see, the need for this latter approximation can be circumvented through the use of a tomographic inverse SAR model.

3. 3-D SPOTLIGHT-MODE SAR AND INVERSE SAR

Figure 2 depicts the imaging geometry for spotlight-mode SAR. The radar is carried on a moving platform, and the antenna is steered to continuously illuminate the same target scene. Thus, spotlight-mode SAR synthesizes high-resolution images using data gathered from multiple observation angles. Furthermore, the processing for this form of SAR can be formulated as a tomographic reconstruction problem and analyzed using the projection-slice theorem from computer tomography (CT) [5]. Specifically, [5] showed that for a linear FM transmitted waveform, the data collected from each viewpoint, after quadrature demodulation, are a slice of the 2-D Fourier transform of the scene along the viewing angle.

Although observing the scene from different perspectives is the key to any form of SAR, a moving platform is not necessary. Instead, if the target scene of interest is moving with respect to a static radar, then the change in aspect required to image the object with high resolution may be provided by the moving target itself. When this occurs and the image is formed using SAR-like processing, the imaging system is called Inverse SAR (ISAR). The imaging of a 3-D planet with an Earth-based radar is a candidate for ISAR processing, since we may assume that Earth is stationary and the target planet is rotating. In exploring this possibility, we will consider, for simplicity, the SAR scenario (rather than ISAR) since the mathematics of SAR and ISAR are basically equivalent.

The tomographic formulation of SAR given in [5] considered the imaging of only planar scenes. This paper considers the scenario in which radar targets are assumed to lie on the surface of a 3-D structure. It is explained, using tomographic principles, that each demodulated radar return signal from a spotlight-mode collection geometry represents a radial set of samples of the 3-D Fourier transform of the target reflectivity density function. Although 3-D SAR imaging was first studied in [6], much of what follows is based on a tomographic formulation by Jakowatz and Thompson [7].

3.1 Three-dimensional tomography

To develop the 3-D tomographic formulation of SAR, it is first necessary to extend the projection-slice theorem from two to three dimensions. This extension produces two alternative forms, depending upon how the third dimension is used. Each version is relevant to the discussion of Section 3.2 and is described below.

Let $g(x,y,z)$ denote the 3-D complex-valued reflectivity density in spatial coordinates. Then define the 3-D Fourier transform of $g(\cdot)$ as

$$G(X,Y,Z) = \int \int \int g(x,y,z) e^{-j(xX+yY+zZ)} dx\,dy\,dz.$$

The two versions of the projection-slice theorem, stated below, relate certain projections of $g(\cdot)$ to corresponding slices of its Fourier transform $G(\cdot)$:

Theorem 1 *The 1-D Fourier transform of*

$$p_2(x) = \int \int g(x,y,z) dy\,dz \qquad (8)$$

is equal to a 1-D ray of $G(X,Y,Z)$ obtained by evaluating $G(\cdot)$ on the X-axis:

$$\begin{aligned} P_2(X) &= \int p_2(x) e^{-jxX} dx \\ &= G(X,0,0). \end{aligned} \qquad (9)$$

This theorem will be referred to as the linear-ray theorem for obvious reasons.

Theorem 2 *The 2-D Fourier transform of*

$$p_1(x,y) = \int g(x,y,z) dz \qquad (10)$$

is equal to a 2-D slice of $G(X,Y,Z)$ taken through the (X,Y) plane:

$$P_1(X,Y) = \int\int p_1(x,y) e^{-j(xX+yY)} dx dy$$
$$= G(X,Y,0). \quad (11)$$

Since it relates planar projection functions to planar Fourier transform sections, this theorem will be referred to as the planar-slice theorem.

By invoking the rotational property of Fourier transforms, both theorems may be generalized to an arbitrary orientation in three-dimensional space.

3.2 Three-dimensional imaging via SAR

For 3-D SAR imaging, we will consider $g(x,y,z)$ to be equal to zero everywhere inside the outer surface of the planet. This assumption is reasonable, since for high carrier frequencies in the gigahertz range, the wavelength of the transmitted waveform is too small to substantially penetrate the planetary surface.

The collection geometry for 3-D SAR is depicted in Fig. 3, where the azimuth angle θ and grazing angle ψ specify a particular orientation in the synthetic aperture from which the scene is being viewed. Following the procedure of [5], a linear FM chirp pulse, given by

$$s(t) = \begin{cases} e^{j(\omega_o t + \alpha t^2)} & \text{for } |t| \le T/2 \\ 0 & \text{else,} \end{cases} \quad (12)$$

is transmitted, where 2α is the FM chirp rate, T is the pulse duration, and ω_o is the RF carrier frequency. The returned echo from orientation (θ, ψ) follows Eq. (6) in [5] as

$$r_{\theta,\psi}(t) = A \cdot Re\left[\int_{-L}^{0} p_{\theta,\psi}(u) s\left(t - \frac{2(R+u)}{c}\right) du\right] \quad (13)$$

on the interval

$$-\frac{T}{2} + \frac{2R}{c} \le t \le \frac{T}{2} + \frac{2(R-L)}{c} \quad (14)$$

where L denotes the radius of the target and R is the distance from the antenna to the origin of the coordinate system at the target. The projection function $p_{\theta,\psi}(\cdot)$ in Eq. (13) involves the integration of the 3-D reflectivity density function $g(x,y,z)$ over the planar surfaces in Fig. 3, as opposed to integration over straight lines as in [5]. In [5] it was shown that the radar return signal, after mixing with the reference chirp (delayed by $\tau_o = 2R/c$) and low-pass filtering, is proportional to the Fourier transform of the projection function. Similarly, demodulating the returned echo in Eq. (13) gives

$$\overline{C}_{\theta,\psi}(t) = \frac{A}{2} \cdot P_{\theta,\psi}\left[\frac{2}{c}(\omega_o + 2\alpha(t-\tau_o))\right]. \quad (15)$$

According to the linear-ray theorem, $\overline{C}_{\theta,\psi}(t)$ must represent a ray at angular orientation (θ,ψ) of the 3-D Fourier transform of the unknown reflectivity $g(x,y,z)$. Since the processed return $\overline{C}_{\theta,\psi}(t)$ is available only on the interval defined by Eq. (14), $P_{\theta,\psi}(X)$ is determined only for $X_1 \le X \le X_2$ with

$$X_1 = \frac{2}{c}(\omega_o - \alpha T)$$
$$X_2 = \frac{2}{c}\left(\omega_o + \alpha T - \frac{2\alpha L}{c}\right) \approx \frac{2}{c}(\omega_0 + \alpha T). \quad (16)$$

At this point, tomographic principles can be exploited further to better understand and interpret the image that is formed. Consider the collection process in Fig. 4 with a straight-line flight path, where the surface swept out in 3-D Fourier space by a sequence of demodulated returns is in fact a plane, referred to as the slant plane. Processing returns in this manner provides samples of $G(X, Y, Z)$ on the polar grid in the annulus segment shown in Fig. 5. (The scaling is exaggerated; generally $X_2 - X_1 << \frac{2\omega_0}{c}$.) The usual procedure for forming a slant-plane image is to inverse Fourier transform the 2-D array of data in this plane, after interpolation from the polar grid to a Cartesian grid. Note that this image formation procedure places no restriction on motion through resolution cells, since our coordinate system was fixed on the target.

Let us now address the question of how this 2-D slant-plane image relates to the 3-D structure of the actual scene. Recall that the planar-slice theorem links a planar projection of a 3-D object to a planar slice of its Fourier transform. Therefore, the inverse 2-D transform of the interpolated slant-plane data will produce an image that is the projection of the 3-D target in the direction normal to the slant plane in the image space. From Eq.(10) this type of projection is a 2-D function in which each value corresponds to a line integral in a given direction. Figure 3 portrays this situation. Notice that the w-axis specifies the direction of integration, while the u- and v-axes define the slant plane. Given the assumption made earlier that the 3-D reflectivity function is nonzero only on the surface, line integration here is equivalent to the projection onto the slant plane of the two points (or in some cases several discrete points) on the surface where it is intersected by the line of integration.

4. CONCLUSION

In this paper we reviewed the standard range-Doppler processing model used for radar astronomy and explained that the image constructed using this technique is the 2-D convolution of the planet's reflectivity density with the radar ambiguity function. This form of processing, however, is based on a stationary coordinate system that is not affixed to the rotating planet. Thus, in this model the reflectivity density is actually a function of time. This temporal dependence (resulting in motion through resolution cells) is not accounted for in the range-Doppler model. The difficulty with the temporal dependence of the reflectivity can be overcome by simply fixing the coordinate system on the rotating planet and using a tomographic ISAR model. The resulting processing involves polar-to-Cartesian interpolation followed by a 2-D FFT, if a linear FM waveform is transmitted. The ISAR processing can be easily modified for the case with an arbitrary waveform [8].

Reference [8] contains considerable additional information on the topic of this paper and shows that for the case where there is no motion through resolution cells, range-Doppler and ISAR imaging will produce nearly the same image. In the future we hope to apply ISAR processing to data acquired at the Arecibo Observatory. In the case of imaging asteroids, it may be necessary to estimate the rotation rate directly from the radar data. We are also interested in exploring true 3-D imaging within the ISAR framework.

References

[1] P. E. Green, Jr., "Radar astronomy measurement techniques," Technical Report 282, Massachusetts Institute of Technology Lincoln Laboratory, Lexington, MA, December 1962.

[2] T. Hagfors and D. B. Campbell, "Mapping of planetary surfaces by radar," *Proc. of the IEEE*, 61, Sept. 1973.

[3] C. E. Cook and M. Bernfeld, *Radar Signals: An Introduction to Theory and Application*, Academic Press, New York, 1967.

[4] R. E. Blahut, *Theory of Remote Surveillance Algorithms*, IBM, 1990.

[5] D. C. Munson, Jr., J. D. O'Brien and W. K. Jenkins, "A tomographic formulation of spotlight-mode synthetic aperture radar, *Proc. of the IEEE*, 71 (1983) 917-925.

[6] J. L. Walker, "Range-Doppler imaging of rotating objects," *IEEE Trans. Aerosp. Electron. Syst.*, AES-16, 23-52, January 1980.

[7] C. V. Jakowatz, Jr. and P. A. Thompson, "A three-dimensional tomographic formulation for spotlight-mode synthetic aperture radar," Sandia National Labs, Albuquerque, New Mexico, 1992.

[8] M. R. Coble, "High-resolution radar imaging of a rotating sphere," M.S. Thesis, University of Illinois at Urbana-Champaign, 1993.

Figure 1: Contours of equal range (solid lines) and equal Doppler shift (dashed lines).

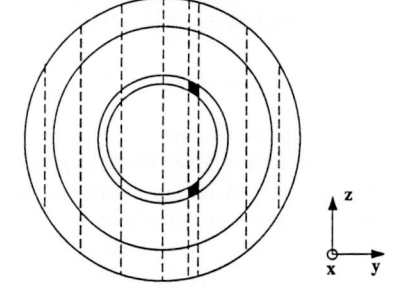

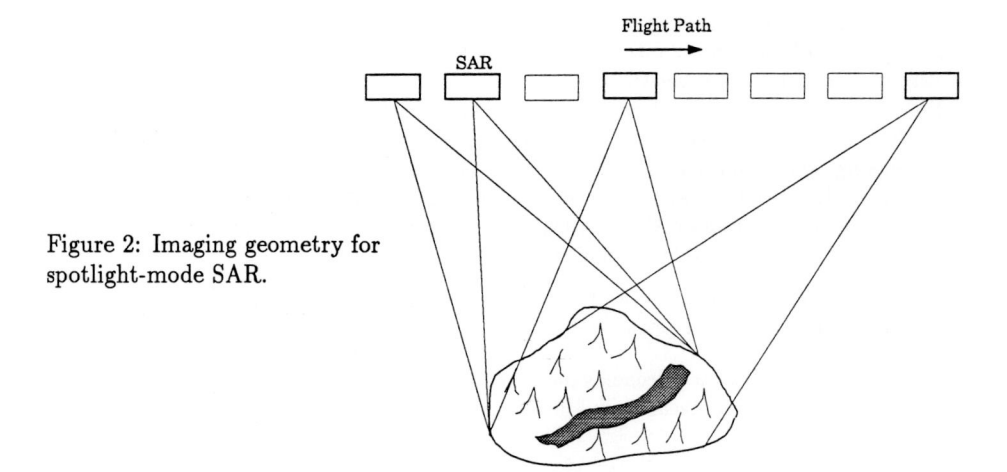

Figure 2: Imaging geometry for spotlight-mode SAR.

Figure 3: Imaging geometry for 3-D SAR.

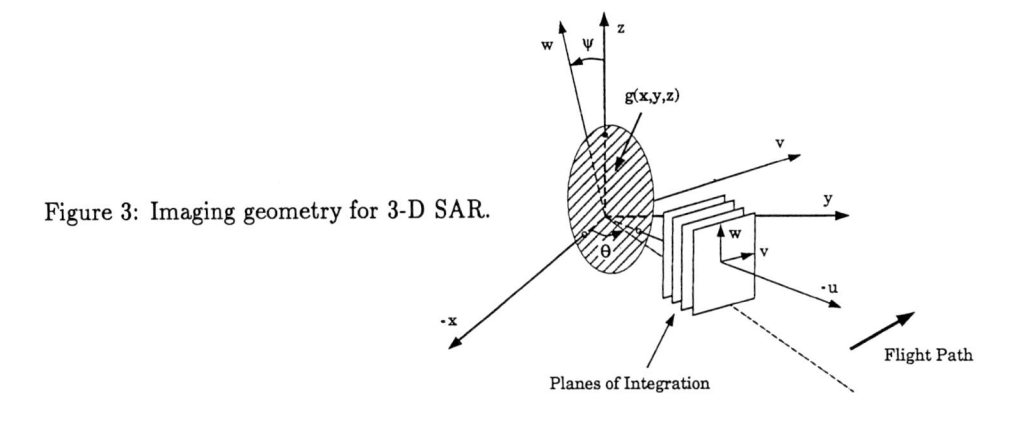

Figure 4: Projection of a 3-D object onto the SAR slant plane.

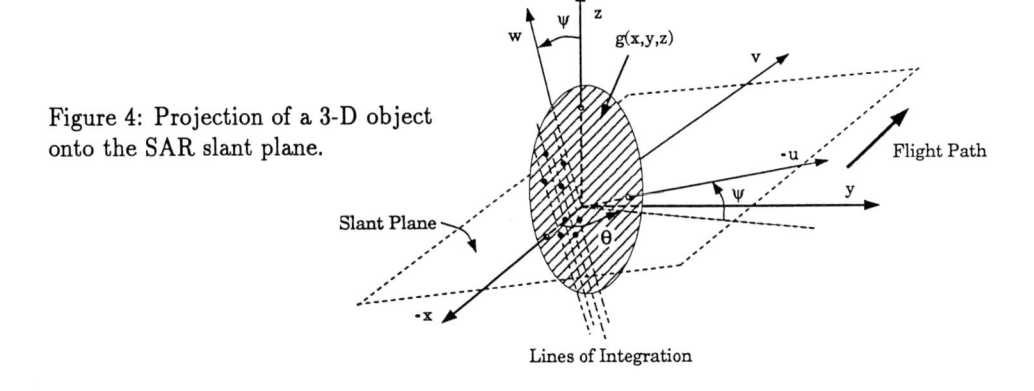

Figure 5: Annulus segment containing returned samples of $G(X, Y, Z)$.

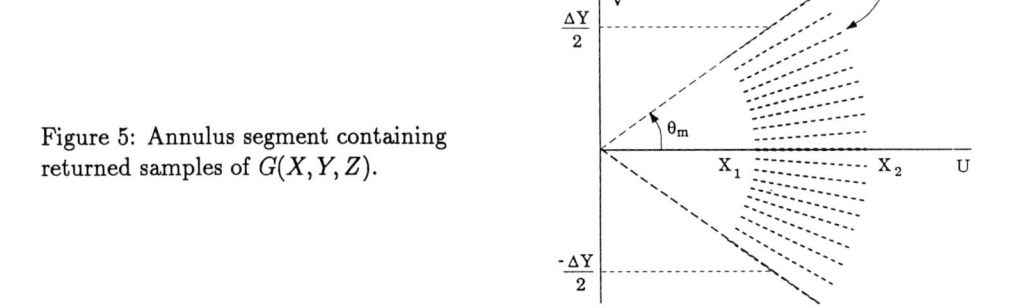

Multispectral Image Processing

P.E. Trahanias and A.N. Venetsanopoulos

Department of Electrical & Computer Engineering, University of Toronto, Toronto, Ontario, Canada M5S 1A4

Abstract

Automatic processing of multispectral (multichannel[1]) images has received increased attention recently due to its importance in processing color images (TV, video), satellite images and multispectral images that result from various multichannel acquisition devices. In this paper we review the approaches developed for multispectral image processing. A new class of filters for multispectral image processing is also proposed and briefly analyzed.

1. VECTOR IMAGE PROCESSING

Conventional approaches to multispectral image processing are based on processing the different image components (channels) separately [1, 2]. However, these approaches fail to utilize the inherent correlation that is usually present in multispectral images. Consequently, *vector processing* of multispectral images is desirable. Recently, this has been recognized by many researchers and has been adopted in many tasks including filtering, restoration and edge detection [3, 4, 5, 6]. The term vector processing indicates the vector nature of multichannel images, where the value at each pixel is an m−dimensional vector (m is the number of image channels).

2. VECTOR MEDIAN FILTERS

An important case of vector image processing operators are the vector median filters (VMF) that have been introduced as extension of scalar median filters to the multichannel case [7, 8]. VMF can be derived either as maximum likelihood estimates when the underlying probability densities are double-exponential or using vector order statistics techniques [9]. In the former case, consider an m-dimensional distribution

$$f(\mathbf{g}) = \gamma e^{-\alpha \|\mathbf{g}-\beta\|} \tag{1}$$

where γ and α are scaling factors and $\beta = (\beta_1, \beta_2, \ldots, \beta_m)^T$ is the location parameter of the distribution. The maximum likelihood estimate $\hat{\beta}$ for β, based on a random sample $\{\mathbf{g}_1, \mathbf{g}_2, \ldots, \mathbf{g}_N\}$ from (1) can be obtained by maximizing the likelihood function

$$L(\beta) = \prod_{i=1}^{N} \gamma e^{-\alpha \|\mathbf{g}_i - \beta\|} \tag{2}$$

---

[1]The terms multispectral and multichannel will be used interchangeably throughout this text

There is no closed form solution of (2); a suboptimal estimate can be found if the additional requirement that $\hat{\beta}$ be one of the g_i is added. This leads to the definition of the vector median as

$$g_{VM} \in \{g_i | i = 1, 2, \ldots, N\} \tag{3}$$

and

$$\sum_{i=1}^{N} \|g_{VM} - g_i\| \leq \sum_{i=1}^{N} \|g_j - g_i\|, \ \forall j = 1, 2, \ldots, N \tag{4}$$

The vector median of a population can also be defined using vector order statistics techniques. In this case, it is defined as the minimal vector according to the aggregate ordering technique, which leads again to equation (4) [9].

3. EXTENSIONS OF VECTOR MEDIAN FILTERS

Based on vector order statistics, extensions or modifications of the vector median filter can also be defined. The *vector α−trimmed mean* [10] is defined as

$$g_{V\alpha TM} = \sum_{k=1}^{N(1-2\alpha)} \frac{1}{N(1-2\alpha)} g_{(k)} \tag{5}$$

where

$$g_{(1)} \leq g_{(2)} \leq \cdots \leq g_{(N)}$$

is an ordering of g_i.

Based on the distances from the population mean a class of estimators can be defined as [6]

$$g_i^D = \begin{cases} g_i & if \ \|\bar{g} - g_i\|_D \leq t \\ g_j \in W^* & : \ \min_g \|g_j - g_i\|_2 \end{cases} \tag{6}$$

where D denotes a distance measure (Euclidean or Mahalanobis) and W^* is a *confidence set*, i.e. the vectors contained in W^* are not outliers and, therefore, they are not replaced by the filter.

4. VECTOR DIRECTIONAL FILTERS

A new class of filters for processing vector-valued signals has been recently proposed by the authors [11, 12]. It is called vector directional filters (VDF); the principle underlying these filters is vector ordering with the novelty that the ordering criterion is the *angle* between the image vectors. VDF actually operate on the direction of the image vectors; as a result, the processing of vector data is separated into "directional processing" and "magnitude processing". This property of VDF (separation of processing) establishes a link between multichannel signal processing and single-channel signal processing. In the later case the signal values are scalars and, therefore, magnitude processing is performed. Such a signal can also be seen as a vector-valued signal in which the direction of all the vectors is exactly the same (parallel vectors) and the magnitude (length) of each vector is equal to the signal value at this point. Directional processing is, therefore, not needed in this case.

In vector-valued signals, however, each vector is uniquely characterized by its direction and its magnitude, which makes the idea of processing these components separately, very attractive. After directional processing, a multichannel signal can locally be seen as a single-channel signal since this processing results in vectors having the same (approximately) direction. Consequently, the bulk of techniques developed for grey-level signal/image processing can be employed in the next step, namely magnitude processing. This is indeed a major advantage of this approach since it facilitates the use of many efficient image processing operators (order statistics, max/median, α-trimmed mean, morphological) to multichannel images.

Let α_i correspond to $\mathbf{g}_i$ and defined as

$$\alpha_i = \sum_{j=1}^{n} \mathcal{A}(\mathbf{g}_i, \mathbf{g}_j), \ i = 1, 2, \ldots, N \tag{7}$$

where $\mathcal{A}(\mathbf{g}_i, \mathbf{g}_j)$ denotes the angle between the vectors $\mathbf{g}_i$ and $\mathbf{g}_j$, $0 \leq \mathcal{A}(\mathbf{g}_i, \mathbf{g}_j) \leq \pi$. An ordering of the α_is

$$\alpha_{(1)} \leq \alpha_{(2)} \leq \cdots \leq \alpha_{(N)} \tag{8}$$

implies the same ordering to the corresponding $\mathbf{g}_i$s

$$\mathbf{g}_{(1)} \leq \mathbf{g}_{(2)} \leq \cdots \leq \mathbf{g}_{(N)} \tag{9}$$

The first term in (9), $\mathbf{g}_{(1)}$, defines the output of the *basic vector directional filter* (BVDF). The first k terms in (9), $\{\mathbf{g}_{(1)}, \mathbf{g}_{(2)}, \ldots, \mathbf{g}_{(k)}\}$, constitute the output of the *generalized vector directional filter* (GVDF).

BVDF chooses as the output vector the one that minimizes the sum of the *angles* with all the other vectors. In other words, it chooses the vector most centrally located, without considering the vector magnitude. It can be proved that this is the least error estimate of the angle location. GVDF generalizes BVDF in the sense that its output is a superset of the (single) BVDF output. GVDF outputs the set of vectors whose angle from all the other vectors is *small*. In other words, the output set of the GVDF consists of vectors centrally located in the population with (approximately) the same direction in the vector space. Consequently, GVDF achieves, in a sense, to produce (locally) a *single-channel* signal, since the set of vectors produced contains samples in the same direction.

The GVDF output should subsequently be passed through a filter $\mathcal{F}$ in order to produce a single output at each pixel. Since the GVDF output consists of vectors with (approximately) the same direction, $\mathcal{F}$ may consider only the magnitudes of the vectors, i.e. it can be any grey-scale image processing filter.

VDF posses many properties that make them appropriate in multispectral image processing. Among them we mention invariance under scaling and rotation, preservation of step edges and convergence to root signals.

References

[1] S. Naqvi, N. Galagher, and E. Coyle, "An application of median filtering to digital television," in *IEEE ICASSP*, pp. 2451–2454, April 1986.

[2] R. Strickland, C. Kim, and W. McDonnel, "Digital colour image enhancement based on the saturation component," *Optical Engineering*, vol. 26, pp. 609–616, July 1987.

[3] R. Machuca and K. Phillips, "Applications of vector fields to image processing," *IEEE Trans. Pattern Anal. Mach. Intell.*, vol. PAMI-5, pp. 316–329, May 1983.

[4] N.P. Galatsanos and R.T. Chin, "Digital restoration of multichannel images," *IEEE Trans. Acoust. Speech Signal Process.*, vol. 37, pp. 415–421, March 1989.

[5] P.E. Trahanias and A.N. Venetsanopoulos, "Color edge detection using vector order statistics," *accepted for publication, IEEE Trans. Image Process.*, August 1992.

[6] R.C. Hardie and G.R. Arce, "Ranking in R^p and its use in multivariate image estimation," *IEEE Trans. Circuits and Systems for Video Tech.*, vol. 1, pp. 197–209, June 1991.

[7] J. Astola, P. Haavisto, and Y. Neuvo, "Vector median filters," *Proc. of the IEEE*, vol. 78, pp. 678–689, Apr. 1990.

[8] M. Gabbouj, E.J. Coyle, and N.C. Gallagher, "An overview of median and stack filtering," *Circuits, Systems, Signal Process.*, vol. 11, no. 1, pp. 7–45, 1992.

[9] V. Barnett, "The ordering of multivariate data," *J. Royal Statistical Society A*, vol. 139, Part 3, pp. 318–343, 1976.

[10] S. Sanwalka and A.N. Venetsanopoulos, "Vector order statistics filtering of colour images," in *13th GRETSI Symp. on Signal and Image Processing*, pp. 785–788, 1991.

[11] P.E. Trahanias and A.N. Venetsanopoulos, "Vector directional filters – A new class of multichannel image processing filters," *submitted for publication, IEEE Trans. Image Process., under review*, July 1992.

[12] P.E. Trahanias and A.N. Venetsanopoulos, "Multichannel image processing using vector-angle ranking," in *accepted for presentation, SPIE Conf., Nonlinear Image Processing IV*, (San Jose, CA), Jan.31 - Feb.5 1993.

Fractal Transforms: Complexity versus fidelity

D. M. Monro

School of Electronic & Electrical Engineering, University of Bath, BA2 7AY, England

Abstract
In this paper the fidelity of approximating images by fractal functions is investigated over a range of complexity options, by combining types of searching with orders of approximation. The implications are discussed for application to picture archiving systems and real time video.

1. BACKGROUND

Two fractal distinct block coding techniques have been described. In Jacquin's ITT-coding [1], each "range block" is encoded by mapping from a larger "domain block". Monro and Dudbridge [2] encoded a block by tiling it with reduced copies of itself, using a least-squares criterion to derive an optimal mapping. Both of these methods are particular cases of the more general Bath Fractal Transform (BFT) [3, 4]. Here we explore some of the complexity options which combine various degrees and types of searching with various orders of approximation.

2. THEORY

To encode an image, define an Iterated Function System (IFS)

$$W = \{w_k; k = 1,..., N\} \tag{1}$$

whose attractor A is a tiling of the image. A fractal function f is defined on A as

$$f(w_k(x, y)) = v_k(x, y, f(x, y)) \tag{2}$$

where the mappings v_k have parameters $\alpha_i^{(k)}$; $k = 1,..., N$; $i = 1, 2, ... M$. Here, N is the order of the IFS and M is the number of free parameters of the BFT. The function f is found by minimizing $d(g, \tilde{g})$ over tile k, where g is the greyscale function of the block, $\tilde{g}$ is its mapping

$$\tilde{g}(x) = v_k(w_k^{-1}(x), g(w_k^{-1}(x))) \tag{3}$$

and d is a suitable metric, with respect to the parameters $\alpha_i^{(k)}$ of v_k. We solve

$$\frac{\partial \, d(g,\tilde{g})}{\partial \alpha_i^{(k)}} = 0 \tag{4}$$

for all i and k to obtain the BFT. If the w_k are affine transforms and the v_k are polynomials, the BFT is the solution of N systems of M linear equations. Our previous coding method [2, 4] tiled a block by four copies of itself ($N = 4$), using a bilinear BFT with four parameters ($M = 4$).

In ITT-coding, the image is tiled by adjoint "range blocks". For each of these a larger "domain block" is mapped onto it, which is selected by searching the domain blocks. The mappings may include rotation, reflection or greylevel scaling, and always shrink the domain onto the range. The ensemble of mappings is an IFS of high order. To study the combined complexity options, the range block is here called a "tile", and a "parent" is the same as the domain block.

3. SEARCHING OPTIONS

In this evaluation, 4 by 4 pixel tiles and 8 by 8 pixel parent blocks are used. The parents align with tile boundaries, hence they overlap one another. A level zero search chooses the parent that a tile is inside without searching, as in [2] and [4]. A level-one search considers each of the four parent blocks which overlap the tile. The BFT is derived for each parent, by computing the metric $d(g, \tilde{g})$, and thus the best parent for each tile is identified. A level-two search includes all the parents which overlap those in level one, and so on. Figure 1 identifies the 16 parent blocks in a level 2 search, the 4 in level 1 and the single parent of the level 0 search, all for the shaded tile.

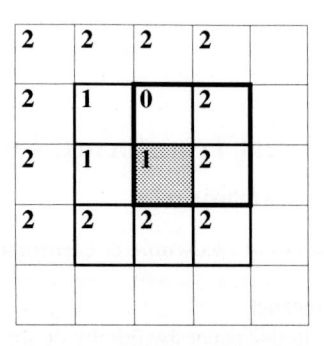

2	2	2	2
2	1	0	2
2	1	1	2
2	2	2	2

Figure 1. Parents labelled, upper left, by search level.

4. COMPLEXITY OPTIONS

A parent block might be rotated to any of four orientations, and there are five possible reflections, giving 20 combinations of both. With the BFT, various orders of polynomial fractal functions can be used. In the simplest (zero-order) case, each mapping v_k has the form

$$v_k(x, y, f) = s_k f + t_k. \tag{5}$$

as in Jacquin's method [1]. Also considered are higher order variants without cross products

$$v_k(x, y, f) = a_3^{(k)}x^3 + a_2^{(k)}x^2 + a_1^{(k)}x + b_3^{(k)}y^3 + b_2^{(k)}y^2 + b_1^{(k)}y + s_k f + t_k. \tag{6}$$

Table 1 (a). RMS Error, Search Level 0, Gold Hill Fragment					Table 1 (b). RMS Error, Search Level 2, Gold Hill Fragment				
BFT Order	NoRots or Refls	4 Rots Only	5 Refls Only	20 Rots & Refls	BFT Order	NoRots or Refls	4 Rots Only	5 Refls Only	20 Rots & Refls
0	16.54	15.48	15.44	15.26	0	13.25	11.86	11.80	11.58
1	11.19	10.63	10.68	10.47	1	9.74	9.06	9.08	8.85
2	9.51	9.17	9.20	9.08	2	8.33	7.90	7.91	7.75
3	9.06	8.79	8.83	8.73	3	8.18	7.72	7.73	7.58

5. RESULTS

A 128 by 128 pixel fragment of the intensity (Y) component from the standard test image "Gold Hill" was chosen for investigation. Table 1 shows the rms errors measured over the image fragment for combinations of complexity at two searching levels. Either rotations or reflections alone give a similar reduction in rms error, which from experience we know will be noticeable in the picture quality. Searching over all 20 combinations provides little additional benefit, for both the searches shown. More is gained by increasing the search level than by consideration of rotations and reflections. A noticeable feature is that in all cases but one the rms error is reduced further by using a higher order approximation than it is by considering rotations and reflections.

The rms errors have also been evaluated for different orders of approximation over a wider range of searching levels. Here the entire Gold Hill image is coded, and the rms errors obtained were generally smaller because the fragment used above contains more detail than other areas. Figure 2 is a graph of the results. A noticeable feature of these graphs is the flatness of all the curves. Figure 3 shows the original fragment of "Gold Hill" along with its approximation using level-zero searching and zero-order BFT, and by level-six searching and cubic BFT, both without rotations or reflections. The rms errors over the fragments are given in each case.

6. DISCUSSION

Unlike many transforms, fractal transforms involve a loss of fidelity, so understanding of these losses is necessary before attempting to use them for compression. All of the complexity options studied here are capable of reducing the error in image approximation by a fractal transform. The improvements offered by considering reflections and/or rotations is relatively small compared to the other options, and in combination little is gained for considerable cost. In nearly every case, the gain in fidelity achieved by using either higher order approximation or more searching is greater. The fidelity of the fractal transform increases with both search level and the order of the approximating function. The flatness of the curves in Figure 2 favours increased order over search level in trading cost against fidelity.

Also in relation to cost, the optimal Minimal Plotting Algorithm (MPA) [5] for decoding of fractal functions can only be applied to the level zero case in which no pixel from a parent is mapped into a different parent. This favours zero searching, which also produces the most symmetrical cost between coding and decoding.

All fractal transforms can be decoded at low cost. In picture archiving systems where images are compressed once and decompressed often, fractal technology could be the preferred method. Low orders of the BFT with little searching are also fast to encode [2, 4] and so could be of value in video mail applications. Finally all fractal methods have scaling properties, so that they can be used to decode at resolutions which are either coarser or finer than the original.

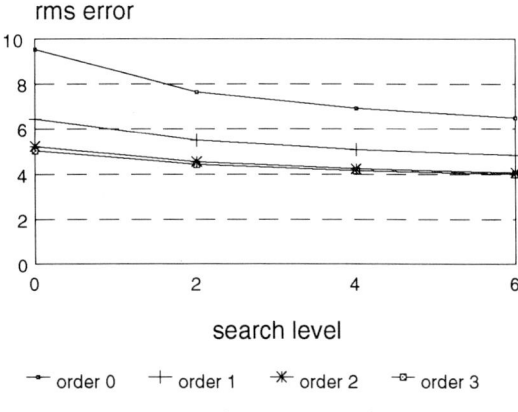

Figure 2. RMS errors as a function of BFT order and search level.

Figure 3 (a). Original image fragment.

Clearly practical use of fractal transforms must take account of a three way trade, between cost, fidelity and compression. All of the complexity options discussed involve a loss of compression. We postulate that the compression/fidelity trade of the BFT will be approximately neutral. If so, cost considerations would favour a second or third order approximation as a more efficient and symmetric fractal transform over the zero order approximation with any or all of the complexities of searching, rotation and reflection.

7. REFERENCES

[1] A. E. Jacquin, "Image coding based on a fractal theory of iterated contractive image transformations", *IEEE Trans Image Processing*, Vol.1 No. 1, pp. 18 - 30, 1992.

[2] D. M. Monro and F. Dudbridge, "Fractal approximation of image blocks", Proc. IEEE ICASSP, pp. III: 485-488, 1992.

[3] D. M. Monro, F. Dudbridge and J. A. Dallas, "Least squares fractal interpolation", Ecole Polytechnique Colloquium on Fractals in Engineering, Montreal, 1992.

[4] D. M. Monro and F. Dudbridge, "Fractal block coding of images," *Electronics Letters*, vol. 28, no. 11, pp. 1053-1054, 21 May 1992.

[5] D. M. Monro, F. Dudbridge and A. Wilson, "Deterministic rendering of self-affine fractals", *IEE Colloquium on Fractal Techniques in Image Processing*, London, 1990.

Figure 3 (b). BFT order 0, no search, e_{rms} 16.54.

Figure 3 (c). BFT order 3, search level 6, erms 6.74

Region-Based Coding of Images Using a Spline Model

R. Baseri and J.W. Modestino

Electrical, Computer and Systems Engineering Department and Center for Image Processing Research, Rensselaer Polytechnic Institute, Troy, New York, 12180

Abstract

This paper describes a second-generation region-based coding scheme operating in a subband environment and based on approximating the lowest-frequency subband of an image with a 2-D spline model. Following a segmentation procedure, parametric polynomials of appropriate orders are fitted to the resulting contiguous regions of the lowest-frequency subband. The residual error term plus the higher-frequency subbands are encoded using conventional pixel-based quantization schemes while we describe an efficient encoding scheme for the 2-D spline representation. An entropy-constrained design procedure is developed for designing these individual encoders to minimize the overall mean-square distortion subject to a constraint on the output entropy. Experimental results are described demonstrating the substantial superiority of this approach relative to more traditional pixel-based entropy-constrained subband encoding schemes.

1. INTRODUCTION

Most current image compression algorithms are pixel-based techniques in that they operate on individual pixels (or small blocks of pixels) without taking advantage of the significant structures and features often present in real-world images. To increase compression ratios beyond those offered by pixel-based techniques, compression schemes must expand the domain of their operation beyond coding individual pixel intensities to higher-level entities. An example is the region-based coding scheme described here.

Some of the previous related work in this area has been presented in [1]-[3]. Unlike any previously published work, this paper describes a new approach to image coding using spline approximations in a subband environment. Since the lowest-frequency band of the image consists mostly of relatively large regions of approximately constant intensity, it is intuitively appealing to model such regions in terms of polynominal functions. Higher-frequency bands, which contain most of the texture information, generally cannot be efficiently modeled by polynomials. In this work higher-frequency subbands are coded using traditional pixel-based coding schemes.

Using a previously-developed segmentation algorithm [4], the lowest-frequency band is decomposed into regions consisting of pixels with nearly constant intensities. The spline approximation is calculated for the image by representing each region with a polynomial of appropriate order. Given the contour map, together with the parameters used for modeling each region of the image, an approximation to the original image can be reconstructed. An overall block diagram of the resulting region-based encoder is illustrated in Fig. 1.

2. SPLINE REPRESENTATION OF IMAGES

The 2-D spline model in this work is constructed by representing each region by a parametric polynomial surface of the form

$$P(x,y) = \sum_{I=0}^{n(k)} \sum_{i=0}^{I} A_{k,I,i} x^i y^{I-i} \quad \text{for} \quad (x,y) \in \Re_k \ ; \ n(k) \in \{0,1,2,\ldots,m\}, \quad (1)$$

where x and y represent the horizontal and vertical locations of pixels within the image and $\Re_k$, $k = 1, 2, \ldots, K$, is the set of pixel locations constituting the k'th region of the image. Here, $n(k)$ represents the order of the polynomial used to approximate the k'th region which we assume is of most order m. The $A_{k,I,i}$ represent the coefficients of the spline model corresponding to the k'th region.

The total number of coefficients in the spline model is thus given by $C = \sum_{k=1}^{K} c(k)$, where $c(k)$ is the number of coefficients used for representing the k'th region. The problem of determining the appropriate $n(k)$ for each segment can be cast as an optimization problem aimed at minimizing the mean-square difference between the actual low-frequency subband and its spline approximation, subject to a given complexity as represented by the total number of polynomial coefficients, C.

This problem is similar to that of optimal bit allocation, given an overall rate constraint. A number of methods for determining the optimal bit allocation have been developed. In this work a variation of the BFOS algorithm [5] was used to determine the $n(k)$.

The coefficients of the spline representation, $A_{k,I,i}$, are calculated using an algorithm similar to that described in [1]. The spline representation thus consists of a contour map defining the boundaries between adjacent regions together with the coefficients of the approximating polynomials. Chain codes are used to represent the boundaries with compression achieved by entropy coding the indices.

3. QUANTIZATION OF SPLINE COEFFICIENTS

Figure 2 shows a histogram of the coefficients of a typical spline approximation to a natural image. The skewed histogram make spline coefficients a suitable candidate for entropy-coding schemes. Uniform threshold scaler quantizers are used to quantize the spline coefficients with the number of quantizer reconstruction levels chosen from the set $\{32, 64, 128, 256\}$. The quantizer indices were then entropy coded.

Figure 3 illustrates typical resulting operational rate-distortion characteristics when a spline model is used to approximate the lowest-frequency band of the image "lena". Each curve corresponds to a specific quantizer characterized by its number of reconstruction levels. To calculate the operational rate-distortion curve for each quantizer, the number of spline coefficients, C, is varied. In particular, for each value of C, the optimum spline approximation is determined using the BFOS algorithm to determine the optimum polynomial order for each region. The coefficients are then quantized and the mean-squared distortion in approximating the actual subband with the spline approximation is then determined together with the first-order entropy of the quantizer indices. The resulting operational rate-distortion functions are then traced out by varying C. The overall operational rate-distortion characteristics, $R(D)$, for operating the quantizers is then defined as the lower envelope of the curves presented in Fig. 3.

4. RESULTS

Experimental results have shown that the spline coding method developed in this work outperforms traditional pixel-based coding techniques. To illustrate this point, operational rate-distortion characteristics of the spline coding method and that of a representative pixel-based subband coding technique [6] are depicted in Fig. 4. The image in this case was manually segmented. Observe that the spline coding method is able to provide a 2 dB peak-signal-to-noise-ratio (PSNR) improvement.

In order to investigate the performance of this method in an unsupervised fashion, the EM segmentation algorithm described in [4] was used. The model for this work consisted of a NSHP (Non-symmetric half-plane) autoregressive model for the pixel intensities, and a Markov random field for modeling the class distributions. The EM algorithm was used to perform a 7-class segmentation of the lowest-frequency subband of "lena". The resulting operational rate-distortion characteristics are illustrated in Fig. 5. It is evident that even though an unsupervised segmentation algorithm is used, the spline coding method is still able to perform considerably better than the pixel-based scheme. An example of results obtained by the spline coding method using the EM segmentation scheme is presented in Fig. 6. This image was coded at 0.25 bits/pixel and resulted in 33.8 dB PSNR.

5. CONCLUSIONS

In conclusion, it has been demonstrated that region-based coding in a subband environment using spline models is a more suitable method than traditional pixel-based schemes. The results clearly show that the region-based coding method introduced in this work provides higher signal-to-noise-ratios than pixel-based methods especially at low bit rates.

6. REFERENCES

1. M. Gilge, T. Engelhardt, and R. Mehlan, "Coding of arbitrary shaped image segments based on a generalized orthogonal transformation," *Signal Processing: Image Communication*, Vol. 1, pp. 153-180, 1989.
2. M. Kunt, A. Ikonomopoulos, and M. Kocher, "Second generation image coding techniques," *Proc. of the IEEE*, Vol. 73, No. 4, pp. 549-574, Apr. 1985.
3. M. Kunt, M. Benard, and R. Leonardi, "Recent results in high-compression image coding," *IEEE Trans. Circuits and Systems*, Vol. CAS-34, No. 11, pp. 1306-1336, Nov. 1987.
4. J. Zhang, J. Modestino, and D. Langan, "Maximum-likelihood parameter estimation for unsupervised stochastic model-based image segmentation," Submitted to *IEEE Trans. Inform. Theory*, Vol. IT-37, pp. 400-402, Mar. 1991.
5. E.A. Riskin, "Optimal bit allocation via the generalized BFOS algorithm," *IEEE Trans. Inform. Theory*, Vol. IT-37, No. 2, pp. 400-402, Mar. 1991.
6. Y.H. Kim and J.W. Modestino, "Adaptive entropy coded subband coding of images," *IEEE Trans. Image Process.*, Vol. IP-1, pp. 31-48, Jan. 1992.

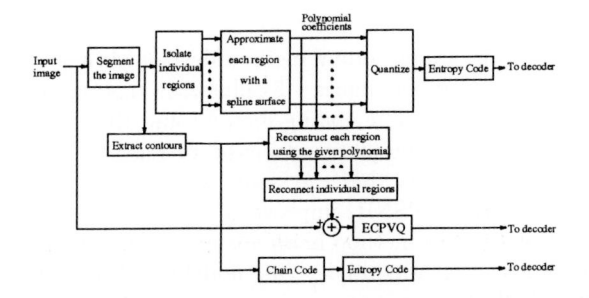

Fig.1 System block diagram.

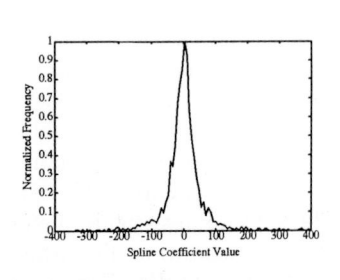

Fig.2 Spline coefficient histogram.

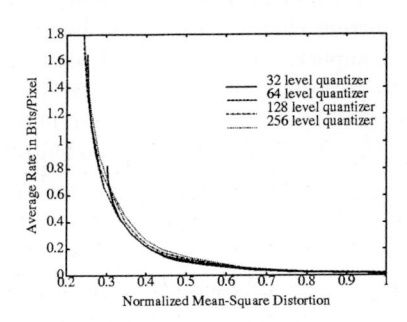

Fig.3 R-D curves for spline approx. operating on "lena".

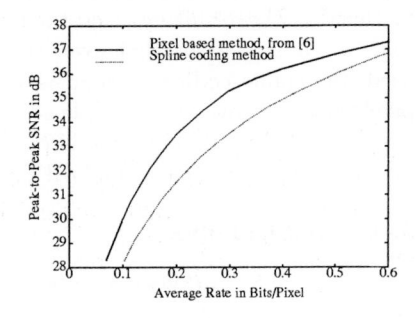

Fig.4 Spline coding vs. a representative pixel-based scheme.

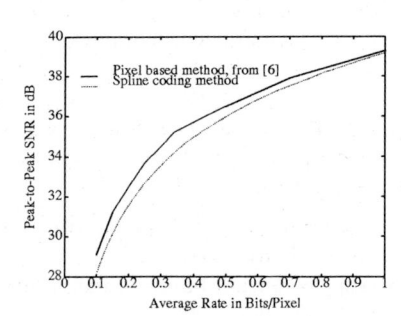

Fig.5 Spline coding vs. a representative pixel-based scheme, using EM-based segmentation.

Fig.6 "lena", Spline coded at 0.25 bits/pixel, 33.8 PSNR.

Compatible HDTV coding using MDPCM and VQ

Xudong Song and Yrjö Neuvo

Signal Processing Laboratory, Tampere University of Technology,
P.O. Box 553, SF-33101 Tampere, Finland

Abstract

In this paper, we present a compatible HDTV coding system based on one-stage nonlinear pyramid structure, which is used to separate an interlaced HDTV signal into a compatible interlaced TV signal and an interlaced HDTV residual signal. A MDPCM scheme with variable-length coding (VLC) is used to code the compatible interlaced TV signal. A high-quality compatible TV signal is obtained at 36 Mbps. A VQ scheme is used to code the HDTV residual signal at 72 Mbps. From the compatible TV signal and the HDTV residual signal, a high-quality HDTV signal can be reconstructed at 108 Mbps.

1. INTRODUCTION

The aim of high-definition television (HDTV) is to provide the viewer with a more realistic viewing experience than is offered by today's television. Having a wider aspect ratio than conventional television displays is one of major features of HDTV. Another distinction is the potential ability to deliver much higher resolution than today's television standards can accommodate. The channels that are available today do not provide the high bandwidth required to deliver these high-quality signals to the consumer. Recent advances in high-speed optical transmission and VLSI signal processing have provided opportunities to create the Broadband Integrated Services Digital Network (BISDN) with seemingly unlimited capacity. One of the most promising applications in the BISDN is the digital transmission of HDTV signals.

The coding of HDTV signal for digital transmission via BISDN has recently drawn considerable attention [1, 2, 3]. Chen et al [1] proposed a subband-based scheme for transporting digital HDTV over BISDN at approximately 130 Mbps. Tzou proposed [2] a DCT-based coding system for 130Mbps HDTV transmission in ATM networks. Sakurai et al [3] proposed an adaptive subband-DPCM and subband-DCT coding system for 156Mbps HDTV transmission in an SDH-based STM network. On the other hand, the compatibility issue between HDTV and standard TV has been the focus of a great deal of research [4, 5]. In a compatible coding scheme, standard TV coding data are embedded in HDTV coding data as their subsets.

In this paper, a HDTV/TV compatible coding system based on one stage nonlinear pyramid structure is introduced, which can achieve high quality HDTV picture compression, while ensuring compatibility with TV.

2. SYSTEM DESCRIPTION

A block diagram of the system is shown in Fig. 1. A one-stage nonlinear pyramid structure is used to realize a compatible HDTV and TV coding system. In the first path, the incoming high-resolution interlaced HDTV signal is first low-pass filtered by detail-preserving multistage median filter (MMF) and then subsampled horizontally and vertically by a factor of two. The subsampled picture is the interlaced TV signal. After the subsampling, the low-pass interlaced TV signal goes through a median DPCM (MD-PCM) encoder. The decoded interlaced TV signal is interpolated back to its original picture size and is then subtracted from the original interlaced HDTV signal. The resulting difference signal in this second path is then coded using vector quantization (VQ). The advantage of this approach is that any residual aliasing present in the interpolated low-pass signal and coding error in the first path will be included in the high-pass signal in the second path. Consequently, during the reconstruction process, the high-pass signal will be able to cancel the aliasing due to the interpolation and coding error at the receiver.

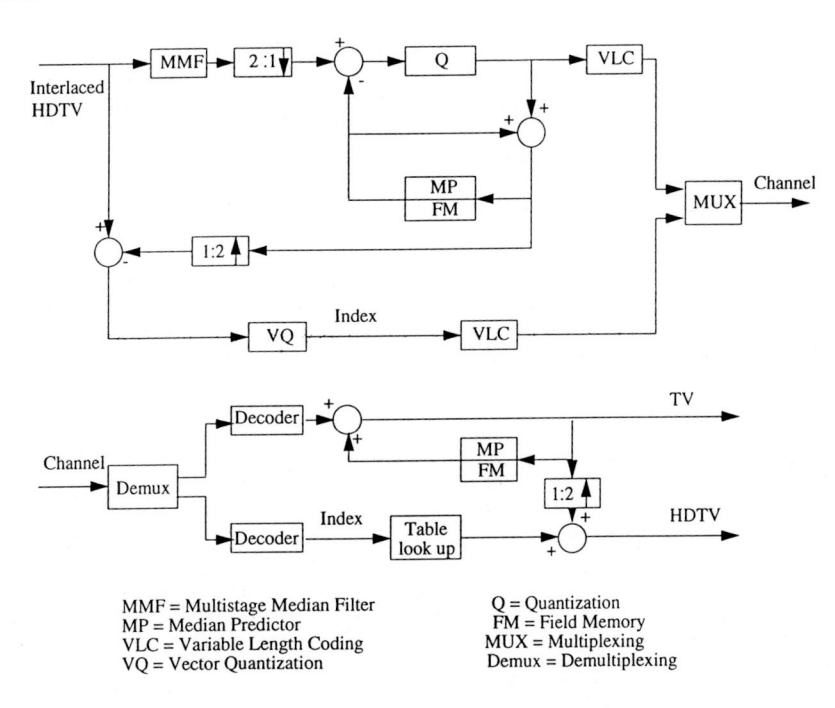

Fig. 1 Block diagram of a compatible HDTV encoder and decoder.

3. MDPCM CODING OF LOW-PASS HDTV SIGNAL

In MDPCM coder, three-dimensional median type predictor is designed in such a way that motion is automatically taken into account [7]. In other words, when there is motion the predictor output is selected from the current field (spatial prediction). In the still parts of the image sequence the previous frame is used for prediction (temporal prediction). The pixel configuration for the three-dimensional median prediction is shown in Fig. 2.

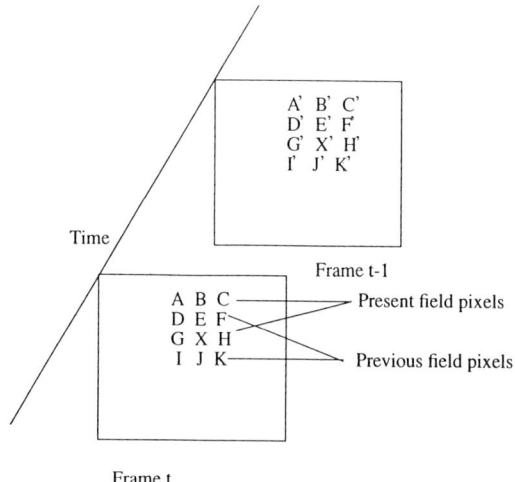

Fig. 2 The pixel configuration for the three-dimensional median prediction.

$$P = MED[G, B - B' + X', X', G + B - A, G - G' + X'] \tag{1}$$

where A, B, and G, are previously reconstructed pixels which are in the same field as pixel X. P is the prediction of the present pixel X. The previously reconstructed pixels B', G', and X' are in the previous frame. Pixel X is in the same spatial position as pixel X'. The prediction error signal is then quantized using a symmetric nonuniform quantizer.

4. VQ CODING OF HIGH-PASS HDTV SIGNAL

Vector quantization (VQ) with a block size of 4×2 is applied to encode high-pass HDTV signal to achieve sufficient bit-rate reduction. The 4×2 is used since vertical correlation is smaller than the horizontal correlation in a field. Due to the presence of large smooth region and high energy edge information in the residual interlaced HDTV signal, the use of a classified VQ schemes is under investigation to improve the performance of this path.

5. SIMULATION RESULTS

Computer simulations have been conducted to evaluate the proposed coding system. The 10 frames (20 fields) of the HDTV sequence UNBELDI were used in the simulations. The sampling frequency for UNBELDI is 54 MHz for Y and 27 MHz for U and V. The UNBELDI sequence is interlaced at 50 fields/sec, each field consisting of 1440×576 pixels. We encoded the Y, U, V separately.

In the simulation, a multistage median filter (2LM+) [8] (3×3) was used to form one-stage nonlinear pyramid. Because of the broad bandwidth of HDTV signals having more random noise than conventional TV signals, the multistage median filter (MMF) is used for random noise reduction while not causing picture degradation such as blurring. To evaluate the performance of the proposed coding scheme, a Burt's linear filter ($a = 0.6$) [6] is used to from one-stage linear pyramid. Table 1. shows the first-order entropy of high-pass HDTV signal on various filters. It is seen from Table. 1 that one-stage nonlinear pyramid gives smaller entropy for an interlaced HDTV residual signal than those obtained with one-stage linear pyramid.

Table 1. The first-order entropy of high-pass HDTV signal on various filters.

Entropy	Filter type	
	2LM+ (3 × 3)	Burt ($a = 0.6$)
Field.3.y	4.307318	4.412960
Field.3.u	3.625636	3.690894
Field.3.v	3.546426	3.608307

The Y-component of the compatible interlaced TV signal was MDPCM encoded using three-dimensional median prediction with 35-level tapered quantizer as shown in Table 2. Table 2 only shows the output level of zero and the positive side of this tapered quantizer. 35-level VLC codeword was derived based on the statistical analysis of the quantized error signal. The chrominance signals U and V were MDPCM coded using 15-level tapered quantizer obtained from Table 2 (top eight codewords). The luminance

Table 2. Quantizing characteristics

Input level	Output level	VLC
-2 ... 2	0	00
2 ... 7	4	01
7 ... 13	10	101
13 ... 21	17	10001
21 ... 30	26	1000010
30 ... 39	35	10000110
39 ... 49	44	100000110
49 ... 60	55	1000001010
60 ... 71	66	1000011100
71 ... 83	77	10000010110
83 ... 95	89	10000111010
95 ... 108	102	10000111011
108 ... 121	115	1000001011110
121 ... 134	128	1000001011101
134 ... 147	141	10000011100100
147... 161	154	10000010111111
161 ... 174	169	10000010111110010
174 ... 255	179	1000001011111001111

signal Y of the high-pass interlaced HDTV signal was divided into 4×2 blocks, each block was vector-quantized using a codebook with 256 codevector. The codebook of the vector quantizer was generated from a set of 10 fields high-pass HDTV signal. The chrominance signals U and V of high-pass interlaced HDTV signal were vector-quantized in the same manner as for the luminance signal. Only the size of codebook is 64. A Huffman coding scheme is applied to achieve further data compression.

Fig. 3 Original picture of the third field of the UNBELDI sequence.

Fig. 4 Reconstructed HDTV picture of the third field

Fig. 5 Reconstructed compatible TV picture of the third field.

Figure 3 shows the original luminance picture of the third field of the UNBELDI sequence. Figure 4 shows the reconstructed HDTV picture of the third field. Figure 5 shows the reconstructed compatible TV picture of the third field. It is seen from Fig. 4 and Fig. 5 that the observed picture quality of the reconstructed HDTV and TV pictures was very good.

6. CONCLUSIONS

In this paper, a compatible HDTV coding system based on one-stage nonlinear pyramid structure is proposed. The advantage of one-stage nonlinear pyramid is that it not only can produce a detail-preserving interlaced TV signal but also give smaller first order entropy for an interlaced HDTV residual signal. Our simulations show very promising results using the proposed scheme.

References

[1] T.C. Chen, P.E. Fleischer and S.M. Lei, "A subband scheme for advanced TV coding in BISDN applications", *Proc. of 3rd Workshop on Signal Processing of HDTV*, L'Aquila Italy, Sept. 1989.

[2] K.H. Tzou, "An intrafield DCT-based HDTV coding for ATM networks", *IEEE Trans. Circuits Syst. Video tech.*, Vol. 1, No. 2, pp. 184-196, June 1991.

[3] N. Sakurai, K. Irie, and R. Kishimo, "A transmission characteristics of subband coded HDTV signal transmission system," *Globecom'92*, pp. 1098-1102 Dec. 1992.

[4] M. Breeuwer and P. H. N. de With, "Source coding of HDTV with compatibility to TV," *SPIE Visual Communications and Image Processing*, Lausanne, pp. 765-776, Oct. 1990.

[5] T.C. Chen, K.H. Txou and P.E. Fleischer, "A hierarchical HDTV coding system using a DPCM-PCM approach," *SPIE Vol. 1001 Visual Communications and Image Processing'88*.

[6] P. Burt and E. H. Adelson, "The laplacian pyramid as a compact image code," *IEEE Trans. Commun.*, Vol. COM-31, N0. 4, pp. 532-540, April 1983.

[7] T. Jarske and Y. Neuvo, "Adaptive DPCM with median type predictors", *IEEE Trans. Consumer Electronics*, Vol. 37, N0. 3, pp. 348-352, Aug. 1991.

[8] A. Nieminen, P. Heinonen, and Y. Neuvo, "A new class of detail-preserving filters for image processing," *IEEE Trans. Pattern Anal. machine Intell.*, Vol. PAMI-9, Jan. 1987.

[9] Y. Linde, A. Buzo, and R. M. Gray, "An algorithm for vector quantizer design," *IEEE Trans. Commun.*, Vol. COM-28, pp. 84-89, Jan. 1980.

[10] X. Song and Y. Neuvo, "Image data compression using nonlinear pyramid vector quantization," *Signal processing VI: theories and applications*, 1992 Elsevier Science Publishers.

Coding of Motion Compensated Prediction Error Images by Morphological Segmentation

Wei Li and Murat Kunt
Swiss Federal Institute of Technology

Abstract

A morphological segmentation algorithm is presented for coding the motion compensated *prediction error images* (PEI). Low correlation of the PEIs favors the adoption of such a segmentation based algorithm. An average compression ratio of 50 is obtained with excellent reconstruction quality on the CCIR 601 test sequence.

1 Introduction

With the increasing importance of digital television, image sequence coding becomes a highly active research domain. Motion compensation is generally used to decorrelate the data along the temporal direction. Although large amount of researches have been dedicated to the motion compensation algorithms, much less efforts have been done to the encoding of the motion compensated prediction error images (PEI). Due to the uncovered background and the simplified motion models, there exist certain high energy zones in the PEIs. Linear transforms are still widely used to code them. However, statistics of the PEIs [1] demonstrate that they have little correlation. This leads us to concluding the inappropriateness of the linear transforms for the PEIs, and to the adoption of a segmentation based algorithm.

In this paper, a morphological segmentation algorithm is presented, which segments a PEI into a small number of segments to be coded. Due to the temporal masking effect of the human visual system, the isolated small segments can be removed without affecting the visual quality. Small interior regions are merged to its adjacent segments so as to reduce the number of small contours. The proposed method has the advantage to keep the sharpness of the moving objects' contours. The de-interlaced CCIR601 test sequence *Table Tennis* is used to test the proposed coding method, and the multigrid block matching algorithm [2] is employed for the motion compensation.

The rest part of the paper is organized as follows. Section 2 deals with the segmentation algorithm. Quantization and entropy coding of the segments are treated in Section 3. In Section 4, simulation results are presented. Conclusions are given in Section 5.

2 Morphological Segmentation Algorithm

The four major steps of the segmentation algorithm can be described as follows.
(1) Dynamic Thresholding
The threshold is determined dynamically for each PEI by computing its histogram and fixing the threshold to separate the pixels into high energy part ($\beta\%$ of the pixels) and

low energy part. A marker image is produced where a white region corresponds to pixels with absolute value higher than the threshold in original image.

(2) Elimination of Small Segments

The isolated white or black pixel in the marker image can be removed by examining the V-4 neighbors of each point and putting its color to the color of majority neighbors. The procedure to remove small white segments is composed of a median filtering, n successive erosions on the white regions, followed by a geodesic reconstruction of the large segments which are not completely removed.

(3) Merging of Small Interior Regions

The small interior regions should be merged to the large segments in order to make the entire coding efficient. This step consists principally of m erosions on the black region followed by a geodesic reconstruction.

(4) Contour Processing

There exist various algorithms to code contour position. A very efficient one is proposed in [3] to code the contours with only three codewords. Nevertheless, it imposes the V-4 connectness of adjacent contour points. An opening is performed on the marker image, giving the final segmentation. The borders between white and black regions in the marker image are kept as contour pixels, leading to V-4 connected closed contours.

3 Quantization and Coding

Pixels of the PEI which belong to the segments are quantized using a uniform quantizer. This quantization is made possible due to the generally lower spatial resolution of human eyes in the moving parts of images. The quantized values are then coded using an adaptive arithmetic coder [4]. Each contour is coded by sending the absolute coordinates of the first contour pixel followed by relative positions of the following contour pixels coded by the same adaptive arithmetic coder. Let R_s be the number of bits to code the pixel values of contours, R_c the number of bits to code contour points, I the number of segments, and $M \times N$ the size of image. Suppose $2^{(b_1-1)} < M \leq 2^{b_1}$, and $2^{(b_2-1)} < N \leq 2^{b_2}$, then the first contour pixel position will cost $b_1 + b_2$ bits. The total bits R needed to code a PEI is thus given by:

$$R = R_s + R_c + I \cdot (b_1 + b_2). \tag{1}$$

4 Simulation Results

The algorithm is tested using PEIs computed from the first 12 frames of the test sequence *Table Tennis* (576 × 720). Parameters in the algorithm are set as given in [1], that is, $n = 4$, $m = 2$, $\beta = 10$. Quantization step size is fixed to 4. Table 1 summarizes the segmentation and coding results. The PSNR of each reconstructed PEI is around 46.8dB. As an example, the original PEI computed from the $3rd$ and $4th$ frames is shown in Fig. 1(a), while the segmented image to be coded with the superposed contours is shown in Fig. 1(b). It can be seen that the segments correspond well the large high energy zones in the original PEI. Figure 2 shows the original $4th$ frame and the reconstructed one assuming that the receiver disposes a perfect previous frame. The reason for this assumption is just to isolate the PEI coding problem. The reconstructed sequence shows little visible defects. In particular, sharp contours of moving objects are well preserved, which lead to very pleasant visual quality.

5 Conclusions

A morphological segmentation based method is described to code the motion compensated prediction error images. This contour-region approach preserves sharpness of the moving objects' contours and makes use of certain characteristics of human visual systems, e.g. the spatial-temporal resolution trade-off. Compression ratio around 50 is achieved with excellent quality of the reconstructed sequence. Such promising results mean that all sequence coding systems incorporating motion compensation can be improved by using the segmentation based coding methods for prediction error images.

Acknowledgment

The authors are grateful to Dr. Henri Nicolas for the proofreading of the present paper and for his helpful suggestions.

References

[1] W. Li and F.X. Mateo. Segmentation based coding of motion compensated prediction error images. In *ICASSP'93*.

[2] F. Dufaux. Multigrid based motion estimation for interframe image sequence coding. In *Proc. of Eusipco'92*, volume III, pages 1323–1326.

[3] M. Eden and M. Kocher. On the performance of a contour coding algorithm in the context of image coding part i: contour segment coding. *Signal Processing*, 8(10):381–386, 1985.

[4] I.H. Witten, R.M. Neal, and J.G. Cleary. Arithmetic coding for data compression. *Communications of the ACM*, 30:520–540, June 1987.

frame	H	I	C	S	R_c	R_s	R	C_r
2	4.4	12	3152	15477	6200	67256	73696	45.0
3	4.4	8	3082	12625	6112	54496	60768	54.6
4	4.4	12	3800	15928	7240	70544	78024	42.5
5	4.4	12	2464	11843	5056	56456	61752	53.7
6	4.8	6	1048	7454	2536	40960	43616	76.1
7	5.5	4	766	5708	2008	32672	34760	95.4
8	5.7	11	2575	9548	4816	51648	56684	58.5
9	4.8	10	3444	15761	6264	77536	84000	39.5
10	4.6	25	3849	15638	7120	71960	79580	41.7
11	4.5	27	3767	14404	7112	65272	72924	45.5
12	4.5	39	4737	16791	8696	72248	81184	40.9
total	4.7	166	32684	141177	63160	661048	726988	50.2

Table 1: Simulation results on *Table Tennis*. H is the estimated entropy of the original PEIs. I is the number of contours, C is the number of contour pixels, S is the number of pixels included in the segments, and C_r is the compression ratio.

(a) (b)

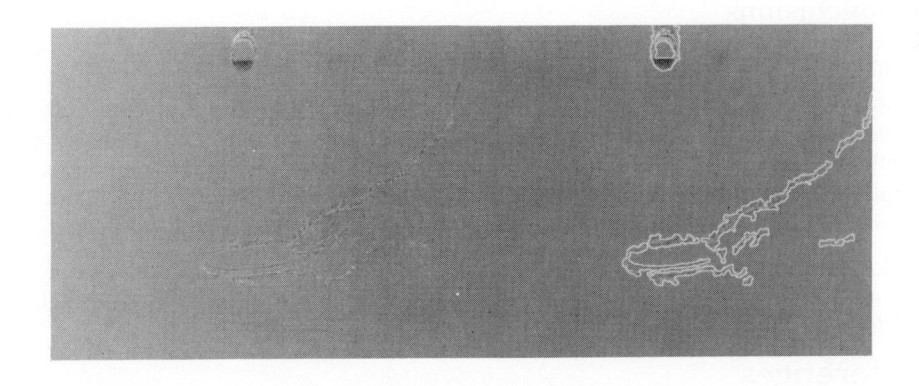

Figure 1: (a)Original PEI, (b) Segmented PEI.

(a) (b)

Figure 2: (a)Original 4th frame of *Table Tennis*, (b) Reconstructed 4th frame.

Two-dimensional delayed-decision lossy image coding via copying and the application to intraframe low-rate image compression

Ei-ichiro Takemasa, Takashi Komatsu, Takahiro Saito

Department of Electrical Engineering, Kanagawa University, 3-27-1 Rokkakubashi, Kanagawa-ku, Yokohama, 221, Japan

Abstract
Recently we have extended one of the conceptions of lossless universal pattern-matching coding, viz. the concept of coding via copying, to multi-dimentional lossy coding, thus developing an efficient coding technique for irreversible image compression. To enhance the adaptability of the coding technique, the work introduces delayed-decision mechanism along with a continuous scanning method, and applies the delayed-decision coding technique to low-rate image compression.

1. INTRODUCTION

In the lossless universal pattern-matching coding developed by Ziv and Lempel[1], its pattern-set is gradually built up in the process of coding incoming data, and hence pattern-set mismatch does not occur. We have recently extended one of the conceptions of lossless universal pattern-matching coding, viz. the concept of coding via copying, to multi-dimensional lossy coding, thus developing a novel universal pattern-matching coding technique for irreversible image compression and demonstrating its superiority over the adaptive DCT coding technique at a low coding rate[2]. In this paper, we refer to this coding technique as PMC for short.

The previously proposed PMC technique employs multiple different coding modes, and adaptively chooses the proper coding mode for a current input subblock by the threshold logic based on the waveform distortion. This adaptive selection of the coding mode is not optimal, because the current selection of the coding mode affects future behavior of the encoder. To enhance the adaptability of the coding mode selection, this paper incorporates a kind of delayed-decision mechanism known as the (M,L) algorithm[3] along with a continuous scanning method instead of the usual raster scanning method into the PMC technique[2], and applies the delayed-decision PMC to low-rate image compression.

2. UNIVERSAL PATTERN-MATCHING CODING VIA COPYING (PMC)[2]

PMC employs the concept of coding via copying. The concept is to encode future subblocks with the waveform distortion less than the allowed distortion value D^* via maximum-length copying from a frame memory containing the recently decoded subblocks. If any copying operation does not yield the waveform distortion less than the allowed distortion value D^*, then the incoming subblock is encoded by some conventional image coding technique, which is referred to as residue coding. Figure 1 illustrates the copying operation, which is identified by both the address of starting point and the number of copied subblocks. Compression is achieved by transmitting them. To alleviate the computational requirement, we restrict a search area used for determining the starting point to some small region. We define all the subblocks contained in the search area as a subblock-template.

The coding algorithm for compression of still images is described below :
(0)Input subblock :
Partition a given input image into nonoverlapping square subblocks of 4 x 4 pels, scan subblocks, and feed them to the encoder as an input subblock.
(1)Copying Mode :
Compare a given input subblock with all the subblock-templates contained in the search area defined in Fig. 2 and determine the best subblock-template yielding the minimum waveform distortion. If the minimum distortion is less than the allowed distortion value D*, then start the copying operation at the location of the best subblock-template; if not, proceed to the step (2).
In the copying operation, the number of the copied subblocks is determined by the following procedure. In Fig. 1, the coder computes the waveform distortion between the input subblock X1 and the past subblock Y1. If the computed distortion is less than the allowed distortion value D*, then the coder continues the copying operation and applies the above procedure to the next input subblock X2; if not, the encoder stops the copying operation.
(2)Residue Coding Mode :
Encode a given input subblock by using the conventional image coding technique. We employ the conventional orthogonal transform coding technique for residue coding, of which uses extrapolative prediction and the discrete sine transform[4].

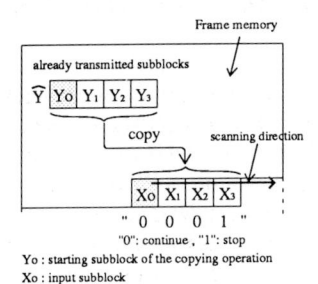

Figure 1 - Copying operation

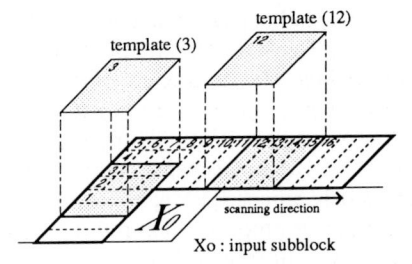

Figure 2 - Definition of the search area

3. DELAYED-DECISION PMC(DD-PMC)

We incorporate an efficient instrumentable multi-path tree search procedure known as the (M,L) algorithm[5] into the coding mode selection of PMC. The (M,L) algorithm keeps M possible branches in contention for a finite period of L consecutively scanned input subblocks, and after the delay of L subblocks determines the fittest path which gives the minimum cumulative cost among M paths in contention. We define the cost C, which is allocated to the branch emanating from the node, as a function combining the squared waveform distortion with the code length :
$$C = D + \lambda \cdot R \qquad (1)$$
D; squared waveform distortion , R; code length , λ; Lagrange multiplier

The delayed-decision PMC encoder employs the three different coding modes:
[1] Lead Copying Mode (C/L-mode)
Compare a given input subblock with all the subblock templates contained in the search area defined in Fig.2, and then start some new copying operations at their respective locations of the good subblock-templates chosen up to a fixed number in the order of increasing the squared waveform distortion value. The number of the selected starting points is limited to Ns. In this coding mode, the multi-path tree has Ns branches emanating from a node, each branch of which corresponds to each selected starting point.
[2] Successive Copying Mode (C/S-mode)
If and only if the previous subblock is encoded via copying, this sub-mode can be applied to a current input subblock. In this sub-mode, the encoder continues the current copying operation.
[3] Residue Coding Mode (R-mode)

As the method for scanning given input subblocks, the usual raster scanning method is not suited to this coding technique because of the repetitive discontinuity. Hence, the work presented here employs a continuous scanning method. Figure 3 illustrates the simplest example of continuous scanning methods; other continuous scanning methods such as the zigzag scanning may be used, but generally speaking complex scanning methods render the coding procedure rather complicated. Hence, we employ the continuous scanning method illustrated in Fig.3.

4. SIMULATIONS

The delayed-decision PMC(DD-PMC) is simulated on test still images with 512 lines and 512 pixels per line. Figure 4 shows the simulation results for various values of M, L, and Ns. The signal-to-noise ratio, SNR, is defined by
$$SNR = 20\log 10(255/R.M.S.E) \qquad (3)$$
The simulation results show that the value of the search parameter L is not critical and the DD-PMC encoder performs close to its best at L=16~32. As the values of M and Ns increase, its coding performance is improved gradually; but the DD-PMC encoder with large values of M and Ns needs a vast computational effort and does not seem practicable. The simulation results suggest values of (M,L,Ns)=(8,16,1) as being adequate for the PMC technique, and the (8,16,1)-based DD-PMC encoder works well to provide satisfactory performance with its comparatively low computational complexity. Figure 5 shows the SNR versus coding rate performance for the test image "Weather Forecast". In Fig.5, the SNR versus coding rate performance of the (8,16,1)-based DD-PMC encoder is compared with that of the other coding techniques :
[1]Instantaneous-Decision PMC(ID-PMC)[2]
Our previously proposed PMC. The delayed-decision mechanism is not used. In addition, the usual raster scanning method is employed.
[2]EP-DST[4]
This coding technique is used as the residue coding technique in PMC.
[3]SAC[5]
Scene adaptive 8 x 8 DCT coder proposed by Chen and Pratt.
In Fig.5, DD-PMC shows the best coding performance, and the incorporation of the (M,L) algorithm along with the continuous scanning method into PMC improves the coding performance by about 2 dB or more at low coding rates under about 0.4 bit per pixel.

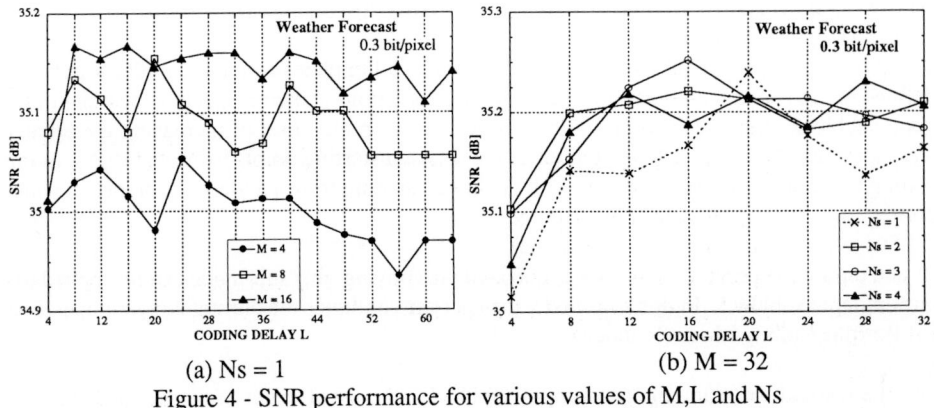

Figure 4 - SNR performance for various values of M,L and Ns

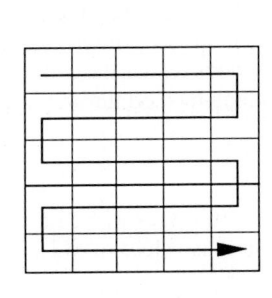

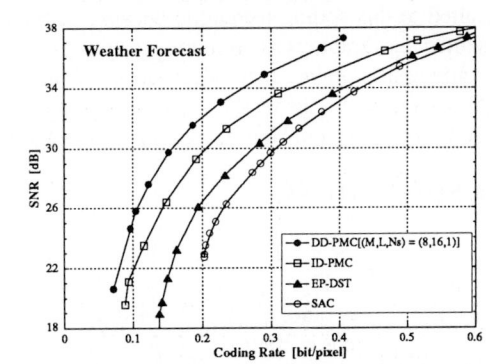

Figure 3 - A continuous scanning method

Figure 5 - SNR vs coding rate performance

5. CONCLUSIONS

The delayed-decision mechanism improves the adaptability of the coding mode-selection in PMC, and provides a significant performance improvement especially at low coding rates.

6. REFERENCES

1 J.Ziv and A.Lempel, A Universal Algorithm for Sequential Data Compression, IEEE Trans. Inform. Theory, IT-23, (1977), 337-412.
2 T.Saito et.al., Self-Organizing Pattern-Matching Coding for Picture Signals ,Proc. IEEE Int. Conf. Acoust., Speech & Signal Process., (1989), 1671-1674.
3 J.B.Anderson and J.B.Bodie, Tree Encoding of Speech, IEEE Trans. Inform. Theory, IT-21, (1975), 379-387.
4 N.Yamane et.al., An Image Data Compression Method Using Extrapolative Prediction-Discrete Sine Transform; In The Case of Two-Dimensional Coding, Trans. IEICE Japan, J71-B, (1988), 717-724, (in Japanese).
5 W.H.Chen and W.K.Pratt, Scene Adaptive Coder, IEEE Trans. Commun., COM-32, (1984), 225-232.

Improving the Error Resilience of Entropy Coded Video Signals

D. W. Redmill and N. G. Kingsbury

Signal Processing and Communications Laboratory, Cambridge University Engineering Department, Trumpington Street, Cambridge, CB2 1PZ UK, Tel: [+44] 223 33 2767

Abstract

This paper addresses the problem of designing video coders that give good performance over channels with unpredictable bit error rate (BER), whilst maintaining the compression achievable with standard coding schemes [1], designed for clean channels. Error correction coding works well if the BER is known in advance to be fairly small and constant, but not well for channels with variable BER or long bursts [2].

The error resilient entropy code (EREC) is introduced as a method for adapting existing entropy coding schemes to give increased error resilience. The EREC is applicable to many types of data and is well suited for video coding as it can give increased priority to important motion vector and low frequency information over less critical high frequency information. The EREC has been applied to produce a modified version of the H261 coding scheme [1]. The resulting scheme has been simulated and shown to have almost identical performance to standard H261 coding in clean channels whilst providing a much improved performance at high bit error rate (BER).

INTRODUCTION

In order to design error resilient coding strategies we need to understand how channel errors affect the performance of video decoders. There are several ways in which channel errors can propagate within the decoder and degrade its performance:
- Bit decoding. In variable length code-word schemes channel errors can effect the decoding of following code-words.
- Code-word decoding. In most schemes the meaning of individual code-words is dependent on previous code-words.
- Spatial domain. In most sub-band schemes a single error in a sub-band will affect a local region of the picture (not just one pixel).
- Temporal domain. In systems that employ temporal prediction, errors in one decoded frame will propagate to all dependent frames.

In this paper we are primarily concerned with minimising the first two forms of channel error propagation by maintaining synchronization between the encoder and decoder.

Traditional methods of maintaining synchronization involve the inclusion of unique synchronization words at regular points. However, these add redundancy to the code and thus degrade the performance in clean channels.

It has been shown [3] that some Huffman codes contain synchronization codewords, after which the bit decoding is resynchronized. These techniques have also been developed [4,5] to resynchronize the code-word decoding with a minimal added redundancy.

Due to the nature of the constraints on synchronizing codewords, they are forced to be longer than average length, and thus in order to maintain low redundancy they must be used relatively infrequently. The result is schemes that maintain reasonable performance at moderate BER, but degrade seriously at high BER.

THE ERROR RESILIENT ENTROPY CODE (EREC)

The algorithm introduced in this paper is a development of the error resilient positional code (ERPC) [6]. The data to be coded is split into a fixed number N of blocks, each consisting of a variable length b_i bits. Each block is constrained to be an **overall prefix code** which means that in the absence of channel errors the decoder knows when it has finished decoding a block without reference to any following information. The total number of bits to be coded is also constrained to be less than or equal to some pre-chosen total T bits that can be transmitted within the EREC structure. The output code comprises of N slots of s_i bits (usually equal length) such that $\sum b_i \leq T = \sum s_i$.

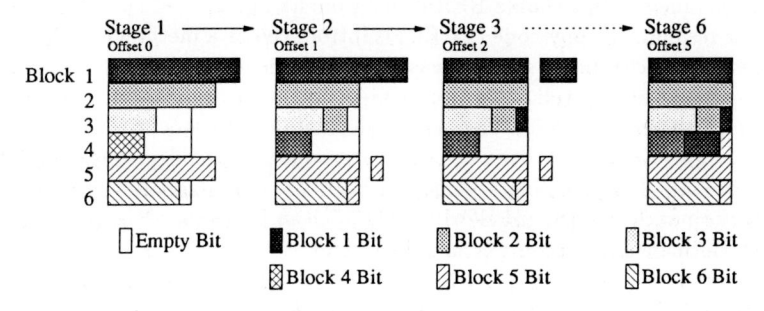

Figure 1: Bit Reallocation Algorithm.

Initially each variable length block of input data is assigned to the corresponding slot, as in figure 1 stage 1. Some input blocks will contain less than the available bits in the corresponding slot; ie $b_i < s_i$ leaving $s_i - b_i$ bits unused. These can be used for the bits that are left over from blocks that contain too many bits $b_i > s_i$. Figure 1 shows how the encoder algorithm operates in order to fill these spaces: it uses a fixed offset sequence ϕ_n (in this case 0,1,2,3,4,5) so that, at stage n, unplaced bits from block i are placed in slot $i + \phi_n$ (modulo N) if space is available. Provided the total number of input bits is $\leq T$, all the bits will be placed within N stages of the algorithm.

In the absence of channel errors the decoder can follow this process by decoding all blocks simultaneously until it finds the end of each block, which is possible because the code is an overall prefix code.

This code is error resilient because the structure of N fixed length slots means that the decoder knows exactly where to start decoding each variable length block of data. Some error extension does occur, but this effect is relatively minor. If important data is kept towards the beginning of blocks then it is less likely to suffer from the propagation of errors.

These ideas can be extended to provide a level of quad-tree decomposition that will perform well with sparse data. We define a macro-block (MB) as a group of four blocks of data. The data to be coded can now be split into two levels, the MB level and the block level. Each macro-block can now be assigned to a code slot. After coding the MB level data, the coder can allocate active blocks to code slots. If there are more active blocks than slots then some blocks will not be allocated a code slot, and these will have to be added onto the end of other data.

APPLICATION TO VIDEO CODING

We have based our experimentation on the H.261 coding scheme [1]. In order to examine error resilient aspects of the code it is essential to include some decay in the feed-back loops. Thus we have slightly modified the H.261 scheme by using a fixed loop filter for the whole frame and removing the intra-frame prediction of motion vectors. This scheme is referred to as the standard scheme. Note that this scheme includes synchronization words (PSC and GBSC), but not the channel coding.

We have then applied the EREC to produce an improved scheme. The synchronization words (PSC and GBSC) have been omitted as they are no longer needed. The MB address words have been replaced with a single activity bit for each MB. The encoder is limited to a total of less than T bits by removing excess high frequency coeficients. The EREC has been applied by using one slot for each MB (16×16 pixels). The number of active blocks has been limited to be less than or equal to the number of slots.

Both schemes have been simulated for the 'Claire' sequence (256×256 pel 8 bit monochrome 10 frames/s) sequence at 75 kbit/s. Note for the standard scheme this is an average bit rate while for the EREC it is fixed for each frame. Figure 2 shows these results. The performance of the EREC in clean channels was found to be within 0.1 dB of the standard scheme which is considered to be insignificant. There is a large improvement in noisy channels. Subjectively the pictures at high BER have accurate low frequency and motion, at the expense of blurring caused by errors in high frequency terms.

If the truncation of high frequency coefficients is not accounted for in the encoder feedback loop, then the system could in principle be used in conjunction with an existing coder to recode the data into an error resilient structure. Experimentation shows a loss in quality of about 0.1 dB with this constraint.

SUMMARY

The EREC coding strategy introduced in this paper has several advantages over conventional variable length coding techniques:
- It is very versatile and could be applied in conjunction with many variable length coding strategies for sources such as still pictures and audio as well as video.
- If designed correctly there is virtually no redundant information.
- Performance is improved and degradation is graceful with increasing BER.
- Important data such as motion and low frequencies can be given greater resilience.

The principal constraint is that the overall code must be of fixed length per frame. This can be overcome by using more complicated hybrid schemes.

REFERENCES

[1] *H261. Specification for p*64 Kbit/s flexible hardware*. CCITT Standard Recommendation H.261, Part II., 1989.

[2] N. MacDonald. Transmission of compressed video over radio links. In *VCIP*, pages 1484–1488, 1992.

[3] T. J. Ferguson and J. H. Rabinowitz. Self-synchronizing Huffman codes. *IEEE Trans on Info Theory*, 30:687–693, 1984.

[4] B. L. Montgomery and J. Abrahams. Synchronizing of binary source codes. *IEEE Trans on Info Theory*, 32:849–854, 1986.

[5] W. M. Lam and A. R. Reibman. Self-synchronizating variable-length codes for image transmission. In *ICASSP*, pages III–477 – III–480, 1992.

[6] N. T. Cheng and N. G. Kingsbury. The ERPC: an efficient error-resilient technique for encoding positional information of sparse data. *IEEE Trans on Comms*, 40:140–148, 1992.

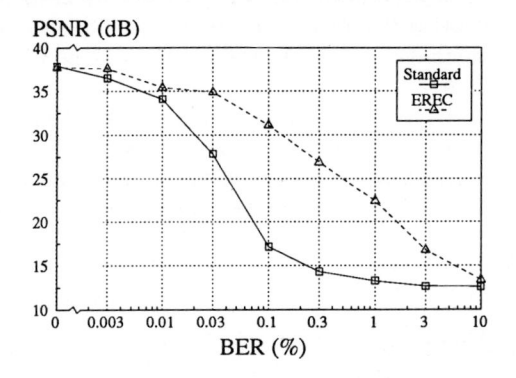

Figure 2: Simulation Results for 'Claire' at 75 kbit/s.

Figure 3: Frame 40 at 0.03% BER (a) Standard (b) EREC

Low bit-rate image VQ with progressive transmission

Giovanni Poggi

Università di Napoli, Dipartimento di Ingegneria Elettronica
Via Claudio,21 80125 Napoli, Italy Fax +39 81 768.31.49 E-mail poggi@nadis.dis.unina.it

Abstract

In this paper an algorithm is presented to perform progressive transmission VQ at a reduced bit-rate. It is based on the use of an ordered tree-structured codebook, and allows one to obtain a significant bit saving (over 50%) in typical PT applications, while preserving the reproduction quality of full-search VQ.

1. Introduction

Vector quantization (VQ) [1] provides, in theory, the best possible performance among all block-based compression techniques. The compression rate achievable in practice, however, is severely limited by the encoding complexity, which increases exponentially with the block size. To improve performance without an explosion of complexity, several encoding schemes have been proposed (e.g., [2-5]) which exploit the statistical dependence among spatially close blocks through some form of prediction and/or joint entropy coding. For a given picture quality these techniques allow one to significantly reduce the bit-rate with respect to memoryless VQ; all of them, however, are hardly compatible with progressive transmission (PT).

The goal of PT [6-7] is to provide the receiver as soon as possible with an approximation of the image, which is then gradually improved as further bits arrive, until the full quality encoded picture is obtained when all the information is transmitted. This procedure gives the receiver the opportunity to quickly identify the image and possibly stop the transmission if desired. Such a feature can largely increase the speed of search through a remote database of images, like radiological images or crime suspects pictures.

It is easy to obtain PT with memoryless VQ, but not so with the far more interesting high-compression VQ schemes. When some form of inter-block prediction is used, in fact, the decoder needs the full quality reconstructed image in order to track the behavior of the encoder. Entropy encoding of codeword indexes, on the other hand, changes the meaning of the transmitted bits, destroying any progressive refinement property.

A simple way to overcome this limitation is proposed in this paper, based on the use of a properly ordered VQ codebook. A new VQ-based coding scheme is then presented, which allows for PT, and retains the quality of full-search encoding, while providing a significantly lower bit-rate than memoryless VQ. After a brief review of PTVQ in Section 2, in Section 3 the coding scheme is detailed, and in Section 4, its performance is assessed, via computer simulations.

2. Progressive transmission vector quantization

The simplest way to perform PTVQ is by resorting to a tree structured (TS) codebook, where the codewords are associated to the terminal nodes (leaves) of a tree. To each internal node a vector is associated as well, which is a weighted average of all the codewords belonging to the subtree originating by that node. As such, these vectors can be regarded as sub-optimal reproduction vectors.

Each time the encoder descends a layer of the tree (supposed binary, here) from the root, down to the selected leaf, a bit is transmitted to indicate which of the children node was chosen. Whit each bit the decoder can address the corresponding sub-optimal codeword and progressively improve the reproduction until the last bit is received. The encoder can transmit first the most significant bit of all the indexes to be sent, then the second one, and so on, allowing the receiver to build immediately a rough reproduction of the image and gradually improve it thereafter.

Usually, TS codebooks are built with a sub-optimal procedure, by recursively splitting the training set in non-overlapping subsets, for each of which a suitable codeword is chosen. Recently a different approach has been proposed [7], where the codebook design is unconstrained (hence better), and the tree structure is built in a subsequent step by properly reassigning the indexes. The basic idea is to assign close indexes to similar codewords, and retain the natural indexing in the tree structure. As a consequence, sibling codewords are similar to each other and to their common parent, obtained as their weighted average. The ordering effect is induced on all the intermediate layers up to the root, producing a well organized TS codebook.

3. Low-rate PTVQ by codebook ordering

Besides allowing for a tree structured codebook, and hence PT, codebook ordering has another useful consequence. Since spatially close image blocks are usually similar to one another so are the codewords associated with them; therefore, if the codebook is ordered, the indexes themselves will exhibit a high degree of correlation (see Fig.1). Such a correlation can be easily exploited by performing a low-complexity linear predictive encoding of the indexes [5] thus reducing the bit-rate at virtually no computational cost and without any increase in distortion.

Although index prediction is not directly compatible with PT, the correlation among indexes implies a similar correlation in the index bit-planes, and so, by compressing these binary images, one can reduce the bit-rate without compromising the PT property. Performing compression at the bit-plane level is usually somewhat less effective than working on the original data. Most of the bit-rate reduction, however, is obtained on the planes formed by the most significant bits, which are the only ones necessary for an early recognition of the image, and therefore the ones most frequently used in PT applications, like the remote scanning of a database.

The encoding scheme can now be summarized:

1. The codebook is designed off-line by means of the Kohonen algorithm [8] which automatically organizes it. The TS is then built as suggested in [7];

2. The current image is vector quantized (full-search VQ is performed) and the VQ indexes are arranged spatially to form an index-image;

3. The index bit-planes are formed and encoded. The i-th bit-plane is sent only after the $(i-1)$-th bit-plane has been completely transmitted, allowing so for progressive reconstruction.

The bit-plane compression, in particular, is carried out as follows: first the binary maximum-likelyhood prediction of the current bit is performed on the basis of some sorrounding bits in order to exploit the bidimensional correlation. Then the modulo-2 prediction error is inserted in either a 'good' or a 'bad' bit stream depending on the expected probability of error: this way the 'good' stream will have long sequences of '0' and will result highly compressible. The resulting streams are finally compressed using the Ziv-Lempel algorithm.

4. Simulation results

Computer simulations were performed using the gray-scale images of the USC database. Some of the images were used as a training set for the codebook design while two others (representing human faces) were used to test the encoding performance. A vector size of 16 was chosen (blocks of 4x4 pels) and a codebook of 256 codewords (hence 8 bit/vector and 0.5 bit/pel) was built.

In Fig.1 the index bit-planes number 1 (the most significant) 2 and 8 are shown (4x4 times magnified) for the test image Lena. Thanks to the ordered codebook much of the original-image correlation has been transferred onto the index-image, and especially the most significant bit-planes exhibit large homogeneus regions and can be efficiently encoded. The least significant bits instead show little correlation, and often it is convenient not to encode them at all.

In Fig.2 the progressive reconstruction of Lena is shown after the transmission of 1, 2 and 8 index bit-planes. It seems safe to say that the second image (at least) is already detailed enough for the receiver to identify it and decide whether to stop the transmission or not.

Fig.1: index bit-planes (4x4 times magnified) arising from the vector quantization of the test image Lena with an ordered codebook. (a) most significant bit-plane; (b) second most significant; (c) least significant.

Fig.2: progressive reconstruction of the image Lena after the transmission of (a) 1 bit-plane, BR=0.025 bit/pel, PSNR=16.77 dB; (b) 2 bit-planes, BR=0.063 bit/pel, PSNR=21.40 dB; (c) 8 bit-planes, BR=0.398 bit/pel, PSNR=30.45 dB.

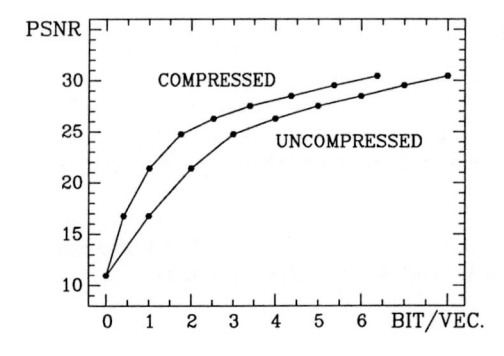

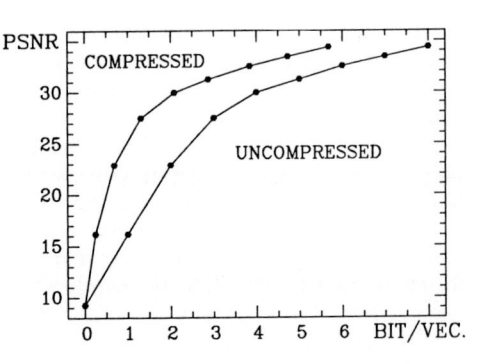

Fig.3: encoding performance, image Lena. **Fig.4:** encoding performance, image Tiffany.

Figg.3 and 4 illustrate the compression performance of the proposed scheme: the peak signal-to-noise ratio (PSNR) is reported as a function of the average number of transmitted bits per vector. In the uncompressed case the ascissa takes on only integer values while using the proposed encoding scheme several bit planes require less than one bit/vector. In order to reach a 22dB PSNR, sufficient for the recognition of the image as seen before, less than 1 bit/vector is required for the first image and just 0.6 for the second, instead of the 2 bit/vector otherwise necessary without compression. In absolute terms this translates to bit-rates of 0.06 and 0.04 bit/pel. The saving on the fully encoded images is not as striking, since the last bit planes exhibit less correlation, but it still goes from 20% to 30% with respect to memoryless PTVQ. Better results can probably be obtained by improving the bit-plane encoding technique. Research on this topic is currently under way.

References

[1] A.Gersho, R.M.Gray, *"Vector quantization and signal compression"*, Kluwer Academic Press, 1992.

[2] J.Foster, R.M.Gray, M.O.Dunham, "Finite-state vector quantization for waveform coding", IEEE Trans. Inform. Theory, vol IT-31, pp.348-359, May 1985.

[3] V.Cuperman, A.Gersho, "Vector predictive coding of speech at 16 Kb/s", IEEE Trans. Commun., vol COM-33, pp.685-696, July 1985.

[4] N.M.Nasrabadi, Y.Feng, "Image compression using address-vector quantization", IEEE Trans. Commun., vol COM-38, pp.2166-2173, Dec. 1990.

[5] E.Cammarota, G.Poggi, "Address vector quantization with topology-preserving codebook ordering", 13th GRETSI Symposium, pp.853-856, Sep.1991.

[6] K.H.Tzou, "Progressive image transmission: a review and comparison of techniques", Optical Engineering, vol 26, pp.581-589, July 1987.

[7] E.A.Riskin, L.E.Atlas, S.R.Lay, "Ordered neural maps and their applications to data compression", IEEE Workshop on Neural Networks for Signal Processing, pp.543-551, Sep.1991.

[8] T.Kohonen, *"Self-organization and associative memory"*, 2nd Ed., Springer-Verlag, 1988.

PERCEPTUALLY BASED CONTRAST ENHANCEMENT ALGORITHM

Tian-Hu Yu and Sanjit K. Mitra

Department of Electrical and Computer Engineering,
University of California at Santa Barbara, California 93106

ABSTRACT

A perceptually based contrast enhancement algorithm is proposed. The proposed algorithm can enhance contrast in low contrast regions and de-enhance contrast in high contrast regions simultaneously to improve the quality of images for human vision.

1. INTRODUCTION

There are two distinct kinds of contrast degradations. One may be due to low contrast with almost all the pixel values closely clustered, and the other may be due to extra high contrast or glare, resulting in low image details. One class of contrast enhancement algorithms are called direct contrast enhancement methods originally proposed by Gordon and Rangayyan [1]. In these approaches, a contrast measure is first obtained and then modified by a mapping function to generate the enhanced version. Dhawan, Buelloni and Gordon [2] noticed that the direct contrast enhancement method using the square-root mapping function proposed by Gordon and Rangayyan [1] introduced too much noise and digitization effects, and they proposed other contrast modification functions, such as the exponential and the logarithm functions. Beghdadi and Negrate [3] defined another contrast measure based on edge information of the image. Since degradations of images may vary from one region to another, Dash and Chatterji [4] proposed an adaptive technique to enhance and de-enhance image contrast simultaneously. However, their method is computationally intensive.

In this paper, we propose a perceptually based direct contrast enhancement algorithm. In our approach, an image can be enhanced in low contrast regions and de-enhanced in high contrast regions simultaneously. In addition, the dynamic range of the enhanced pixel values depends on that of the input pixel values.

2. THE DIRECT CONTRAST ENHANCEMENT METHOD

In the direct method, for each pixel of value x of the input digital signal, a neighborhood edge pixel value x_e is first estimated. Then the enhanced value x' of the input pixel is made much greater than x when $x > x_e$ or much smaller than x if $x < x_e$. To determine x', a contrast measure c based on its original value x and x_e,

$$c = \frac{|x - x_e|}{x + x_e} \quad (1)$$

is employed. A modified contrast measure c' is then obtained using a nonlinear mapping function $\Psi(c)$:

$$c' = \Psi(c). \tag{2}$$

Finally, the enhanced pixel value x' is calculated as follows:

$$x' = \begin{cases} \frac{x_e \cdot (1-c')}{(1+c')} & \text{if } x \leq x_e, \\ \frac{x_e \cdot (1+c')}{(1-c')} & \text{if } x > x_e. \end{cases} \tag{3}$$

As indicated above, in the direct contrast enhancement method, x_e is a threshold for modifying the pixel value x, and it should be a good approximation to a local edge pixel value. A gradient weighted mean within a window [3] or the average pixel value inside a window [1] has been proposed as the local edge pixel value x_e. We have found that the former is more appropriate than the latter as x_e. Several nonlinear functions have been proposed as the mapping function for contrast enhancement; such as (i) the root function: $\Psi_1(c) = c^\alpha$ where $0 \leq \alpha \leq 1$, (ii) the exponential function: $\Psi_2(c) = (1 - e^{-kc})/(1 - e^{-k})$, (iii) the logarithmic function: $\Psi_3(c) = [\ln(1+kc)]/[\ln(1+k)]$, (iv) the hyperbolic function: $\Psi_4(c) = [\tanh(kc)]/[\tanh(k)]$, where k is a constant. It should be noted that the formula for enhancement given by (3) ensures that the enhanced pixel value is the same as the original if the contrast measure is not modified.

3. THE PROPOSED APPROACH

We propose a new contrast enhancement algorithm based on the modification of the direct method. We use here the gradient weighted mean within a window as x_e [3]. For contrast mapping, we employ a polynomial mapping function given by

$$c'_k = (1-a) \cdot c'_{k-1} + a \cdot c'^2_{k-1}, \tag{4}$$

where c'_k is a modified contrast measure after a k-th iteration of the polynomial mapping function and $c'_0 = c$, and a is a parameter within -1 and 0. We have found that the proposed mapping function has low contrast enhancement when c is small and high contrast enhancement when c is large. Thus, it makes a better trade-off between contrast enhancement and background noise effect than the others.

For image enhancement, it is not necessary to keep the enhanced pixel value same as the original in the absence of contrast modification. Therefore, the pixel generation formula should be such that if $x < x_e$, then $x' < x_e$, and if $x > x_e$, then $x' > x_e$. In this case, either when $x < x_e$ and $x' < x$, or when $x > x_e$ and $x' > x$, the process is to enhance contrast; otherwise, it is a de-enhancement process. Based on the above analysis, we propose a new formula to generate enhanced pixel values from modified contrast measures as follows,

$$x' = \begin{cases} x_e \cdot [1 + \frac{2Mc'}{(1+c')}] & \text{if } x \leq x_e, \\ 255 - (255 - x_e) \cdot [1 + \frac{2Mc'}{(1+c')}] & \text{if } x > x_e, \end{cases} \tag{5}$$

where M is a prefixed positive integer number. Thus, we can split up the range of the input contrast measure into two parts: contrast enhancement region if $c' < \frac{c}{M-(M+1)\cdot c}$,

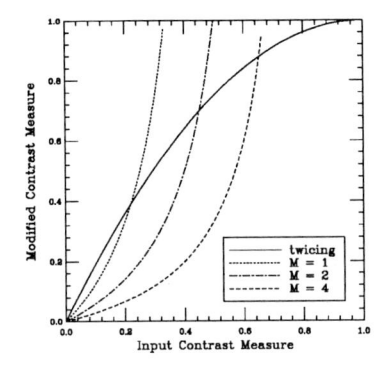

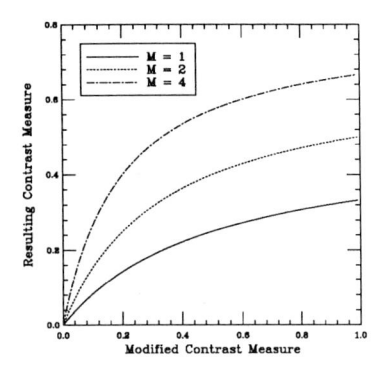

Figure 1: The critical points for $M = 1, 2$ and 4.

Figure 2: The relationship between c" and c'.

and de-enhancement region if $c' > \frac{c}{M-(M+1)\cdot c}$. It can be seen that the effect of the proposed algorithm on the image contrast depends on c' and M, and for a given M and mapping function it depends only on the input contrast measure. For example, if $c' = c \cdot (2.0 - c)$, the critical values of input contrast measures are the crossing points as shown in Figure 1 for the cases of $M = 1, 2$ and 4.

In addition, the contrast modification of the proposed formula can be shown using a resulting contrast measure c'':

$$c'' = \frac{|x' - x_e|}{x' + x_e}. \tag{6}$$

The relationship between c'' and c' is given by

$$c'' = \frac{M \cdot c'}{[(M+1) \cdot c' + 1]}, \tag{7}$$

and is shown in Fig. 2. As can be seen, when c' is small, c'' is proportional to c', and when c' is large, c'' is bounded by $\frac{M}{M+2}$.

In our proposed algorithm, the dynamic range of the enhanced pixel values is dependent on that of the input image. For example, if the input pixel values are in the range of 0-255, the enhanced pixel values are restricted within the range of $[-M \cdot 255, (M + 1) \cdot 255]$. Because of this property, we can employ a contrast stretching technique to transfer this range to the display range.

4. ILLUSTRATIVE EXAMPLES

To test the performance of the proposed algorithm, we have applied it to an image corrupted by impulse. The original image is a part of the image "Lena" as shown in Figure 3(a). Its noisy version is shown in Figure 3(b) in which there are 1409 randomly distributed negative impulses and 1291 positive ones. The probability of impulse occurrence is 6.75 percent. The enhanced image is shown in Figure 3(c). Note

(a)　　　　　　　　　　(b)　　　　　　　　　　(c)

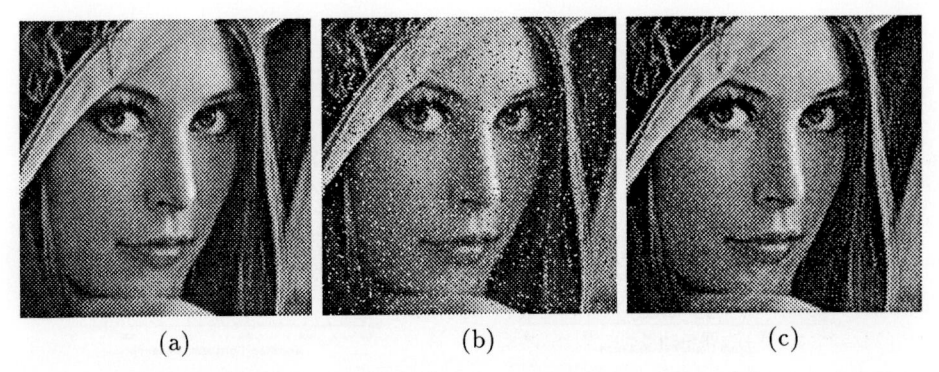

Figure 3: Example of the proposed algorithm. (a) Original image. (b) Input image corrupted by impulse noise. (c) Enhanced image obtained using the proposed algorithm.

that the edges of the original image have been enhanced while impulse noise has been suppressed.

5. CONCLUDING REMARKS

We have proposed a new direct contrast enhancement approach. It has been shown that in the proposed approach, an image can be enhanced in low contrast regions and de-enhanced in high contrast regions simultaneously. The critical point for enhancement and de-enhancement depends on the contrast mapping function and a prefixed integer number M. In addition, the dynamic range of the enhanced pixel values can be obtained based on the the input dynamic range and the integer M.

References

[1] R. Gordon and R. M. Rangayyan, "Feature enhancement of film mammograms using fixed and adaptive neighborhoods," *Applied Optics*, vol. 23, no. 4, pp. 560-564, February 1984.

[2] A. P. Dhawan, G. Buelloni and R. Gordon, "Enhancement of mammographic features by optimal adaptive neighborhood image processing," *IEEE Trans. Medical Imaging*, vol. MI-5, no. 1, pp. 8-15, March 1986.

[3] A. Beghdadi and A. L. Negrate, "Contrast enhancement based on local detection of edges," *Computer Graphics and Image Processing*, vol. 46, no. 4, pp. 162-174, 1989.

[4] L. Dash and B. N. Chatterji, "Adaptive contrast enhancement and de-enhancement," *Pattern Recognition*, vol. 24, no. 4, pp. 289-302, February 1991.

Restoration of incomplete contour images using Markov random fields.

Sabine URAGO, Marc BERTHOD and Josiane ZERUBIA

INRIA, 2004 Route des Lucioles, BP 93, 06902 Sophia Antipolis Cedex, France.
Fax : (33) 93 65 76 43, email : urago@sophia.inria.fr.

Abstract :
In this paper, we describe an algorithm which performs the restoration of images with incomplete contours. We use Markov Random Fields (MRF) and a global measure : a Gibbs distribution. In order to restore the contours we define some criteria which have to be optimized. A deterministic (Iterated Conditional Mode) or a stochastic (Gibbs sampler) algorithm generates a configuration in which the contours are completed. Our study shows ways to modify and extend a method initially proposed by J. L. Marroquin to enable the processing of real and noisy images. To illustrate this algorithm, several examples have been tested including synthetic, noisy and real images. We show herein the results obtained on an indoor image.

1. INTRODUCTION

In this paper, we present an algorithm which allows the **restoration of images that consist of incomplete contours**.
To restore the images a **Markov Random Field** model associated with a **Gibbs distribution** is used. The method implements a stochastic process, based on the **Gibbs sampler algorithm**. One modification made to the method proposed by J. L. Marroquin [4], [5], is the implementation of the deterministic **ICM** (Iterated Conditional Mode) algorithm [1]. These algorithms generate an equilibrium configuration in which the contours are reconstituted.

We have modified and extended (see [6] for more details) an algorithm initially proposed by J. L. Marroquin in 1989 [4], [5], with the goal of restoring real incomplete contour images (in particular indoor and SPOT images) which are sometimes noisy.

2. RESTORATION OF IMAGES

To restore images of incomplete contours, we first determine a classification of boundary elements.

2.1. Markov Random Fields of contour elements

At each site, we store two quantities :
• The **first component of the Markov Random Field** specifies the **state** of contour elements. The boundary elements take values from a set of 21 states (state=0 : no contour ; 1<=state<=4 : straight element ; 5<=state<=20 : turning element). These different states are used to determine the macroscopic directions.

• The **second component** corresponds to a more global information, the **macroscopic direction** of the line segments, i.e. the direction in which the line is supposed to go (see [3], [4] for more details).

In order to extend the contours, it is necessary to know the connection with the neighboring elements. We determine different classes of neighbor interactions [5] (indifferent connection "o", termination of lines "e", T-junction "t", T-junction "t_1" , sharp turn "s", straight continuation "a", right turn "b, f, j, r", left turn "d, g, k, l", complete a turn "c", forbidden connection "-").

J. L. Marroquin [4], [5], has not defined the type "t_1" (T-junction) of connections and he considers this interaction as "forbidden". Nevertheless, we have to include this type of T-junction ("t_1") in order to get all the possible connections between the contour elements.

To define the energy function, we need to have a few informations :

When we extend the contours, we complete a line by a **state compatible with the macroscopic direction**. The direction is determined by a number which takes values between (-1) and (11). For a given site, **the number given to the macroscopic direction** can be **negative** (-1) when there is no contour or when a corner is detected.

We select subintervals of orientations, so that one number associated with a subinterval is compatible with a unique set of states and connections (see [4], [5]).

We specify also a value **"alpha"** which represents the probability of selecting a turn (i.e. state > 4) divided by the probability of selecting a straight continuation (i.e. 1<=state<=4). **The method used to determine "alpha" is different from the one proposed by J. L. Marroquin** [4], [5] (see [6] for more details).

2.2. Energy function

Both potential functions V1 and V2 are used to determine the energy (see [4], [5] and [6]).
First, to define these potential functions, we need to introduce some notation :

$l_{i,j} = l(q_i, q_j)$ describes the connection of two neighboring states q_i and q_j.

At a given site i, a direction number (noted d_i) is compatible with a unique set of connections divided into three **classes** :
- $SC_i = SC(d_i)$ **Straight continuation**
- $T_i = T(d_i)$ **Turn**
- $CT_i = CT(d_i)$ **Double turn**

$C_i = C(d_i)$ represents the **states compatible with the direction d_i**.
$a_i = a(d_i)$ is the **probability to turn divided by the probability to go straight.**

The **function V1** depends only on the state (q_i, d_i), in which q_i represents the state of the site i, and d_i the direction.

$$V1(q_i, d_i) = \begin{cases} \infty & \text{if } (d_i \geq 0 \text{ and } q_i \notin C_i) \\ \ln(1+a_i) & \text{if } (d_i \geq 0 \text{ and } q_i \in [1,4] \cap C_i) \\ \ln(\frac{1+a_i}{a_i}) & \text{if } (d_i \geq 0 \text{ and } q_i \in [5,12] \cap C_i) \\ 0 & \text{otherwise} \end{cases} \quad (1)$$

V2 is determined for two states (q_i,d_i) and (q_j,d_j), where i and j are two neighboring sites. For some constants VE, VT and VC so that $|VC|<VT<VT_1<<VE$, we can define V2 as follows:

$$V2(q_i,d_i,q_j,d_j) = \begin{cases} 0 & \text{if } (l_{i,j} = \text{"o"}) \\ VE & \text{if } (l_{i,j} = \text{"e"}) \\ VC & \text{if } (d_i=d_j) \text{ and } (q_i,q_j \in C_i) \text{ and } (l_{i,j} \in SC_i \cup T_i) \\ VC-\ln(\frac{1+ai}{ai}) & \\ & \text{If } (d_i=d_j) \text{ and } (q_i,q_j \in C_i) \text{ and } (l_{i,j} = CT_i) \\ VT & \text{If } (l_{i,j} \notin \{\text{"o"},\text{"e"},\text{"-"}\} \text{ and } d_k<0 \text{ and } q_k>0) \\ & \text{for some } k \in \{l,j\} \text{) or if } (l_{i,j} = \text{"t"}) \\ VT_1 & \text{if } l_{i,j} = \text{"}t_1\text{"} \\ \infty & \text{else} \end{cases} \quad (2)$$

The total energy "U" is the sum over all sites of V1 and the sum over all cliques of size 2 of V2:

$$U = \sum_{i,j:\,||i-j||<2} V2(q_i,d_i,q_j,d_j) + \sum_i V1(q_i,d_i) \quad (3)$$

i,j represent two neighboring sites (belonging to the same clique).

According to the Hammersley-Clifford theorem, at a given site i, the probability of selecting the vector (q_i,d_i) is proportional to:

$$P = \exp\left(\frac{-1}{T}\left(\sum_{i,j:\,||i-j||<2} V2(q_i,d_i,q_j,d_j) + \sum_i V1(q_i,d_i)\right)\right) \quad (4)$$

where T characterizes the temperature.

2.3. Initialization

Before running an algorithm (ICM or Gibbs sampler) minimizing the non-convex energy function described previously, we have to initialize the vector (q_i,d_i), in order to model the initial image. The **initialization of the second component** of the MRF, the direction, **is different from the description proposed by J. L. Marroquin** [4], [5], and this completely modifies the behaviour of the algorithm. The initialization algorithm looks at the state lattice in order to determine the macroscopic direction at every site. The propagation algorithm enables us to obtain the largest set of neighboring sites having the same direction number. Then, we can extend a line in the best direction (see [6] for more details).

2.4. Results

The restoration of incomplete contour images has been implemented on a **Connection Machine** (CM200). An edge detector is applied to an image of a real scene. Canny-Deriche algorithm [2], [3] and a thresholding for non-maxima suppression have been used for this purpose. We then apply the proposed method in order to complete the contours. Tests have been conducted on synthetic as well as real images (SPOT, indoors, etc...).

Figures [1-3] show the results on an indoors image of size (256*256).

Figure 1. Original image.

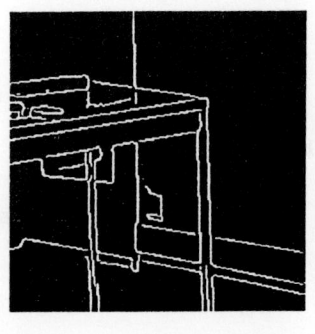

Figure 2. Contours obtained after a Canny-Deriche filtering with non-maxima suppression.

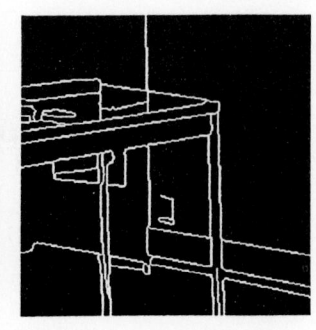

Figure 3. Restoration of incomplete contour image.

3. CONCLUSION

The method proposed in this paper to restore incomplete contour images enables the filling of large gaps. A classification of contour elements (states) and of different connection types is defined in order to model the image. Only local interactions are memorized, but every site stores a more global information : the macroscopic direction. To complete the images, we successively perform extension and shortening of contour lines, to obtain an equilibrium configuration in which "T-junctions" appear. Two algorithms have been implemented : a stochastic algorithm (Gibbs sampler) and a deterministic method (ICM). The deterministic algorithm is faster when the initial image is not noisy, i.e. when there is only correct information in the image.

4. REFERENCES

[1] J. Besag, "On the statistical analysis of dirty pictures", *JL of Royal Statistical Society*, series B, Vol. 68, pp 259-302, 1986.

[2] J. Canny, "A computational approach to edge detection", *IEEE Transactions on Pattern Analysis and Machine Intelligence*, Vol. PAMI-8(6), p 679-698, 1986.

[3] R. Deriche, "Using Canny's criteria to derive a recursively implemented optimal edge detector", *International Journal of Computer Vision*, p 167-187, 1987.

[4] J. L. Marroquin, "A Markovian Random Field of Piecewise Straight Lines", internal report, *"Centro de Investigacion en Matematicas"*, Mexico, 1989.

[5] J. L. Marroquin, "A Markovian Random Field of Piecewise Straight Lines", *Biological Cybernetics* , p 457-465, 1989.

[6] S. Urago, M. Berthod and J. Zerubia, "Restauration d'image de contours incomplets par modélisation par champs de Markov", INRIA research report n° 1688 (in french), 1992.

a Coarse-to-Fine Scheme for Region Growing using Statistical Tests

G.J. Brelstaff

Perceptual Systems Research Centre, Dept. of Psychology, University of Bristol, BS8 1TN, UK.

Abstract

This paper illustrates a coarse-to-fine image segmentation scheme that applies selected statistical tests to decide whether or not to split or merge regions. A parametric model is selected and fitted to the grey-level surface within each putative region. The deviation of the model from the surface is assessed by statistical tests appropriate to the state of knowledge of the image noise. The deviation is thus either accepted as noise variation or interpreted as due to a real region boundary. A coarse-to-fine solution is proposed to the logistic problem of applying such tests in order to obtain a reasonable segmentation.

1. INTRODUCTION

An image's underlying structure can be thought of as a patch-work of homogeneous regions—within each of which the signal can further be assumed to be approximated by a chosen parameterised model. For example, a smooth grey-level patch might be described by a constant, planar or biquadratic function and a textured patch by a grating or a plaid. Once chosen the model can be least-squared fitted so to closely represent the local grey-level surface. The actual grey-levels within the patch will, in general, deviate from those predicted by the fit due to the statistical effects of imaging noise. Knowledge of the noise characteristics allows appropriate statistical tests to be applied to determine whether a grey-level deviation is accounted for by noise variation or should be interpreted as a change in the image's underlying structure. Here we address the logistic problem of applying such knowledge to obtain a reasonable segmentation. Fig. 1. illustrates what can be achieved when the noise is known to be additive Gaussian and has amplitude that varies unpredictably between regions. The underlying image structure is of planar and biquadratic regions. Images with different structural and noise characteristics can be similarly segmented provided that enough is known about them to formulate the appropriate statistical tests.

2. METHOD

For segmenting the image in Fig. 1.(a) two statistical components are particularly useful. Both are briefly described below—see Brelstaff and Ibison [2] for comparison with

Chi-squared [1] and Maximum Likelihood [7] approaches:

(1) The F-statistic that can be used to determine whether two adjacent grey-level surfaces may be regarded as part of the same region. It is computed for each candidate merge as follows:

$$F = \frac{(e_{1,2} - e_1 - e_2)/v_b}{(e_1 + e_2)/v_w} \quad (1)$$

where e_1, e_2 and $e_{1,2}$ are the total squared fitting errors in the first, second, and combined areas, and $v_b = d_1 + d_2 - d_{1,2}$, and $v_w = N - d_1 - d_2$. The d's are the numbers of parameters in the models fitted to these areas (for a biquadratic: 6), and N is the number of pixels in the combined area. F obeys an F-distribution pdf: $p_{v_b,v_w}(F)$ having v_b, v_w degrees of freedom. Over the range $[F, \infty]$ this pdf integrates to a finite probability P. At a given level of significance α, the two areas are significantly different if $P < \alpha/2$, and they should not be merged. This is an F-test [8,6].

(2) Spatial-randomisation can be used to determine whether the individual deviations of the fitted model from the grey-level surface have a sufficiently random spatial organisation so as to regard the area as part of a single region. The test proceeds as follows: (1) Create a spatial map of the sign of the deviation at each pixel. (2) Count T the number of crossings from positive to negative between neighbouring pixels. (3) Test whether T takes on a statistically unlikely value [4,9]. This Spatial test can be applied at the same level of significance α as the F-test (we use $\alpha = 0.01$ throughout).

Before outlining the coarse-to-fine scheme, we describe how these two components are combined into an iterative scheme that acts to merge areas of a given scale into putative regions.

Iterative merging: Initially the image is tessellated into a rectangular grid of tiles and the biquadratic model is fitted within each tile. Tiles spanning region boundaries should be prevented from merging. Some can be identified beforehand by either (a) failing the Spatial test, or by (b) having significantly larger fitting error than a neighbouring tile (failing a local F-test). These are marked in grey in Fig. 1. We take a precaution to avoid other, unidentified boundary tiles in subsequent merging operations. Our tactic is to seed region growth from a tile that is as far a possible from any pre-identified boundary tiles. Once chosen a seed is iteratively grown until none of its adjacent tiles will merge with it. At each iteration a tile is chosen to be merged with the seed on the basis of the F-statistic. The tile providing the maximum value of the probability density $p_{v_b,v_w}(F)$ is selected. However, this merge is rejected if the F-test fails or if the combined area fails the Spatial test. Merging stops when no further seed tiles can be grown.

Coarse-to-fine strategy: The iterative scheme outlined above is applied at each individual stage in our coarse-to-fine segmentation strategy. At each successive stage the tile size is reduced and the scheme is applied again—progressing towards a high resolution segmentation. Regions that begin to grow at coarse resolution are preserved and allowed to continue growing at the next stage. Thus at any particular stage only those tiles that remain unmerged, or were pre-identified as spanning a boundary, are re-tessellated (split) at the next stage. However, it also proves necessary to re-tessellate any region grown in a linear formation: i.e. any that do not exceed one tile in width along their entire extent. This prevents adjacent tiles that span the same boundary, and were not pre-identified, from growing along that boundary (e.g. see those in Fig. 1.(d)).

3. RESULTS & DISCUSSION

The coarse-to-fine strategy is illustrated in Fig. 1.: the 64x64 pixel image (a) is first tessellated into coarse, 16x16 tiles and (b) segmented by iterative merging. At this scale, only two regions are grown (all other tiles marked grey were pre-identified as spanning boundaries). Neither of these regions, however, is preserved, as neither exceeds the 16 pixel tile width. Thus at the next scale, the entire image is re-tessellated into 8x8 tiles. At this stage, iterative merging (c) creates two large regions that are preserved at the next stage, where 4x4 tiles are used and resultin a segmentation (d) containing four preservable regions. At the final stage (e), the tile size becomes a single pixel and iterative merging acts to assign these single pixels to adjacent regions preserved from the previous stage. Note single pixels are not used here as seeds and do not require the Spatial test.

The segmentation boundaries marked in Fig. 1.(e) closely align with the known region boundaries in the test image. This compares well to the Canny operator [3]—that marks false edges within the high gradient area (on the left of the image) and leaves incomplete, broken edges (in the centre). Haralick's F-test based, local, facet-fitting strategy [5] does better but still marks many false edge points within each region. Furthermore, both of these approaches require more than a single input parameter to tune their segmentation process (α is all that we need). The grey areas remaining in our final segmentation represent localities where our chosen bi-quadratic model was shown to be inappropriate: there was not enough space to fit it (a planar model might be fitted there). Where our segmentation boundaries do deviate from the known region boundaries they do so for two reasons: (1) because of local ambiguity where the imaging noise allows the boundary to be statistically misplaced by a pixel or so (e.g. around the left blob area), and (2) because of global ambiguity: e.g. the central area of the image can equally be interpreted as part of the background patch or part of the vertical strip region. To reliably resolve this kind of ambiguity requires a higher-level model of the image structure than that provided by the low-level patch model considered here. The model would need to embody such concept as occlusion and 3-D surfaces.

4. REFERENCES

1. P.J. Besl, and R.C. Jain, "Segmentation Through Variable-Order Surface Fitting," *IEEE Trans. PAMI*, 10, 2, 167-192, 1988.
2. G.J. Brelstaff, and M.C. Ibison, "Statistical Recipes for Region Growing", In prep., 1993.
3. J.F. Canny, *Finding edges and lines in images.* S.M. thesis, MIT, Cambridge, USA, 1983.
4. A.D. Cliff, and J.K. Ord, *Spatial Processes: Models and Applications.* Pion, London, 1981.
5. R.M. Haralick, "Edge and Region Analysis for Digital Image Data", *CGIP* 12, 60-73, 1980.
6. Laprade, and Doherty, "Split and merge segmentation using an F-test criterion", *Image Understanding and the Man-Machine Interface*, SPIE Vol. 758, 1987.
7. J.F. Silverman, and D.B. Cooper, "Bayesian Clustering for Unsupervised Estimation of Surface and Texture Models," *IEEE Trans. PAMI* 10, 482-495, 1988.
8. M.R. Spiegel, *Probability and Statistics.* Schaum's Outline Series, Mc Graw Hill, 1982.
9. G.J.G. Upton, and B. Fingleton, *Spatial Data Analysis by Example*, Vol 1. John Wiley, 1985.

Acknowledgements: This work was supported through grant D/ER/1/9/4/2034/102 from the DRA RARDE Fort Halstead Sevenoaks UK, and by the IBM UK Scientific Centre.

Fig. 1. The coarse-to-fine segmentation scheme that adopts the biquadratic model to segment the test image (a) at a progressively higher resolution—i.e. at tile sizes: (b) 16x16, (c) 8x8, (d) 4x4, and (e) single pixels. Level of significance is 0.01 throughout.

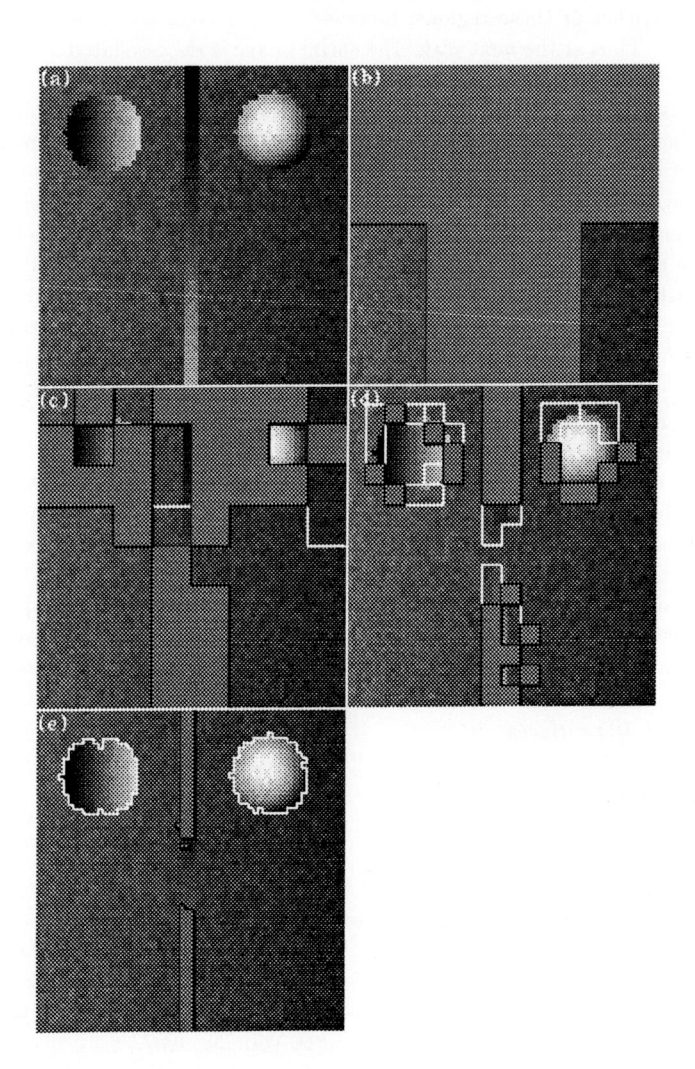

Computational Considerations on Randomized Hough Transform and Motion Analysis

Heikki Kälviäinen

Lappeenranta University of Technology, Department of Information Technology, P.O. Box 20, SF-53851 Lappeenranta, Finland, E-mail: Heikki.Kalviainen@it.lut.fi

Abstract

A new and efficient version of the Hough Transform, the Randomized Hough Transform (RHT), has recently been suggested. The RHT has been applied to motion detection by the author and his coworkers and a novel method to calculate 2-D motion in a sequence of time-varying images has been proposed. The method, called Motion Detection using Randomized Hough Transform (MDRHT), utilizes the main ideas of the RHT, random sampling of, e.g., edge points of the original image, and a converging mapping of randomly sampled points into one point in a dynamically-linked accumulator. In this paper the generalized algorithm is suggested and computational behavior of this generalization is examined. Probability mechanisms of the MDRHT are considered on both translation and rotation.

1 Introduction

Below two main approaches in motion estimation, the feature-based and optical flow based [1], there is many motion detection methods using the Hough Transform (HT) [2]. The aim of this work is to develop a non-model-based 2-D motion detection method using edge pixels as features. Motion estimation is done first, and only after that the pattern recognition of a rigidly moving object. One important objective is to examine motion as a movement of pixels in images instead of a movement of some wider segmented regions. Many earlier applications of motion detection with the HT suffer from both the problems of the HT, e.g., time and memory consumption and the general problems of motion analysis, e.g., need of relatively short motion displacements. A new approach to the HT has been recently developed, called the Randomized Hough Transform (RHT) [5]. When used for curve segment detection in static images, it has been shown that the RHT decreases dramatically the time consumption and memory usage compared to the standard HT. Ideas of the RHT have recently been applied to motion analysis, and a novel algorithm, called the Motion Detection using Randomized Hough Transform (MDRHT) [3], has been developed by the author and his coworkers. In this paper, basic mechanisms of the MDRHT are studied and concepts behind computational behavior are examined through definitions.

In Section 2, the kernel of the MDRHT algorithm is presented. Generalization of the algorithm is shown in Section 3. Probabilistic mechanisms are also considered. There are versions to calculate both translation and rotation (Section 4).

2 Motion Detection using Randomized Hough Transform

The basis of the RHT algorithm for line detection was the fact that each parameter space point could be expressed with, e.g., two points from the original binary edge picture. The point pairs

(p_i, p_j) were picked randomly from the binary edge picture. The parameter point (a, b) was solved from the curve equation with each pair (p_i, p_j), and the cell $A(a, b)$ was accumulated in an accumulator space. The RHT was run long enough to detect the global maximum in the accumulator space. The parameter space point (a, b) of the global maximum described the parameters of the detected curve.

The idea of RHT can be applied to motion detection. Assume that there are two grey-level images. Two binary edge pictures are formed from them. Then correspondences of the edge points in the two sets are studied in the following way: Two points are picked randomly from the first picture and a corresponding point pair is sought from the second picture. A corresponding point pair is any point pair which has the same relative displacement vector as the point pair in the first picture. When a corresponding point pair is found, the translation (dx, dy) is calculated between points of the first and the second picture and a cell $A(dx, dy)$ is accumulated in the accumulator space. It is obvious that there are more hits in those accumulator cells which correspond to the real translation of a moving object. Random picking is continued until a global maximum is detected in the accumulator space. The global maximum can be determined with a threshold. This global maximum gives the detected motion. The kernel of the MDRHT algorithm is the following:

Algorithm 1 : *Kernel of the MDRHT*

1. Form the sets B and C of edge pixels in two pictures.
2. Pick point pairs (b_i, b_j) and (c_i, c_j) randomly from sets B and C.
3. If the point pairs correspond, calculate the x- and y-translations $dx = c_{ix} - b_{ix}$ and $dy = c_{iy} - b_{iy}$ and goto step 4; otherwise, goto step 2.
4. Accumulate the cell $A(dx, dy)$.
5. If a global maximum in the accumulator space is detected, motion (dx, dy) has been found; otherwise, goto step 2.

3 Generalization and Analysis of the Algorithm

The MDRHT algorithm can also be used with three points instead of two; then motion is detected as a movement of three pixels, with obvious generalizations. The generalization of the MDRHT is shown in Algorithm 2. N point pairs (b_i, c_i) are picked from the sets B and C, i.e., a set of N pixels from both sets B and C are randomly chosen. Motion is studied as a movement of N points. The correspondences of the point pairs (b_i, c_i) are checked requiring that all of them have the same translation.

Algorithm 2 : *Generalization of the MDRHT*

1. Form the sets B and C of edge pixels in two pictures.
2. Pick N point pairs (b_i, c_i), $i = 1, \cdots, N$, randomly from sets B and C.
3. If the point pairs correspond, calculate the x- and y-translations $dx = c_{ix} - b_{ix}$ and $dy = c_{iy} - b_{iy}$ and goto step 4; otherwise, goto step 2.
4. Accumulate the cell $A(dx, dy)$.
5. If a global maximum in the accumulator space is detected, motion (dx, dy) has been found; otherwise, goto step 2.

Effects of this generalization are studied next. We start with some definitions:

Definition 1 : A set B has points $b_i = (b_{ix}, b_{iy})$ of the first picture and a set C has points $c_i = (c_{ix}, c_{iy})$ of the second picture, respectively.

Definition 2 : C_r is a set C shifted with translation $r = (dx, dy)$, thus C_r has points $c_i = (c_{ix} - dx, c_{iy} - dy)$, $(c_{ix}, c_{iy}) \in C$.

Definition 3 : $D(B \times C)$ is a set of point pairs (p_1, p_2), $p_1 \in B$ and $p_2 \in C$, having the same coordinates (x, y).

Note that the number of points in the set $D(B \times C)$ is the same as the number of points in the set $B \cap C$. This is denoted $nr(B \cap C)$. Assume now that one pair of points is randomly picked from the two pictures. Then the probability $p_1(r)$ that their mutual displacement is r is given by

$$p_1(r) = \frac{nr(B \cap C_r)}{\sum_s nr(B \cap C_s)}, \qquad (1)$$

where $nr(B \cap C_r)$ is the number of matching points in the pictures when the second picture is translated with r and $\sum_s nr(B \cap C_s)$ is the number of all possible matches.

If two point pairs are randomly picked, the probability $p_2(r)$ is as follows:

$$p_2(r) = \frac{(nr(B \cap C_r))^2}{\sum_s (nr(B \cap C_s))^2}. \qquad (2)$$

In general if N point pairs (not necessarily different) are picked from the set $D(B \times C_r)$, then $nr(B \cap C_r)^N$ picks can be formed. Equation (2) can be extended to the generalized form so that the probability $p_N(r)$ of translation r is

$$p_N(r) = \frac{(nr(B \cap C_r))^N}{\sum_s (nr(B \cap C_s))^N}. \qquad (3)$$

The more points are picked, the sharper the probability distributions will become. It means that the maximum probability value is emphasized. This effect is illustrated in Fig. 1 which shows all possible hit values of (dx, dy) cells in the accumulator space of a test Pic1 introduced in [3]. Only the cells having at least 43 hits when $N = 1$ are displayed. In practice, however, an increasing need of computing time with more random sampling elements can ruin the advantage of growing N. This problem can be avoided by parallel computing.

Let n be the number of different translations r. From Equation (3) probabilities of different translations $p_i, i = 1, \cdots, n$ can be formed. They define a discrete probability distribution which has probability values p_i. Random sampling of the MDRHT is done from this distribution. With the distribution, two main interests can be solved: what is the expectation of random picks until we get t times same translation, and what is, in this case, the probability to get some known translation t times?

4 Rotational Motion

Algorithm 1 can be extended to detect rotation. In this case the object, hence the point pair, has been assumed to rotate rigidly with some unknown rotation point and an angle. The problem is to find these parameters by sampling from the two images. After the correspondence of the point pairs using some characteristics of them is checked, the rotation point (x_r, y_r) and the angle θ can be calculated. The accumulator space A is now three-dimensional. When a global maximum is found, the parameters of the accumulator cell of the maximum $A(x_r, y_r, \theta)$ define detected motion. The kernel of the extended algorithm is presented in [4]. A rotated point set can be defined as follows:

Definition 4 : C_{rot} is a set C shifted with rotation $rot = (x_r, y_r, \theta)$, thus C_{rot} has points $c_i = ((c_{ix}(cos\theta) - c_{iy}(sin\theta) - x_r(cos\theta - 1) + y_r(sin\theta)), (c_{ix}(sin\theta) + c_{iy}(cos\theta) - x_r(sin\theta) - y_r(cos\theta - 1)))$, $(c_{ix}, c_{iy}) \in C$.

In the rotational algorithm two point pairs are selected. The probability $p(rot)$ that their mutual displacement is rot is given for point pairs not necessarily different by

$$p(rot) = \frac{(nr(B \cap C_{rot}))^2}{\sum_s (nr(B \cap C_s))^2}. \tag{4}$$

References

[1] Aggarwal, J.K., Nandhakumar, N., "On the Computation of Motion from Sequences of Images - A Review", *Proceedings of the IEEE*, vol. 76, no. 8, 1988, pp. 917-935.

[2] Illingworth, J., Kittler J., "A Survey of the Hough Transform", *Computer Vision, Graphics, and Image Processing*, vol. 44, 1988, pp. 87-116.

[3] Kälviäinen, H., Oja, E., Xu, L., "Motion Detection using Randomized Hough Transform", *Proc. of 7th Scand. Conf. on Image Analysis*, Aalborg, Denmark, August 1991, pp. 72-79.

[4] Kälviäinen, H., Oja, E., Xu, L., "Randomized Hough Transform Applied to Translational and Rotational Motion Analysis", *Proc. of 11th Int. Conf. on Pattern Recognition*, The Hague, The Netherlands, August-September 1992, pp. 672-675.

[5] Xu, L., Oja. E., Kultanen P., "A New Curve Detection Method: Randomized Hough Transform (RHT)", *Pattern Recognition Letters*, vol. 11, no. 5, 1990, pp. 331-338..

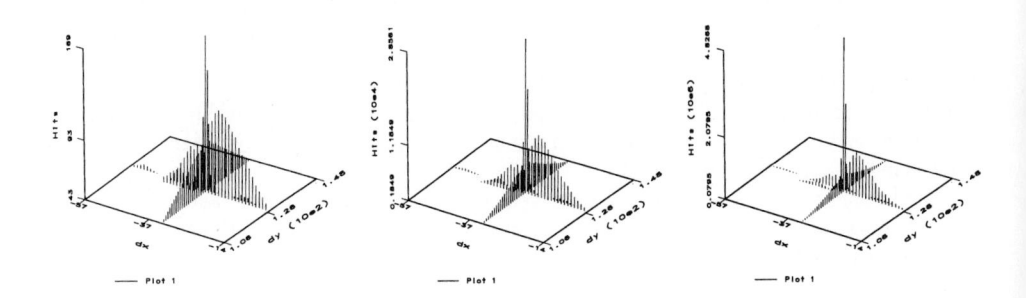

Figure 1: Hit histogram in a test Pic1: $N = 1$, $N = 2$, and $N = 3$.

A New Algorithm for Active Contours

Arnaldo J. Abrantes[a]   Jorge S. Marques[b]

[a]INESC/ISEL   [b]INESC/IST

R. Alves Redol 9, P-1097 Lisboa Portugal

Abstract

This paper presents a novel algorithm for updating the snake contours which allows a more flexible description and improved control of the contour internal forces. Specifically, internal forces are interpreted in this paper as a filtering operation performed on the control points of the snake. This new approach also suggests the use of tailored strategies in the choice of the internal forces, adapted to the detection of particular classes of objects.

1. INTRODUCTION

Active contours have been proposed to circumvent the well known difficulties observed in classical edge detection algorithms: contour breaks, missing data, presence of spurious edge points. Many of these difficulties arise from the fact that boundary detection is an ill-posed problem which can only be solved by imposing additional constraints. A way of tackling with these difficulties is the one adopted in the snake algorithm proposed by Kass et al. in [1]. A snake is an elastic string whose evolution depends on internal forces, which try to give some coherence to the contour, and on external forces which depend on the image field. The continuity of the contour is imposed by the elastic model and therefore snakes can interpolate missing data and adapt to object boundaries.

Unfortunately, several difficulties are found when the snake algorithm is used to update the position of an elastic contour: the contour evolution is strongly dependent on the initialization and sometimes converges to incorrect shapes; contour motion is hard to control; convergence is slow. Experimental tests show that the choice of the snake internal forces is critical and it is not flexible enough to cope with several contour shapes. Several attempts have been made to overcome some of the above mentioned problems [2,3] but none has addressed the improvement of internal forces.

This paper presents a novel algorithm for updating the snake contours which allows a more flexible description and improved control of the contour internal forces. Specifically, internal forces are interpreted in this paper as a filtering operation performed on the control points of the snake. This new approach also suggests the use of tailored strategies for the choice of the internal forces, adapted to the detection of particular classes of objects.

2. CLASSIC SNAKE ALGORITHM

The snake algorithm was proposed by Kass et al. in [1]. The basic idea is to approximate the object contour by an elastic string acted by internal forces, which try to keep the snake units together and to avoid significant curvature, and by image forces which attract the snake units to image features (e.g., high gradient intensity points). The snake contour is defined as a parametric curve in the complex plane (or in R^2) depending on a continuos arc length parameter, or as a set of discrete points in the complex plane if one adopts a discrete formulation. In both approaches, the snake contour is obtained by the minimization of a potential cost functional

$$P = P_{int} + P_{img} \qquad (1)$$

where P_{int} is an internal potential and P_{img} is the image potential which attracts the contour to the desired image features. Replacing the internal potential by its classic expression one obtains

$$P(z(0), z(1), \ldots, z(N-1)) = \sum_{n=0}^{N-1} \frac{\alpha_1}{2} \left| \Delta^{(1)} z(n) \right|^2 + \frac{\alpha_2}{2} \left| \Delta^{(2)} z(n) \right|^2 + P_{img}(z(n)) \qquad (2)$$

where $z(0), \ldots, z(N-1)$ are the control points of a closed active contour, and $\Delta^{(i)}$ denotes the i-th difference operator. Equation (2) is an image dependent nonlinear cost functional which can be minimized by standard optimization techniques. In most works a steepest descent algorithm is usually adopted. Therefore, the new snake shape is given by

$$\mathbf{z}^{t+1} = \mathbf{z}^t - \gamma \left(\mathbf{A} \mathbf{z}^t - \mathbf{f}_{img}(\mathbf{z}^t) \right) \qquad (3)$$

where

$$\mathbf{z}^t = \begin{bmatrix} z_0^t & z_1^t & \cdots & z_{N-1}^t \end{bmatrix}^T \qquad (4)$$

is a vector of snake control points at t-th iteration, $\mathbf{A}$ is a real pentadiagonal matrix given by

$$\mathbf{A} = \begin{bmatrix} a_0 & a_1 & a_2 & 0 & 0 & \cdots & a_2 & a_1 \\ a_1 & a_0 & a_1 & a_2 & 0 & \cdots & 0 & a_2 \\ a_2 & a_1 & a_0 & a_1 & a_2 & \cdots & 0 & 0 \\ 0 & a_2 & a_1 & a_0 & a_1 & \cdots & 0 & 0 \\ \cdots & \cdots & \cdots & \cdots & \cdots & \cdots & \cdots \\ a_1 & a_2 & 0 & \cdots & 0 & a_2 & a_1 & a_0 \end{bmatrix} \quad \text{with} \quad \begin{aligned} a_0 &= 2\alpha_1 + 6\alpha_2 \\ a_1 &= -\alpha_1 - 4\alpha_2 \\ a_2 &= \alpha_2 \end{aligned} \qquad (5)$$

$f_{img}(z(n)) = -\nabla P_{img}(z(n))$ is an image force field and γ is the adaptation gain. Equation (3) can be rewritten in the form

$$\mathbf{z}^{t+1} = \mathbf{M} \mathbf{z}^t + \gamma \, \mathbf{f}_{img}(\mathbf{z}^t) \qquad (6)$$

where $\mathbf{M} = \mathbf{I} - \gamma \mathbf{A}$. Another strategy which is often used to optimize cost functional (2) consists of measuring the gradient at the next snake configuration $\mathbf{z}^{t+1}$. In this case, equation (6) holds with $\mathbf{M} = (\mathbf{I} + \gamma \mathbf{A})^{-1}$.

3. ACTIVE CONTOURS IN THE TRANSFORM DOMAIN

In order to derive the new algorithm described in section 4, we will first rewrite the classic snake algorithm in a transform domain defined by the eigenvectors of matrix $\mathbf{M}$. Since matrix $\mathbf{M}$ is real and symmetric, there is a unitary eigenmatrix $\mathbf{V}$, of rank N, which diagonalizes the matrix $\mathbf{M}$ according to

$$\mathbf{D}_\lambda^M = \mathbf{V}^H \mathbf{M} \mathbf{V} \qquad (7)$$

where $\mathbf{D}_\lambda^M$ is a diagonal matrix containing the eigenvalues of $\mathbf{M}$ and $\mathbf{V}^H$ denotes the conjugate transpose of $\mathbf{V}$. Fortunately, the diagonalization performed in equation (7) can be accomplished in a very efficient manner. In fact, since $\mathbf{M}$ is a circulant matrix, its eigenvectors are complex exponentials with harmonically related frequencies and its eigenvalues can be efficiently computed by the DFT of M's first column [5].

Multiplying both sides of (6) by $\mathbf{V}^H$ and using (7) one obtains a recursive algorithm for the updation of the snake parameters in the transform domain

$$\mathbf{Z}^{t+1} = \mathbf{D}_\lambda^M \mathbf{Z}^t + \gamma \mathbf{F}^t \qquad (8)$$

where $\mathbf{Z} = \mathbf{V}^H \mathbf{z}$ and $\mathbf{F} = \mathbf{V}^H \mathbf{f}$ are the snake and image force vectors in the transform domain, respectively.

Interpreting equation (8) as a state space formulation of a dynamical system, we have managed to decouple the different modes. The N-th order system can therefore be implemented as a set of N independent first order systems. This property does not necessarily mean that equation (8) provides a faster implementation of the snake updation scheme since we now have to transform the image force field into the transform domain.

4. NEW SNAKE ALGORITHM

In this section we will propose an interpretation of the internal forces of the previous algorithms as a filtering operation on the snake control points. Therefore, the choice of the internal forces becomes a problem of filter design. We will then generalize the class of filters (internal forces) which may be used in the updation scheme. This generalization allows a more flexible choice of the internal force model which can be modified during the adaptation process and can also be designed according to the class of objects one is trying to locate.

Fig. 1 shows the block diagrams of the snake updation algorithms. The computation of a new vector of snake control points is the sum of an image force vector which attracts the snake to desired image features (e.g., edges) and a term which depends on the snake internal forces. The internal force term is obtained multiplying the snake vector by matrix $\mathbf{M}$ in the classical updation scheme (see Fig. 1a) and corresponds to a simple product of the matrix eigenvalues by the corresponding vector components in the transform domain algorithm (see Fig. 1b). This operation can be interpreted as a filtering operation as follows. Since matrix $\mathbf{M}$ is circulant, it corresponds to a circular convolution of the snake positions by a FIR filter with impulse response given by the first column of matrix $\mathbf{M}$. In addition, since the snake is a description of a closed contour, it can be periodically extended and the circular convolution is nothing but a linear convolution of the snake positions by the impulse response of an FIR filter. The FIR filter performs a regularization of the previous contour shape by a lowpass filtering operation.

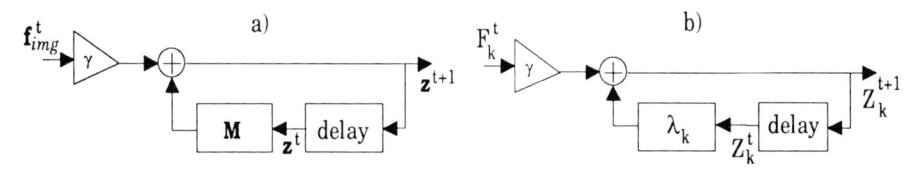

Figure 1 - Block diagram of snake algorithm in a) space domain, b) transform domain.

Since matrix $\mathbf{M}$ is circulant, its eigenvalues are obtained by the DFT of the first column of matrix $\mathbf{M}$, i.e., from the DFT of the filter impulse response. Therefore, the eigenvalues are samples of the filter impulse response and if one changes the eigenvalues of matrix $\mathbf{M}$ one modifies the filter frequency response accordingly. This property provides a practical way of controlling the lowpass filtering operation or, stated in other words, the internal forces of the snake algorithm.

Unfortunately, in the classic formulation, there is little flexibility in the choice of lowpass frequency response because all eigenvalues are parametrized by α_1, α_2 (see eq. 2). This results in a limited class of filter shapes shown in Fig. 2a, which is not well suited to deal with many object contours.

In the algorithm proposed in this paper we allow every eigenvalue to be independently specified therefore allowing greater flexibility in the choice of the snake internal forces. Strategies to design the set of eigenvalues will not be discussed in this paper but they can be object dependent and they can vary along the adaptation process tailoring the internal forces to different stages of the updation process. Figure 2b shows an example of a filter frequency response which can be used to regularize the snake contour in the proposed algorithm.

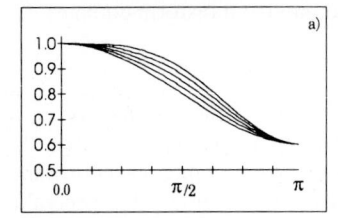

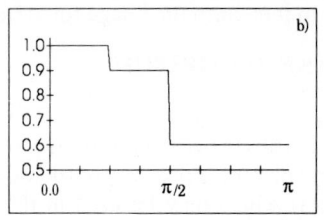

Figure 2 - Filter frequency responses: a) classic algorithm; b) example of a general design.

The updation of the snake parameters with independently specified eigenvalues can either be performed by recursion (6) or (8). In the first case one has to compute matrix **M** everytime the eigenvalues are modified. This can be done easily by computing the inverse DFT of the eigenvalue sequence to obtain the first column of matrix **M** which defines the matrix. In the transform domain, there is no need to update any matrix. All we have to do is to use the appropriate eigenvalues in the adaptation scheme (8).

An interesting property is that although we have removed all restrictions on the eigenstructure of matrix **M**, the resulting algorithm is still a steepest descent algorithm for the optimization of an extended cost functional

$$P(z(0),...,z(N-1)) = \sum_{n=0}^{N-1} \frac{\alpha_1}{2}\left|\Delta^{(1)}z(n)\right|^2 +......+ \frac{\alpha_{N/2}}{2}\left|\Delta^{(N/2)}z(n)\right|^2 + P_{img}(z(n)) \qquad (9)$$

where the weights α_i (i=1, .., N/2) can be easily related with the chosen eigenvalues by closed form expressions. When α_i=0 (i=3, .., N/2) we obtain the classic algorithm described in section 2. Eigenvalue design and experimental results will be presented in a forthcoming paper.

References

[1] Michael Kass, Andrew Witkin, Demetri Terzopoulos, "Snakes: Active Contour Models", First Int. Conf. on Computer Vision, June, London, 1987.
[2] Laurent D. Cohen, "On Active Contour Models and Balloons", CVGIP: Image Understanding, Vol.53. nº 2, pp.211-218, March, 1991.
[3] Donna J. Williams, Mubarak Shah, "A Fast Algorithm for Active Contours and Curvature Estimation", CVGIP: Image Understanding, Vol. 55, pp.14-26, January, 1992.
[4] Anil K. Jain, "Fundamentals of Digital Image Processing", Prentice-Hall, 1989.

Cooperation between edge detection and region growing: the problem of control

M. Salotti and C. Garbay

Groupe SIC, Lab. TIMC/IMAG,
 Batiment CERMO BP 53X
38041 Grenoble CEDEX -FRANCE-

Abstract
In the cooperation task between edge detectors and region growing, the goal is to take the best from each technique. The problem is very complex and the cooperation is often much simplified. We discuss the importance of control in such systems and propose new ideas. We present two algorithms that have a flexible information management convenient for a powerful cooperation.

1. INTRODUCTION

In low-level vision, a cooperation between an edge detector and a region growing method has often been used [1-4]. Our aim is to show that specific properties are needed to perform a powerful cooperation. In the first part, we present the problem of control. Then, we propose two algorithms that have suitable properties for designing a cooperative system. Finally, we present the perspectives of our approach.

2. COOPERATION: A DIFFICULT TECHNIQUE

2.1 A suitable approach
The most important problem in the segmentation task is to have a powerful information management [5]. Miscellaneous information has to be taken into account to achieve this goal and a cooperation between different techniques seems to be a suitable approach. Edge detectors are sensitive to local variations while region growing methods are more sensitive to more global ones. It seems therefore logical to make them cooperate in order to get the best of each technique.

2.2 Complex problems
The common approach [1-4] consists in two steps. Edges are detected first. Then, using information relative to the edge map, a region growing method is performed. The main idea of the cooperation is that, if no significant gradient is present at the frontier of two regions, the two regions have to be merged. This cooperation is not reciprocal because the edge detector do not use information on regions to find edges. Thus, the results depend much on the quality of the edge detector and the best of each technique has not been used in an efficient way. However, how to get new information coming from other methods ? When ? How to take it into account ? What information does other methods need ? To avoid losing efficiency in the new information management, a powerful control has to be used.

2.3 How to control cooperation ?

In order to control efficiently the cooperation between edge detection and region growing and to proceed carefully step by step, the main stages and therefore the main decisions have to be clearly identified. We propose the following decomposition.
- The choice of the focus of attention is the first important stage. It consists in choosing a partial goal, by identifying the situation (context) and by making a synthesis of any available local information.
- The second stage is the exploration of the neighbourhood. New information is collected around the focus of attention.
- Finally, the last stage is a partial segmentation result. A synthesis has to be made here in order to progress and to increase quality and quantity of information. After the last stage, it is possible to come back to the previous stages, depending on the last decision and the control management.

This decomposition implies that each method should give partial results, after each important stage. Moreover, in order to be adapted to a specific situation, a cooperating method should have a variety of controllable routines. In this case, an expertise on the use of each routine is needed.

3. TWO COMPLEMENTARY APPROACHES

3.1 A region growing based cooperation

We present in figure 1 a region growing method that tries to respect the main stages presented in the previous chapter. This algorithm has already been presented but not in details [5]. A more complete description will be given here.
- First, for all pixels, the algorithm computes local information. All values are stocked in arrays. A lot of information is thus available.
- Then, four connex pixels are chosen (focus of attention). Actually, in the set of pixels that do not belong to any region, the pixels chosen have small gradient values. A region growing model is chosen. This model will help in determining which adapted criteria have to be used when evaluating new pixels for a specific region growing. A region is a dynamic record with several attributes: the number of pixels, the grey level average, the standard deviation, the average of the local 3x3 standard deviation, the average gradients, the direction of the average gradient, the percentage of pixels that have the same direction of gradient and the type of the region. The type of the region can be "very homogeneous region", "homogeneous region", "small uniform gradation" if there is a small gradation in a constant direction, "strong uniform gradation", "non uniform gradation" if there is a gradation in several different directions, or "irregular" if it does not belong to the previous types. All the attributes are taken into account to compute the type of the region.
- When a new pixel belongs to a region, its four neighbours are examined. If they are not marked, they are evaluated and ordered into the stack "Possible new pixels of the region". The evaluation takes into account local information in a 5x5 neighbourhood and information stocked in the region growing model. This stage corresponds to the exploration around the focus of attention.
- Then, the first pixel of the stack is reexamined. According to criteria previously presented, the pixel is rejected (specific mark) or accepted (another specific mark). All attributes are updated and this processing continue until all pixels around the region have been rejected and the stack is empty.
- When all pixels belong to a region,, the region growing ends.

```
┌─────────────────────────────────────────────────────────────────┐
│  Precalculations: accumulation of available information:        │
│  local standard deviation (3x3), local average (3x3), masks 5x5->gradients │
└─────────────────────────────────────────────────────────────────┘
                              │
                              ▼
       ┌──────────────────────────────────────────┐   All pixels belong
    ──▶│ 1) Focus of attention:                   │──▶ to a region,
       │ choice of the first 4 pixels for region growing │   end of region
       │ choice of the region growing model       │       growing
       └──────────────────────────────────────────┘
                              │
                              ▼
       ┌──────────────────────────────────────────┐
    ──▶│ 2) Exploration of the neighbourhood:     │
       │ neighboring pixels ordered into the stack according │
       │ to local information and the region growing model │
       └──────────────────────────────────────────┘
                              │
                              ▼
       ┌──────────────────────────────────────────┐
       │ 3) Partial result, synthesis:            │
       │ evaluation of the first pixel of the stack, update of the │
       │ region growing model and update of the region attributes │
       └──────────────────────────────────────────┘
   stack empty
   region completed
```

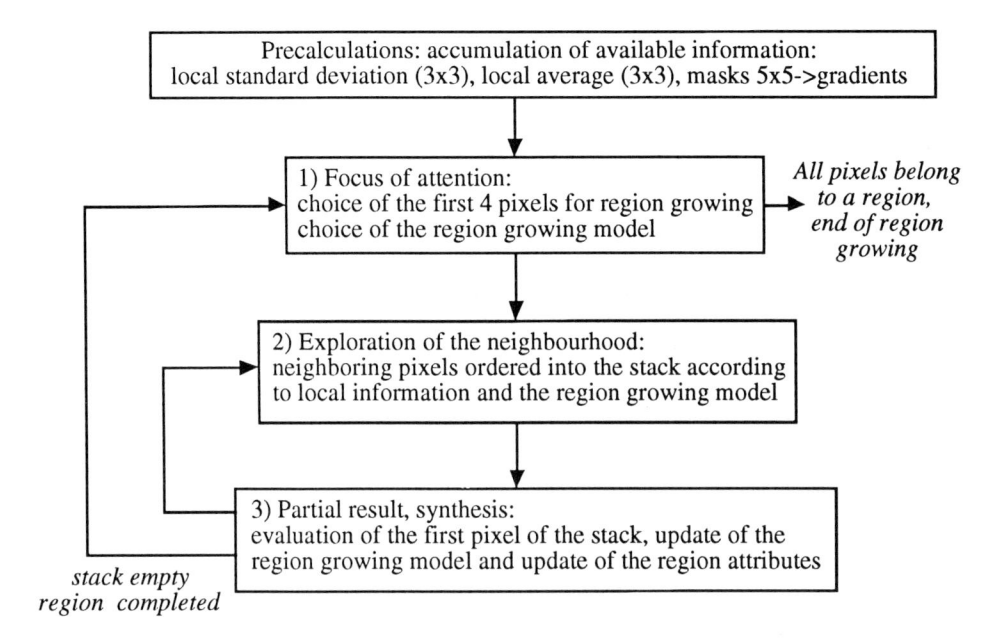

Figure 1 : A region growing algorithm

Discussion: because the seed of the region growing is entirely independent, there are several advantages. First, it is possible to have partial results. We just have to stop choosing a new pixel for region growing. Second, it is possible to impose the first pixel of the region, either the exact position, or just the properties it has to respect. By this way, the aggregation process can be strongly controlled.

3.2 An edge detector based cooperation

We present here an edge growing technique that uses information on regions to find edges. It is therefore a cooperative system. The goal is to form chains of edges that have a significant "relative gradient" value. By relative, we mean that the neighboring regions have to be relatively homogeneous compared to the gradient of the edge pixel. To get information on regions, given an edge pixel, a region growing technique starts with a pixel located farther in the gradient direction, and aggregates the first six pixels that have the smallest local standard deviation in a 3x3 window. Two regions are considered: one in front of the pixel, and another one behind.

• After some calculations (as in figure 1), the first important stage consists in choosing the first pixel of the edge growing (focus of attention). Actually, each pixel of the image is examined in turn. If some conditions are satisfied, high gradient value, local neighboring regions relatively homogeneous, a seed of edge is validated and a model of edge growing is chosen, according to all available information, in particular the type and value of the gradients (four masks have a strong response to a "step edge gradient", and four other masks have a strong response to a "line gradient", according to four possible directions).

• The next stage is "the exploration of the focus of attention". First, a pixel is chosen

among the possible next pixels at both ends, according to the model of edge growing. Then, information on the two neighboring regions (on the right and on the left of the edge) is required to compute the "relative gradient". To get this information, local information (precalculated) is used and if the result is not sure, a short region growing technique is undertaken (actually limited to 6 pixels).
• The third stage is a synthesis and a partial result: the new edge pixel is accepted or rejected and the edge growing model is updated. Then, the processing continue to the previous stage. If all pixels have been rejected at both ends, the edge map is updated and a new edge pixel is chosen (first stage). When all pixels have been examined, the edge growing ends.

Discussion: our results are very encouraging, but we do not have enough place to present them (an article on this edge detector will be soon published). Our approach has many advantages:
• Since the focus of attention is an independent stage, the seed of the edge can be imposed, according to the needs of the cooperation. This technique can be therefore strongly controlled.
• The short region growing process is actually very simple. However, if the region has remarkable properties (very homogeneous for example), it would be suitable to continue the region growing processing until the entire region has been found. Information on this region would be used later to help another edge growing. A cooperation with the region growing technique presented in the previous section will be soon implemented.
• New models of edge growing can be easily added. By this way, specific edges can be detected. For example, the characteristics of the neighbouring region can be imposed, or the shape can be constrained.
To resume, the main quality of this algorithm is a flexible information management. Each important stage can be controlled and a close cooperation is thus possible.

4. CONCLUSION

We have presented new ideas to manage a cooperation between edge detection and region growing, illustrated with two algorithms. The main point is that the methods should be prepared to cooperate. The main stages (control cycle) have to be determined in order to control each decision by using adaptive routines in a well defined context. We are now intending to develop a complete cooperative system that will integrate the region growing method and the edge detector. We hope to present our results in a near future.

References
[1] C. Chu and J.K. Aggarwal. The integration of region and edge-based segmentation, Proc. 3rd Int. Conf. Comp. Vision, Osaka (Japan), 101-105. (1990)
[2] J.F. Haddon and J. Boyce. Image segmentation by unifying region and boundary information, IEEE Trans. PAMI , Vol. 12, N°10, 929-948. (1990)
[3] A.M. Nazif and M.D. Levine. Low-level image segmentation: an expert system, IEEE Trans. PAMI , Vol. 6, N°5, 555-577. (1984)
[4] T. Pavlidis and Y.T. Liow. Integrating region growing and edge detection, IEEE Trans. PAMI , Vol. 12, N°3, 225-233. (1990)
[5] M. Salotti and C. Garbay. A new paradigm for segmentation. Proc. 11th Int. Conf. on Pattern Recog. (ICPR 92), The Hague, Vol. 3, 611-614. (1992)

Motion-Compensated Wavelet Coding of Colour Images Using Adaptive Vector Quantisation

S. Yao[†] and R.J. Clarke

Dept. of Computing and Electrical Engineering
Heriot-Watt University
Riccarton, Edinburgh EH14 4AS
Scotland

ABSTRACT

In this paper, a compression technique for colour image sequences using adaptive vector quantisation based on motion compensated difference frame wavelets and multiscale motion estimation is proposed. The wavelet transform is used to decompose an image into a group of subimages with varying resolutions corresponding to the different frequency characteristics. Since the motion activity of the image can be characterised at different resolutions, multiscale motion compensation is adopted. The chrominance signals are compressed by only coding the "low-low" frequency bands of the wavelet decomposition. Finally, adaptive vector quantisation is used to achieve good overall performance.

1. Introduction

The wavelet transform has recently received considerable attention for image compression[1]. The representation provides a multiresolution interpretation of an image signal with localization in both space and frequency, justified by relevance to human visual system (HVS) models, and is therefore appropriate for image coding. The goal is to exploit redundancy as much as possible while preserving good image quality. Using the wavelet transform, an image can be decomposed into a set of subbands on orthogonal bases and allow us to achieve good performance.

In coding an image sequence, motion compensation can reduce the entropy of the error signal. Since the wavelet decomposition characterizes the motion activity over different scales and frequency ranges, a multiscale motion compensation scheme can be adopted. After motion compensation, the residual error signal has lower energy and is easily coded by adaptive vector quantisation, as reported earlier[2]. Here, we extend the technique to the processing of chrominance signals. These have less energy than the luminance signals[3], and are filtered and downsampled to reduce the data rate. Two coding schemes are proposed and their efficiencies compared.

In section 2, the wavelet transform is reviewed. Multiscale motion compensation is described in section 3 and the coding algorithms shown in section 4. Section 5 contains simulation results.

[†] S.Yao is now with Dept. of Electrical and Electronic Engineering, Changsha Inst. of Tech., Hunan, P.R. China

2. The Wavelet Transform

The wavelet transform expands a signal onto a family of orthogonal bases which are the dilation and translation of a unique function $\psi(x)$ and scaling function $\phi(x)$ [4]. The decomposition results in details and an approximate version of the signal at different resolutions. It can be implemented by quadrature mirror filters[5], and here a simple four tap Daubechies' filter is used. The filtered signals can be written as

$$S_{j,k} = \sum_p S_{j+1,p} h(2k-p) \; ; \; D_{j,k} = \sum_p S_{j+1,p} g(2k-p)$$

where $g(n) = (-1)^{1-n} h(-n+1)$, $S_{j,k}$ is a representation of signal $S(x)$ at scale 2^j and $D_{j,k}$ gives a representation of the details between the scale 2^{j+1} and 2^j ($S_{j+1,p}$ is the higher resolution version). The wavelet transform is carried out by separably filtering row and column components of the image to give a pyramid structure, each level of which contains three upper and one lowest frequency subbands.

3. Multiscale Motion Detection and Compensation

Motion compensation improves coding efficiency, and one recent technique which has been reported is a coarse to fine method[6]. In the wavelet transform domain, motion detection can be carried out on the "low-low" frequency band by evaluation of the luminance difference between blocks. The criterion used here is as follows: a pixel is considered "moving" if its brightness level changes by more than a certain value. If the number of "moving" pixels in a block exceeds a predetermined threshold, the whole block is classified as moving.

Let $S_{j,(m,i)}^k$ represent a block in the mth frame. $S_{j,(m-1,i)}^k$ represents the corresponding block in the previous frame and the absolute pixel difference R_i, is defined as

$$R_i = |S_{j,(m,i)}^k - S_{j,(m-1,i)}^k|, \; i = 1, 2, ..., N$$

where N is the number of pixels within the block. Set

$$O(R_i) = \begin{cases} 1 & \text{if } R_i > T_1 \\ 0 & \text{if } R_i \leq T_1 \end{cases}$$

Then, a block is considered to be motion block if and only if $\sum_{i=1}^{N} O(R_i) \geq T_2$. T_1 and T_2 are perceptually-based thresholds determined experimentally. Due to the smaller size of subimage S_j the calculation is easy and quick. The blocks in other bands at the same location are similarly classified. After motion detection, most blocks (especially in the background) have been classified as still. For moving blocks, variable-size block matching corresponds with the pyramidal structure of the transform. Motion vectors calculated on the lower resolution level are now refined by using conjugate direction search over a small area of the next level. If $V_{d2}^s(x,y)$, $s = GH, HG, GG$ represents the motion vectors centered at (x,y) for the second level subimage blocks of the various frequency bands, then the motion vectors for the blocks on the first level are given by $V_{d1}^s = 2 V_{d2}^s + \Delta^s(\delta_x, \delta_y)$ where, $\Delta^s(\delta_x, \delta_y)$ is the incremental motion vector.

4. Coding

Fig.1 and Fig.2 show the two coding schemes. In the first, the present frame is conventionally predicted by searching for the closest block in the previous frame and the error signal is then decomposed into wavelet subbands. In the second, the motion vectors are obtained using the multiresolution estimation method. The motion-compensated low-low frequency band (S_2) is entropy coded to preserve low frequency information. The residual difference signals are coded using adaptive vector quantisation[7] where signals derived from the first two frames are used as a training sequence to generate an initial multiresolution codebook. For the current frame, input vectors are assigned to the nearest partition using the shortest distance rule. The current codewords are compared with the previous set; if the difference is significant, the codebook is replenished and new codewords sent.

For the chrominance signals, we downsample the U and V components by four horizontally and vertically with a wavelet low-pass filter, i.e. only low-low frequency bands are coded and transmitted. At the decoder, the compressed chrominance signals are upsampled again. Degradation in the colour information is almost negligible.

5. Simulation Results

The experimental results are derived from the test colour image sequence "Claire" which has 256 × 256 pixels/frame with 8 bits per pixel. A 4 × 4 block size is used for the first pyramid level and 2 × 2 for the second. The sizes of the codebooks are 128 for the second level in horizontal and vertical directions and 64 diagonally. Half these sizes are used for the first level. Coding results using the two schemes are listed below.

	\multicolumn{2}{c	}{Y}	\multicolumn{2}{c	}{U}	\multicolumn{2}{c	}{V}
\multicolumn{7}{	c	}{Experimental Results}

	Y bit rate	Y PSNR	U bit rate	U PSNR	V bit rate	V PSNR
Scheme 1	0.47bpp	36.8dB	0.20bpp	40.09dB	0.18bpp	45.26dB
Scheme 2	0.32bpp	36.5dB	0.20bpp	40.09dB	0.18bpp	45.26dB

6. References

[1]. Antonini, M., Barlaud, M., Mathieu, P., and Daubechies, I.:'Image coding using vector quantisation in the wavelet transform domain', IEEE Proc. ICASSP,1990,pp.2297-2300.

[2]. Yao, S. and Clarke, R.J.:'Motion-compensated wavelet coding using adaptive vector quantisation', IEE Colloquium on "Application of Wavelet Transforms in Image Processing", London, Jan. 20th 1993.

[3]. Limb, J.O., Rubinstein, C.B. and Thompson, J.E.:'Digital coding of color video signals - a review', IEEE Trans. on Comm., Vol.COM-25, No.11, pp.1349-1385, Nov. 1977.

[4]. Daubechies, I.:'Orthonormal bases of compactly supported wavelets', Communications on Pure and Applied Mathematics,1988,Vol.XLI, pp.909-996.

[5]. Mallat, S.G.:'A theory for multiresolution signal decomposition: the wavelet representation', IEEE Trans., 1989,PAMI-11, (7) ,pp.674-693.

[6]. Wang, Q. and Clarke, R.J.:'Motion estimation and compensation for image sequence coding', Signal Processing: Image Communication 4 (1992), pp.161-174.

[7]. Yao, S. and Clarke, R.J.:'Image sequence coding using adaptive vector quantisation in wavelet transform domain', Electron. Lett., 1992, 28, (17), pp.1566-1568.

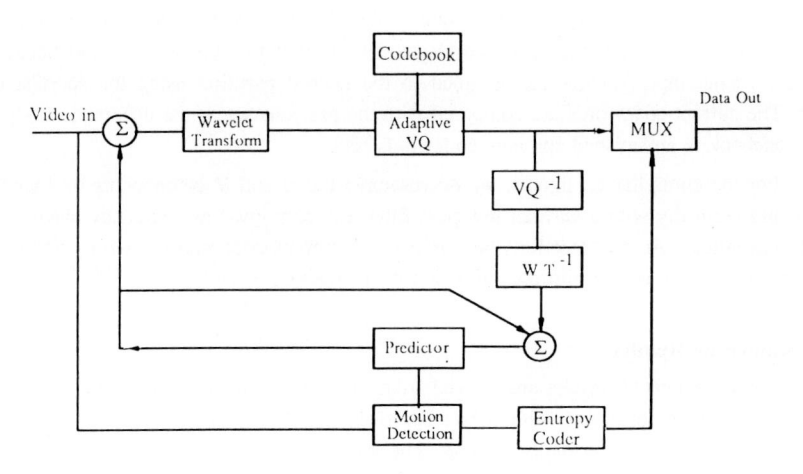

Fig.1 Scheme One

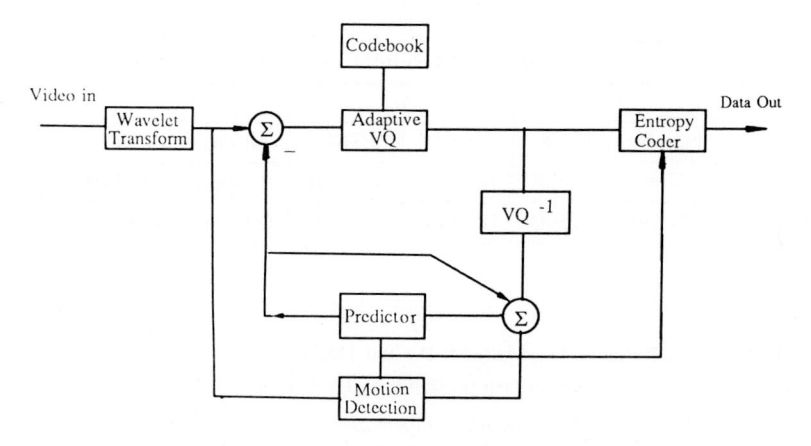

Fig.2 Scheme Two

> # High–rate still–image compression by DWT coding

M. Corvi and E. Ottaviani

Elsag Bailey, R.&D., Via Puccini 2, 16154 Genova, Italy

Abstract
A compression algorithm for still images based on the discrete wavelet transform (DWT) is described. Hige–rate, quality compressions are obtained through an efficient coding of the DWT coefficients. The results of numerical experiments are presented.

1. INTRODUCTION

Image compression is currently motivated by the ever–growing need for transmission of images over band–limited channels and digital storage. Digital images usually contain a huge amount of data, which are often highly redundant, and several techniques have been proposed exploiting this redundancy to reduce the number of bits per image.

The JPEG algorithm [1], based on DCT and statistical coding allows for fast implementations with good compression ratios. It has become a standard term of comparison for other techniques. Recently subband coding technique [2] have shown very promising in terms of bit rate vs. distortion. Image compresion relies on the data decorrelation achieved through the decomposition of the image into frequency bands.

Wavelets prove particularly useful for the design of subband codings in which the conditions for perfect reconstruction are met [3]. The regularity properties of the wavelet underlie the effectiveness of the coding technique. The discrete wavelet transform (DWT) [4-5] performs a bi–orthogonal multi–scale decomposition of the image. Our work relies on the structure of DWT coefficients for an efficient coding. Compression is obtained by keeping only the most relevant coefficients. They are then quantized with a scale–dependent bit number, through uniform scalar quantization, and their value is coded with standard statistical techniques. Their identification is also coded with statistical techniques by exploiting the pyramidal structure inherent to the DWT coefficients.

The algorithm has been implemented in FORTRAN on a VAX 6000-410 computer. On the 8–bit 512×512 image "LENNA", its performances are comparable to those of the best techniques currently available, like adaptive DCT [6] and SBC, with entropy–coding [7]. Image quality is good, showing several details present in the original image up to compression ratios as large as 40.

2. THE DWT CODING ALGORITHM

Our image compression algorithm is based on an efficient coding of the DWT coefficients. The coding part consists of four blocks. The first one carries out the DWT of the image. Next the DWT coefficients are normalized and quantized. The compression block selects the *relevant* coefficients. Finally, the relevant information is fed into the coded–data stream (file or transmission channel). The reconstruction part does these operations in reverse order. The coded data are inputed first, a matrix of DWT coefficients is formed, and the recostructed image is obtained with an inverse DWT.

In view of a potential parallel implementation the time performances of the algorithm are determined by the slowest block, which is, in our case, the DWT one. Therefore it is crucial to operate with efficient transforms.

We employ the iterative DWT algorithm described in [4]. The input matrix F_N, of size 2^N, is the original image. At each iteration a matrix F_n, of size 2^n, is filtered to obtain four matrices of size 2^{n-1}; three of them HF_nG, GF_nG, and GF_nH are the details at scale $N+1-n$. The fourth one, HF_nH, is a coarse image of size 2^{n-1} and forms the input of the next iteration. Here H and G denote the low–pass and the band–pass filters, respectively. The process ends with a coarse image of size 2^0, and a sequence of triplets of details with increasing size from 2^0 to 2^{N-1}. These are the DWT coefficients and can be arranged into a matrix of size 2^N like the initial image, see [4].

The original image is reconstructed stepwise by applying the conjugate filters H^* and G^*, according to $F_n = H^*(HF_nH)H^* + H^*(HF_nG)G^* + G^*(GF_nH)H^* + G^*(GF_nG)G^*$. To have perfect reconstruction the filters H, G, H^*, and G^* must satisfy suitable conditions [3]. A particular class of these filters is obtained from the theory of wavelets. In this case the decomposition of the image is related to its projections on the multiresolution spaces associated with the wavelet.

Our implementation employs finite length filters arising from compactly supported bi-orthogonal wavelets [8]. The specific filters H and H^* in our experimentation are derived from the spline $_2\phi$ and $_{2,2}\tilde{\phi}$, see [9]. The compactness of the support and the regularity of the wavelets are important for the suppression of the ringing effect at the edges in the reconstructed image. These filters are described in details in [8–9], here we remark only that their lengths are three and five taps, respectively, and their values are small multiples of 2^{-m} with m=1, 2, 3. This underlies the fast computational speed of our implementation.

For the compression the entries of the coefficient matrix must be quantized with a small number of bits, therefore some image degrade is unavoidable in the process. However the information in the DWT coefficients is usually more decorrelated than in the original image, and, granting a mild degrade, a high compression can be achieved by retaining the most important coefficients only. Large coefficients correspond to important projection components in the image multiresolution decomposition, and are necessary to obtain a reconstructed image close to the original one. On the other hand small coefficients can be ignored without affecting the image reconstruction appreciably.

The coefficients are selected by the magnitude, keeping into account that, because the coefficients are usually larger at the lowest scales, it is useful to compensate this effect with a scale–dependent gain factor, evaluated adaptively from the image data. Denoting by $y(k)$ the logarithm of the maximum of the coefficients at scale k, the gain rate is defined as $\alpha = \frac{12}{N^2-1}E[y^2] - \frac{2}{N+1}E[y]$, where $E[x] = (1/N)\sum_{k=1}^{N} x(k)$. The coefficients at scale k are then scaled by $2^{1+\alpha k}$.

The quantization–bit numbers for the normalized DWT coefficients vary with the scale and can be adjusted to the image. We have obtained the best results with bit numbers ranging from nine, for the largest scales, to four, for the smallest scales.

The set of *relevant* coefficients is characterized by the lower bound on the magnitude of the *relevant* coefficients. This is evaluated from the desired bit/pixel ratio estimating the number of bits required to code both the coefficients values and their positions in

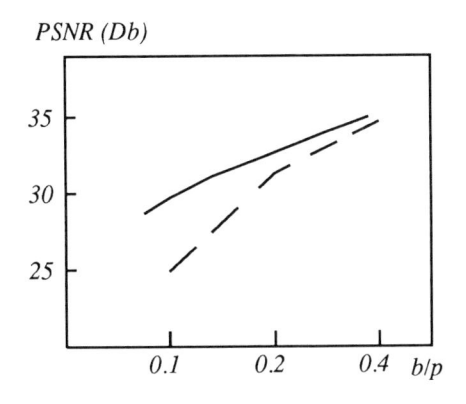

Figure 1. PSNR curves for the DWT algorithm and for JPEG.

Figure 2. Image "LENNA" compressed at 51700 bits.

the DWT coefficient matrix.

The positions coding exploits the pyramidal structure of this matrix. A given detail entry belongs to all the detail blocks that lie on the segment joining the entry to the coarse–image corner. Therefore, it is assumed that all the coefficients directly above a *relevant* coefficient are also *relevant*. According to this principle the *relevant* coefficients form a tree structure. We have found experimentally that a minor percentage of coefficients which would otherwise be discarded is added to the coded stream. The transmission of some small coefficients is strongly compensated by the fewer bits required to code a tree structure than a generic set of locations.

The last block feeds the data into the coded stream. First, it records general information about the image: size, bit–quantization numbers, rate of the normalization exponent α, and the maxima of the details at the various scales. Then the coarse image is recorded in the stream, and the three details at the scale 2^0 are put in the relevant set. This set is structured as a LIFO queue. Next, while the queue of *relevant* coefficients is not empty, a coefficient is taken out from it; its value and the information about which of the four underlying coefficients is *relevant* are recorded in the stream; the appropriate coefficients are also added to the queue of *relevant* coefficients. Statistical coding is used both for the coefficient values and for the position information.

3. NUMERICAL RESULTS

The program has been developed on a VAX 6000–410 computer using FORTRAN language. This general–purpose code takes about 15 CPU seconds to code, and 10 to decode a 512×512 image. However, the process is composed of blocks in pipeline, and lends itself to a custom hardware parallel implementation [10]. In our implementation the DWT blocks take about 6 CPU seconds for a 512×512 image. A dedicated implementation can provide for much higher speeds, making this approach competitive for

Figure 3. Edges curves for Figure 2 (left) and for the original "LENNA" (right).

many potential industrial applications.

DWT subband coding achieves in general very good performances with respect to JPEG. Figure 1 shows the peak signal to noise ratio (PSNR) obtained with our algorithm on the 512×512 test image "LENNA", the JPEG curve (dashed) is displayed for reference. The proposed algorithm yields PSNRs of 32.5 dB at 0.25 bits/pixel and of 29.7 dB at 0.125 bit/pixel. These are comparable with the performances reached by the best techniques presently available, like adaptive DCT [6] and SBC, with entropy coding [7].

Image quality after reconstruction is generally good, showing several fine details present in the original image. This is due to the presence of coefficients at all scales, whose importance is emphasized by the scale–dependent normalization. Figure 2 shows the image "LENNA" at about 0.20 bit/pixel.

The bit/pixel ratio performances are even better on images of indoor scenes recorded by a mobile robot for navigation purposes. In this case we get an image quality sufficient for navigation with a bit/pixel ratio lower than 0.1. To evaluate also a perceptional distorsion measure, we tested the application of standard edge–detection techniques to reconstruction images at high compression ratios, figure 3. Experiments show that only a few edges of minor importance are missing, keeping essentially unchanged the primal sketch of the image itself.

REFERENCES

1 G.K.Wallace, Comm. A.C.M. No. 34 (1991) 30.
2 P.H.Wasterink, et al., IEEE Trans. Comm. 36 (1988) 713.
3 M.Vetterli, and C.Herley, IEEE Trans. SP 40 (1992) 2207.
4 S.G.Mallat, IEEE Trans. PAMI 11 (1989) 674.
5 R.A.DeVore, B.Jawerth and B.J.Lucier, IEEE Trans. IT 38 (1992) 719.
6 W.A.Pearlman, IEEE Trans. Comm. 38 (1990) 698.
7 N.Tanabe and N.Farvardin, IEEE Jour. of Sel. Areas in Comm. 10 (1992) 926.
8 A.Cohen, Preprint AT&T Bell Labs, Murray Hill, NJ, 1992.
9 A.Cohen, I.Daubechies, J.C.Feauveau, Comm. Pure Appl. Math. 45 (1992) 485.
10 J.D.Hoyt and H.Wechsler, 11 IAPR Proc. Vol.4 (1992) 19.

Motion Compensation for compatible Multiresolution coding schemes

F. Bellifemine, A. Chimienti, R. Picco

Centro di Studio per la Televisione of C.N.R. - Strada delle Cacce no.91 - Torino (Italy)

1 Introduction

Subbands, Orthogonal Wavelets transform and, more generally, all techniques for the multiresolution (MR) representation of signals seem to open a new way for the coding of signals. The space of signals is represented by a new orthonormal basis which is convenient for easily extracting approximations of signals with increasingly lower resolutions. An image coding scheme compatible for classes of receivers of a different size and resolution may be then designed. Suitable requirements for the video part of a transmission scheme [1], such as progressive coding, multi-layer coding, bitstream scalability, compatibility between different picture formats and graceful degradation, can all be adapted naturally to well known MR technique capabilities.

Even if this seems to be the path to follow in order to develop a worldwide image coding scheme breaking the barriers of applications, MR techniques have not yet given the expected results and they still perform worse than traditional DCT schemes for video coding. The translation invariance problem has already been recalled as one of the primary weaknesses of the MR transform in the sense that the content of the subbands is unstable under translations of the input signal and this characteristic does not allow the temporal correlation of a video sequence to be fully exploited. Therefore, the motion compensation technique (MC), which has been proved to be quite effective in DCT schemes, does not have a simple and effective application in MR schemes.

When the input signal is translated, the transform coefficients do not behave in the same way [2]: they change drastically in a complex manner. In general, $W(T(f)) \neq T(W(f))$ where $W(\bullet)$ is the MR transform operator and $T(\bullet)$ is the translation operator. On the other hand, MR representations are based on filtering and subsampling. Filtering (i.e. convolution) is a linear operator and is shift-invariant, while subsampling is a linear operator but, in general, is not shift-invariant apart from when the shift value is an integer multiple of the subsampling factor. Furthermore, MR schemes employ Quadrature Mirror Filters (QMF) that allow the Nyquist criterion in subsampling to be violated and the aliasing error when the subbands are recombined to be cancelled. This means that, under translation of the input signal, the information content within a subband moves into other subbands.

The translation invariance problem is a great weakness of these schemes as it does not allow a fully compatible scheme to be designed (where the subbands are independent of one other) and the MC to be simply exploited within the subbands. On the other hand, if the movement is compensated before subdividing the signal into subbands, the MR scheme loses its flexibility and it no longer allows compatibility between different picture formats: whatever the necessary signal resolution, all the subbands must be received in order to invert the motion compensation process at the receiving end. Therefore, a translation operator T or a set of operators must be found to compensate the movement after the signal has been MR represented, that is, if the translation value x_0 and the transform coefficients $c_{n,j}^f$ are known, the operator is defined as $c_{n,j}^g = T(x_0, c_{n,j}^f)$ where $g(x) = f(x - x_0)$.

The aim of this paper is to analyse the translation invariance problem and to define an operator which exploits the detected movement information and compensates the signal translation directly within the content of each subband.

2 Translation operator

According to the construction and definition of an MR scheme:

$$\mathbf{f}(x) = 2^J \sum_{m=-\infty}^{+\infty} A^{-J} f_d(m)\, \phi^{-J}(x - 2^J m) + \sum_{k=-J}^{-1} 2^{-k} \sum_{m=-\infty}^{+\infty} D^k f_d(m)\, \psi^k(x - 2^{-k} m) \quad (1)$$

where $A^{-J} f_d(m)$ and $D^k f_d(m)$ are the transform coefficients for the spaces V_{-J} (the set of all the signals approximated at resolution 2^{-J}) and O_k (the set of the detail signals necessary to go from V_k to V_{k+1}), respectively, and $\phi(x)$ and $\psi(x)$ are the scaling function and the wavelet function of the MR scheme (with the notation $w^i(x) = 2^i w(2^i x)$), i.e. the mother functions to construct the orthonormal basis of the spaces in the scheme. The transform coefficients of the function $g(x) = f(x - x_0)$ for the space generated by the mother function $w^j(x)$ are then:

$$\mathbf{c_{n,j}^g} = <g(u), w^j(u - 2^{-j} n)> = \int_{-\infty}^{+\infty} g(x) w^j(x - 2^{-j} n) dx = \int_{-\infty}^{+\infty} f(y) w^j(y + x_0 - 2^{-j} n) dy =$$

$$= 2^J \sum_{m=-\infty}^{+\infty} A^{-J} f_d(m) \int_{-\infty}^{+\infty} \phi^{-J}(x - 2^J m) w^j(x + x_0 - 2^{-j} n) dx +$$

$$+ \sum_{k=-J}^{-1} 2^{-k} \sum_{m=-\infty}^{+\infty} D^k f_d(m) \int_{-\infty}^{+\infty} \psi^k(x - 2^{-k} m) w^j(x + x_0 - 2^{-j} n) dx \quad (2)$$

Now, let us consider separately the two cases where $w^j(x) = \phi^{-J}(x)$ and $w^j(x) = \psi^j(x)$, that is compensation in the V_{-J} space and compensation of the detail signals. Formula (2) becomes then, respectively, (3) and (4):

$$\mathbf{A^{-J} g_d(n)} = \sum_{m=-\infty}^{+\infty} A^{-J} f_d(m) ACF(\phi, 2^{-J} x_0 + m - n) +$$

$$+ \sum_{k=-J}^{-1} 2^{-k} \sum_{m=-\infty}^{+\infty} D^k f_d(m) CCF(\psi^k, \phi^{-J}, x_0 + 2^{-k} m - 2^J n) \quad (3)$$

$$\mathbf{D^j g_d(n)} = 2^J \sum_{m=-\infty}^{+\infty} A^{-J} f_d(m) CCF(\phi^{-J}, \psi^j, x_0 + 2^J m - 2^{-j} n) +$$

$$+ \sum_{k=-J}^{j-1} 2^{-k} \sum_{m=-\infty}^{+\infty} D^k f_d(m) CCF(\psi^k, \psi^j, x_0 + 2^{-k}m - 2^{-j}n) +$$

$$+ \sum_{m=-\infty}^{+\infty} D^j f_d(m) ACF(\psi, 2^j x_0 + m - n) +$$

$$+ \sum_{k=j+1}^{-1} 2^{-k} \sum_{m=-\infty}^{+\infty} D^k f_d(m) CCF(\psi^k, \psi^j, x_0 + 2^{-k}m - 2^{-j}n) \quad (4)$$

where ACF and CCF are the autocorrelation and crosscorrelation functions of the base functions:

$$\mathbf{ACF(u, \tau)} = \int_{-\infty}^{+\infty} u(x) u(x + \tau) dx \; ; \quad \mathbf{CCF(u, v, \tau)} = \int_{-\infty}^{+\infty} u(x) v(x + \tau) dx \quad (5)$$

Formulas (3) and (4) represent the searched for translation operator T: a convolution operation that gives the transform coefficients of a signal $g(x)$ which is the translated copy of $f(x)$. The new coefficients are simply a linear combination of the old ones with the combination coefficients given by the ACF and CCF functions of the base functions.

In an image coding scheme $A^{-J} f_d(m)$ and $D^k f_d(m)$ are the wavelets transform coefficients of the previous coded image, x_0 is the estimated displacement (for the whole image or, better, for independent areas of the image), while $A^{-J} g_d(n)$ and $D^j g_d(n)$ are the predicted transform coefficient values for the actual image. The differences between the predicted values and the effective values are to be quantized and transmitted over the channel to the receiving end.

If all the terms of the formulas are used, the receiver must decode all the subbands to be able to invert the MC process; in this case, the scheme loses its flexibility and is no longer compatible. Three different compensation schemes which retain flexibility have instead been analysed. The first one uses only the ACF terms so it compensates the movement directly within the content of each subband without considering the contribution from other subbands. The second, instead, for each subband j, uses terms from all the subbands k with a lower resolution, $-J \leq k \leq j$; this scheme maintains compatibility with the realistic hypothesis that if a receiver decodes the detail signal at resolution j, it has previously decoded all the detail signals at coarser resolutions $k \leq j$. Finally, the third scheme interpolates the content of each subband, regardless of the used QMF, in order to evaluate the transform coefficients translated by the value $x_0/2^{-k}$, where 2^{-k} is the subsampling factor of each subband; the interpolating functions must then be the ideal half-band interpolators.

Formulas (3) and (4) may in fact also be considered as interpolating functions matched upon the orthonormal basis of the spaces of the MR scheme. In practice, they are true interpolating filters as the following properties hold:

$$ACF(u, \tau) = 1 \quad , \tau = 0 \quad (6)$$
$$ACF(u, \tau) = 0 \quad , \tau \in Z, \tau \neq 0 \quad (7)$$
$$CCF(u^i, v^j, \tau) = 0 \quad , \tau = k 2^{-j}, k \in Z \quad (8)$$

for $u, v = \phi, \psi$ as, by definition of an MR scheme, $\{\phi(x - n)\}_{n \in Z}$ is an orthonormal basis of V_0 and $\{\psi(x - n)\}_{n \in Z}$ is an orthonormal basis of O_0. These properties simply state

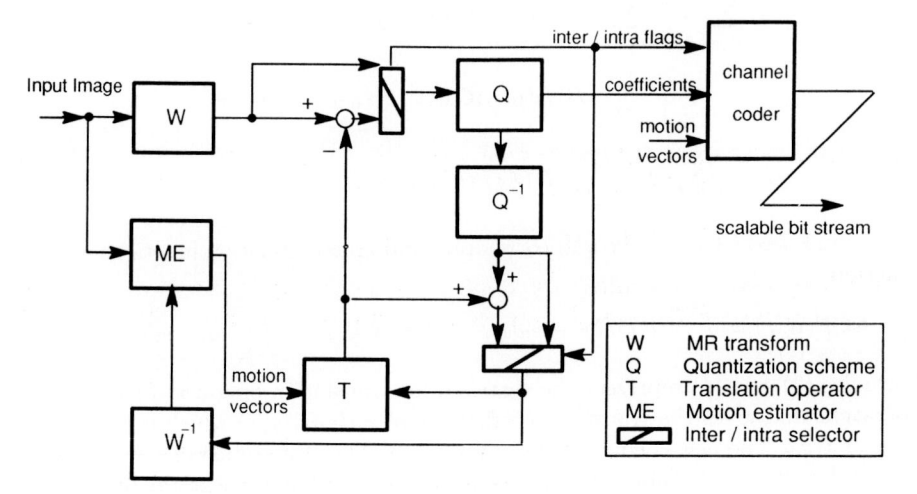

Figure 1: Architecture of an MR video coding scheme

that the interpolated values do not change if the translation value is an integer multiple of the subsampling factor of the system.

Experimental results have been obtained which confirm the validity of the approach and which show the obvious superiority of the second compensation scheme. Its performance is strongly dependent on the coupling between the subbands, however the QMF filters used (32C and 32D of Johnston [3]) show that only adjacent lower resolution subbands substantially contribute to the compensation performance.

3 Conclusions

A translation operator has been defined which compensates the movement after the signal has been MR represented. A condition has been given in order to partially solve the problem of MC within subbands while preserving compatibility, and three different proposals for an operator which maintains flexibility have also been suggested. An architecture is also described in the figure showing the simple coding structure which is derived from the operator.

Results have been obtained which show the effectiveness of this operator by simulating a coding algorithm with the three schemes proposed for interpolation.

References

[1] *List of requirements for MPEG-2 Video.* Technical Report N0159, ISO/IEC JTC1/SC29/WG11, March 1992.

[2] E.P. Simoncelli, W.T. Freeman, E.H. Adelson, and D.J. Heeger. Shiftable multiscale transforms. *IEEE Trans. on Inf. Theory*, 38(2):587–607, March 1992.

[3] J.D. Johnston. A filter family for use in quadrature mirror filter banks. In *Proceeding ICASSP*, pages 291–294, April 1980.

ENTROPY-CONSTRAINED DESIGN OF AN IMAGE CODER WITH A COMPOSITE SOURCE MODEL

C. F. Harris and J.W.Modestino

Electrical, Computer and Systems Engineering Department and Center for Image Processing Research, Rensselaer Polytechnic Institute, Troy, New York, 12180

Abstract

We demonstrate an efficient image compression scheme based on a composite source model. Most current coding techniques attempt to make a global fit of a stochastic model to the entire image. Natural images, however, are generally non-homogeneous and can be better modeled by individually fitting appropriate models to locally homogeneous regions of the image. Specifically, we model the image as a causal 2-D autoregressive mixture with a Markov random field assumption for the state map. The intensity process is coded adaptively using previously developed entropy-constrained predictive quantization schemes. Results show excellent rate-distortion performance gains, and typical PSNR improvements of 3 dB at rates of approximately 1 bpp.

1. INTRODUCTION

We have developed a highly efficient coding technique for images based on a composite source model. Composite models have been used before in other application areas such as speech coding. In the area of image coding, however, virtually no work has been done towards using composite source models. In our work, the subsources are all modeled as 2-D Gaussian autoregressive (GAR) random processes, with each having a distinct set of parameters. The 2-D causal GAR models all hav enonsymmetric half-plane (NSHP) regions of support. It is also assumed that a random process governs the switching between the various subsources. This switch is modeled as a Markov random field (MRF).

For the underlying state process, a MRF model was assumed in order to impose region continuity constraints on the image. The particular MRF used here has the following local characteristic:

$$f(z(i,j) = k|z_{\eta_{ij}})) = \frac{\exp(\beta \delta_{ij}(k))}{\sum_{l=1}^{K} \exp(\beta \delta_{ij}(l))} \quad ; \quad k = 1, 2, ..., K. \tag{1}$$

The value K is the number classes present in the image, and the MRF takes on integer values in the range $1, 2, ...K$. The quantity $\delta_{ij}(k)$ is the number of neighbors of (i,j) (i.e., elements of η_{ij}) in the state k. The coefficient β is a parameter which controls the degree of clustering. For $\beta = 0$, the states are independent and equally probably, while for $\beta \geq 1$, the probability of a pixel being in the same state as its nearest neighbors, is very high. A sample realization of the MRF for $K = 3$ and $\beta = 1.5$ is provided in Fig. 1(a).

The first step of the design procedure requires the image to be segmented into a finite number of regions, each associated with a corresponding 2-D GAR subsource. An unsupervised stochastic model-based approach is used for the image segmentation [1]. In particular, the model parameter estimation problem is formulated as an incomplete data problem, with the expectation-maximization (EM) algorithm being used to determine maximum-likelihood estimates. The EM-based segmentation algorithm determines estimates of the class-conditioned GAR parameter values as well as the class status of each pixel.

Ordinary DPCM quantization generally does not account for the composite nature of the source. Instead, it uses a single fixed quantizer and globally optimized predictor for the entire image. This quantizer is ordinarily designed on the basis of global statistics or a set of training data. The quantizer and linear predictor coefficients for a particular subsource in our scheme are designed using the image data from the regions classified as belonging to a fixed class.

The type of quantizers used determines to a large degree the system performance. For our work, we use scalar quantizers designed under an entropy constraint. These quantizers are non-uniform and therefore add to the complexity of the system. However, they have operational rate-distortion performance better than either uniform quantizers or level-constrained quantizers. The design algorithm for quantizers under an entropy constraint is described in detail in [2] for a memoryless source. This approach uses a Lagrangian formulation to optimize the quantizer problem by minimizing the functional $J = D + \lambda H$, where D is the average distortion and H is the entropy of the ouput codeword indices. The multiplier λ can be viewed as corresponding to the slope at the point (D, H) on the operational R-D curve and can be varied to solve the problem for different average rates.

For sources with memory, the quantizer is embedded within a predictor feedback loop which complicates the design problem. The algorithm employed here solves this problem by using a doubly-iterative design procedure for the design of optimal predictive quantizers under an entropy constraint. This algorithm has been shown to converge to a local minimum of distortion, and a detailed treatment can be found in [3].

After determining the maximum-likelihood (ML) segmentation for the image under the assumption of K existing classes, we design a separate codebook for each class. The image segmentation indicates which subsource is acting at each pixel location and is also called the class map. The image data belonging to a class is used as training data in the design of a quantizer for that class. We used entropy-constrained predictive scalar quantizers (ECPSQ's) for quantization [3].

Optimal bit allocation is achieved among classes by choosing quantizers at points of equal slope on each class' operational R-D curve [4]. This was done by designing quantizers for each class using the same Lagrange multiplier λ.

The class map is used to switch between quantizers and predictors as the image is encoded. Both the receiver and transmitter must have the estimated class information available in order to choose the quantizer and linear predictor for the current pixel. This data takes the form of side information which must be transmitted separately from the intensity data. Typical class map realizations can be efficiently coded using a lossless entropy-coded generalized chain coding technique.

2. RESULTS

To demonstrate the efficacy of this procedure, we applied the coding algorithm to synthetically generated 2-D data. More specifically, we considered a MRF for the unobservable 3-class switch process as shown in Fig. 1(a). Each subsource was generated using a GAR process with a unique set of AR parameter values. Fig. 1(b) shows the resulting synthetic composite source used in this experiment.

Unsupervised segmentation of the synthetic image, followed by coding, was performed for different assumptions on the number, K, of classes acting. The EM segmentation for 10 iterations is shown in Fig 1(c). The performance was compared to entropy-constrained DPCM using a single globally optimized codebook and linear predictor. Figure 3(a) shows the operational R-D performance for this synthetic source. On typical synthetic images, we obtain performance gains of up to 5 dB.

The procedure was also applied to several real-world images to obtain the corresponding operational R-D performance. Performance gains of approximately 3 dB were achieved for representative real-world images. In the natural image case, the true number of classes is unknown, so we must make an assumption as to the number of classes. Results indicate that, for the images tested, increasing the number of classes beyond 3 or 4 classes does not significantly improve image reconstructions. We include R-D performance for the image Lena in Fig. 3(b) under different number of class assumptions. The original grayscale image and the resulting 3-class EM segmentation after 10 iterations are shown in Fig.'s 2(a) and 2(b), respectively. Figures 2(c) and 2(d) show reconstruction results of the image Lena coded at 1.0 bpp for nonadaptive ($K = 1$) ECPSQ and for the assumption $K = 3$, respectively.

3. SUMMARY AND CONCLUSIONS

We have described the application of a composite source modeling approach to the encoding of images. This included the use of an unsupervised stochastic model-based ML segmentation procedure and an optimum quantizer design algorithm. This model proved to be effective for real-world images. Performance was improved over standard entropy-constrained 2-D DPCM in both a rate-distortion sense and in a subjective sense.

4. REFERENCES

1. J. Zhang, J.W. Modestino and D.A. Langan, "Maximum-Likelihood Parameter Estimation for Unsupervised Stochastic Model-Based Image Segmentation," Submitted to IEEE Trans. Signal Proc., July 1991.
2. P.A. Chou, T. Lookabough, R.M. Gray, "Entropy-constrained vector quantization," *IEEE Trans. on Acoustics, Speech and Signal Processing*, Vol. ASSP-37, pp.31-42., Jan. 1989.
3. Y.H.Kim and J.W.Modestino, "Adaptive Entropy-Coded Predictive Vector Quantization of Images," *IEEE Transactions on Signal Processing*, Vol. SP-40, pp. 633-644, March 1992.
4. T.Berger, *Rate-Distortion Theory*. Englewood Cliffs, NJ, Prentice Hall, 1971.

(a) 3-Class MRF Realization   (b) Observed Image   (c) Resulting Segmentation

Figure 1: Synthetic Test Images.

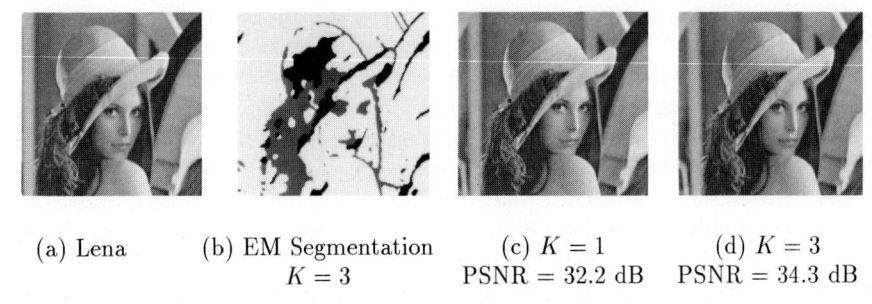

(a) Lena   (b) EM Segmentation $K = 3$ (c) $K = 1$ PSNR = 32.2 dB   (d) $K = 3$ PSNR = 34.3 dB

Figure 2: Original Lena Image, Resulting EM Segmentation, and Reconstructions at $R = 1.0$ bpp.

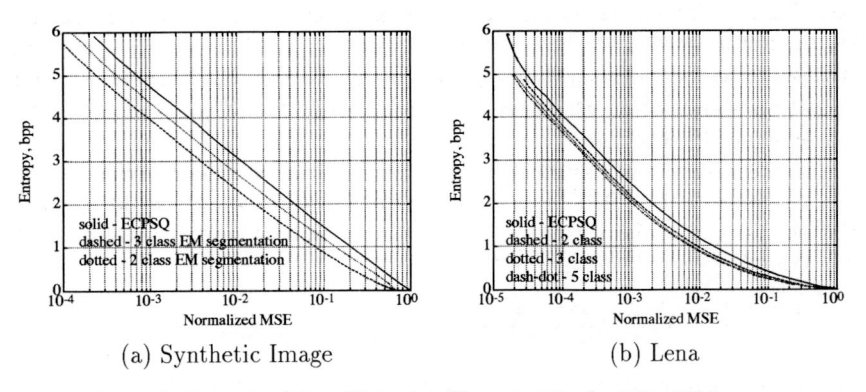

(a) Synthetic Image

(b) Lena

Figure 3: Operational Rate-Distortion Characteristics for 256×256 Images.

Layered and packetized subband coding for compatible digital HDTV

S. Simon and I. Bruyland

Communication Engineering Laboratory, University of Ghent, St.-Pietersnieuwstraat 41, B-9000 Gent, E-mail : simon@lem.rug.ac.be

1 INTRODUCTION

Normal *TV compatibility* is one of the important issues in digital HDTV coding. It should be possible to receive full resolution digital HDTV bitstreams on lower resolution TV receivers, without first decoding the whole incoming bitstream and then subsampling it, as this would lead to the unnecessary incorporation of expensive HDTV circuitry into normal TV receivers. One of the ways to go for is layering : the basic signal layer consists of the coded low resolution TV signal, while some extra "upgrade" layers contain the additional information needed for HDTV.

A second issue is the possibility for *graceful degradation*. In the presence of noisy transmission channels, bit errors should occur preferably in the less important parts of the bitstream thus protecting the more important parts of the bitstream. In this way, the bit error visibility is reduced, and the quality of the decoded signal degrades proportionally to the quality of the transmission channel. Important layers could use e.g. more sophisticated error correction.

Also, there is a trend towards very low bitrates for HDTV. If one wants to distribute HDTV over *existing terrestrial and cable channels*, the signal has to be compressed down to bitrates as low as *60 Mbit/s*, or even *30 Mbit/s*. In e.g. a 16-QAM modulation scheme, this corresponds to two resp. one 8MHz TV channels.

2 SYSTEM FEATURES

As uncompressed interlaced HDTV (1440x1250, 50Hz) originates at a bitrate of more than 670 Mbit/s, it is clear that very high compression has to be applied. To reach the desired compression ratios of about 11 resp. 22, high-performance techniques have to be applied to reduce temporal, spatial and statistical redundancy. Our codec is based on a combination of full frame motion compensation and subband coding, as presented in fig. 1. Subband coding is already firmly established in HDTV coding [1, 2], because it does not cause annoying blocking effects at low bitrates and it makes possible "natural" compatibility with lower resolutions by simple selection of the lower subbands.

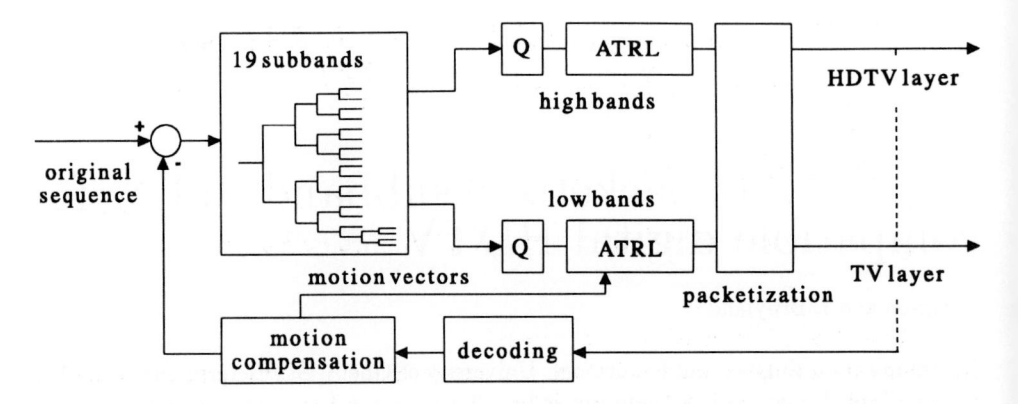

Figure 1: Layered HDTV subband coder with motion compensation

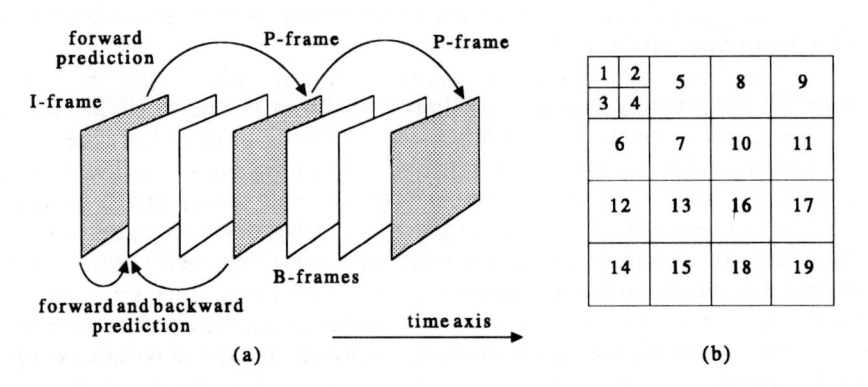

Figure 2: (a) I-, P- and B-frames for GOF=3; (b) 19-band splitting

2.1 Motion compensation

An MPEG-like technique is used with a division of the incoming sequence in so-called groups-of-frames (GOFs), as in fig. 2(a). Inside one GOF, one predictive frame (P-frame), which is n frames away, is calculated using single-sided forward motion compensation starting from the current frame. The bidirectional frames (B-frames) in between are interpolated with forward (starting from the current frame) *or* backward motion compensation (starting from the predicted frame, n frames away) on a block by block basis. After every few GOFs an intra frame is calculated (I-frame). A hierarchical approach is used, where the HDTV picture is predicted on the basis of the normal TV resolution layer, in order to ensure compatibility with TV.

2.2 Subband splitting

The prediction error frames are then split into nineteen subbands, using classical separable one dimensional 14-tap QMF filter banks, applied vertically and horizontally in six successive stages. The bands are grouped into two layers as in fig. 2(b). Bands 1 to 7 make up the TV layer, while bands 8 to 19 contain the "upgrade" information for HDTV.

	Mean entropy (bpp)	Actual rate ATRL inc. packet overhead (bpp)
bands 1-4	2.70	3.95
bands 5-7	3.40	4.05
bands 8-19	1.81	1.94

Table 1: ATRL coder performance for a typical B-frame

2.3 Quantization and ATRL coding

After a weighting stage, where all bands are multiplied by a weighting factor according to their visibility to the HVS (Human Visual System), the bands are quantized using a fixed linear 255-level quantizer. The resulting values are converted to a signed-magnitude representation where the magnitude is represented by the highest-order bits and the sign is put into the least significant bit.

First, the more important bitplanes are scanned. The runlengths of zeros are coded using an ATRL code [3]. In this code, only codewords of length 1 (for runlength M) or of length $m + 1 = log_2 M + 1$ (for the runlengths 0 to $M - 1$) can appear. m is optimized using a statistic of the runlengths in the next 2048 bits. The optimal m is put into a 4-bit *prefix*, which is enclosed in the bitstream. When a non-zero bit is encountered, the remaining lower significant bits of the pixel are encoded.

Then, succesively, the lower bitplanes are tackled. Pixels which were already coded because they had a higher order bit set, are skipped, and the zero runlength is increased by 1. This ensures independent decodability of lower bitplanes, thus making the code more robust in case of loss of parts of the bitstream. The code we use has acceptable efficiency, as table 1 shows.

Higher or lower quantization precision can be selected, simply by increasing or decreasing the number of bitplanes one wants to transmit. This is in fact equivalent to variable thresholding : the more planes are transmitted, the lower the threshold, and the better the reconstruction quality.

2.4 Packetization and layering

The quantized and ATRL-coded outputs of the different bands, and additional information like the motion vectors, are divided up into packets of a fixed size of 2048 bits. Each of the packets contains a 32-bit identification header which ensures its independent decodability. The packets are divided into several classes according to their relative importance for the total image quality. This importance is measured by the contribution of a packet loss to the total mean squared error between the original frames and the decoded frames. In order to reach the desired bitrate, packets are successively selected in descending order of importance till the bitrate is reached. Packets that originate from the lowest 4 bands are always selected for transmission.

Except for the loss of very important packets like those with the motion vectors, the system can stand loss of individual packets without severe degradation in image quality.

	15Mb/s Error variance	7.5Mb/s Error variance
GOF=1 : P	110.3	260.5
GOF=2 : P,B	66.9	191.3
GOF=3 : P,B,B	62.0	196.5

Table 2: Luminance error variances for different bitrates, sequence MOBILE, frame 68 to 79, packet size 2048 bits

3 RESULTS

The whole coding system was simulated in software. Subjective and objective quality assessments were carried out on the well-known TV test sequence "MOBILE". This sequence is very critical, as it contains a lot of motion, color and high spatial detail. Fig. 2 shows some results for the bitrates 15 and 7.5Mb/s, which are equivalent to the HDTV bitrates of resp. 60 Mb/s and 30 Mb/s. Only luminance error variances between the original frame and the reconstruction are given, as the luminance is much more critical than the chrominance. About 4/5 of the bitrate is allocated to the luminance and 1/5 to the chrominance. At 15 Mb/s, no coding artefacts were visible. The picture quality at 7.5 Mb/s is acceptable. Remark that, mostly, the larger the GOF, the smaller the error variance is, and the better the visual picture quality. Especially for 15 Mb/s, the gain of using larger groups-of-frames is high. At lower bitrates, increasing the number of GOFs to more than two has a negative effect for this sequence.

4 CONCLUSION

This paper proposes a novel combination of subband coding with motion compensation for low bitrate HDTV, featuring a layered structure, predictive and interpolative frames and packetization. The coding scheme allows for graceful degradation. Partial decoding of HDTV bitstreams by lower resolution digital TV receivers is possible. Very good image quality has been reached for 60 Mb/s, and acceptable distribution quality at 30 Mb/s.

References

[1] S. Simon, B. De Canne, M. Van Bladel, F. Fan, "A General Subband Coding Scheme for 140 Mbit/s HDTV", IPA 92, Maastricht, The Netherlands, pp. 401-404, 7-9 April 1992

[2] J. De Lameillieure et al., "Subband Coding at 140Mbit/s of interlaced HDTV signals", EUSIPCO 92, august 25-28, 1992, Brussels, Belgium

[3] H. Tanaka, A. Leon-Garcia, "Efficient Run-length Encodings", IEEE Trans. on Information theory, vol. IT-28 no6, pp.880-890, November 1982

Image Processing: Theory and Applications
G. Vernazza, A.N. Venetsanopoulos, C. Braccini (Editors)

Pyramid structures for hierarchical content-driven image coding

B. Aiazzi[a], L. Alparone[b], S. Baronti[a], A. Casini[a], F. Lotti[a]

[a]Istituto di Ricerca sulle Onde Elettromagnetiche IROE - CNR,
via Panciatichi, 64, 50127 Firenze, ITALY

[b]Dipartimento di Ingegneria Elettronica, University of Florence,
via di S. Marta, 3, 50139 Firenze, ITALY

Abstract
An adaptive scheme employing pyramid structures is proposed for multiresolution encoding of still pictures. Efficiency is increased by designing a low-entropy pyramid decomposition and also by giving encoding priority to important features through a content-driven decision rule. Lossless reconstruction capability is ensured. Significant results are achieved.

1. INTRODUCTION

Progressive image coding is gaining attention for transmission over low and medium bit-rate channels. A coarse but global version of the image is transmitted first; refinements are achieved by encoding progressively finer details. Compacting information is a crucial point, since performance increases when the image decomposition is designed to reduce the source entropy at each resolution level. Coding results depend on the entropic efficiency of the encoder, and on the capability to select image details relevant from a subjective point of view.

We propose a complete image coding system, based on a modified Laplacian pyramid (LP) designed in order to possess low correlation between adjacent levels and low entropy at each level, in comparison with the classical LP [1]. Innovations concern the adoption of different filters for reduction and expansion, and the delivery of quantization errors through the pyramid levels, maintaining the advantages of a content-driven decomposition [2][3]. The complete outline may be regarded as a multiresolution interpolative DPCM, with the advantages of a differential scheme and a source decorrelation better than that of a spatial predictive method.

2. GENERALIZED PYRAMIDS

Let $G_0(i,j)$, i,j=0,...N-1, be the original image with N a power of 2. The set:

$$G_k(i,j) = \sum_{m=-M_1}^{M_1} \sum_{n=-N_1}^{N_1} W_1(m,n) G_{k-1}(2i+m, 2j+n) \tag{1}$$

of down-sampled images, is defined as Generalized Gaussian Pyramid (GGP) for $i,j=0,...,(N/2^k)-1$ and $k=1,...,T$, where k identifies the current level of the pyramid and $T \leq \log_2 N$ the top level. From the GGP a Generalized Laplacian Pyramid (GLP) is defined as:

$$L_k(i,j) = G_k(i,j) - \sum_{m=-M_2}^{M_2} \sum_{n=-N_2}^{N_2} W_2(m,n) G_{k+1}(\frac{i+m}{2}, \frac{j+n}{2}) \qquad (2)$$

for $i,j=0,...,(N/2^k)-1$, and $k=0,...,T-1$; summation terms are null for odd values of $i+m$ and $j+n$. $W_1(m,n)$ and $W_2(m,n)$, transformation kernels, are taken separable and symmetric as $W_1(m,n)=W_{1y}(m)W_{1x}(n)$ and $W_2(m,n)=W_{2y}(m)W_{2x}(n)$. In the following it is assumed that $W_{1x}(n)=W_{1y}(m)=w_1(n)$ and $W_{2x}(n)=W_{2y}(m)=w_2(n)$. For both (1) and (2), the parametric odd kernel of size 5, introduced by Burt & Adelson, has been widely used [1][2][3].

Better results are generally achieved by using half-band selective filters for both (1) and (2), as suggested in [4]. However selective filters are expensive to implement and the half-band requirement is tighter for interpolation than for decimation. Therefore a trade-off has been adopted in this work by employing the filter proposed in [1] for (1):

$$w_1(0)=a, \ w_1(\pm 1)=0.25, \ w_1(\pm 2)=0.25-a/2 \qquad (3)$$

that is not half-band except for $a=0.5$. For interpolation, a half-band parametric kernel of size 7, with same computational cost as (3) (5 nonzero coefficients), has been applied:

$$w_2(0)=1, \ w_2(\pm 1)=b, \ w_2(\pm 2)=0, \ w_2(\pm 3)=0.5-b. \qquad (4)$$

The DC gain equals 2, since samples to be interpolated are zero-interleaved. Suitable values of b range in 0.5÷0.625, resulting in linear and cubic interpolation, respectively.

Adopting the kernels (3) and (4) in (1) and (2), makes reduction and expansion independent of one other, introducing a separate adjustment that significantly decreases the entropy and the variance at each level of the resulting zero-mean GLP. In any case the frequency response of the cascaded two filters must be as flat as possible when the frequency is close to the DC.

3. CONTENT-DRIVEN ENCODING ALGORITHM

The basic encoding scheme consists of transmitting an intermediate level $T' < T$ of the GGP (root image) followed by the quantized and coded GLP at levels from $T'-1$ to 0 [1]. In order to introduce a selection criterion for the information to be coded (*entropy decomposition*), each level of the GLP is divided into adjacent subimages of 2×2 pixels (or nodes if $k>0$), that are the information atoms to be considered for transmission. A subimage at level k is taken to have one father node at level $k+1$, and each element of a subimage at level $k>0$ is considered to be the father node of a subimage at level $k-1$.

Content-driven transmission [2][3] consists of a single breadth-first scanning step driven by a set of thresholds, one for each level. The objective of these thresholds is to identify those parts of the GLP which are most meaningful. On each 2×2 subimage an *entropy* function is computed. If this measure is lower than the current threshold, the subimage, as well as all its descendants (if any), are not transmitted. Otherwise, if the entropy exceeds the threshold, GLP 2×2 subimages (interpolation errors) are quantized and Huffman coded. Since not all subimages are selected for transmission, flag bits must be added for each expandable quartet to synchronize the transmitter and the receiver.

A crucial point of the above outline is the choice of an efficient *entropy* measure. Several functions have been reported in [3]. In the present work the maximum absolute value over the 2×2 GLP subimage has been employed, representing a trade-off between velocity and efficiency, both in terms of visual quality and objective errors (MSE).

4. QUANTIZATION AND ERROR RECOVERY

LP encoding may be regarded as a differential spatially interpolative method, in which interpolation errors are quantized and coded. Quantization is particularly critical in this scheme, since errors at higher levels can be spread down over lower ones. Conversely, large quantization steps over the whole pyramid are necessary to improve the compression performance. Since differential schemes must ensure that the predictor/interpolator produces the same output both in compression and in reconstruction, quantization errors at upper levels must be considered by delivering them to lower levels.

To take into account errors from the root downward, the actual GLP that is quantized and coded is given by (2), in which G_{k+1} is replaced by $\hat{G}_{k+1}$, recursively defined as

$$\hat{G}_{k+1}(i,j) = [L_{k+1}(i,j)]_{\Delta_{k+1}} + \sum_{m=-M_2}^{M_2} \sum_{n=-N_2}^{N_2} W_2(m,n) \hat{G}_{k+2}(\frac{i+m}{2}, \frac{j+n}{2}) \qquad (5)$$

in which Δ_{k+1} denotes the quantization step at level $k+1$ and the quantity in square brackets represents a value rounded to an integer multiple of Δ_{k+1}.

This strategy has the effect of damping the propagation of quantization errors at upper levels and allows the user to control the reconstruction error (up to *lossless* performance, in case of complete decomposition) by acting only on the quantization step at level 0. Moreover, the entropy functions are computed on the actual data to be coded, thus improving the efficiency of the entropy decomposition.

5. EXPERIMENTAL RESULTS

Results have been evaluated by considering the total bit-rate (BR) in bits per pixel. Distortion has been measured by means of the mean square error (MSE) and the peak signal-to-noise ratio (PSNR). Figure 1 shows the 512x512 test image "Lena" (8 bpp., entropy H_0=7.45 bpp) and the reconstruction after the application of the proposed scheme with entropy decision, error recovery and different kernels (a=0.65625, b=0.578125, in a 7 bit representation).

BR, MSE and PSNR values are shown in Table 1, where the first entry refers to lossless reconstruction. As results show, this scheme is more efficient than other pyramidal methods [1][3], and comparable with subband coding. The observer's judgement also confirms that significant visual features, such as contours and step edges, are well preserved.

BR	MSE	PSNR
4.5 bpp	0.0	∞
0.8 bpp	12.9	37.0 dB
0.6 bpp	17.8	35.6 dB
0.4 bpp	27.7	33.8 dB

Table 1. Lossless and lossy encoding performance for picture "Lena"

Figure 1. Original (left) and coded version of "Lena", at 0.4 bpp (right).

6. ACKNOWLEDGMENT

Work partially supported by CNR within the Telecommunications Project.

7. REFERENCES

1 P.J. Burt and E.H. Adelson, "The Laplacian Pyramid as a Compact Image Code", *IEEE Trans. Communications*, Vol. COM-31, April 1983, pp. 532-540.
2 G. Mongatti, L. Alparone and S. Baronti, "Entropy Criterion for Progressive Laplacian Pyramid-Based Image Transmission", *Electronics Letters*, Vol.25, 7, 1989, pp. 450-451.
3 G. Mongatti, L. Alparone, G. Benelli, S. Baronti, F. Lotti and A. Casini, "Progressive Image Transmission by Content Driven Laplacian Pyramid Encoding", *IEE Proceedings-I, Communications, Speech and Vision*, Vol.139, N. 5, 1992, pp. 495-500.
4 P. Meer, E.S. Baugher and A. Rosenfeld, "Frequency Domain Analysis and Synthesis of Image Pyramid Generating Kernels", *IEEE Trans.* Vol. PAMI-9, N. 4,1987, pp. 512-522.

Image Processing: Theory and Applications
G. Vernazza, A.N. Venetsanopoulos, C. Braccini (Editors)
© 1993 Elsevier Science Publishers B.V. All rights reserved.

Fast VQ Codebook Design with the Kohonen Algorithm

Giovanni Poggi

Università di Napoli, Dipartimento di Ingegneria Elettronica
Via Claudio,21 80125 Napoli, Italy Fax +39 81 768.31.49 E-mail poggi@nadis.dis.unina.it

Abstract

This paper proposes a fast Kohonen algorithm for VQ codebook design. To speed up the selection of the best-matching codevector several tricks are adopted, which allow one to exploit the structure of the codebook despite its high-rate evolution. Simulations show that the resulting complexity is comparable to that of well-known fast-GLA techniques.

1. Introduction

The design of effective and robust codebooks plays a central role for the performance of any Vector Quantization (VQ) [1] system. The most popular technique to this end is the Generalized Lloyd Algorithm (GLA) which, given a suitable training set, alternately optimizes the codebook and the input-space partition until a local minimum of the encoding distortion is reached.

The Kohonen Algorithm (KA) [2] represents an interesting alternative to the GLA. It was originally proposed in the neural network field as a tool to produce auto-organized nets of neurons where, thanks to interactions during the training phase, spatially close cells are sensitive to similar stimuli. The design of VQ codebooks, however, has probably become now its major application, and several researchers [3, 4] have already investigated its performance for this task.

Although the KA does not pursue any optimality criterion, it often produces better codebooks than the GLA, since the interaction among cells (codewords in this context) helps reducing the bias due to the particular training set [4].

The main argument in favor of the KA, however, is its inherent ability to produce ordered codebooks, where codewords having close indexes (with the assigned index metric) are similar to each other. This property can be exploited in several ways. One can perform linear predictive coding of the indexes [5] thus reducing the bit rate with virtually no additional computation or storage cost. Also, it is possible to exploit the ordered structure of the codebook to perform progressive transmission [6] while retaining the quality of full-search encoding. Finally, by assigning the codebook a hypercubic structure, rather than linear, a high degree of tolerance to channel errors can be obtained [7].

A major problem presented by the KA is its considerable running time, mostly due to the selection, for each training vector and each pass, of the best matching (BM) codevector. Many algorithms are known (e.g., [8, 9]) for the fast selection of the BM; the most efficient ones, however, exploit some side information on the codebook that is not available with the KA, since the codewords are continuosly updated.

In this paper an efficient technique for the BM selection is presented which only exploits some low-cost side information and the correlation among training vectors to speed up the search, and is therefore compatible with the KA. In next Section the KA is described in some detail, then the fast search technique is presented in Section 3, and numerical experiments and their results in Section 4.

2. The Kohonen algorithm

Let $\{\mathbf{y}_n(0), n = 1, \ldots, N\}$ be an arbitrary initial codebook. For each vector of the training set $\{\mathbf{x}_s, s = 1, \ldots, S\}$ two processing steps are performed:

- The BM is selected as the codevector minimizing the assigned distortion measure $D(\cdot, \cdot)$

$$\mathbf{y}_{\mathrm{BM}}(s) : D(\mathbf{x}_s, \mathbf{y}_{\mathrm{BM}}(s)) \leq D(\mathbf{x}_s, \mathbf{y}_n(s)), \qquad n = 1, \ldots, N \qquad (1)$$

- The BM, and possibly some other codewords are updated according to the relation

$$\mathbf{y}_n(s+1) = \mathbf{y}_n(s) + \alpha(s)\beta(n - \mathrm{BM})(\mathbf{x}_s - \mathbf{y}_n(s)) \qquad \forall\, |n - \mathrm{BM}| \leq \mathcal{N}(s) \qquad (2)$$

Here, $\alpha(s) = \alpha_0 exp(-s/\sigma_\alpha)$ is a learning factor, which controls the speed of adaptation of the codebook and its convergence. At the beginning it is quite high, so that the codewords quickly match the training set, then it decreases allowing so the codebook to converge. The function $\beta(n) = 1 + \cos(n/\mathcal{N}(s))$ instead, with $\mathcal{N}(s) = \mathcal{N}_0 exp(-s/\sigma_\mathcal{N}) + \mathcal{N}_\infty$, controls the extent of the mutual interaction among codevectors. $\mathcal{N}_0$ is chosen pretty large, so that in the early phases of learning many codewords are updated for each training vectors, and an organized codebook is readily obtained. Then the mutual interaction weakens as $\mathcal{N}(s) \to \mathcal{N}_\infty$ (a much smaller value) and the codewords are free to reach a near-optimal final state.

The second step has little weight on the overall complexity. Indeed, the algorithm's parameters can be reasonably updated at a low rate, and therefore, for each training vector only $\mathcal{N}(s)K$ multiplications and twice as much additions are required, K being the vector dimension.

The first step instead, assuming the squared error distortion measure

$$D(\mathbf{x}, \mathbf{y}) = \sum_{k=1}^{K} (x_k - y_k)^2 \qquad (3)$$

calls for NK multiplications, $2NK$ additions and N tests for each training vector. As $\mathcal{N}(s)$ is typically from one (at the beginning) to two orders of magnitude smaller than N, the BM selection is clearly responsible for almost all the computational complexity.

3. Fast BM selection

Many techniques have been devised to speed up the search for the BM, most of them, however, require some expensive side information on the codebook like, for example, the distances between all couples of codevectors [9]. While this is reasonable with GLA, which updates the codebook only once for each pass of the training set, it is out of the question with KA, which updates the codebook at each new training vector.

Partial Distance Elimination (PDE) [8] is perhaps the only fast search technique which does not require side information. It simply evaluates the distortion sample by sample and discards the codevector under analysis as soon as the partial distortion for the first $K' < K$ samples

$$D_{K'}(\mathbf{x}, \mathbf{y}_n) = \sum_{k=1}^{K'} (x_k - y_{nk})^2 \qquad (4)$$

exceeds the minimum total distortion available so far, say $D_{\min}$. As it is, however, PDE provides quite a limited complexity reduction. In the following three independent additional steps are proposed to help speed up the search.

Average Value Elimination (AVE)

Let $\mathbf{x}$ be the current training vector and $\mathbf{y}_n$ the generic codevector, (to simplify notations, the dependence on s is not shown), and let

$$\xi = \frac{1}{K}\sum_{k=1}^{K} x_k, \quad \eta_n = \frac{1}{K}\sum_{k=1}^{K} y_{nk} \qquad (5)$$

be the average values of their components. It can be easily proved that $D(\mathbf{x}, \mathbf{y}_n) \geq K|\xi - \eta_n|^2$
As a consequence, the following implication holds

$$|\xi - \eta_n| \geq \sqrt{D_{\min}/K} \implies D(\mathbf{x}, \mathbf{y}_n) \geq D_{\min} \qquad (6)$$

namely, whenever the absolute difference between the averages exceeds the value $\sqrt{D_{\min}/K}$ the codevector can be discarded without further analyses.

For a relatively large value of K the AVE allows one to immediately reject many codewords that would have required several processing steps with the PDE. The PDE itself can be used, then, to single out the BM among the remaining ones. The advantage obtained is well worth the evaluation and updating of the average values, which only requires additions (the sum value instead of the average can be used).

Best Matching Prediction (BMP)

The actual efficiency of both PDE and AVE strongly depends on how fast, during the search, a codevector close 'enough' to the input vector is found. As this happens, in fact, $D_{\min}$ drops to such a small value that most remaining codevectors can be quickly rejected.

Most of the times, a good starting guess is readily available. Let us suppose, in fact, that successive training vectors $\mathbf{x}_{s-1}$ and $\mathbf{x}_s$ are highly correlated. Then, using the best matching relative to $\mathbf{x}_{s-1}$ we obtain a very small $D_{\min}$ at the first try already, and consequently a significant complexity reduction. The high-correlation hypotesis, on the other hand, is almost always satisfied in image coding, since the training vectors are sequentially drawn from a collection of natural (smooth) images. The overall saving, therefore, can be remarkable.

Jump Codebook Scanning (JCS)

Even when the BMP is used one can sometimes start the search with a wildly mismatched codevector. In such cases, scanning the codebook sequentially is a bad idea. Due to the ordering property of KA, in fact, codevectors having close indexes are quite similar to each other and hence will have comparable distances from the training vector. Therefore, a linear scanning of the codebook will very slowly reduce a large initial $D_{\min}$.

This drawback is easily overcome by adopting a large-step scanning of the codebook so as to quickly try quite different codevectors and obtain, within a few trials, a reasonably low $D_{\min}$. Such a scanning, on the other hand, cannot hurt when the initial try is already satisfactory.

4. Numerical experiments

To evaluate how much the proposed BM search method speeds up the KA when compared to the brute-force search, or even the PDE, several codebooks have been built using the various methods. The training set was composed by 16-dimensional vectors (4x4 blocks) drawn by the gray-scale images of the USC database, and the initial parameters of the KA were set to $\alpha_0 = 0.2, \mathcal{N}_0 = 16, \mathcal{N}_\infty = 1$. Both $\alpha(s)$ and $\mathcal{N}(s)$ were then reduced at each pass of the training set by a factor 0.9. Of course, all of the resulting codebooks were identical, since all the search algorithms always find the BM, no matter how.

The performance (in terms of total number of operations per training vector) is reported in Tab.1. The new method largely outperforms the simple PDE (not to mention the brute-force method), especially when considering that a fixed share of operations is required for the mere updating of the codevectors (step two of the KA). To gain some insight on the relative effectiveness of AVE, BMP, and JCS, they were also used separately in combination with the PDE. It appears that AVE has the greatest impact on complexity followed in order by JCS and

BMP, each one providing, separately, at least a 40% improvement over PDE for multiplications which are the heaviest operations (square roots are in negligible number).

Fig.1 shows the number of multiplication per vector as a function of the pass over the training set, also reporting the individual contributions of the BM selection and the vector updating. Both contributions are decreasing, the first because the codevectors become increasingly well organized, the second because there are less and less vectors to update as $\mathcal{N}(s)$ goes to one. The BM search accounts, now, for little more than half the overall complexity, and therefore, a possibly more efficient method would provide little reduction anyway in the KA running time.

In conclusion, with the proposed algorithm for the fast selection of the BM, the running time of the KA is much reduced, and appears to be comparable with that of the fast GLA algorithms. In addition, since the KA is less affected by the amount of training data [4], a smaller training set can be used making it even more convenient than the GLA for the VQ codebook design.

Table 1
Performance (operations/vector/1000)

METHOD	MULT	ADDT	TEST
STANDARD	95.0	182.3	5.6
PDE	45.7	91.6	40.5
PROPOSED	11.3	38.7	11.6
AVE+PDE	15.9	47.5	16.0
BMP+PDE	27.8	55.7	22.6
JCS+PDE	22.2	55.7	17.0
UPDATING	4.9	9.9	0.0

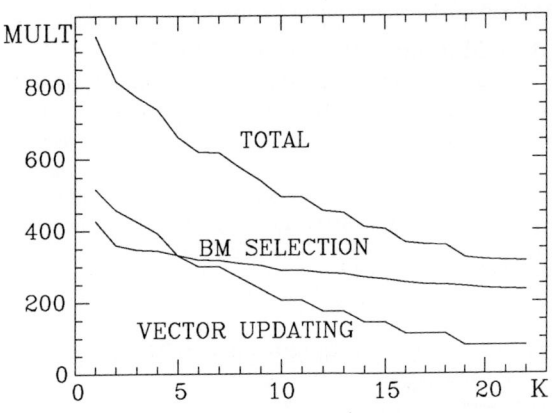

Figure 1. multiplications/vector vs pass

References

[1] A.Gersho, R.M.Gray, "Vector quantization and signal compression", Kluwer Academic Press, 1992.

[2] T.Kohonen, "Self-organization and associative memory", 2nd Ed., Springer-Verlag, 1988.

[3] N.M.Nasrabadi, Y.Feng, "Vector quantization of images using Kohonen self-organization", IEEE Conf. on Neural Networks, vol.I, pp.101-108, July 1988.

[4] J.D.Mc Auliffe, L.E.Atlas, C.Rivera, "A comparison of the LBG algorithm and Kohonen neural network paradigm for image VQ", ICASSP-90, pp.2293-2296, 1990.

[5] E.Cammarota, G.Poggi, "Address vector quantization with topology-preserving codebook ordering", 13th GRETSI symposium, pp.853-856, Sep.1991.

[6] E.A.Riskin, L.E.Atlas, S.R.Lay, "Ordered neural maps and their applications to data compression", IEEE Workshop on Neural Networks for Signal Processing, pp.543-551, Sep.1991.

[7] G.Poggi, "VQ codebook design for noisy channel by the Kohonen algorithm" PCS-93, Lausanne, Mar.1993.

[8] C.D.Bei, R.M.Gray, "An improvement of the minimum distortion encoding algorithm for vector quantization", IEEE Trans. Commun., vol COM-33, pp.1132-1133, Oct. 1985.

[9] S.H.Chen, W.M.Hsieh, "Fast algorithm for VQ codebook design", IEE Proc. I, Commun., Speech & Vision, vol 138, pp.357-362, Oct. 1991.

ADVANCED MOTION ESTIMATION FOR SUBBAND CODING OF IMAGE SEQUENCES

L. Böröczky *, K. Fazekas ** and P. Csillag **

* KFKI Research Institute for Measurement and Computing Techniques, P.O. Box 49, H-1525 Budapest, Hungary, Phone: +36 1 169-9499, Fax: +36 1 169-5532, E-mail: h2136bor@ella.hu

** Technical University of Budapest, Dept. of Microwave Telecommunications, Goldmann ter 3, H-1111 Budapest, Hungary, Phone and fax: +36 1 181-2968

Abstract

In this paper an advanced multiresolution motion estimator is presented for a motion-compensated subband coder. The motion estimator is based on a modified version of the well-performing general pel-recursive algorithm, which is applied at various levels of the image pyramid created from low frequency subbands. The proposed hierarchical motion estimator outperforms the algorithms utilizing pure full resolution images.

1. INTRODUCTION

Encoding of time-varying image sequences for the Broadband Integrated Services Digital Network (BISDN), which is expected to become the prime multimedia communication network, is a research field of growing interest.

Recently, subband coding has been proved to be a suitable method for coding of image sequences. In subband coding the signal frequency band is split up into subbands and each band is encoded with a coder and a bit rate accurately matched to the statistics of that particular band [1]. It is common to use predictive or transform coding of the highly correlated low frequency subband, which preserves the features of the original image. Higher frequency subbands have much less correlation and are often encoded using scalar quantization and a subsequent noiseless code. Since, in typical image sequences the temporal activity is high due to the object motion, motion compensation can improve coding efficiency of the low frequency subbands.

Frame-to-frame object motion in an image sequence can be represented by a 2-D time-varying vector field consisting of the local displacement vectors. This vector field is often referred to as motion field, which is unknown in general and has to be estimated based on the given time-varying image sequence. Several motion field estimation algorithms applied to image coding have been developed and generally they are based on the assumption of smoothness of the estimated motion field [2-4]. According to their computation methods, they are known as block-matching and adaptive/recursive type of estimators. Generally, each type of smoothness-based motion estimation algorithms, which uses two consecutive frames of full resolution, exhibits very slow convergence rate or is unable to estimate large displacement vectors of the motion field. To overcome this problem, hierarchical motion estimation scheme has been proposed in [5], that is especially suitable for motion-compensated (MC) subband coding.

In this paper a multiresolution motion estimator is proposed to the subband coding environment, which is presented in Section 2. At various levels of the image pyramid created from the low-frequency subbands, a modified version of the general pel-recursive motion estimation algorithm is applied and described in Section 3. Experimental results are presented in Section 4. Finally, a conclusion including the main directions of future research is provided.

2. MULTIRESOLUTION MOTION ESTIMATION FOR SUBBAND CODING

Subimages, which are lowpass-filtered and subsampled versions of the original image and available during subband decomposition, can conveniently be used for motion estimation. For illustration, one can consider an image decomposition into 13 subbands with an intermediate 4-band splitting. In this 13-subband decomposition the lowest subband of the first 4-band splitting and the lowest subband of the 13-splitting can be retained for subsequent motion estimation as creating a subband-image pyramid with lowest level containing the original image [1].

The use of a multiresolution image representation, regardless of the type of the pyramid transform, introduces two main questions in the context of motion estimation, such as selection of the motion estimation algorithm applied at pyramid levels and design of a sophisticated propagation strategy for the estimated motion fields between the different pyramid levels. In order to maintain low computational complexity of the motion estimation, a coarse-to-fine propagation strategy is proposed for the MC subband coding scheme. In this strategy the estimated motion field obtained at a lower level of the image pyramid is considered as the initial estimate for the next level of the multiresolution motion estimation. The chosen motion estimation algorithm for the levels of subband pyramid is a well-performing, general pel-recursive algorithm proposed in [2].

3. PEL-RECURSIVE GENERAL MOTION ESTIMATION

The chosen pel-recursive general motion estimator outperforms the existing pel-recursive algorithms, however, the original algorithm of [2] in subband coding environment and multiresolution representation needs some modifications. It operates in a scanwise causal way through the images and it is based on a prediction/update principle. At each pixel the following iteration is performed:

$$\mathbf{d}^{i+1}(\mathbf{x}, t) = \mathbf{d}^{i}(\mathbf{x}, t) + \lambda^{i}\mathbf{u}^{i}, \tag{1}$$

where $\mathbf{d}^i(\mathbf{x}, t)$ is the current estimate for the motion field vector at spatio-temporal location $(\mathbf{x},t)$, $\lambda^i \mathbf{u}^i$ is an update vector, $\mathbf{d}^{i+1}(\mathbf{x}, t)$ is the new estimate and i is the iteration index. Three strategies need to be specified, such as determination of the initial estimate $\mathbf{d}^0(\mathbf{x}, t)$ to start the iterations, the computation of the update vector $\lambda^i \mathbf{u}^i$ at every iteration, and the control of the iteration process. Originally, the general estimator contains a motion-compensated spatio-temporal motion vector predictor, an adaptive update equation based on a MC spatio-temporal image sequence model and an estimator control strategy.

In the proposed multiresolution scheme the initial estimates of the motion field vectors are obtained by the coarse-to-fine propagation strategy incorporating a spatial vector predictor, which relies on the spatial smoothness of the motion field and mainly increases the robustness of the motion estimator. The update term $\lambda^i \mathbf{u}^i$, consists of a vector $\mathbf{u}^i$ that represents the direction of the update, and a scalar λ^i that determines the length of the update vector. For estimation of $\mathbf{u}^i$, instead of using a complex estimator proposed in [2], an efficient adaptive Wiener-based update vector [4] is applied:

$$\mathbf{u}^i = [G^T G + \mu I]^{-1} G^T \mathbf{z}, \qquad (2)$$

where G is a matrix containing the first order derivatives of the intensities calculated in an estimation support around and including the actual pixel, $\mathbf{z}$ is the displaced-frame-difference vector estimated in the estimation support and I is a 2x2 identity matrix. The damping parameter μ is calculated adaptively from pixel to pixel according to the specific image contents and estimation accuracy [4]. The scalar multiplier λ^i in (1) is determined by a linear search applied along the direction of the estimated update vector. The control of the estimation procedure is carried out by utilizing a simplified divergence check and a termination rule of the iteration process, as proposed in [2].

4. EXPERIMENTAL RESULTS

In order to investigate the performance of the proposed hierarchical motion estimator, experiments were carried out on synthetic motion, where the motion fields are known a priori. The experiments were performed on the middle part of 64x40 pixels of the image "Mickey", shown in Fig. 1. This image was translated uniformly by (3,-2) pixels and together with the original image, they were considered as two consecutive frames from a time-varying image sequence. For subband decomposition of the images QMF12a filter [1] was used and 3-level image pyramids were created using two low frequency subbands and the original image. The general pel-recursive motion estimator includes a coarse-to-fine propagation strategy with 4-pixel spatial vector predictor, an adaptive Wiener-update considering a second order non-symmetrical causal support and a termination rule as maximizing the allowable length of the estimated motion vectors ($\mathbf{d}^{max} = 10$). This estimator is applied at each level of the subband-image pyramid.

Fig. 2 shows a reference motion field, which was estimated using pure full resolution images and applying 4-pixel spatial vector predictor. It can be observed, that the estimation diverges at a large number of rows. The motion field estimated by the hierarchical scheme is presented in Fig. 3, where a main improvement is obtained.

Fig.1 The "Mickey" image

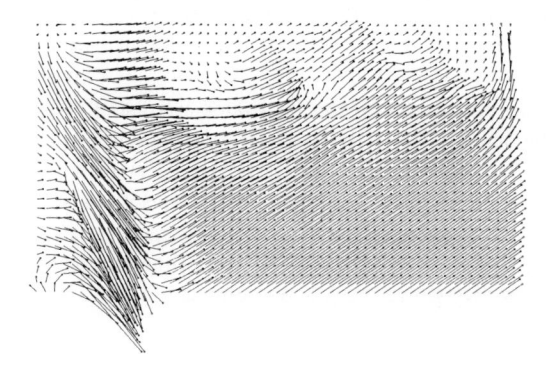

Fig. 2 Motion field estimated using only full resolution images

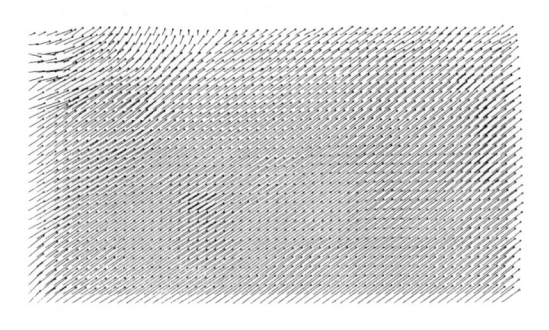

Fig. 3. Motion field estimated using 3-level subband pyramid

5. Conclusion

An advanced multiresolution pel-recursive motion estimator was proposed for the subband coding of time-varying image sequences. The initial experimental results proved that the proposed hierarchical motion estimator outperforms the algorithm using only full-resolution images. In future research, the main attention will be focused on developing a more sophisticated propagation strategy exploiting the temporal smoothness of the motion fields and on investigation of the proposed motion estimator in a total subband coding scheme.

References

[1] P.H. Westerink, "Subband Coding of Images", Ph.D. Thesis, 1989, TU Delft, The Netherlands.
[2] J.N. Driessen, L. Böröczky, J. Biemond, "Pel-recursive motion field estimation from image sequences", Journal On Visual Communications And Image Representation Vol. 2, No. 3, Sept. 1991, pp. 259-280.
[3] L. Böröczky, "Pel-Recursive Motion Estimation for Image Coding", Ph.D. Thesis, 1991, TU Delft, The Netherlands.
[4] L. Böröczky, J.N. Driessen, J. Biemond, "Adaptive algorithms for pel-recursive displacement estimation", SPIE Vol. 1360 Visual Communications and Image Processing'90, pp. 1210-1221.
[5] W. Enkellmann, Investigations of multigrid algorithms for the estimation of optical flow fields in image sequences" CVGIP, Vol 43, 1988, pp. 150-177.

3D Modeling of Human Heads from Multiple Views

C. Braccini, S. Curinga, A. Grattarola and F. Lavagetto

DIST, Department of Communication, Computer and Systems Science - University of Genova - Genova, I-16145, Italy

Abstract

In this paper we propose a new technique for modeling human heads based on an integrated 2D-3D approach capable to faithfully reproduce their volumetric and pictorial characteristics. The 2D representation is used to model the face region that exhibits the highest pictorial detail and time-varying deformations, while the 3D representation adds true depth information necessary to synthesize profile views of the head. The experimental results obtained by applying the modeling technique to an artificial head show significant improvements with respect to simplified approaches where synthetic depth information is pasted on the 2D model. Through the use of a well-suited regularization procedure, acceptable results have been obtained also with live heads where a certain amount of error due to the head motion has to be taken into account. Promising fields of application basically concern very low-bitrate videophone coding for real-time visual communication and synthetic face animation for interactive virtual reality.

1. INTRODUCTION

Very low bitrate coding of "head-and-shoulder" videophone sequences can be obtained through complex 3D modeling of the scene where a priori knowledge is fully exploited for efficient frame prediction [1-6]. As the most relevant component of the scene is the speaker's head and the basic information is associated to his facial mimics, the availability of efficient and reliable procedures for modeling and animating human faces evidently represents a fundamental issue.

Since only one video camera is typically integrated in the videophone set, no stereo information is available to reconstruct the head structure, which is usually synthesized, exploiting a priori knowledge. As long as frontal or nearly frontal views are synthesized, even a coarse approximation of the depth information does not affect too much the perceived quality of the reconstructed sequence. Conversely, in case of slightly rotated or profile views, a reliable 3D information on the speaker's head is necessarily required.

Within the approach described in this paper the speaker's head is modeled by means of an object-oriented data structure organized as a wire-frame mask. The mask is initially fitted on a frontal view of the speaker's face and then mapped onto a 3D solid specifically adapted to the geometry of the speaker's head. The head-shaped solid is obtained by analyzing a set of calibrated views of the speaker's head corresponding to different orientations with respect to the camera position: the higher is the information conveyed by the views, i.e. the more scattered they are in space, the more precise is the volumetric representation.

As the calibration of the views can be performed explicitely only within a controlled environment with fixed references, different solutions must be devised for general applications such as videophone communications. In our experimental setup the speaker is seated at a constant distance from the camera on a chair capable to perform discrete rotations of fixed and predefined angles, as described in Figure 1, so that the corresponding views are implicitly and roughly calibrated. The procedure for the volumetric representation of the speaker's head is based on the "occluding contours" technique [7]. The head silhouette is extracted from all the views and the corresponding generalized cones, whose vertices coincide with the camera, are back-projected in space, so that their intersection identifies an excess approximation of the head volume.

As the views used for the estimation of the head volume are not taken all at the same time, some motion usually occurs from one to the other, so that a certain amount of uncertainty is present. In our approach the calibration error is automatically corrected by applying a regularization procedure [8] which iteratively modifies the extrinsec camera parameters corresponding to each view in order to minimize an error functional. The resulting head-shaped solid is then finely carved, on the basis of a priori knowledge, in correspondence of the eyes, where the "occluding contours" technique fails in reproducing the head concavities.

The 2D wire-frame modeling of the speaker's face and the 3D head shaping are briefly reported in Sections 2 and 3, respectively, while the achieved results are presented and discussed in Section 4.

2. 2D WIRE-FRAME MODELING

A frontal view of the speaker is processed in order to segment his silhouette from the background and to adapt a wire-frame structure on his face. Silhouette segmentation is obtained through the motion-based analysis of a given number of frames extracted from the videophone sequence in the hypothesis of slight movements of the speaker's head against a stationary background.

The procedures for the construction and adaptation of the wire-frame structure, described in [1-2], take advantage of further simplifying assumptions on the class of human faces that can be successfully modeled: in particular only white-skinned faces without bear and eyeglasses are considered in this paper.

Primary facial features like eyes and mouth are finely approximated by means of suitable wire-frame submasks whose geometry closely matches the actual human anatomy. Each of these submasks can be animated through knowledge-based deformation rules acting on a predefined domain of triangles whose vertices are accordingly and temporarily relocated. The domain of activity of each deformation rule affects a circular or rectangular subregion of the wire-frame structure whose size and shape depend on the specific mimics to be reproduced.

Several approaches oriented to the anatomical modeling of human faces have been recently proposed [9-11] showing very promising results, with significant improvements with respect to less sophisticated approaches [12] where no assumption is made on the underlying muscle structure.

The model we employ falls somewhere in between: no tridimensional muscle modeling is performed but only a simplified single-layer bidimensional muscle representation is used, capable to yield satisfactory results with low complexity. The addressed field of application is in fact that of very low bitrate video communication with image format of 1/9 CIF, roughly 100x100 pixels, where the face region covers an average area of a few thousand pixels with a resolution poverty which does not justify the use of a too sophisticated model.

Through the analysis of the real image sequence a set of facial control parameters is evaluated [3-6] by means of suitable algorithms operating on different image subregions for the extraction of specific somatic features. Some a priori knowledge is exploited, specifically:

- facial features usually exhibit high contrast with respect to the surrounding skin;
- the skin region around facial features is usually rather homogeneous in color and texture;
- facial features are symmetrically distributed with respect to a face longitudinal axis.

3. 3D HEAD MODELING FROM MULTIPLE VIEWS

Among the many 3D reconstruction methods proposed in the literature, those based on the occluding contours [7,13] are considered for the application proposed in this paper, because they may provide quite satisfactory approximations to the shape of objects by means of relatively simple and efficient algorithms. An inherent limit of the occluding contour techniques is the impossibility of recovering holes and concavities not visible in the available views. In the proposed system the output of the 3D reconstruction module is used to modify a generic head model built from a-priori knowledge. In this way a good approximation of the existing concavities can be introduced.

Another major weak point is the sensitivity to errors in the input data and, in particular, in the intrinsic (focal length, optical center, lens distortion) and extrinsic (space position and orientation) sensor parameters. The sensitivity is due to the ill-conditioned nature of the inversion of the perspective equations, and affects all the 3D reconstruction methods.

A typical solution to the sensitivity problem is based on a calibrated space, i.e. a background where a great number of reference markers allows a reduction of the errors in the estimated sensor parameters [14]. Unfortunately, this kind of solution is not easily applicable in the proposed system.

A different solution is using a new regularization technique [8] that exploits simple constraints induced by the acquisition system geometry. This technique provides good reconstructions from a set of poorly calibrated views, without imposing any constraint on number, position and orientation of the viewpoints.

To summarize the regularization procedure, sketched in Figure 1, let us recall that the 3D object reconstruction from views based on the occluding contours starts from a set of 2D perspective images of an object. In each view the contour which separates

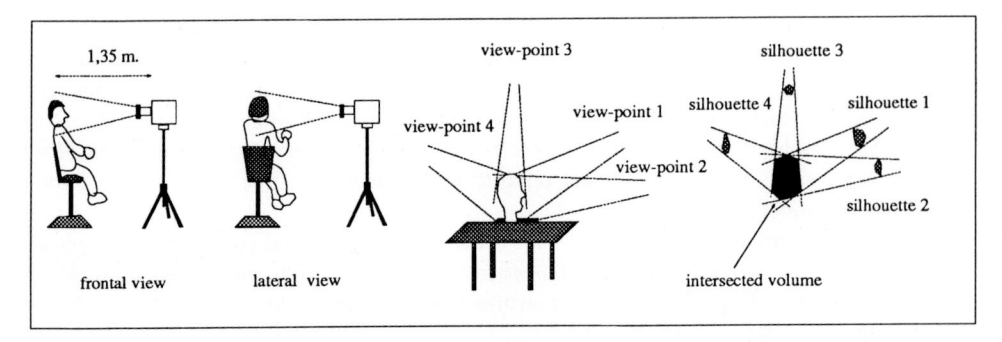

Figure 1: (Left) Experimental set-up for the image acquisition. (Right) Description of the occluding contours procedure used for the head 3D modeling.

the object image from the background is identified. Each occluding contour can then be projected in space from the corresponding viewpoint, so defining an infinite volume (a half space, denoted c-volume in the following) which is surrounded by a conic-like surface.

The intersection of these volumes approximates the unknown object better and better, as the number of available views increases. The regularization procedure, starting from a rough estimate of the extrinsic parameters, exploits a simple physical constraint induced by the geometry of the system to progressively refine their value as comprehensively discussed in [2].

The 3D volumetric model built from the input views is then converted to a 3D surface model and, after the integration of the pictorial information (as described, e.g., in [15]), it is used both to correct the depht information of the wire–frame mask of the speaker face and to model the remaining part of the speaker's head.

4. EXPERIMENTAL RESULTS

The modeling technique described above has been initially applied to a clay made head employing multiple views differing in orientation. In this case the head can be definitely considered as a rigid body and only miscalibration errors have to be corrected by the regularization. In Figure 2 the original views employed to build the 3D model are shown while in Figure 3 the three basic steps of the volumetric procedure are described. Some facial expressions synthesized from the 3D model are then presented in Figure 4. Live head modelling, presently under experimentation, has provided promising preliminary results showing that the employed regularization procedure can be applied effectively to compensate the non rigid motion of the head.

5. REFERENCES

1. C.H. Chien and J.K. Aggarwal, Model Construction and Shape Recognition from Occluding Contours, IEEE Trans. on PAMI, vol. 11, n. 4, 1989, pp. 372-389.

2. A.A. Grattarola, Volumetric Reconstruction from Object Silhouettes: a Regularization Procedure, Signal Processing, vol. 27, n. 1, 1992, pp. 372-389.

3. F. Lavagetto, A.A. Grattarola, S. Curinga and C. Braccini, Muscle Modeling for Facial Animation in Videophone Coding, Proc. IEEE Int. Workshop on Robot and Human Communication, Tokyo, 1992, pp.369-375.

4. F. Lavagetto and S. Curinga, Videophone Coding Based on 3d Modeling of Facial Muscles, Proc. SPIE Int. Conf. on Visual Communication and Image Processing, Boston MA, 1992, pp. 1366-1374.

5. D. Terzopoulos and K. Waters, Analysis of Facial Images Using Physical and Anatomical Models, Proc. IEEE 3rd Int, Conf. on Computer Vision, Osaka, 1990, pp. 727-732.

6. N. Magnenat-Thalmann, E. Primeau and D. Thalmann, Abstract Muscle Action Procedures for Face Animation, The Visual Computer, vol. 3, 1988, pp. 290-297.

7. K. Waters, A Muscle Model for Animating Threedimensional Facial Expression, Computer Graphics, vol. 22, n. 4, 1987, pp. 17-24.

8. F.I. Parke, Parameterized Models for Facial Animation, IEEE Computer Graphics and Applications, vol. 2, n. 9, 1982, pp. 61-68.

9. R. Forchheimer and T. Kronander, Image Coding - from Waveforms to Animation, IEEE Trans. on ASSP, vol. 37, n. 12, 1989, pp. 2008-2023.

10. K. Aizawa, H. Harashima and T. Saito, Model-Based Analysis/Sysnthesis Image Coding (MBASIC) System for Person's Face, Image Communication, vol. 1, n. 2, 1989, pp.139-152.

11. C.S. Choi, H. Harashima and T. Takebe, Analysis and Synthesis of Facial Expressions in Knowledge-Based Coding of Facial Image Sequences, Proc. ICASSP-91, S.Francisco CA, 1991, pp. 2737-2740.

12. C. Braccini, S. Curinga and F. Lavagetto, Model-Based Coding of Facial Images: from Analysis to Synthesis, PCS-93, Lausanne, March 17-19, 1993.

13. W.N. Martin and J.K. Aggarwal, Volumetric Description of Objects from Multiple Views, IEEE Trans. on PAMI, vol. 5, n. 2, 1983, pp. 150-158.

14. R.Y. Tsai, An Efficient and Accurate Camera Calibration Technique for 3D Machine Vision, Proc. IEEE Comp. Soc. Conf. on CVPR, Miami FA, 1986, pp. 364-374.

15. C. Braccini, A.A. Grattarola and F. Lavagetto, A Tool for Artificial Reality: from Views to Object Representation, Proc. IEEE Int. Workshop on Robot and Human Communication, Tokyo, 1992, pp. 222-226.

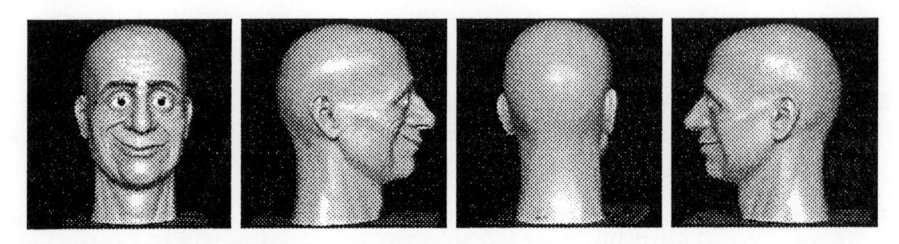

Figure 2: Views employed for the 3D modeling of the clay made head.

Figure 3: 3D voxel representation of the clay made head before (left) and after (center) applying the regularization procedure. (Right) Surface head representation.

Figure 4: Facial expressions synthesized onto the 3D model.

A novel approach to data interpolation for image-data compression

F.G.B. DeNatale, G.S. Desoli, S. Fioravanti and D.D. Giusto

Signal Processing and Understanding Group, DIBE, University of Genoa, Italy

Abstract

A novel image interpolation algorithm is presented, which uses an adaptive least-square interpolation by means of a spline-like scheme. Such interpolation is utilized for image-data compression in a two-source decomposition scheme. In the paper, conventional linear-interpolation approaches are first described, then a least-square bilinear interpolation is proposed based on a fixed grid size. After, the problem of an adaptive implementation is addressed, and a sub-optimal quadtree segmentation for the least square approach is proposed.

1. THE INTERPOLATION PROBLEM

Many interpolation algorithms have been used for image coding [Yan77]. The one proposed in [Farrelle86] is very interesting, as it offers many advantages. The basic idea is a two-source decomposition in which the prediction (related to the stationary part of the image) is obtained through a linear predictor based on a 2-D interpolation. A simplified version of interpolant can be derived by using an approximation for the solution (for the least square problem) in the transformed domain through the Discrete Sine Transform (DST), the image being modeled as a 1st-order Markov process. This procedure represents an extension of a previous work [Jain76], where a fast Karhunen-Loeve (KL) transform for a 1st-order Markov process is proposed.

From the technical point of view, the interpolation of a source requires the definition of a model, which can be deterministic or stochastic, depending on whether it uses random variables or not. It is often assumed that a source can be derived from an auto-regressive (AR) model [Derin89]. This type of model is characterized by the use of a minimum-variance representation (MVR) for a stochastic sequence. Therefore, the actual sequence value can be expressed as the sum of a prediction and a prediction error, as follows

$$x(n) = x'(n) + \varepsilon(n) \tag{1}$$

where the prediction $x'(n)$ must be taken such as to minimize the mean square value of the prediction error $\varepsilon(n)$. The minimum value of this error is given by

$$v^2 = \min_{x'(n)} E\{\varepsilon^2(n)\} \qquad (2)$$

In general, the prediction x'(n) can be obtained through a nonlinear interpolation. However, for simplicity, one can first consider a linear model, and after transform it into a nonlinear one by introducing a least-square model. For the linear model, the prediction can be written as

$$x'(n) = \sum_{r=1}^{\infty} h(r)\, x(n-r) \qquad (3)$$

where the prediction coefficients h(r) can be found by minimizing the square error between the prediction and the original image. The solution can be obtained by means of a set of equations associated with the orthogonality condition

$$E\{\varepsilon(n)\, x(n-r)\} = 0; \quad r = 1,2,... \qquad (4)$$

The above equation points out that the prediction error is not correlated with the previous values of the sequence. Then

$$R_\varepsilon(r) = E\{\varepsilon(n)\,\varepsilon(n-r)\} = v^2\,\delta(r) \qquad (5)$$

where $R_\varepsilon(r)$ is the autocorrelation of the error $\varepsilon(n)$. The spectrum energy density function (SDF) of this error is

$$G_\varepsilon(z) = v^2; \quad z = e^{j\omega} \qquad (6)$$

showing that the prediction error is represented by white noise. If the MVR model cannot be realized by using a finite number of coefficients (i.e., the SDF contains some zeros), an approximation for the SDF of x(n) is utilized. Random sequences are often modeled by a pth-order AR model

$$x(n) = \varepsilon(n) + \sum_{r=1}^{p} a(r)\, x(n-r) \qquad (7)$$

In a data-compression application, it is often possible to use a linear MRV model, in which the prediction depends on the future as well as on the past

$$x(n) = \varepsilon(n) + \sum_{r \neq 0} a(r)\, x(n-r) \qquad (8)$$

In this case, the solution in the transformed domain is given by

$$A(z) = \frac{v^2}{G_x(z)} \qquad (9)$$

where

$$A(z) = -\sum_{r=-\infty}^{+\infty} a(r) z^{-1}; \quad a(0) = -1 \tag{10}$$

Moreover, by considering ε(n) and x(n) as input and output, respectively, of a linear system with a transfer function of the type 1/A(z), it is possible to demonstrate that the relation between ε(n) and x(n) is given by

$$G_x(z) = \frac{G_\varepsilon(z)}{A(z) A(z^{-1})} \tag{11}$$

from which, after substituting Gx(z) from (9), it derives

$$G_\varepsilon(z) = v^2 A(z^{-1}) = v^2 A(z) \tag{12}$$

In this case, ε(n) is a sequence of colored noise.

2. BILINEAR INTERPOLATION

Bilinear interpolation of an image block is based on a simple prediction that uses the original four corners values of the block to be interpolated. This approximation simplifies the optimum 2D interpolation scheme (given a 1st order Markov process as model for the source). It works as follows

$$X_c(i,j) = \frac{A\,ij + B\,(N-i)\,j + C\,(N-j)\,i + D\,(N-i)\,(N-j)}{N^2} \tag{13}$$

where A,B,C,D are the values of the corner pixels and $X_c(i,j)$ is the interpolant value inside each block (where N is the block size).
This type of interpolation can be utilized for N≤7; above this threshold, the approximation for high correlation decays. Then to improve quality, a further coding operation must be performed on the residual.

3. LEAST SQUARE BILINEAR INTERPOLATION

The algorithm proposed in this paper is based on the bilinear interpolation previously described. The significant innovation lies in the fact that the corner values are not the ones of the pixels in the original image; they are computed in order to minimize the square error between the prediction and the original data. The prediction scheme works as follows

$$X'(n) = \varepsilon(n) + \sum_{r=1}^{p} a(r) Y(n-r) \tag{14}$$

where Y(n) is a sequence of values (as a function of X(n)) which minimizes the square error between X(n) and the prediction X'(n)

$$\sigma^2 = \min_{Y(n)} E\{[X(n) - X'(n)]^2\} \tag{15}$$

If $\{X_{i,j}, X_{i+1,j}, X_{i,j+1}, X_{i+1,j+1}\}$ are the corner values of the block considered, the expression for $X'_{i,j}(u,v)$ is

$$X'_{ij}(u,v) = \alpha(u,v) X_{ij} + \beta(u,v) X_{i+1,j} + \gamma(u,v) X_{i,j+1} + \eta(u,v) X_{i+1,j+1} \tag{16}$$

$$\alpha(u,v) = \frac{(N-v)(N-u)}{N^2} \qquad \beta(u,v) = \frac{u(N-v)}{N^2} \tag{17a}$$

$$\gamma(u,v) = \frac{v(N-u)}{N^2} \qquad \eta(u,v) = \frac{uv}{N^2} \tag{17b}$$

By replacing $\{X_{l,m}\}$ with $\{Y_{l,m}\}$ in the previous equation

$$X'_{ij}(u,v) = \alpha(u,v) Y_{ij} + \beta(u,v) Y_{i+1,j} + \gamma(u,v) Y_{i,j+1} + \eta(u,v) Y_{i+1,j+1} \tag{18}$$

where $\{Y_{l,m}\}$ are the unknown values for which:

$$\frac{\partial \sigma^2}{\partial Y_{h,k}} = \sum_{i,j} \sum_{u,v} [X_{ij}(u,v) - X'_{ij}(u,v)]^2 = 0 \quad \text{for each } \{h,k\} \tag{19}$$

In this equation, the elements being not zero in the first summation are those for which $(i,j) \in \{(h,k),(h,k-1),(h-1,k),(h-1,k-1)\}$ with exceptions at image boundaries, where some of these elements may turn out to be null. Then

$$\frac{\partial \sigma^2}{\partial Y_{hk}} = -S_{hk} + Y_{hk}A_0 + Y_{h+1,k}A_1 + Y_{h,k+1}A_2 + Y_{h+1,k+1}A_3 + Y_{h-1,k}A_4 + \\ + Y_{h-1,k+1}A_5 + Y_{h,k-1}A_6 + Y_{h+1,k-1}A_7 + Y_{h-1,k-1}A_8 \tag{20}$$

where $S_{hk} =$

$$\sum_{u,v}[X_{hk}(u,v)\alpha(u,v) + X_{h-1,k}(u,v)\beta(u,v) + X_{h,k-1}(u,v)\gamma(u,v) + X_{h-1,k-1}(u,v)\eta(u,v)] \tag{21}$$

and

$$A_0 = \sum_{u,v}[\alpha^2(u,v) + \beta^2(u,v) + \gamma^2(u,v) + \eta^2(u,v)] \ldots A_8 = \alpha(u,v)\eta(u,v) \tag{22}$$

$S_{h,k}$ is constant for each block, and the coefficient $\{A_0 \ldots A_8\}$ remain constant as h and k vary. In the boundary regions, some terms are eliminated; however, the basic form of equation (20) does not change. Such equation can be written in matrix form

$$[T] \underline{Y} = \underline{S} \tag{23}$$

where $\underline{Y}$ is the vector containing the unknowns, and $\underline{S}$ is the vector of known terms. The matrix [T] is then a symmetric Toeplitz matrix with 9 non null diagonals. The solution of the system through the inverse matrix is

$$\underline{Y} = [T]^{-1} \underline{S} \qquad (24)$$

If the block dimensions are not too small (e.g., ≤10), the matrix dimensions leads to an acceptable computational load. In any case, an approximate solution can be simply obtained by using an iterative gradient descent algorithm (e.g., the Gauss-Jacobi or the Gauss-Sidel ones). After initializing $\{Y_{l,m}\}$ with $\{X_{l,m}\}$ (i.e., the original values of the block corner), the iterative procedure gives a nearly exact solution after few iterations (e.g., <5). It has to be noted that this approach does not require the generation of the matrix [T], not the computation of its inverse; as a matter of fact, such matrices are constant for any given value of the block dimensions, hence they can be previously computed.

5. ADAPTIVE INTERPOLATION

In order to apply such interpolation to coding an image, adaptive criteria should be introduced, as the prediction obtained by using a fixed size grid is not efficient. To obtain a good interpolation and a low energy residual, the grid size must be too small, even in areas of image showing low intensity variations. In order to avoid such a problem, an adaptive segmentation of the image should be used. A possible choice is the use of a quad-tree segmentation, which is driven by the interpolation error obtained through bilinear interpolation within each block. However this process cannot be performed in a least-square way as easy as in the case of a fixed grid size. This is mainly due to the difficulty of deriving an explicit form for the derivative of the energy function; this function now depends on an inconstant number of nodal variables for a given block. Since this drawback cannot be avoided, a sub-optimal procedure has been developed, which gives a solution well approximating the optimal one.

6. SUB-OPTIMAL QUAD TREE SEGMENTATION

The sub-optimal segmentation divides the image into squared blocks at the first step, and then, after performing the algorithm previously described to achieve the least-square solution, performs a further split in those blocks whose mean square error (MSE) is above a predefined threshold. After splitting a plane of the quadtree, each nodal variable, previously computed for the least-square solution, is left unchanged only if it does not belong to split blocks. This is mainly for avoiding annoying side effects that arise if the nodal variables, interested by block splitting in the previous quadtree plane, are left unchanged.

By this way, the quadtree segmentation is performed very easily. Such procedure retains most of the properties of robustness against noise that are featured by the fixed-grid-size least square solution, even if it is sub-optimal.

7. ADVANTAGES OF THE LEAST SQUARES APPROACH

By comparing the results obtained by the least-square bilinear interpolation with those yielded by other interpolation approaches, which do not take into account the pixel values inside blocks, one can notice that our scheme achieves a marked increase in the visual quality of the reconstructed images, both in objective (SNR) and subjective terms; in particular, edges are better defined. A comparison of such a proposed technique with the 2-D recursive block coding technique presented in [Farrelle86] points out that a higher quality can be obtained by the former, without any increase in the bit rate for the prediction. A lower bit rate is required for coding the residual, as it is characterized by smaller energy value. In addition, the assumptions about a minimum tile effect and the uncorrelation between adjacent blocks in the residual domain are easily verified. Traditional block-coding methods (e.g., VQ and DCT coding) consider adjacent blocks in an independent way, and show lower efficiencies when compared to our approach. Moreover, the proposed approach reduces the tile effect; this can be explained by observing that the tile effect is due to different coding of different blocks. This give rise to a type of noise that is correlated within each block, but uncorrelated between adjacent blocks.

In conclusion, the least square bilinear interpolation for image coding presents the following major advantages, as compared with the traditional block-coding methods: reduction in the tile effect, both by means of the continuity constraints imposed on the reconstruction and through the uncorrelation of the residual; lower computational complexity; more effective exploitation of interblock correlation.

Acknowledgments

This work has been partly supported by the Commission of the European Communities (contract: CEC - RACE II - R2042; project title: *Europublishing*).

References

[Derin89]   H.Derin, A.K.Patrik, "Discrete-index Markov-type random processes", Proc. IEEE, vol. 77, pp. 1485-1510, 1989.

[Farrelle86] P.M.Farrelle, A.K.Jain, "Recursive block coding: A new approach to transform coding", IEEE Transactions, vol. COM-34, pp. 161-179, 1986.

[Jain76]   A.K.Jain, "A fast Karhunen-Loeve transform for a class of stochastic processes," IEEE Transactions, vol. COM-24, pp. 1023-1029, 1976.

[Yan77]   J.K.Yan, D.J.Sahrison, "Encoding of images based on a two-component source model", IEEE Transactions, vol. COM-25, pp. 1315-1322, 1977.

Contour detection using the valleys of a similarity histogram

R. OULAD HAJ THAMI and J.P ASSELIN de BEAUVILLE

Laboratoire d'Informatique, Ecole d'Ingénieurs en Informatique pour l'Industrie, Parc de Grandmont, Avenue Monge 37000 TOURS, FRANCE.
e-mail : oulad@univ-tours

Abstract
This article proposes a non-supervised algorithm of image segmentation. The approach chosen is based on the extraction of the contours of the homogeneous regions. The algorithm detect the anti-modes (or the valleys) of a local histogram. We give examples of application of this algorithm to artificial and natural images.
Key words : segmentation, contours, similarity histogram, valley.

1. INTRODUCTION

In this paper, we are interested in the contours detection which are primitives of low level. They are very much used in systems of artificial vision. The contours techniques detection which are best known are based on gradient operators or "Lapalcien" type which consists in extracting the local extrema of the gradient or the zero of the "Laplacien" [3].

The proposed approach is based on the transformation of the image so that, systematically, its regions and its contours be, respectively, the plateaus or the modal regions and the valleys of the transformed image. This transformation brings the segmentation problem to calculation of modal regions or valleys of the transformed image .The basic idea is to associate each pixel with a vector of attributes and to define a similarity distance on the set of these vectors of attributes. The affectation of a pixel to a region or a contour is decided by measuring the number of pixels in its local neighborhood which are similar to it.The algorithm is sequential and linear. It is characterized by its simplicity and rapidity. The results obtained are very encouraging.

2. DEFINITIONS AND ALGORITHM

Let I be an image with a NxN size and a grey level function f. To each pixel $p(i,j)$ of I is associated the grey level $f(i,j)$ within $[0..255]$ and let F be a window with dimension LxL (L=3 or 5 or 7...).

2.1. Definitions
Definition 1 : We call neighbours of a pixel p (i,j) every pixel belonging to the window F which is centered on p(i,j). Let V(i,j) be this set, it is given by :

$$V(i,j) = \{p(x,y) \in I \ / \ i-l \leq x \leq i+l \ et \ j-l \leq y \leq +l\} \quad (1)$$

Where *l*=L div 2 (div being the operator of integer division).

Definition 2 : To each pixel p(i,j) of I, a vector of attributes is associated with it :

$$A(i,j) = [a_1, a_2, ..., a_n]^T \quad (2)$$

Where a_i, $1 \leq i \leq n$ are measures (grey-level variance, average ...) calculated in V(i,j). Let λ be the set of these vectors :

$$\lambda = \{A(i,j) \ / \ 1 \leq i,j \leq N\} \quad (3)$$

and let d(.,.) be some distance defined on λ.

For each pair (p(i,j) ; p(i',j')), d(A(i,j) ; A(i',j')) reflects the closeness of the attributes of p(i,j) and p(i',j').

Definition 3 : A pixel p(i',j') is said to be locally similar to p(i,j) if and only if :

(1) $p(i',j') \in V(i,j)$ (4)
(2) $d(A(i,j), A(i',j')) < S$ (5)

Where S is a pre-defined threshold (S>0). Both pixels p(i,j) and p(i',j') are as similar as the distance between their associated vectors of attributes are close to 0.

Let E(i,j) be the set of the pixels locally similar to p(i,j) ; E(i,j) is given by :

$$E(i,j) = \{p(i',j') \in V(i,j) / d(A(i,j), A(i',j')) < S\} \quad (6)$$

E(i,j) is the set of pixels, within the window F centered on p(i,j), whose calculated attributes are close to those of p(i,j) :

Let N(i,j) be the number of elements in E(i,j) :

$$N(i,j) = card(E(i,j)) \geq 1 \quad (7)$$

N(i,j) is a measure of the number of pixels similar to p(i,j) within its local neighborhood. The higher N(i,j) is, the higher the probability that the corresponding pixel p(i,j) belong to a homogeneous region is. Inversely, the smaller N(i,j) is, the lower the probability that p(i,j) belong to a homogeneous regions is.

2.2 Algorithm
The algorithm proceeds in two steps. The first one is the generation of the transformed image or histogram of similarities. The second step is the detection of the histogram valleys which are associated with the image contours.

2.2.1 Generation of the histogram of similarities

Here, an image is associated to a histogram called histogram of similarities. Let H be this histogram. A class of H of coordinates (i,j) is noted H(i,j), i,j=1..N, and its size h(i,j). To each H(i,j) is therefore associated the number of pixels similar to the pixel p(i,j) giving:

$$h(i,j) = N(i,j) \text{ for } 1 \leq i,j \leq N \qquad (8)$$

The histogram H can be interpreted as a "cartography" of the image I in the case of an ideal image without any noise, the homogeneous regions of I will generate inside of H some plateaus and their contour wil generate valleys separating these plateaus (see image B). The existence of algorithms for detecting modal regions or valleys [1] [2] [5] [4] makes the exploitation of this histogram very easy. The subjacent relation between the valleys of H and the contours of the image I allows extration of these frontiers. The use of a histogram of type H, instead of the classical histogram of grey levels, allows a better separation of the histograms associated wich each regions.

2.2.2 Valley detection of the histogram of similarities

The methods of valley detection used in this work is based on the following observation: a few exceptions apart, characterizing school hypothesises rather than real cases, the value h(i,j) of a valley's point, corresponding to a contour, is strictly inferior to the values of its adjacent neighbours, in the sense of the 8-connexity. According to the above observation, we calculate, for each element H(i,j) of H the following indicator:

$$\Delta(i,j) = \sum_{i',j'}[h(i',j') - h(i,j)] \qquad (9)$$

With $1 \leq i-1 \leq i' \leq i+1 \leq N$ et $1 \leq j-1 \leq j' \leq j+1 \leq N$.

The affectation of a pixel p(i,j) to a "region" or to a "contour" category is then done according to the Δ (i,j) sign. The decision rule is the following:

(D1) if Δ (i,j)<0 decide that p(i,j) belongs to a contour.

(D1) if Δ (i,j)≥0 decide that p(i,j) belongs to a region.

In fact, this rule of decision allows the detection of the transition points between a modal region and a valley.

3. APPLICATION

In this paragraph, we present a particular application of this algorithm in order to segment some images with non texturized levels of grey. The only attribute which is taken into account in this application is the pixels grey level.

$$A(i,j) = [f(i,j)]^T = f(i,j) \text{ for } 1 \leq i,j \leq N \qquad (10)$$

The distance measuring the closeness between the attributes of two pixels is, very simple, i.e. the absolute value of the difference between their levels of grey.

$$d(A(i,j),A(i',j')) = |f(i,j) - f(i',j')| \qquad (11)$$

The window choice is important for the quality of the contours. A large window's drawback is the thickening of the detected contours. In this application, the best window's size is 3x3. The interest of a threshold S is to avoid the detection of a too large number of contours. We give one result for a natural image which size is 128x128 and for which the grey level is coded on 8 bits as well as the calculation time T on a 25 MHz machine in seconds, the "threshold S's value and the window F's size.

4. CONCLUSION

We have presented an algorithm of image segmentation based on a histogram of similarities. This histogram is calculated by counting the similar pixels in a local neighborhood. This algorithm reduce the image segmentation to the detection of the histogram of similarities' valleys and to that of the modal regions (or plateaus). We are now studying this algorithm's behaviour on strongly texturized or very noisy images and look for the calculation of an adaptive similarity-threshold S.

Figure 1. Original image

Figure 2. Resulting edges image : L=3, S=15, T=3"

References

1 ASSELIN de BEAUVILLE J.-P., "L'estimation des modes d'une densité de probabilité multidimensionnelle. Statistique et Analyse de données, Vol. 8, N° 7, pp. 16-40, 1983.
2 KOONTZ W.L.G., NARENDRAA P.M., FUKUNAGA K., A graph-theoretic to nonparametric cluster analysis. I.E.E.E. Trans. on Computer., Vol. C-25, pp. 936-944, 1976.
3 MONGA O., Segmentation d'images : où en sommes nous ?., INRIA-Rocquencourt, Rapport de recheche N° 1216, Avr 1990.
4 TOUZANI A., POSTAIRE J.-G., Mode detection by relaxation, I.E.E.E. Trans. on Pattern Analysis and Machine Intelligence. Vol. 10, N° 6, pp. 970-978, 1988.
5 VASSEUR C., POSTAIRE J.-G., A convexity testing method for cluster analysis. I.E.E.E Trans. Syst. Man. Cyber. Vol. SCM-10, N° 3, pp. 145-179., 1980.

Expectation-driven segmentation: a pyramidal approach

A.Broggi and G.Destri [1]

Dipartimento di Ingegneria dell'Informazione, Università di Parma, Viale delle Scienze, 43100 Parma, Italy

Abstract

One of the characteristics of low-level processing of images is that it is a *data-driven* process, this means that generally no global knowledge about the image is required to process it. In this work, the pyramidal approach has been used both to decrease the amount of data to be processed, working at coarse resolution, and to increase the performance of the segmentation algorithm, allowing the use of a sort of *a-priori* knowledge about the image.

1. INTRODUCTION

The effort done in this work is mainly devoted to the extension of the use of massively parallel architectures, generally restricted to low-level aspects of image processing, in order to solve medium-level tasks, such as image segmentation. This extension is made possible by the use of an architecture based on a multiresolution Cellular Automata [4] computational model. The algorithm presented in this paper has been conceived to be implemented on PAPRICA (PArallel PRocessor for Image Check and Analysis) [1], a special-purpose SIMD architecture developed at the Politecnico di Torino, with a two-dimensional interconnecting mesh for interprocessors communication. The special structure of the hardware board and the processing array virtualization mechanism implemented on PAPRICA allow also an efficient simulation of a pyramidal architecture with no constraints at all on the intra-layer interconnections.

One of the fields of application of PAPRICA architecture is within the PROMETHEUS Eureka Project: more precisely, it will be the target architecture which will run the low-level portion of the image processing task of a complex Computer Vision system. The specific application in which PAPRICA is involved is related to the problem of automated navigation: in this case, a camera grabs image sequences from a moving vehicle, and pipelines them to PAPRICA system which preprocess and prepare them for a higher level processing, aimed to the identification of the road and the lane. Since the main goal is to achieve real-time performances, the reduction of the processing time and the performance analysis play a fundamental role in the algorithms design stage.

---

[1]This work was partially supported by CNR Progetto Finalizzato Trasporti in the frame of the Eureka PROMETHEUS project.

2. THE MULTIRESOLUTION APPROACH

The typical image captured by a camera placed on a moving car consists of a portion of the street, with moving and standing obstacles, surrounded by any kind of background depending on the specific environment (fig. 2.a). Since there is a strong correlation between the images grabbed by the moving vehicle and a sort of "expected" image, the segmentation process can be driven by this kind of synthetic image. The synthetic image (fig. 2.b) consists of two gray *plateaux*, representing the street and the sky, while an area of smooth varying brightness, fading from black to white, tracks the position of interest.

The problem of matching a natural and a synthetic image can be solved decreasing their resolution, in order to eliminate the presence of small details in the first. In fact, the algorithm subsamples the input image I and stores it into a higher and coarser level of the pyramid; an average operation is performed before subsampling. The average and subsampling operations are iterated until the image gets smaller enough to loose its details and looks like the synthetic image (typically the number of iterations is $k = 3$, starting from a 512×512 pixel image and decreasing to a 64×64 pixel image):

$$I_{x,y}^{l+1} = \frac{\sum_{i=0}^{+1} \sum_{j=0}^{+1} I_{2x+i,2y+j}^{l}}{4} , \qquad \text{with} \quad l = 0, 1, ..., k-1 \tag{1}$$

where $I_{x,y}^{l}$ denotes the luminance of the pixel at coordinates (x, y) belonging to the pyramid layer l. As shown in the block diagram in figure 1, each intermediate image is kept in memory for further usage. Also the synthetic image S is subsampled until it reaches the dimensions of the original subsampled image. At this point the iterative clustering process begins.

2.1. The clustering algorithm

The clustering algorithm [5] is an enhanced version of the *Adaptive Smoothing* [3], which is basically an edge-preserving average filter, iterated a small number of times. The intensity discontinuities that are preserved do not depend on a global threshold, as in its original formulation, but they are a function of a 5×5 neighborhood of each pixel. The great advantage that pushed to the choice of this kind of filter is represented by its simplicity and its low computational cost when compared to its performances.

Considering a single input image, the algorithm can be sketched as follows. Let's denote with $I_{x,y}^{(n)}$ the luminance intensity of the pixel at coordinates (x, y) and iteration number n, belonging to image I; the value $I_{x,y}^{(n+1)}$ is thus defined as follows:

$$I_{x,y}^{(n+1)} \triangleq \frac{\sum_{i=-1}^{+1} \sum_{j=-1}^{+1} I_{x+i,y+j}^{(n)} \times w_{x+i,y+j}^{(n)}(x,y)}{\sum_{i=-1}^{+1} \sum_{j=-1}^{+1} w_{x+i,y+j}^{(n)}(x,y)} \tag{2}$$

where $w_{u,v}^{(n)}(x,y)$ represents the weight of the pixel at coordinates (u, v) at iteration number n, when computing the new value of $I_{x,y}^{(n)}$. The weight $w_{u,v}^{(n)}(x,y)$ is thus defined as follows:

$$w_{u,v}^{(n)}(x,y) \triangleq e^{-\frac{|g_{u,v}^{(n)}|^2}{2|k_{x,y}^{(n)}|^2}} \qquad \text{with} \quad k_{x,y}^{(n)} = M \cdot \max_{i,j=-1,0,+1} g_{x+i,y+j}^{(n)} \tag{3}$$

where $g_{x,y}^{(n)}$ represents the magnitude of the gradient computed at pixel (x, y). The parameter M measures the deviation from the classical average filter ($M \to \infty$): the lower is M, the higher is the ratio between the weights $w_{x+i,y+j}$, with $i, j = -1, 0, +1$.

Since the computation of nine exponential values for each pixel is very time-consuming and not easily implementable on the extremely simple processing elements of massively parallel architectures, some approximations are needed in order to improve the performances of the algorithm, and to render it more suitable for real-time processing.

After a few manipulations [5], the approximated expression of $I_{x,y}^{(n+1)}$ becomes:

$$I_{x,y}^{(n+1)} = \frac{\sum_{i=-1}^{+1}\sum_{j=-1}^{+1} I_{x+i,y+j}^{(n)} \times \left(G_{x,y}^{(n)} - g_{x+i,y+j}^{(n)}\right)}{\sum_{i=-1}^{+1}\sum_{j=-1}^{+1} \left(G_{x,y}^{(n)} - g_{x+i,y+j}^{(n)}\right)} \quad , \quad \text{with } G_{x,y}^{(n)} = \max_{u,v=-1,0,+1} g_{x+u,y+v}^{(n)} \quad (4)$$

2.2. Driven clustering

Now let's consider the case of *driven* clustering, namely a clustering in which, together with the input image, a second *synthetic* image is used to produce the output. The values $g_{x,y}^{(n)}$ in this case of *driven* clustering are defined as a weighted average between the gradient of the original image I and the gradient of the synthetic image S:

$$g_{x,y}^{(n)} = W_I^{(n)} \text{grad}(I^{(n)}) + W_S^{(n)} \text{grad}(S) \quad (5)$$

with $W_I^{(n)} + W_S^{(n)} = 1$ and $W_I^{(h)} = 1$, where h is the number of clustering iterations. After the clustering process, the resulting image C is transferred to a higher resolution layer of the pyramid ($C_{2x,2y}^l = C_{x,y}^{l+1}$); the values of pixels at odd coordinates are computed with a simple clustering principle using the neighboring even values of C and the values of the original subsampled image I:

$$C_{2x+1,2y}^l = \begin{cases} C_{2x,2y}^l & \text{if } \left|C_{2x,2y}^l - I_{2x+1,2y}^l\right| \leq \left|C_{2x+2,2y}^l - I_{2x+1,2y}^l\right| \\ C_{2x+2,2y}^l & \text{if } \left|C_{2x,2y}^l - I_{2x+1,2y}^l\right| \geq \left|C_{2x+2,2y}^l - I_{2x+1,2y}^l\right| \end{cases} \quad (6)$$

The values of $C_{2x,2y+1}^l$ and $C_{2x+1,2y+1}^l$ will be computed following the same clustering rule.

The final image C^0 (fig. 2.c) is thus a segmented version of the input image I^0 (fig. 2.a) and contains homogeneous gray-tone regions. This prefiltering operation simplifies the typical following edge detection procedure (fig. 2.d), that can now be performed using a fast and simple gradient based filter [2].

3. CONCLUSIONS

In this paper a real-time oriented algorithm for expectation-driven segmentation has been presented: it has been conceived to be implemented on PAPRICA, taking advantage of its multiresolution Cellular Automata computational paradigm. The current ongoing research is based on a simple extension of the algorithm for image sequences segmentation: the synthetic image is replaced by the output image obtained processing the previous frames. This simple improvement, now under development, decreases the problem of matching pairs of images which can be fairly different (i.e. in a junction or in a curve) and, at the same time, since during the computation the resolution is sensibly reduced, there is no need for two subsequent frames to be temporarily very close to each other, allowing longer processing time.

REFERENCES

1. A.Broggi, G.Conte, F.Gregoretti, L.Reyneri, L.Rigazio, C.Sansoè, and C.Zamiri. *PAPRICA*, in *CAD and Architectures: Reports on Architectures and Algorithms for VLSI Design*, pages 21–141. CNR - Progetto Finalizzato MADESS, Rome, 1990.
2. A.Broggi, V.D'Andrea, G.Destri, "Cellular Automata Machines as a Computational Paradigm for low-level Vision," in International Journal of Modern Physics C, Singapore, World Scientific, Volume 4, Number 1, February 1993.
3. P.Saint-Marc, G.Medioni, "Adaptive smoothing for feature extraction," in *Proceedings of Image Understanding Workshop*, Boston, MA, pp.1100-1113, 1988.
4. T.Toffoli,N.Margolus,*Cellular Automata Machines.* Cambridge, MA, MIT Press, 1987.
5. A. Broggi, "Parallel and Local Simple Features Extraction: Street Boundary Detection in Traffic Images", Technical Report, University of Parma, CS91-2, 1991.

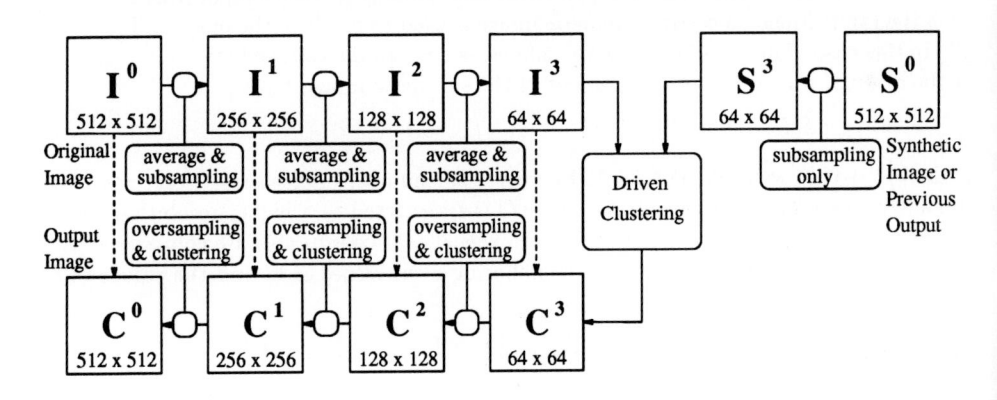

Figure 1. Block diagram of the algorithm

Figure 2. a) Original Image; b) Synthetic Image; c) Output Image.

Definition and some properties of the watershed of a continuous function

Laurent Najman[a,b] and Michel Schmitt[a]

[a]L.C.R, Thomson-CSF, Domaine de Corbeville, 91404 ORSAY-Cedex, France.

[b]CEREMADE, Université Paris Dauphine, 75775 Paris-Cedex, France.

Abstract

The notion of watershed [1, 7], used in morphological segmentation, has only a digital definition. In this paper, we propose to extend this definition to the continuous plane. Using this continuous definition, we present the watershed differences with classical edge detectors. We then exhibit a metrics in the plane for which the watershed is a skeleton by influence zones and show its lower semi continuous behaviour. Finally, this theoretical approach suggests a new algorithm for solving the eikonal equation: $\|\nabla f\| = g$.

1 The watershed: from discrete to continuum

Classical digital algorithm: We follow here the presentation of L. Vincent [7].

Let $B = \cup B_i \subset A$, where B_i are the connected components of B.

Definition 1. The *geodesic influence zone* $iz_A(B_i)$ of a connected component B_i of B in A is the set of the points of A for which the geodesic distance to B_i is less than the geodesic distance to other connected components of B. The points of A which do not belong to any influence zone make up the skeleton by influence zones of B in A, noted $SKIZ_A(B)$:

$$SKIZ_A(B) = A \setminus IZ_A(B) \text{ where } IZ_A(B) = \bigcup_{i \in [1,k]} iz_A(B_i). \tag{1}$$

The watershed algorithm on digital images by recurrence on grey level is:

Definition 2. The set of the *catchment basins* of the numerical image I is the set $X_{h_{max}}$ obtained after the following recurrence:

$$\begin{array}{l}(i) \quad X_{h_{min}} = T_{h_{min}}(I) \\ (ii) \quad \forall h \in [h_{min}, h_{max} - 1], \ X_{h+1} = Min_{h+1} \cup IZ_{T_{h+1}(I)}(X_h).\end{array} \tag{2}$$

where: - $h_{min} \in \mathbb{Z}$ (*resp.* h_{max}) is the lowest (*resp.* the greatest) grey level of image I.
- $T_h(I)$ is the threshold of the image I at height h : $T_h(I) = \{p \in D_I, I(p) \leq h\}$
- Min_h is the set of the regional minima of I at the height h.

The *watershed* of the image I is the complement of this set.

Continuous generalization: From now on, the image f is supposed to be regular enough (C^2) to allow the use of differential operators. The gradient ∇f is the plane vector of the first order derivatives of f, and the Hessian H_f is the symmetric matrix of the second order derivatives.

Definition 3. A path $\gamma :]-\infty, +\infty[\to \mathbb{R}^2$ is called a *maximal line of the gradient* if

$$\forall s \in]-\infty, +\infty[, \dot{\gamma}(s) = \nabla f(\gamma(s)) \neq 0 \text{ and } \lim_{s \to -\infty} \dot{\gamma}(s) = \lim_{s \to +\infty} \dot{\gamma}(s) = 0 \tag{3}$$

Definition 4. We denote by $\mathcal{P}(f)$ the subset of the singular points a of f such that there exists at least two maximal lines of the gradient starting from a and ending respectively in b_1 and b_2, where b_1 and b_2 are two distinct minima of f.

We have the following convergence theorem which will be used in the following as our continuous definition of watershed:

Theorem 5. *Let f a C^2 function, with a connected support. Suppose that f has only isolated singular points, and that, on the singular points, the Hessian has two non zero eigenvalues. We construct a sequence f_n of step functions which converges pointwise towards f. More precisely, we put $f_n(a) = \frac{E(2^n f(a))}{2^n}$. Then, the watershed of f, seen as the limit of the watershed of f_n, is the set of the maximal lines of the gradient linking two points of $\mathcal{P}(f)$.*

This suggest that we can add some lines to the watershed, by adding points to $\mathcal{P}(f)$. If we carefully choose these new points, the result exhibits end points and is a way to tackle the notion of grey-tone skeleton.

Proposition 6. *Let a a point of the support of f such as $\nabla f(a) \neq 0$. Let $\mathcal{V}_a$ a neighbourhood of a which does not contain any singular point. Let γ a path containing a and parallel to the gradient of f on $\mathcal{V}_a$. Then it exists a function f_0, equal to f on $\mathcal{V}_a$, such as γ is in the watershed of f_0.*

In other words, there is no local characterization of the watershed. This is due to the C^2 regularity of f. If f is less regular, there is a local characterization: the watershed is the set of the points of discontinuities of the gradient. The best example is the watershed of $f(a) = d(a, X)$ where X is a binary image. In this case the watershed of f is equal to the skeleton by influence zones of X. As it was shown by Matheron [6], it is locally characterized by the set of the points of non differentiabilty of f.

2 Comparison with the edge detectors

Let f a smooth function. Two second order differential operators are commonly used to detect edges. The first one is the Laplacian $\Delta f = f_{xx} + f_{yy}$ and second one is the non-linear Canny's detector [2] which looks for the maximum of the gradient in the direction of the gradient. The Canny's detector, or more exactly the extrema of the gradient in the direction of the gradient, has the same zero crossings as $Q(f) = \langle H_f \nabla f, \nabla f \rangle$

In [4] the authors exhibit a characterization of lines extracted by Canny's extractor. Its results are useful to point out the differences between the watershed of the gradient and the second order differential operators.

We made a comparison of the action of these operators on the image $I(x,y) = \delta_1 \chi_{\{x>0\}}(x,y) + \delta_2 \chi_{\{y>0\}}(x,y)$ where χ_A is the characteristical function of the set A: $\chi_A(a) = 1$ if $a \in A$ and $\chi_A(a) = 0$ if not. I is regularised by a gaussian kernel G and we obtain $f = G * I = \delta_1 \Psi(x) + \delta_2 \Psi(y)$ where $\Psi(x) = \frac{1}{\sqrt{\pi}} \int_{-\infty}^{x} e^{-s^2} ds$.

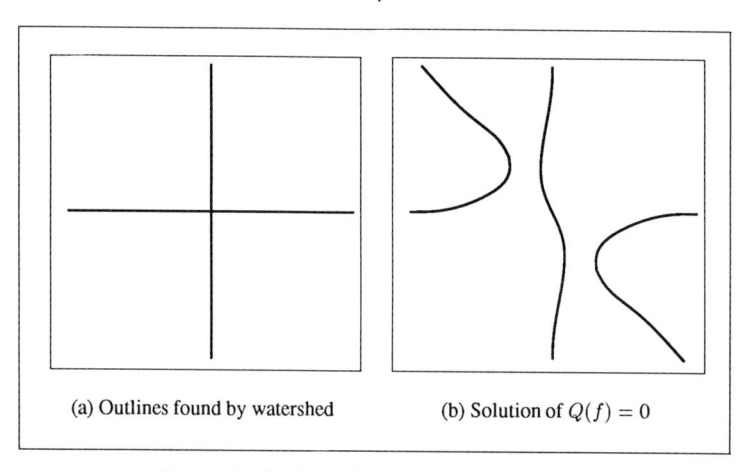

(a) Outlines found by watershed

(b) Solution of $Q(f) = 0$

Figure 1: Outlines found by the algorithms

Fig. 1 shows the results: the second order operators cannot find the multiple point, while the watershed of the gradient modulus can. This is due to the geometric behavior of Canny's operator and zeros of the Laplacian. Both are the intersection of a function $z = g(x,y)$ with $\{z = 0\}$. So, they do not have multiple points. On the other hand, the watershed has multiple points which are necessarily in $\mathcal{P}(f)$.

So if we are interested in finding multiple points, the classical differential crest extractor (local maxima of the modulus of the gradient in the direction of the gradient) has to be replaced by a watershed procedure on the image of gradient modulus.

3 Metrical approach of the watershed

The aim of this section is to exhibit the strong link between the skeleton, one of the notions of the binary mathematical morphology, and the watershed, notion of the grey-level mathematical morphology.

Definition 7. The *image distance* on a C^1 function f with a connected support $\text{supp}(f)$, is:

$$\forall (a,b) \in \text{supp}(f)^2, d_f(a,b) = \inf_{\gamma_{ab}} \left| \int_{\gamma_{ab}} \|\nabla f(\gamma_{ab}(s))\| ds \right| \quad (4)$$

We restrain the choice of f to the C^2 functions which have only isolated singular points. d_f is then a distance. For technical reasons, we suppose that the minima of f are on the same level. Moreover, we suppose that on the singular points, the Hessian has two eigenvalues non zero. We have the following result:

Theorem 8. *The set of points which are at equal d_f-distance of two distinct minima of f is the set of the maximal lines linking two points of $\mathcal{P}(f)$, and thus coincides with the watershed of f.*

A similar result has been stated in [3], but with a much more complex metric used as a definition for the continuous watershed. The advantage of our metric is to allow the statement of new results we present hereafter.

With the results of the previous theorem, we can expect the watershed to have properties similar to those of the skeleton. We state one of those properties.

Definition 9. We define the *skeletal structure* of f as the set of centres of the maximal open d_f-balls contained in $\text{supp}(f) \setminus \mathcal{M}$.

Theorem 10. *The mapping $f \to \mathcal{S}(f)$, where $\mathcal{S}(f)$ is the skeletal structure of f, is lower semicontinuous, if we use the convergence of distributions on the set of functions and the induced hit or miss topology [5] on the set of watersheds.*

This result shows why the watershed is very sensitive to noise, and justify in a way the various smoothing and marking [7] techniques.

4 The eikonal equation

As a side effect, the algorithm of the watershed can be adapted to solve an equation widely used in Shape from Shading, the eikonal equation: finding f such as $\|\nabla f\| = g$. The idea is that, on each catchment basin of f, we have $f(a) = d_f(a, b) + f(b)$, where b is the minima of the catchment basin. As determining $d_f(a, b)$ only depends on g (see formula 4), we can generalize this result: let $\{b_i\}$ a set of points with their associated values $f(b_i)$. A continuous solution to the eikonal equation is given by

$$f(a) = \inf_i \{d_f(a, b_i) + f(b_i)\} \tag{5}$$

References

[1] S. Beucher. Segmentation d'images et morphologie mathématique. Thèse Ecole Nationale Supérieure des Mines de Paris, June 1990.
[2] J.F. Canny. A computational approach to edge detection. In M.A. Fischler and O. Firschein, editors, *Readings in Computer Vision: Issues, Problems, Principles and Paradigms*, pages 184–203. Morgan Kaufmann Publishers, Inc., 1986.
[3] F. Prêteux et N. Merlet. New concept in Mathematical Morphology: the Topographical and Differential Distance functions. In *Image Algebra and Morphological Image Processing II*, volume 1568, pages 66–77, San Diego, California, July 1991.
[4] W.M. Krueger and K. Phillips. The Geometry of Differential Operators with Application to Image Processing. *IEEE PAMI*, 11:1252–1264, December 1989.
[5] G. Matheron. *Random Sets and Integral Geometry*. John Wiley and Sons, New York, 1975.
[6] J. Serra, editor. *Image Analysis and Mathematical Morphology, Volume 2: Theoretical Advances*. Academic Press, London, 1988.
[7] L. Vincent. Algorithmes morphologiques à base de files d'attente et de lacets : Extension aux graphes. Thèse Ecole des Mines de Paris, May 1990.

Detecting Elliptic Objects Using Inverse Hough–Transform

Joachim Hornegger *and* Dietrich W. R. Paulus

Lehrstuhl für Informatik 5, (Mustererkennung), Martensstraße 3, Universität Erlangen-Nürnberg, Germany

Abstract
We describe a fast method for detecting one circular or elliptical object in an image. Based on the well known Hough–Transform for lines a new method with low complexity is developed to compute the centre of gravity and the focal points of an ellipse without knowing the exact contour. The experiments yield satisfactory results both with synthetical images and real scenes like an image of a gastric ulcer. We will also study the robustness of our method with regard to noise. The algorithms are integrated in an object–oriented programming environment for image analysis.

1 Introduction

Objects of circular or elliptical shape will yield an approximately elliptical object when projected on the two dimensional image plane. The estimation of the parameters of the ellipse can be a valuable feature for the recognition and localization of these objects. Examples are medical images of tumors [1] or the problem of finding cylinders in range images. Classical approaches to this problem use the Hough–technique with a five–dimensional parameter [1]; others fit a general second–order curve with a subset of the given data and determine whether or not the result is an ellipse or choose five given points and calculate a conic section fitting to these points [2]. These methods have their own advantages and disadvantages and they are fairly time consuming. Our new approach has the property that the positional parameters of one object with fuzzy contours and roughly elliptical shape can be estimated using only edge elements by applying ideas of the Hough–Transform for lines. The experimental results show that this approach to the given problem offers an efficient and robust solution.

2 Hough–Transform for Lines from Edges

The following methods for detecting elliptic objects are based on edge–images. For each pixel in the graylevel–image the discrete values for orientation and absolute values of the gradients can be computed using an edge operator, e.g. the Sobel–operator. The orientation computed from edge masks (e.g. the operator of Nevatia and Babu) is perpendicular to the gradient. Both methods result in a uniform edge image object in the object–oriented programming environment [4], with edge orientation aligned parallel to the edge.

A straight line in a (x,y)–coordinate system, non–parallel to the y–axis, can be represented using the formula: $y = a\,x + b$, where the parameter a is called the *slope* and b the *translation* of the given line. Obviously, we can associate with each line in the (x,y)-plane one point in the (a,b)-parameter space. The detection of lines based on a gradient image is done in the following manner: For each pixel p_i we compute the slope a_i and the translation b_i of the line using the given information of the edge image. In the (a,b)–array, the *accumulator*, we increment the entry (a_i,b_i) which is initialized with zero. After all pixels of the image are visited, we utilize the values of the entries in the accumulator and conclude which lines occur in the given image. Using the strength of the edges and some given thresholds for the entry in the accumulator, the strength and the length of the lines we get lines corresponding to edges. We do not treat lines of infinite slope here (see for instance [1] on this topic).

3 Inverse Hough–Transform for Circles

The new idea of the *Inverse Hough–Transform* is to use the classical Hough–Transform the other way round. With a given line in the parameter space $b = -x\,a + y$ we can associate a point in the (x,y)-plane in a unique manner, i.e. the point (x,y).

Lemma 1 *For all straight lines $g_i : y = a_i\,x + b_i$ ($1 \leq i \leq N$) which intersect in one point $M = (x_M, y_M)$, we can associate the points (a_i, b_i) ($1 \leq i \leq N$) in the parameter space. All points (a_i, b_i) ($1 \leq i \leq N$) are element of a straight line s in the (a,b)-parameter space satisfying the following equation:*

$$s: \quad b = -x_M\,a + y_M. \tag{1}$$

Proof:
Let $M = (x_M, y_M)$. All straight lines g_i ($1 \leq i \leq N$) intersecting in M fulfil the following equations $g_i : y = a_i\,(x - x_M) + y_M$. Therefore we associate the point $(a_i, -a_i\,x_M + y_M)$ in the parameter space with each line g_i. Obviously, all such points in the parameter space are elements of the line $s : b = -x_M\,a + y_M$. □

The lemma can be used to compute the centre of an image containing one circle. The image is segmented into a edge–image for this purpose. Using the edge information and the coordinates of each point we calculate the slope and the translation of the straight line, which is perpendicular to the tangent line. Ideal conditions assure that all points in the parameter space associated with the slopes and translations of the line bundle satisify one linear equation of the form: $b = -x_M\,a + y_M$, where $M = (x_M, y_M)$ is the centre of the circle. Suppose that a noisy image of a circle is given. Analogously, we can compute the gradient image and finally the lines perpendicular to tangent lines. Certainly the majority of these lines will intersect close to the centre of the circle. Consequently the associated points in the parameter space will not fit with exactly one straight line. Using linear regression analysis we fit a line $b = -x_R\,a + y_R$ through these points (a_i, b_i) with minimal quadratic error, where

$$x_R = -\frac{n\sum_i a_i b_i - \sum_i a_i \sum_j b_j}{n\sum_i a_i^2 - \sum_i a_i \sum_j a_j} \quad \text{and} \quad y_R = \frac{\sum_i a_i^2 \sum_i b_i - \sum_i a_i \sum_j a_j b_j}{n\sum_i a_i^2 - \left(\sum_i a_i\right)^2}. \tag{2}$$

The associated point $R = (x_R, y_R)$ in the image plane will be an approximation of the centre (x_M, y_M) (see Figure 1) The average distance between the centre and the edge-points determines the radius of the circle. We may remark, that one normal line intersects the circular arc twice. Both points have identical normals. Consequently half of the circle yields the same set of normals as the complete circle. The described method is therefore very robust with respect to partial occlusion.

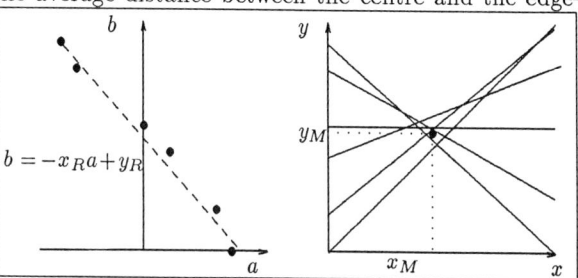

Figure 1: (Inverse) Hough–Transform

The equation 2 can be computed during the analysis of the image with a computational effort depending linearly on the number of edge points. In contrast to the classical Hough-Transform, neither storage space for an accumulator array nor computationally expensive search for local maxima are required.

4 Elliptical Shapes

The lemma 1 can also be used for computing the features of ellipses in general. In the ideal case, where one principal axis is parallel to the x-axis, the method follows immediately from the twofold application of the lemma:

Let the point (x_0, y_0) be the centre of gravity of a given ellipse with the mentioned restriction. Then, the ellipse can be specified by four parameters using the equation

$$\frac{(x-x_0)^2}{a^2} + \frac{(y-y_0)^2}{b^2} = 1. \qquad (3)$$

All normals of the elliptic line intersect close to the centre of gravity (x_0, y_0). If the equations of the normals are known, and indeed they are, the centre will be calculated using linear regression and the lemma 1. Consequently, the squares $(x-x_0)^2$ and $(y-y_0)^2$ can be determined for each contour point. The equation for the ellipse can be written in the following manner:

$$\frac{1}{a^2} = -\frac{(y-y_0)^2}{(x-x_0)^2} \cdot \frac{1}{b^2} + \frac{1}{(x-x_0)^2}. \qquad (4)$$

This term can be interpreted as an affine function in the variables $\alpha := \frac{1}{a^2}$ and $\beta := \frac{1}{b^2}$ with the slope $-\frac{(y-y_0)^2}{(x-x_0)^2}$ and the translation $\frac{1}{(x-x_0)^2}$. Obviously, the above equation is statisfied for each point of the elliptic line. If α and β are unknown parameters, we will associate with each contour point (x, y) one straight line (4) and finally one point in the paramter space assigned to slope and translation. All these lines should intersect in one point $(\frac{1}{a^2}, \frac{1}{b^2})$ determined by the length of principal axis. In noisy images these lines will just intersect close to this point. Using linear regression we can approximately compute this point in the parameter space; the principal axis, focal points and the radius of the ellipses follow directly.

Our issue is to detect a gastric ulcer in medical images. Due to experts in medicine those tumors have approximately elliptical shape. The above ideas can be useful to find an ellipse enclosing the tumor. Indeed the ellipse will not have principal axis parallel to the axis of the underlying coordinate system. First of all, the centre of gravity can be calculated with linear regression. A shift of the origin of the coordinate system into the centre of gravity splits the image into four areas. For each region, the lemma can be used to compute a point. The centres of gravity in the second and the third quadrant are weighted with the number of edge points used for their calculation. The average weighted sum over the x- and y-coordinates results in one focal point. The second focal follows by reflection at the origin of the new coordinate system.

5 Experimental Results

Focal Points	(calc.)	Centre of Gravity	(calc.)	Radius	(calc.)
(50, 100), (150, 100)	(50, 100), (149, 99)	(100, 100)	(99, 99)	160	159
(50, 130), (100, 180)	(57, 149), (92, 149)	(75, 155)	(75, 155)	100	94

Table 1: Parameters for synthetic ellipses and calculated values

The results of the algorithm on synthetic images are shown in Table 1. The first values are used for the generation of the images, the second values are calculated. These methods for the calculation of features of elliptical objects can be used for applications in medical imaging. Figure 2 shows the computed elliptic line overlayed to the red channel of a colour endoscopic image. In a further processing step, the edges close to the elliptic line are used to compute the fractal dimension of the border line to decide, whether the ulcer is malicious or not ([3]).

Figure 2: Endoscopic Image of Gastric Ulcer

References

[1] D.H. Ballard and C.M. Brown. *Computer Vision*. Prentice-Hall, Englewood Cliffs, NJ, 1982.

[2] Illingworth J. and J. Kittler. The adaptive Hough transform. *IEEE Trans. on Pattern Analysis and Machine Intelligence (PAMI)*, 9(5):690–698, 1987.

[3] Christian Lenz. *Fraktale Dimension der Kontur endoskopisch ermittelter Farbbilder von Geschwüren*. Technical Report, Diplomarbeit, Lehrstuhl für Informatik 5 (Mustererkennung), Universität Erlangen–Nürnberg, Erlangen, 1992.

[4] Dietrich W. R. Paulus. *Objektorientierte und wissensbasierte Bildverarbeitung*. Vieweg, Braunschweig, 1992.

Multiresolution detection of texture defects for surface inspection

Dmitry Chetverikov

Computer and Automation Institute, Hungarian Academy of Sciences, Budapest, P.O.Box 63, H-1518 Hungary   E-mail: h1180cse@ella.hu

Abstract

The problem of defect detection in image textures is considered in connection with the task of surface defect detection for industrial inspection. A new multiresolution top-down procedure is proposed for fast and reliable detection of texture defects. The algorithm is based on the data- and model-driven refinement of the initial defect candidates using the principle of *weak consensus* between the candidates extracted at the consecutive levels of resolution.

1. TEXTURE DEFECT DETECTION AND INDUSTRIAL INSPECTION

The task of surface defect detection is of crucial importance in many application problems related to industrial inspection using machine vision [1]. In many cases, e.g. in textile, leather, or wood industry, the defects appear on textured surfaces and should be indicated and classified according to their properties such as size, shape, or colour. From the theoretical viewpoint, this task can often be treated as detection of small non-textural regions ("defects" or "imperfections") within homogenous texture fields whose homogeneity is locally destroyed by the defects. In [2], the specific features of this task were emphasised as compared to the traditional segmentation by texture, and an initial categorisation of the texture defects was introduced. Later, a technique was proposed for detection of defects in image textures [3].

The technique [3] is a framework algorithm that can incorporate various types of texture descriptors depending on the task at hand. It is based on the operator that compares the value(s) of the selected descriptor(s) in a sliding window to the value(s) in a number of the neighbouring windows. A particular configuration of the windows was proposed that enhances the detecting power and the isotropy of the operator. Due to the local character of the comparison, the operator introduced is insensitive to a limited global texture gradient that may result from uneven illumination, tilted surfaces, or slow parameter variation of the process that generated the texture pattern. (For example, the underlying technological process.) The initial experiments have demonstrated that the proposed approach can cope with a limited texture gradient while a conventional method based on a comparison of the current patch to a reference patch fails to discriminate between acceptable and defective patches. It has also been shown that for optimal performance the size of the operator should match that of the defects.

The sensitivity of the texture defect detector [3] is selected by tuning a threshold parameter that discriminates between the response values for defective and non-defective patches. The initial experiments discussed in [3] indicate that this parameter can be easily set in a wide range of values.

However, several important problems are still to be addressed if one wishes to apply the algorithm [3] in a real industrial environment. Most of the industrial tasks related to visual surface inspection require very high processing rates that amount to millions of pixels per second. (See [1].) The technique discussed above is suitable for implementation on a parallel image processing hardware. Nevertheless, it is very desirable to significantly reduce the amount of computation by using a data- and task-driven multiresolution approach that concentrates on the properly selected regions of interest in high-resolution images of the inspected surfaces. Another problem to be solved is the automatic adaptive selection of the operator window to match the size of the defects that may vary within the same image. In practice, the defects are often multisize, with only the expected range of sizes being known *a priori*.

2. MULTIRESOLUTION APPROACHES TO COMPACT OBJECT DETECTION

This study attempts to deal with these two problems in a framework of a multiresolution approach using the principle of attention control in computer vision. (See e.g. Burt [4].) Attention control mechanisms try to emulate the capability of the human vision to focus attention on those relevant parts of the viewfield that are important for solving a given task, e.g. driving a car or detecting particular objects. In addition to higher processing speed, the selectivity combined with a more detailed, full-resolution examination of the relevant subimages improves the reliability of the vision system, which is of crucial importance for many applications.

Multiresolution image data structures such as image pyramids (i.e. stacks of digital images of decreasing size) and the related algorithms are natural tools for flexible attention control. Several versions of pyramids have been proposed and a significant number of multiresolution algorithms have been developed. (See, for example, [5].) Jolion et.al [6] reported on a multiresolution algorithm for detection of compact objects such as blobs. An important aspect of the multiresolution blob detection is the reduction, on a higher level of the pyramid, of the blob size to a few pixels that are readily detected by an appropriate spot detector. This is possible because blobs are simple compact objects of approximately uniform intensity. This property can be preserved at lower resolutions by properly selecting the type of the pyramid. Extending this idea to an arbitrary local pattern, e.g. a texture defect, would need a universal method of reliably representing (the presence of) any local feature at low resolutions. This problem has been discussed by Burt [4] who proposed solutions for a few particular applications. Yet, no universal solution exists. Clearly, one should attempt to find the multiresolution representation that best matches the features used, but one should not necessarily expect that detection becomes more reliable as the resolution is reduced.

3. THE PROPOSED ALGORITHM

In this study, we assume that at higher resolution more information is available for reliable detection. This especially applies to texture defects that may have fine details that disappear when the resolution is reduced. Consequently, our multiresolution approach to detection of texture defects is based on the following general principles:

- *robust image features* are only used;
- the features are examined at the *appropriate levels of the hierarchy*;
- only *relevant regions* are examined at each level;
- when the minimum necessary information is extracted, *concurrent solutions* are proposed that are refined or rejected when more information is available;
- the *final decision* is taken when all relevant information had been considered.

Similar principles were adopted in [7] for multiresolution rotation-invariant matching of planar shapes. There is certain analogy between the two tasks that stems from the assumption that the model of the object considered (i.e. defect or shape) is refined with the resolution.

The proposed multiresolution algorithm for defect detection in textures is a top-down procedure based on a *weak consensus* of the decisions taken on the successive levels of the pyramid. This may be perceived as a specific, relatively simple way of fusion of the data obtained at the pyramid levels. By weak consensus we mean that a defect should be detected at all levels, but the criteria of detection at lower resolutions are less strict than at higher resolutions when the defect is refined.

To initiate the algorithm, the appropriate texture features and pyramid are selected for the particular defect detection task considered. The algorithm starts at the apex of the pyramid. It is assumed that the range of defect sizes is known. The resolution of the apex is selected so as to allow for the detection of the smallest defects. At the apex, the top-down detection procedure is initiated by applying the original uniresolution defect detection algorithm [3]. The size of the operator is varied to match the expected range of defect sizes projected to the lowest resolution selected. The sensitivity of the operator is set to a high value by choosing a low threshold parameter. The candidate defects obtained for different operator sizes are accumulated and used as masks for detection at the next level. A defect is represented by a binary blob with an operator size assigned to the blob. If a defect is detected with several operator sizes resulting in overlapping blobs, the largest blob and the corresponding operator size are selected. This mechanism provides a way of adaptive selection of the optimal operator size locally tuned to the size of each defect.

At any subsequent level, the masks obtained at the previous level are used to guide the process of detection thus significantly reducing the search area. The masks for the next level are computed by AND-ing the previous masks and the current masks. As the process proceeds top down, the threshold parameter of the detector is increased for less permissive decision upon the selection of the candidates. The size of the operator is varied around the value set at the previous level, and an updated optimal value is computed.

Due to the consensus required (i.e. the AND-ing of the candidates) and the increasing strictness of the decision concerning the selection of the candidates, the number of the candidates is gradually reduced. Only a limited number of candidates arrive at the base of the pyramid where the final decision is always made. The computation time of the algorithm is significantly reduced while providing an option for adaptive local selection of the optimal operator size for multisize defect detection. This option is a sort a *tolerance mechanism* whose action is opposite to the reductive action of the consensus mechanism. The tolerance mechanism tries to keep the candidates and guarantees that the defects will not be missed in spite of the gradual reduction of the number of candidates. The parameters of the algorithms should be set so as to find the optimal balance between the two mechanisms.

4. CONCLUSION

We have proposed and briefly described a multiresolution algorithm for defect detection in texture. We hope that the algorithm can be efficiently applied in industrial tasks related to surface inspection. The proposed approach is currently being tested on industrial imagery. The results of the experiments will be published in a forthcoming paper.

5. ACKNOWLEDGEMENT

This work was supported in part by the Hungarian National Science Foundation (OTKA) grant No.2579.

6. REFERENCES

[1] R.T.Chin, "Automated Visual Inspection", CVGIP, vol.41, pp.346-381, 1988.

[2] D.Chetverikov, "Texture Imperfections", Patt. Rec. Letters, vol.6, pp.45-50, 1987.

[3] D.Chetverikov, "Detecting defects in textures", Proc. 5th Scand. Conf. on Image Analysis, Stockholm, Sweden, 1987, pp.427-433.

[4] P.J.Burt, "Attention Mechanisms for Vision in a Dynamic World", Proc.9th ICPR, Rome, Italy, 1988, pp.977-987.

[5] W.G.Kropatsch, "Image Pyramids and Curves - An Overview", Technical Report PRIP-TR-2, Technical University of Vienna, Institute for Automation, March 1991.

[6] J.-M.Jolion, P.Meer, and A.Rosenfeld, "Border delineation in image pyramids by concurrent tree growing", Patt. Rec. Letters, vol.11, pp.107-115, 1990.

[7] D.Chetverikov and A.Lerch, "A multiresolution algorithm for rotation-invariant matching of planar shapes", Patt. Rec. Letters, vol.13, pp.669-676, 1992.

The contribution of region segmentation maps to the localization of geometrical structures in real scene images

Sandra DENASI and Giorgio QUAGLIA

Istituto Elettrotecnico Nazionale "GALILEO FERRARIS"
Strada delle Cacce 91, 10135 Torino, Italy
E-mail: denasi@ien1.to.cnr.it

Abstract

This paper proposes an approach to point out geometric structures by analyzing the drawing of a region segmentation map. An experimental test has provided useful suggestions for the development of a strategy that allows the localization of salient structures in these maps. Results are presented, showing that this technique can be used to integrate the structures detected by other segmentation algorithms and to formulate more reliable object hypotheses.

1. INTRODUCTION

The interpretation of real scenes involves several information processings at different levels of abstractions. In particular, different approaches have been so far carried on by analyzing the contribution of edge and region segmentation maps in order to formulate hypotheses about objects or structures.

The present work proposes the evaluation of an additional contribution that can be inferred from region segmentation by analyzing the drawing of the boundaries of each region, without taking into account any further information about luminance or other associated parameters. The suggestion of investigating this kind of contribution arised from ability of human beings, without previous experience about image segmentation, to understand region maps as if they were line drawings. Indeed, they can easily find out from these maps the most representative structures that give suggestions for the detection and description of the represented objects.

2. EXPERIMENTAL SUPPORT

Split-and-merge [1] and histogram based algorithms [2] have been analyzed because both are able to supply accurate segmentations and to change their resolution in accordance to the level of abstraction required in the interpretation. An

---

[0]This work has been carried on funded by a grant from EUREKA-PROMETHEUS Project (CNR 91.01067.PF93).

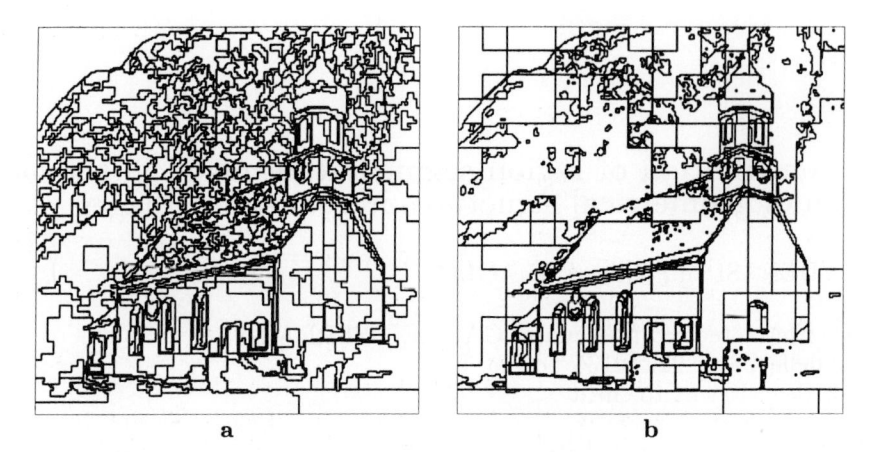

Figure 1: Region segmentation maps.

example of such segmentations is shown in figure 1.

Although these algorithms segment the image into regions that correspond roughly to the objects expected in the scene, human beings are able to neglect the boundaries that are meaningless for the object perception and to override the region distribution produced by the strategy used by the segmentation algorithms. Therefore human beings can see objects thanks to the ability of visual perception to organize the region boundaries so that more complex structures are built, having properties of regularity and closure of their shapes.

In order to find out the criteria that guide the interpretation of these drawings a set of region segmentation drawings were presented to unskilled people. They were asked to outline with a pencil the main objects (if any) that they were able to see. During the experiment the order followed in sketching the lines of the drawing was recorded.

Results obtained from the experiment suggest that the main elements which stimulate visual perception are the following (the order expresses the salience of the structures):

- long straight slanting lines, belonging to the boundary of a single region or composed by joining several boundaries belonging to adjacent regions (figure 2a);

- long narrow regions, in correspondence of strong changes in luminance (figure 2b);

- curved lines originated from boundaries of adjacent regions (figure 2c);

- boundaries of large homogeneous regions.

It is very important to notice the order followed by each observer in drawing the objects or structures, in fact it is evident that the structures that delineate the global shape of the things are drawn first, then the structures of the details are added.

3. THE PROPOSED STRATEGY

According to the previous considerations and following the suggestions derived from the reported experiment, a strategy has been developed to reach the following key goals: detection of salient structures in a region segmentation map and their geometrical organization in order to formulate object hypotheses.

In order to provide a concise representation suitable to represent the noticed structures, an initial linearization has been performed that transforms each region boundary into a list of straight segments. The resulting strokes are analyzed in order to extract the most salient slanting, horizontal and vertical edges. Regular repetition of patterns of elementary strokes drives the localization of slanting edges, while long adjacent parallel horizontal and vertical segments are replaced by single horizontal and vertical edges. Then the meaningful structures are built starting from the longest segments, because they have a higher probability of belonging to an object rather than to the background. These segments are joined to their neighbours to improve their saliency, and then related each other to form more complex structures, that will be the basic elements of objects. Criteria of continuity, collinearity and proximity rule the aggregation processes, according to the principles of Gestalt psychology [3].

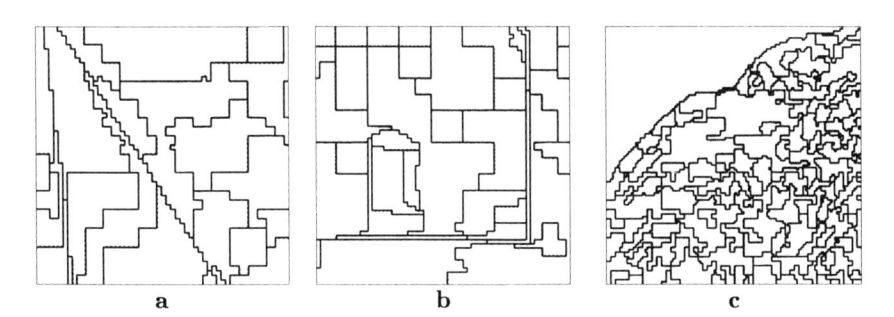

Figure 2: Perceptual suggestions.

We tested the proposed strategy on several segmented pictures. Referring to the region segmentation map shown in figure 1a, all the detected structures are shown in figure 3, and they constitute the main structures from which the search for closed forms and the formulation of object hypotheses start.

In a previous work [4], we proposed the formulation of structure hypotheses starting from edge segmentations and using the same perceptual grouping criteria. The comparison of the results supports our hope to obtain more reliable object hypotheses by merging the two contributions and emphasizing the peculiar meaning of both of them.

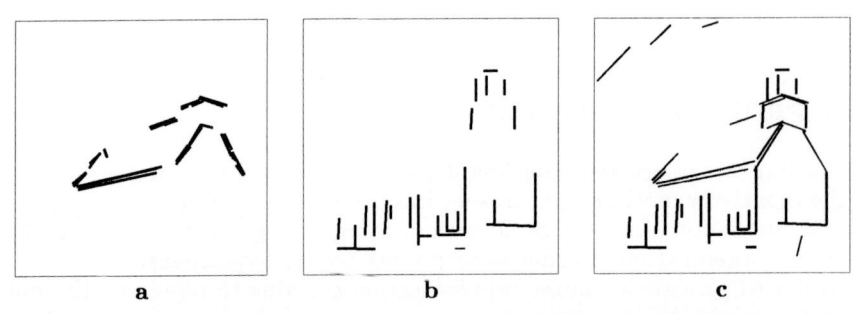

Figure 3: Structures detected in the region map of figure 1a: (a) slanting lines, (b) horizontal and vertical segments, and (c) their resulting integration.

4. CONCLUSION

Comparing figure 3c with the drawings of the experimental test, we observe that the detected structures correspond well to the most salient ones pointed out by the interviewed people. So they can be successfully used to guide the search for other less salient structures and their organization in closed shapes. However, our opinion is that the promising results of this experiment suggest further investigations to discriminate the contribute of perception from the a-priori knowledge of the depicted objects.

Moreover, even if humans are able to perceive the main objects in a region segmentation drawing, many other structures are too much confused to be pointed out. But, as already mentioned, this procedure is part of a scene understanding system in which the structures extracted from the region map will be integrated with the structures extracted from the edge segmentation of the image. This integration can provide more reliable and complete structures to allow a correct formulation of object hypotheses.

REFERENCES

[1] S. Horowitz and T. Pavlidis, "Picture segmentation by a tree traversal algorithm," *J. ACM*, vol. 23, pp. 368–388, April 1983.

[2] J. R. Beveridge, J. Griffith, R. R. Kohler, A. R. Hanson, and E. M. Riseman, "Segmenting images using localized histograms and region merging," *International Journal of Computer Vision*, vol. 2, pp. 311–347, 1989.

[3] M. Wertheimer, "Principles of perceptual organization," in *Readings in Perception*, (D. Beardslee and M. Wertheimer, eds.), New York: Van Nostrand, 1958.

[4] S. Denasi, G. Quaglia, and D. Rinaudi, "The use of perceptual organization in the prediction of geometric structures," *Pattern Recognition Letters*, vol. 13, no. 7, pp. 529–539, 1992.

A Markov Random Field approach to grouping of descriptive primitives

Vittorio Murino, Carlo S. Regazzoni, and Gian Luca Foresti

Department of Biophysical and Electronic Engineering (D.I.B.E.)
University of Genoa, Via All'Opera Pia 11A, 16145 Genova, Italy

Abstract
A probabilistic method for grouping edge-based descriptive primitives is proposed. First, straight segments are organized into a graph representing geometrical relations between segments (i.e., parallelism, collinearity, convergence). then, the grouping process consists in estimating the label and the extremes of each graph segment. These variables are modelled as a Markov Random Field (MRF). A Simulated Annealing algorithm is used to find the best configuration, consisting of closed subsets of segments. The capabilities of the method are shown for a synthetic scene.

1. INTRODUCTION

Image segmentation aims at obtaining a synthetic image representation in terms of descriptive primitives, like edges and regions. Even for such a reduced representation, the task of matching descriptive primitives extracted from an image with those used to describe an object model remains [2] a computationally hard problem, due to the combinatorial explosion of possible association pairs data-object. The goal of grouping methods is to drastically reduce the complexity of computational problems, such as indexing of a model library and matching, by ranking possible subsets of descriptive primitives according to their global properties [2]. For this reason, grouping assumed great importance for the development of robust and efficient visual recognition systems. The work by Lowe [3] is one of the first examples in this sense: Gestalt laws [4] are used as the basis for his SCERPO system. The present paper describes a probabilistic image model for grouping purposes which considers as input rectilinear segments, modelled as nodes of a graph. Links between nodes correspond to relations between segments (i.e., collinearity, parallelism, and convergence). The goal of the proposed grouping method is twofold: 1) to find subsets of segments which are likely to belong to a single object; 2) to complete the missing parts of segments of a certain group on the basis of good-continuation Gestalt laws [4]. In section 2, the representation of segments is described. In section 3, the MRF model for the grouping process is presented, and, in section 4 results are reported.

2. STRAIGHT-SEGMENT GRAPH (SSG)

Segments extracted from an image by a Direct Hough Transform (DHT) algorithm [5] are organized into a graph, denoted as SSG = $\{s_m: m=1..M\}$, where M is the number of edges contained in the image and the node s_m represents the m-th edge. Each graph node is provided with a measurement vector, expressed as

$$g(s_m) = g_m = [\rho, \theta, x_r, y_r, x^{min}, x^{max}] \tag{1}$$

whose components represent intrinsic features of a segment. Each segment is characterized by the parameter $\pi=(\rho,\theta)$, by the coordinates (x_r,y_r) of a reference point in the image reference system, and by the extremes x_{min} and x_{max} of the segment in a one-dimensional reference system associated with π and centered in (x_r,y_r) (see Fig.1).

Graph nodes are linked through collinearity, parallelism and convergence relations. The relations between two nodes are expressed as:

$$H(k,j) = \{h(s_k,s_j) = h_{kj}: k,j \in [1..M], j \neq k\} \tag{2}$$

where h_{kj} is an array of relational features of an edge pair:

$$h_{kj} = [status, \Phi_c, d_c, \rho_p, D_j, D_k, x_c, y_c, \Psi_c, W_c, W_c^{max}] \tag{3}$$

where:
status= {1 parallelism, 2 convergence, 3 collinearity};
Φ_c = angle $\theta_k(=\theta_j)$ of the two collinear segments j and k;
ρ_c, d_p = distance of collinear segments from the origin and distance between two parallel segments, respectively.
D_j, D_k = asymmetry factors of parallel edges;
$\Psi_c, (x_c, y_c)$ = angle and coordinates of the convergence point;
W_c=weakness factor: this parameter indicates the minimum distance of a convergent segment to the convergence point. If $W_c=0$, the segment intersects the convergence point;
W_c^{max} = maximum weakness factor; it refers to the extreme of a convergent segment at the maximum distance to the convergence point.

3. THE MRF GROUPING MODEL

The grouping process is modelled here as a stochastic estimation process to be performed on the SSG. It consists in estimating the optimal values for the label and the extremes of each node. Optimization is based on a non-linear functional. The functional weights a solution (i.e., the complete labelling of all the graph nodes) depending on intrinsic and relational segment attributes. The configurations of the graph are assumed to be described by Gibbs distributions. The energy functional is computed by means of different local terms, separately evaluated for subsets of graph nodes (i.e., cliques). Each local term has a precise geometrical significance: for example, a term may favour the assignment of the same label to highly symmetric pairs of segments, while another penalizes a T-shaped grouping of segments. Modelling the field probability by a Gibbs distribution is equivalent to modelling the grouping process as a Markov Random Field (MRF) (see the Hammersley-Clifford theorem [1,6]). An MRF has the peculiarity of being characterized by a multiple neighbourhood system. The SSG is viewed as an irregular lattice where each node s_m is associated with a line segment. The vector g_m and the vectors $h_{m,j}$ of intrinsic and relational properties form an observed vectorial field $G=\{g_m, h_{mj}: m=1..M, j=1..J_m\}$. A set of random variables to be estimated is assigned to each graph node; such variables represent an identification label and the coordinates of the extremes of a segment in a monodimensional reference system, respectively. Extremes are represented by a vectorial random field $F=\{f_m=[x_m^1, x_m^2]$:

m=1..M}, where $x_m^i \in [0,...,\sqrt{2}N]$ (N by N is the image size). A label field $\mathbf{R}=\{r_m: m=1..M\}$, $r_m \in [1..M]$ is associated with the SSG for grouping edges, to mean that segments having the same label belong to the same group. $\mathbf{F}$ and $\mathbf{R}$ are modelled as coupled Markov Random Fields (MRFs). This means that, according to the Hammersley-Clifford theorem [6] and to the assumption that extreme measures are affected by Gaussian noise [7], the probability of obtaining a configuration ($\mathbf{F}=F, \mathbf{R}=R$) of the coupled field can be expressed as:

$$P(\mathbf{F}=F, \mathbf{R}=R/G) = P(F, R/G) = \frac{1}{Z}\exp\left\{-\frac{U(F,R/G)}{T}\right\} \quad (4)$$

where Z is the so-called partition function, U(F,R/G) is the energy function of the field, and T is the temperature.

Grouping is a process that tries to find the values of the $\mathbf{R}$ and $\mathbf{F}$ fields that are characterized by the minimum values of the energy function. The algorithm searching for the minimum-energy configuration is a stochastic optimization method (i.e., Simulated Annealing [8] with the Metropolis sampler). According to this method, the global energy of each node at step k, U_k is iteratively evaluated. Then, the variables of the current site are changed in a random way. The energy of the resulting new configuration, say U_{k+1}, is computed and is either accepted or not, according to the Metropolis scheme. Thanks to the Markov property, the energy function U(F,R/G) can be expressed as the sum of local terms, V_c, which represent the potentials associated with different configurations of cliques of field variables. A clique c is defined as a subset of sites that are neighbours of each other. A first-order neighbourhood system (i.e., cliques containing only two nodes) is considered. The neighbourhood system $\mathcal{N}=\{\mathcal{N}_m : m=1..M\}$ (where $\mathcal{N}_m=\{s_k: s_k \neq s_m$ and $s_m \in \mathcal{N}_k\}$) associated with S is multiple. This is consistent with the possible relational properties between nodes, i.e., parallelism, collinearity, and convergence. If one denotes by C^i the set of cliques of a different neighbourhood system, the MRF energy of the coupled field can be expressed as:

$$U(F,R/G) = \sum_{m \in S}\sum_{k=1,2} V_{clos}(F/G) + \sum_{m \in S}\sum_{i=1,2,3}\sum_{c \in C^i} V_{c^i}(F,R/G) =$$

$$= \sum_{m \in S}\left[\sum_{k=1,2}\min\left(\left[x_m^k - x_m^{min}\right]^2, \left[x_m^k - x_m^{max}\right]^2\right)\right] + \left[\sum_{j \in N_m^1}\delta(r_m,r_j)\cdot ps(s_m,s_j) + [1-\delta(r_m,r_j)]\cdot K_1\right] +$$

$$+\alpha\cdot\left[\sum_{j \in N_m^2}\delta(r_m,r_j)\cdot cs(s_m,s_j) + [1-\delta(r_m,r_j)]\cdot K_2\right] + \beta\cdot\left[\sum_{j \in N_m^3}\delta(r_m,r_j)\cdot co(s_m,s_j) + [1-\delta(r_m,r_j)]\cdot K_3\right]$$

where δ is the Kroenecker function, i=1 stands for parallelism, i=2 for convergence, and i=3 for collinearity. α and β are model parameters necessary to normalize the energy contributions related to the different cliques, and V_{clos} is a potential term which favours solutions characterized by values of the extremes that are closer to measured ones. This grouping model favours uniform label configurations at neighbourhood sites associated with symmetric or close segments.

3.1. Parallelism

The potential related to parallelism has the form:

$$V_{c^1}(F,R/G) = \begin{cases} ps(s_m,s_j) & \text{if } r_m = r_j \\ K_1 & \text{otherwise} \end{cases} \tag{6}$$

The function *ps* scores the degree of symmetry of a parallel straight line and is defined by the expression:

$$ps(s_m,s_j) = D_m^2 + D_j^2 \tag{7}$$

where D_m and D_j are the relative displacements (with respect to the case of perfect symmetry) between the two segments. The function *ps* penalizes the grouping of segments that are not symmetric, and is used if the same value is assigned to both labels belonging to a clique of two parallel edges. The term K_1 is a cost that is paid whenever different labels are assigned.

3.2. Convergence

The potential term related to the convergence relation is similar to the previous one:

$$V_{c^2}(F,R/G) = \begin{cases} cs(s_m,s_j) & \text{if } r_m = r_j \\ K_2 & \text{otherwise} \end{cases} \tag{8}$$

The function $cs(s_m,s_j)$ has a more complex form:

$$cs(s_m,s_j) = \begin{cases} MNC & \text{if } (W_{c,m} = 0 \text{ AND } W_{c,j} > 0) \text{ OR } (W_{c,j} = 0 \text{ AND } W_{c,m} > 0) \\ [(W_{c,m} - W_{c,j})^2 + (W_{c,m}^{max} - W_{c,j}^{max})^2] & \text{if } W_{c,j} > 0 \text{ AND } W_{c,m} > 0 \end{cases} \tag{9}$$

MNC is a constant that operates whenever a T-like configuration is present; it penalizes T-shaped junctions grouped together. This function aims at penalizing the union of asymmetric convergent edge pairs.

3.3. Collinearity

The potential for collinear edges can be expressed in a similar way:

$$V_{c^3}(F,R/G) = \begin{cases} co(s_m,s_j) & \text{if } r_m = r_j \\ K_3 & \text{otherwise} \end{cases} \tag{10}$$

where

$$co(s_m,s_j) = \|x_m - x_j\| \tag{11}$$

is the Euclidean distance between the neighbouring extremes of the two collinear segments. In this way, the closer the collinear edges, the more likely they are to be considered as a single group.

3.4. Closureness

The potential related to closureness refers to the distance between estimated and measured extremes. It can be expressed as:

$$V_{clos}(x_m^i) = \min\left([x_m^i - x_m^{min}]^2, [x_m^i - x_m^{max}]^2\right) \tag{12}$$

Extremes concur, in a indirect way, to the estimation of costs related to the degrees of symmetry of parallel and convergent segments. Consequently, the obtained solution balances between such two forces.

4. RESULTS

Several experiments on a synthetic image were performed to check on the functioning of the algorithm. Figure 2 gives an example of a synthetic scene containing a rectangle and a triangle. Figure 3 shows (a) the results of the Canny algorithm, (b) the results of the DHT algorithm, and (c) closed groups of segments. It is possible to observe that the extremes of each segment are modified in order to reach the ideal closure point. Two kinds of errors were analyzed: (a) lacunarity errors and (b) spreading errors. Figures 4a and 4b show the behaviours of the lacunarity and spreading errors for different numbers of iterations of the algorithm applied to Fig. 2. The lacunarity error (Fig. 4a) decreases as the numbers of iterations increases; one can notice that, after the minimum number of 20 iterations, the process converges to a minimum value. The spreading error (Fig. 4b) is already acceptable for a small number of iterations: the minimum error value is obtained after a smaller number of iterations than for the lacunarity errors.

5. CONCLUSIONS

An MRF model that takes into account the symmetry and closureness labels of groups of segments has been proposed. The approach can be regarded as an example of extension of classical restoration and segmentation methods, which are used at a higher interpretation level.

6. REFERENCES

[1] S.Geman, D.Geman, "Stochastic Relaxation, Gibbs Distribution, and the Bayesian Restoration of Images", *IEEE Trans. on Pattern Analysis and Machine Intelligence*, Vol.PAMI-6, No.6, pp.721-741, 1984.
[2] W.E.L.Grimson, *Object Recognition by Computer: The Role of Geometric Constraints*, The MIT Press, 1990.
[3] D.G.Lowe, *Perceptual Organization and Visual Recognition*, Kluwer Academic Publishers, 1985.
[4] G. Kanisza, *Organization in Vision*, New York, NY, Praeger, 1979.
[5] G.L.Foresti, C.S.Regazzoni, G.Vernazza, "An Improved Hough-Based Approach to Straight-Segment Detection", submitted to *IEEE Transactions on Pattern Analysis and Machine Intelligence*, 1992.
[6] J.Besag, "Spatial Interaction and the Statistical Analysis of Lattice Systems", *Journal of Royal Statistical Society, B34*, pp.75-83, 1972.
[7] J.Marroquin, "Probabilistic Solution of Ill-Posed Problems in Computational Vision", *Journal of the American Statistical Association,* Vol.82, No.397, March 1987, pp. 76-89.

[8] S.Kirkpatrick, C.D.Gelatt, M.P.Vecchi, "Optimization by Simulated Annealing", *Science*, Vol. 220, No. 4598, pp. 671-680, 1983.

Figure 2. Original synthetic image.

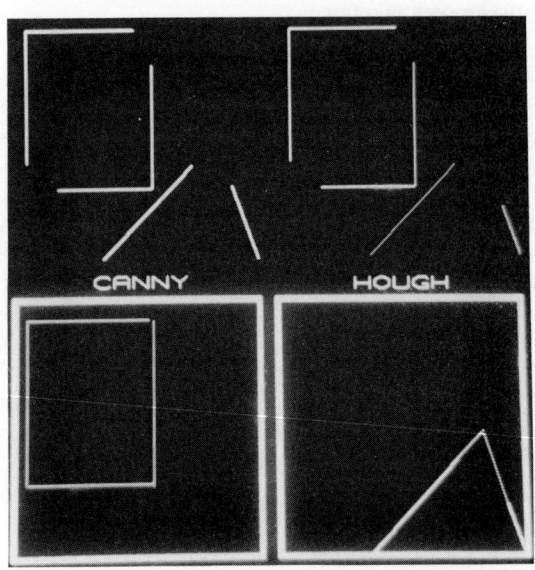

Figure 3. Resulting images from the edge extraction process (top left), the DHT process (top right), and the resulting groups with closure of segments (bottom).

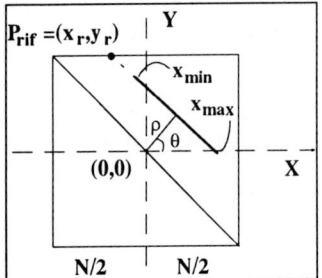

Figure 1. The one-dimensional coordinates of the segment in image reference system.

Figure 4. Profile of the lacunarity error vs. the number of iterations.

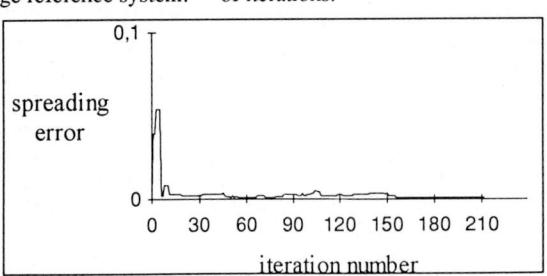

Figure 5. Profile of the spreading error vs. number of iterations.

A New Perspective on Segmentation: Token Grouping at Multiple Abstract Levels

Qian Huang, George C. Stockman
Pattern Recognition and Image Processing Laboratory
Department of Computer Science
Alvin J. M. Smucker
Department of Crop and Soil Sciences
Michigan State University
East Lansing, MI 48824-1027
U.S.A.

Abstract

The problem of segmentation is examined from a new perspective. A paradigm, called *token grouping at multiple abstract levels* is proposed as an alternative methodology for segmentation. A *token* is defined as a general unit of perception. *Token grouping* is aggregation of a set of tokens at some visual level(s) into a new token at another visual level. Under this paradigm, the task of segmentation can be described as a process of continuously grouping tokens into a more abstract one until a token, representing an object, is constructed. The perceptual organizations are determined by grouping criteria which allow both bottom-up grouping principles and top-down expectations to participate in the segmentations at different levels. As an illustration, the proposed paradigm is applied to the problem of segmenting elongated objects.

1 Introduction

Segmentation is an essential problem in the field of machine vision. The capability of a vision system often depends heavily on the quality and the reliability of the segmentation result. Conventionally, segmenting objects from background is through a region-based method, a boundary-based method, or an combination of the two [2]. Most segmentation techniques do not distinguish the regions occupied by different instances of objects if they overlap because the object to be recognized is not the concern of these approaches. To segment instances of objects, therefore, high level analysis is needed. This is usually where we draw the boundary between high level and low level computer vision. In general, due to the fact that object model is not used at the lower levels, a large number of hypotheses will be generated, making the high level verification costly. Utilizing object models earlier may eliminate spurious hypotheses which, ultimately, leads to a more efficient segmentation.

The concept of *token grouping* refers to the functionality of a machine vision system in organizing perceptual entities ([6], [5]). Most of the recent work in token grouping has concentrated on reconstructing lines or curves ([7], [1]). Our view toward token grouping is broader. We characterize the tasks performed by a machine vision system as a process of continuous *groupings across multiple abstract levels*. In this view, only a single operation, namely *grouping*, performs all necessary processing across the entire perceptual hierarchy. Token grouping is a tool that enables the implementation of such a viewpoint.

With this perspective, we propose *token grouping at multiple abstract levels* as an alternative paradigm for segmentation. Such a framework provides a homogeneous environment where information can be integrated and different techniques can be implemented with a uniform principle: *grouping by proximity*, here proximity can be specified by both bottom-up and top-down influences. In Section 2, we formally describe the paradigm of token grouping across multiple abstract levels. The problem of segmentation is examined from the perspective of token grouping in Section 3. As a case study, it is shown, in Section 4, how the theory is applied to the problem of segmenting elongated objects from the background. The concluding remarks is given in Section 5.

2 Token Grouping At Multiple Abstract Levels

Human perception appears to be constructive and hierarchical. Different visual tasks are concerned with various perceptual entities at different levels of abstraction. The tasks performed by a machine vision system can be described as a process of continuous *groupings across multiple abstract levels* in which a constructive and hierarchical machine perception can be achieved.

2.1 Generalized Perception Unit: Tokens

A *token* is defined as a meaningful unit of perception. It can be a point at the finest resolution such as a pixel, or an entire object described at a more abstract conceptual level. Each token can be described by some features that can be intrinsic, relational, or even functional. Tokens are scattered in an $n + 1$ dimensional space, called *token space* with n *spatial* dimensions and 1 *level* dimension, where n is the dimensionality of the input image. Along the dimension of visual level, tokens form a hierarchy of abstractions at different levels of perception. Formally, if t_i^k represents a token at level k, it can be defined as $t_i^k = \{T_i^k, F_i^k\}$, where T_i^k specifies its *organization* and F_i^k is its associated set of features.

2.2 Token Grouping

Token grouping is aggregation of a set of tokens at some visual level(s) into a new token at another (may be the same) visual level. Grouping is an *action* taken by a vision system following the *decision* made based on some predefined grouping criteria. Grouping criteria determine which and how tokens are grouped, and, ultimately, specify the perceptual organization of the data. Grouping operations are constructive due to the nature of aggregation. The only difference between grouping at different levels is the grouping criteria used.

2.3 Grouping Criteria

By what organizational principles should tokens be grouped? Gestalt theory has provided a set of general bottom-up grouping rules. Much of the research in human perception has shown that humans use high level knowledge even in low level visual processing. Perception is influenced by factors besides the physical nature of the stimulus. The *learned perceptual patterns* and *expectation* affect how information might be perceived.

The organizational principles in token grouping allow both perceptual regularities which are data-driven, and a top-down *expectation* to play a role in the visual process. The proposed paradigm offers the opportunity of effectively integrating different sources of information, including decomposed *a prior* knowledge (e.g., object models), in the grouping criteria at any level. This participation of high level knowledge assists in capturing only the useful tokens and resolving the ambiguities in data. The grouping criteria of different levels are the functions of the tasks to be accomplished at these levels and the nature of the tokens involved. The grouping criteria can be numerical, symbolic, or topological. They may appear as an mathematical expression (e.g., an optimization function), a template, a symbolic description, or even a graph.

3 An Alternative Paradigm for Segmentation

Under the framework of token grouping, the task of segmentation can be achieved by grouping tokens into more and more abstract ones until new tokens formed represent instances of objects of interest. The grouping process gradually combines local and non-local information so that the global regularities can be found in a constructive way.

Tokens can be classified into three categories based on their geometric dimensionality. *Dot tokens* have dimensionality of zero; for example, points in a 3D space or pixels in an image. Their features can be intensity, color, or gradient magnitude. *2D tokens* have dimensionality of two. Examples are lines, curves, regions, or texels. The intrinsic features describing a 2D token can be area, length, or orientation. The relational features can be distance, or certain features determined by decomposed object models. *3D tokens* represent 3D perceptual entities which can be surfaces, parts, or objects. Intrinsic features of a 3D token can be its shape, volume, or position. Tokens from different categories can be grouped. A closed curve (a 2D token) can be grouped with its interior pixels (dot tokens) to form a region (another 2D token). On the other hand, the same type of tokens can also be grouped to form a more significant token of the same type. The hierarchy of tokens and the formulation of grouping criteria are the most important issues in applying this methodology.

4 A Case Study: Segmenting Elongated Objects

The proposed methodology has been applied to the problem of segmenting elongated objects in 2D images. We use piecewise 3D cylinders as the object model. We decompose this object model in the following way. The token at the highest level is an object which is formed by a set of tokens, called *tubes*, representing local cylinders. The projection of an object generates a pair of parallelly symmetric curves which corresponds to a *ribbon* token. A ribbon is grouped from two curve tokens which may be approximated as a collection of straight lines, hence line tokens. A line token is formed by a set of pixel tokens that have similar properties such as high magnitude or orientation.

Starting from an input image, various intrinsic properties of pixels are detected first. The segmentation is a process that continuously constructs new tokens from the available ones until the tokens formed represent elongated objects. At each level, the grouping criteria consider both perceptual regularities such as collinearity and symmetry as well as the proximities between data and the decomposed model. Currently, 53 images of elongated objects from various application domains have been segmented and quantified, including plant root images from agriculture, bacteria images from biology, blood vessels images from medical imaging, wire images from industry, and images of synthetic elongated objects. The results have empirically proved the effectiveness of this paradigm ([4], [3]). Figure 1 and 2 show the results for plant root and bacteria images. Compared with the results from simple region-based method (in (c)), the results from token grouping method (in (d)) are improved, especially because instances of objects are also segmented. We regret that only two experimental results are shown in this paper due to the space limit.

5 Conclusions

The problem of segmentation is viewed from a new perspective. The paradigm of token grouping at multiple abstract levels is proposed as an alternative for segmentation. A case study briefly

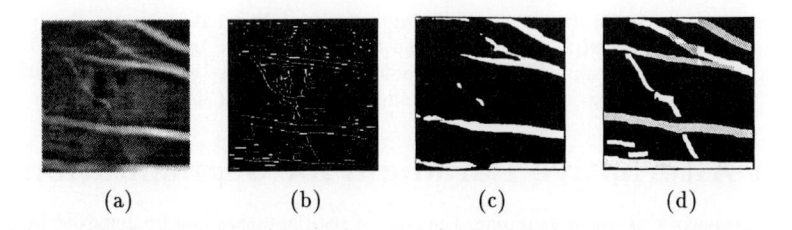

Figure 1: Experimental result for a plant root image. (a) input image, (b) edge map for (a), (c) segmentation result from simple region-based method, (d) segmentation result from token grouping method.

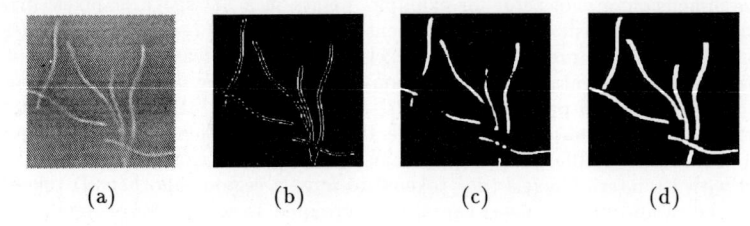

Figure 2: Experimental result for a bacteria image. (a) input image, (b) edge map for (a), (c) segmentation result from simple region-based method, (d) segmentation result from token grouping method.

illustrates how segmentation can be implemented under this proposed framework.

References

[1] J. Dolan and E. Riseman. Computing curvilinear structure by token-based grouping. In *Proc. IEEE CVPR*, pages 264–270, 1992.

[2] Robert M. Haralick and Linda G. Shapiro. Survey: Image segmentation techniques. *Computer Vision, Graphics, and Image Processing*, 29:100–132, 1985.

[3] Qian Huang and G. C. Stockman. 3d elongated object recognition. Technical report, Michigan State University, December 1992.

[4] Qian Huang and G. C. Stockman. Token grouping at multiple abstract levels. Technical report, Michigan State University, October 1992.

[5] David G Lowe. *Perceptual Organization and Visual Recognition*. Kluwer Academic Publishers, 1985.

[6] David Marr. *Vision*. McGraw-Hill Book Company, 1982.

[7] E. Saund. Labeling of curvilinear structure across scales by token grouping. In *Proc. IEEE CVPR*, pages 257–263, 1992.

Seismic image segmentation techniques

I. Pitas and C. Lialios

Department of Electrical Engineering, University of Thessaloniki, Greece

Abstract

This paper presents five techniques for seismic image segmentation that use texture orientation and/or frequency content criteria. The first two use local texture orientation information. The other two use frequency content criteria for texture discrimination, while the last one employs local entropy as an alternative method segmenting chaotic seismic texture patterns.

1. MARKOVIAN IMAGE MODELING

The first method is concerned with the segmentation of a seismic image using the Markovian modeling technique [1]. Each seismic image region corresponds to one Markov chain *state*. Seismic image segmentation is performed by assigning a state to each image pixel. As in most images, the state (region) of each pixel is related to the states of neighbor pixels in the probabilistic sense. The transition from the one state of a pixel to the possible states of neighbor pixels has some *transition probabilities*. The segmentation is performed by calculating which is the most probable pixel state (region) by taking into account the transition probabilities, the states of the neighbor pixels and the pixel value itself. The method described in [1] is used for this purpose. This method provides a suboptimal line-by-line solution to seismic image segmentation. This method has been applied to the segmentation of the seismic image according to its texture orientation. The problem which has been examined can be stated as follows: for a given pixel (i,j) of a seismic image x, calculate which is the most likely local texture orientation by taking into account the texture orientation of the neighbor pixels. Five orientations have been examined for this method: angles of 0, 30, 60, -30, -60 degrees were accepted as possible state labels. The a-priori probabilities that a pixel belongs to the state k are estimated by the following algorithm: The pixels that belong to a local pixel (i,j) neighborhood are examined. The percentage of them that is collinear to (i,j) along direction $k = 0$, 30, 60, -30, -60 and have almost the same gray- level to this pixel gives the apriori probability of pixel (i,j) belonging to the state k is obtained. The segmentation result is rather sensitive to transition probabilities. The segmentation is performed on a line-by-line basis by using a Viterbi-like algorithm [1].

2. HOUGH TRANSFORM

The Hough transform is used for the classification of seismic horizons with respect to their orientation. The seismic image is transformed to a binary one, e.g. by following the seismic horizons. The parametric straight line equation used is $y = m \cdot x + c$ (for a line passing through a pixel (x, y)). m is the slope coefficient and c is the y-coordinate offset. No polar representation is needed, because seismic horizons have slopes close to 0. The Hough transform accumulation array was used for the segmentation of seismic horizons. A two-dimensional histogram is created, which represents how many points of the line with parameters (m, c) occur in the entire binary image of the seismic horizons. Segmentation is performed by identifying peaks in the slope coordinate m or by segmenting its range in equirange or variable-range segments. In most cases, the central values of the slope segments of our interest were 0, -30, -45, -60, 30, 45 and 60 degrees which correspond to certain values of m. The corresponding image regions were obtained by the inverse procedure: the horizon pixels belonging to a specific range of m can be easily calculated by the inverse Hough transform. The seismic image pixels in their vicinity form an image region having a specific texture orientation.

3. DIRECTIONAL FILTERING

The segmentation of a seismic image can be accomplished very successfully using the directional filtering method. *Directional filter* is a filter whose frequency response covers a wedge in the frequency domain [2]. In this work, a set of K directional filters has been used with constant response in the passband, linear decrement to zero in the transition band and zero frequency response in the stop band. These filters cover the entire frequency domain. The segmentation is performed by first calculating all directional filter outputs. A pixel is considered to belong to the region $k = 1, \ldots, K$ if the output of the the k-th directional filter has the largest magnitude at this pixel. In our experiments, the directional filters have been implemented in the frequency domain. In most cases, three filters have been used corresponding to texture with horizontal orientation, negative orientation (-30 degrees) and positive orientation (+30 degrees). The directional filters have given very good segmentation of the seismic image texture with respect to its orientation. The only drawback is the large computational complexity used for their implementation.

An example of the use of directional filters is shown in Figure 1. Figure 1a is the original seismic image. The detected image region having texture orientation with negative slope (-30 degrees) is shown in Figure 1b.

4. GABOR FILTERING

The image decomposition to various subimages with certain orientation and frequency content can be achieved using the Gabor filters. A set of Gabor filters allow the isolation of any narrow range spatial frequencies and provide minimized uncertainty

[3]. The complex 2-D Gabor functions can be tuned over any localized area of the frequency plane. The Gabor filters used in this work were designed and implemented in the frequency domain. Gabor filters are used to distinguish areas of the same orientation but different frequency content. In our study we were interested in three frequency wedges corresponding to texture with horizontal orientation, negative orientation (-30 degrees) and positive orientation (+30 degrees). We have used three Gabor filters per frequency wedge for low-pass, band-pass and high-pass frequency content respectively. They were tuned in different regions inside the wedge of corresponding directional filter. Each Gabor filter defines one texture class. All Gabor filter outputs can be calculated in parallel. Each image pixel is assigned to the class of the Gabor filter which produces the maximal local output. Gabor filters are efficient tools in studying seismic image texture. Their only disadvantage is that they require large computational complexity (much larger than directional filters).

5. ENTROPY-BASED METHODS

Seismic image entropy can give important information with respect to he homogeneity if its texture. A homogeneous seismic texture region will provide a histogram having one strong peak. Consequently, this region will provide low entropy values, whereas chaotic seismic textures will have high entropy values. Thus, chaotic textures that are typical for certain geologic formations can be very easily segmented and identified by using the local entropy. This segmentation method needs some user-defined thresholds for the discrimination of the various seismic image regions. This method provides segmentation of a seismic image with respect to uniformity and regardless to orientation.

6. CONCLUSIONS

Most of the above-mentioned segmentation techniques produce spurious pixels within homogeneous regions in the segmented image. Such pixels can be removed by applying the *majority filter*: for each pixel (i,j), the majority filter calculates the region where the majority of the neighbor pixels belong to. Then the pixel (i,j) is assigned to the local 'majority' class.

Directional filters have been proven to be an efficient tools in seismic texture analysis and segmentation, especially when texture orientation is of importance. Gabor filters are also of importance when the frequency content is important together with the texture orientation. The major drawback of both these methods is their large computational complexity. Hough transform can give good results in seismic image segmentation with respect to orientation. It is also a relatively fast operation. Its only drawback is that it heavily relies on correct seismic horizon following. Markovian techniques give also interesting segmentation results. Their major disadvantage is their sensitivity with respect to the estimation of the transition probabilities. Finally, local entropy is a fast and efficient tool in segmenting chaotic seismic textures.

References

[1] I. Pitas, "Markovian Models for Image Labeling and Edge Detection", *Signal Processing 15,* pp.365-374, North Holland, 1988.

[2] A.Ikonomopoulos and M.Cunt, "High compression of image coding via directional filtering", *Signal Processing 8,* 1985, pp. 179-203.

[3] A.C.Bovik, "Analysis of Multichannel Narrow-Band Filters for Image Texture Segmentation", *IEEE Trans. on Signal Processing,* Vol. 39, No. 9, pp. 2025-2043, Sept. 1991.

[4] M.D.Levine, *Vision in Man and Machine,* McGraw-Hill, 1985.

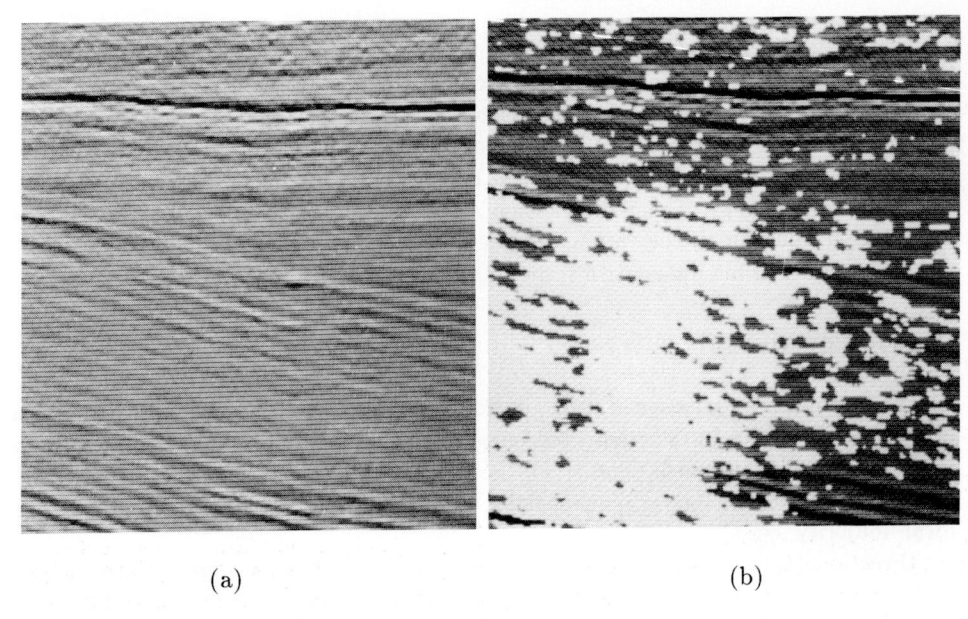

(a) (b)

Figure 1 (a) Seismic section; (b) The white region corresponds to negative texture slope.

EDGE AND REGION INTEGRATION FOR IMAGE SEGMENTATION

Giovanni Venturi

Department of Biophysical and Electronic Engineering -DIBE, University of Genoa, Via Opera Pia 11a I-16145 Genova, Italy

Abstract
   A segmentation method based on stochastic edge-region integration is presented. A first image transformation associates with each pixel, in addition to its gray value, the pixel probability to be an edge point and the pixel probability to be a region-boundary point. The transformed image constitutes a Gibbsian Markov Random Field. A combination of two procedures known as stochastic relaxation and simulated annealing, is used to obtain the final image segmentation according to the MAP criterion.

1. INTRODUCTION

   In this paper, a segmentation method based on statistical integration of edge detection and region growing is presented.
   The approach follows the works by Bajcsy et al [1], Anderson et al [2] and Pavlidis and Liow [3], who demonstrated that the combination of two complementary methodologies, like edge detection and region growing, produces better results than a single methodology.
   Edge detection looks for dissimilarities inside an image, and region growing looks for similarities. If both techniques operated perfectly, edges would be declared at the borders of all regions and, consequently, regions would be the areas surrounded by the edges. Unfortunately, both methods cause errors. According to Pavlidis and Liow, the application of a region growing algorithm may produce three types of errors: a) a boundary is not an edge and there are no nearby edges; b) a boundary corresponds to an edge but does not coincide with it; c) there exist edges with no boundaries near them. On the other hand, the results of an edge detection algorithm are too sensitive to noise and are affected by two types of errors: d) there are edge points due to noise; e) edge segments are not closed or not continuous.
   In our method, by considering the image characteristics and the Signal-to-Noise ratio (SNR) an optimal threshold can be selected twhich is congruent with edge detection and region growing algorithms. A Gibbsian Markov Random Field (GMRF) is built, in which each element is an image pixel with a probability to be an edge point (P_e) or a region-boundary point (P_b). A combination of two procedures known as stochastic relaxation and simulated annealing is utilized to get the final image segmentation, according to the Maximum A Posteriori probability (MAP) criterion[4].
   The GMRF approach allows one to classify a point according to its properties and to some typical configurations (cliques). So, it is possible to correct completely errors of the

type a) (it means high Pe, low Pb), and type c) (high Pb, low Pe). It is also possible to partially correct errors of types b), d) and e), with some limitations.

The method is a general purpose one. It exploits only numerical image information, without utilizing semantic knowledge about the contents of images, so it can be applied to a wide class of images. With respect to other integration methods presented in the literature, in which edges, roughly extracted, are utilized to control or correct of a region growing algorithm, we exploit and integrate, in a statistical way, both edge and region information, thus developing a single segmentation framework.

2. A COMMON FRAMEWORK FOR EDGE DETECTION AND REGION GROWING

According to the classical definition, the segmentation of an image I(x,y) results in the minimum complete, connected and disjoint set of regions r_i that satisfy an uniformity predicate $P(r_i)$. For the region-growing algorithm, the predicate states that contiguous pixels, belonging to a single region, differ from one another by a value smaller than a selected threshold.

$$P(r_i)_{r.g.} \equiv |I(x_1) - I(x_2)| < \tau \qquad \forall\, x_1, x_2 \in r_i,\ x_1, x_2 \text{ contiguous} \qquad (1)$$

In a classical edge detector, a point is considered to be an edge point if:

$$|F(G(x,y))| > \xi \qquad (2)$$

where $G = \nabla I$ is the image gradient, F is a function of the gradient and indicates the edge strength.

The image gradient between two adjacent pixels may be approximated as $\nabla I = (I(x_2) - I(x_1)) / (x_2 - x_1)$; as a consequence, according to (1) and (2), a pixel is an edge point if $|F(|I(x_2)-I(x_1)|)| > \xi$ and, by analogy to (1), we now state that

$$\xi \cong F(\tau) \qquad (3)$$

Equation (3) allows one to select a threshold that is congruent with both edge detection and region growing, and to process homogeneous information. Threshold selection depends on the image characteristics, in particular on the signal amplitude (SR) and the SNR. Under the hypothesis that the image contains regions of uniform gray level, corrupted by additive Gaussian noise, we may define the SNR as:
SNR=20Log(SR/σ_n). Therefore, the optimal threshold for region growing is τ=SR/2; consequently, the optimal threshold for edge detection is ξ=F(τ).

3. STATISTICAL DESCRIPTION

Equations (1) and (2) give a condition to establish whether a pixel is a region or an edge point in a hard way. Since we want to perform a statistical integration, we need to associate with each pixel a probability to be an edge point or a region point. Regarding edges, a point has a much higher probability to be a real edge point if the first member of (1) is greater than the threshold. So the probability to be an edge point may be expressed as:

$$P_e = \frac{1}{K}|F(G(x)) - \xi| \qquad (4)$$

where K is a normalizing constant to assure that the probabilities are in the range [0, 1].

To have a statistical description of the probability of a pixel to be a region pixel, we gradually increase and decrease the region growing threshold around the optimum threshold. In particular, we select the thresholds obtained by increasing and decreasing for 10 times the SNR in a range of 10% of the real value. In this way, for each image point the probability to be a boundary point depends on the threshold and the repetitions.

$$P_b = \frac{1}{Z}\sum \lambda_{\tau_i} \tau_i \qquad (5)$$

where Z is a normalizing constant, and $\lambda_{\tau i} = 1$, if the point is a region boundary point for $\tau = \tau i$, or $\lambda_{\tau i} = 0$ otherwise.

4. EDGE AND REGION INTEGRATION

On the basis of available information, we can classify a pixel as a boundary pixel or a region pixel. We have a set of observed variables Y=[P_e, P_b]; let $s_{k,l}$ be the state of a pixel with coordinates (k,l). $s_{k,l}$ can take only the values 0 and 1, corresponding to the previous two choices. Let S be the entire set of states. We want to perform the classification on the basis of the MAP estimate for the states: $p_{y|s}(Y|S) \cdot Pr[S]$. Since the problem satisfies the GMRF conditions, we may exploit the potentiality of the method described in [4] and find iteratively a maximum with the stochastic relaxation and simulated annealing procedures.

To do this, we have defined a set of cliques and an "ad hoc" energy function for the problem. The energy function U(S) is:

U(Y) = f(gray levels input image)+f(p_e, p_b)+f $_3$(cliques) \qquad (6)

5. EXPERIMENTAL RESULTS AND CONCLUSIONS

Synthetic images and aerial photographs were used to evaluate the system behaviour vs. noise and the system robustness. All images had a resolution of 256x256 pixels and up to 256 different gray levels.

Figure 1.a shows a LANDSAT image of the cultivated fields; the estimated SNR value for this image is equal to 3. Figure 1.b shows the visual results the edge-extraction module. Figure 1.c shows the results obtained by the region extraction module. In these figures, pixels with high probability of being boundary points are white, pixels with low probability are gray, and pixels with medium probability to be boundary points are black. Finally, Figure 1.d shows the final processing results.

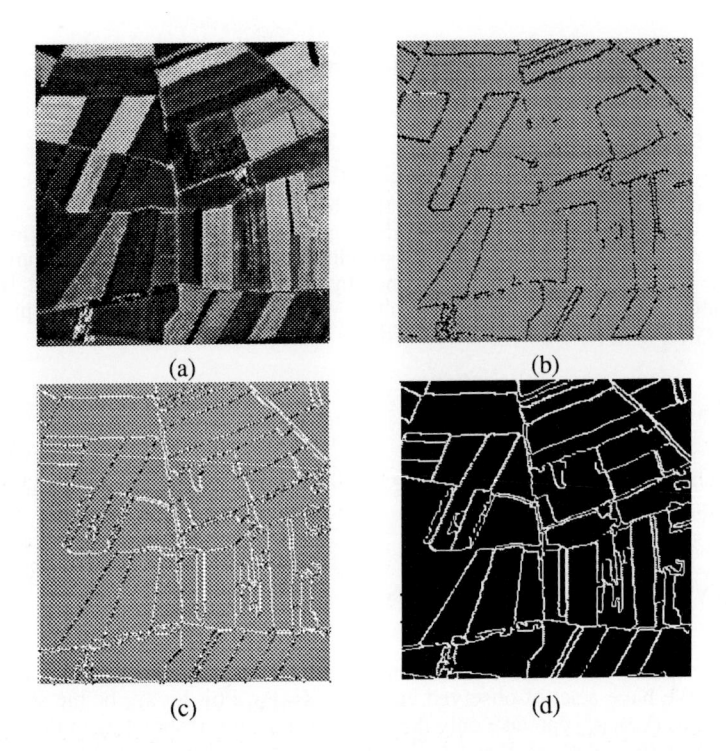

Figure 1. Segmentation results. 1.a Original LANDSAT image. 1.b Edge-extraction result. 1.c Region growing result. 1.d Integration result.

Results may be considered satisfactory. The statistical edge region integration make it possible to preserve more information than the single techniques. Some problems remain to be solved regarding open boundaries and loss of directional information. These shortcomings might be easily avoided by introducing into the system semantic information about the boundary behaviour.

ACKNOWLEDGEMENTS

This work was supported by the CNR Target Project on Biotechnology and Bioinstrumentation

REFERENCES

[1] R. Bajcsy, M. Mintz. A common framework for edge detection and region growing. Un. Pennsylvania GRASP, LAB 61, 1986.
[2] A. L. Anderson, R. Bajcsy and M. Mintz. A modular feedback system for image segmentation. Un. Pennsylvania GRASP, LAB 110, 1987.
[3] T. Pavlidis and Y. T. Liow. Integration of region growing and edge detection. IEEE Trans. Pat. An. Mach. Int. PAMI -12, 225-233,1990.
[4] D. Geman, S. Geman. Stochastic relaxation, Gibbs distributions, and Bayesan restoration of images. IEEE Trans. on PAMI, vol PAMI -6, 721-741, 1984.

Image Matching for Underwater 3D Vision[1]

José Santos-Victor and João Sentieiro

CAPS/DEEC, Instituto Superior Técnico, Av. Rovisco Pais, 1096 Lisboa Codex, Portugal
e-mail : D2760%beta.ist.pt@ptearn.fc.ul.pt

Abstract

This paper presents a matching strategy tailored for 3D vision applications in the underwater environment. Underwater images are usually low contrasted and present poor dynamic range. The illumination often changes from one image to the subsequent. Most matching algorithms do not perform well under these conditions. We propose a method that includes a local equalization procedure, leading to improved results on the processing images. Results are shown for comparison.

1. INTRODUCTION

The design of autonomous systems, able to perform a wide range of tasks in unknown environments, without any human intervention, has received an increasing attention during the last decade, and has become an important field of research in the present days. Autonomous (or with a certain degree of autonomy) mobile robots play already an important role in a large number of applications, like operation in hazardous environments such as nuclear plants, in space or underwater.

To accomplish their missions successfully, these systems must be able to sense the environment, continuously avoiding obstacles, and planning new mission targets. Hence, sensing plays a key role for the success of a given mission scenario.

In this paper, we will discuss the use of autonomous vehicles (AUVs - Autonomous underwater vehicles) in underwater applications, focusing in the use of computer vision as the prime sensory system. Vision is a powerful perceptual sensor and is likely to provide the amount of information necessary to accomplish a certain task.

As the ocean is becoming a resource of increasing economical importance, also the application of autonomous underwater robots, is deserving more and more attention. AUVs can be used to perform tasks such as pipeline laying, equipment repair and retrieval, undersea visual survey, recollection of geological samples, inspection, etc.

In this paper we address the problem of depth estimation for a particular underwater application. The underwater vehicle is equipped with a camera and acquires images during its motion. Each pair of successive images can be used to estimate a dense depth map. There is also the interesting possibility of recursively integrating multiple depth

---

[1]This work has been supported in the context of the MOBIUS project of the EEC MAST programme.

measurements over time, as new images are acquired. The authors have described a system with such characteristics in [1].

Due to the nature of the medium, underwater images are often diffuse and blurred, and may suffer illumination changes along an image sequence. All these issues seriously degrade the usual matching strategies. The use of equalization techniques are usually not suitable, as they perform global changes in the image and, therefore, do not assure that, locally, the brightness is constant from one image to the next. We have used a matching technique which tends to take these problems into account.

2. PROBLEM FORMULATION

The goal of the system being described is the reconstruction of the 3D structure of the seabed and any visualized object. This reconstruction is to be accomplished by using images acquired by a camera moving in an unknown static environment. The camera motion, which is related to the vehicle motion, is assumed to be known.

In order to recover 3D information, the main problem to be solved is that of matching, i.e. establishing the correspondence between image projections that correspond to the same 3D point or feature. Once this is done, by using the knowledge of the camera motion and a pinhole camera model, the 3D point can be reconstructed. We will discuss the matching problem while details on the 3D reconstruction can be found in [2].

2.1. Matching

The matching algorithm described in this paper falls in the class of the multiresolution correlation-based methods. The basic assumption is that corresponding regions have similar gray-level distributions.

To match two images we started by using a derivation of the SSD (Sum of Squared Differences [3]), that embodies a constraint due to the vehicle motion (that is, we assume that the vehicle motion is known, and consider the epipolar constraint)

$$ESSD(u,v,x,y) = \sum_{\alpha,\beta} \phi(\alpha,\beta)[I_{(t,\alpha,\beta)} - I_{(t+\tau,\alpha+u,\beta+v)}]^2 + \lambda_{ep} d_{ep}^2(x,y,u,v) \quad (1)$$

where $I_{(t,x,y)}$ stands for pixel (x,y) of the image acquired at time t, $\phi(\alpha,\beta)$ is a weighting mask, and α,β run across the correlation window. The term $d_{ep}^2(x,y,u,v)$ refers to the distance of a matching candidate pixel to the epipolar line, and is included in the matching cost functional as a penalizing term. The optimal disparity vector (u,v) is the one that minimizes equation (1).

Even though the use of the epipolar constraint significantly improves the matching accuracy, there are often slight changes in illumination which disturb the matching cost functional and, as a result, there are large image areas, where the ESSD does not exhibit a clear minimum, and matching cannot be performed.

The proposed method is based on the following idea. We basically partition the image in two sets of pixels. A pixel is part of set $\mathcal{A}$ if in the neighbourhood of the pixel, the image presents significant gray level changes. The pixel is included in region $\mathcal{B}$ if the gray level is approximately constant in a neighbourhood of the pixel.

In the situation $\mathcal{A}$, the pixel will be matched based on the changing component of the gray level, even though there might be some differences in the average gray level. This is accomplished by using the following cost functional which performs a sort of *local equalization* procedure:

$$ESSD_{(eq)}(u,v,x,y) = \sum_{\beta,\alpha} \phi(\alpha,\beta)[I_{(t,\alpha,\beta)} - \frac{I_t^{dc}}{I_{(t+\tau)}^{dc}} I_{(t+\tau,\alpha+u,\beta+v)}]^2 + \lambda_{ep} d_{ep}^2(x,y,u,v) \quad (2)$$

where I_t^{dc} is the average gray level in the matching window in image $I(t)$. In situation $\mathcal{B}$, where the image has a flat brightness pattern, the matching is done using the ESSD cost functional described by equation (1).

As a result, we are able to match a large number of areas in the image provided that there is some signal variation in the neighbourhood of a pixel, even though there might be some illumination changes. To determine whether a pixel is in situation $\mathcal{A}$ or $\mathcal{B}$, we compare the gray-level variance to a specified threshold.

2.2. Regularization

The disparity vector field is naturally noisy and sparse as the disparity computation may have failed in some image areas (for example if the brightness pattern is constant). In order to overcome these problems, we use a regularization procedure which, by introducing prior knowledge to constrain the disparity vector field to smooth solutions, allows the reformulation of the ill-posed matching procedure as a well posed variational principle. Standard Tikhonov regularization is used with a thin membrane stabilizing functional [4] and the Gauss-Seidel relaxation algorithm [5].

2.3. Coarse-to-fine control structure

The complete algorithm runs in a coarse-to-fine control strategy, based on a gaussian pyramid, obtained by low-pass gaussian filtering and subsampling. The matching is initiated in the coarsest level of resolution. At this level, estimates of the disparity vectors are computed, and the resulting vector field is regularized.

The algorithm then proceeds to the next level of (higher) resolution by performing matching in a small neighbourhood of the disparity predicted by the previous level. The same regularization procedure is then applied. The final estimates are obtained once the highest level of resolution is reached.

The coarse-to-fine control strategy leads to large computational gains, since the most costly computations are performed at the coarsest level of resolution. Moreover this strategy improves the speed of the spatial propagation of the regularization constraints. The results are often better than the single resolution algorithms, since the search-space and spatial support of the matching are extended [5].

3. RESULTS

To conclude, we will present results obtained with real underwater images acquired in an experimental tank, clearly showing the advantages of the described technique. Figure 1 shows an underwater stereo image pair of a test object. Notice the low contrast of the image. Although it is not evident visually, there are illumination changes from one image to the next. The same figure also shows a depth map obtained by the *ESSD* algorithm and the depth map obtained by the proposed algorithm. The improvement obtained is quite significant.

Figure 1: Left : An underwater stereo image pair. Right : On top, a depth map obtained by using the ESSD matching algorithm, while the bottom one results from the proposed method. Darker pixels represent closer regions, and unmatched areas are shown in white. The variance threshold is 0.01 and the matching window size is 9x9.

REFERENCES

[1] J. Santos-Victor and J. Sentieiro. "Generation of 3D Dense Depth Maps by Dynamic Vision". In *Proceedings of the British Machine Vision Conference - BMVC92*, pages 137-147, Springer-Verlag, September 1992.

[2] J. Santos-Victor and J. Sentieiro. "A 3D Vision System for Underwater Vehicles: an Extended Kalman-Bucy Filtering Approach". In José M. F. Moura and Isabel M. G. Lourtie, editors, *Acoustic Signal Processing for Ocean Exploration - NATO Advanced Study Institute*, Kluwer Academic Publishers, 1992. To be published.

[3] P. Anandan. "A Computational framework and an algorithm for the measurement of visual motion". *International Journal of Computer Vision*, vol.4, no. 2, pp. 283-310, May 1989.

[4] M. Bertero, T. Poggio and V. Torre. "Ill-posed problems in early vision". *Proceedings of the IEEE* , vol. 76, no. 8, pp. 869-889, 1988.

[5] D. Terzopoulos. "Regularization of inverse visual problems involving discontinuities". *IEEE Transactions on Pattern Analysis and Machine Intelligence*, vol. 8, no. 4, pp. 413-424, 1986.

Image Processing: Theory and Applications
G. Vernazza, A.N. Venetsanopoulos, C. Braccini (Editors)
© 1993 Elsevier Science Publishers B.V. All rights reserved.

Identification of Multichannel and Multidimensional Systems Using Cumulants: Application to Colour Images

Y. Zhang, D. Hatzinakos and A. N. Venetsanopoulos

Dept. of Electrical and Computer Engineering, University of Toronto, Canada, M5S 1A4

Abstract A higher-order cumulant method for identification of non-causal and non-Gaussian MA-AR models for multichannel and multidimensional systems using output data only is described. The MA-AR model is allowed to be non-minimum phase, asymmetric non-causal or non-separable. The problem is transformed to one of estimating the parameters of a pair of MA models. Then, an algorithm for identifying multichannel and multidimensional non-causal MA models is introduced. The feasibility of the method is examined with computer simulations.

1 INTRODUCTION

Multidimensional and multichannel statistical models and phase reconstruction are important in colour image processing for restoration, spectral estimation, coding, and classification. Higher-order cumulants or spectra (H.O.S.) techniques in one and two dimensional signals has been widely investigated over the last number of years. It is now well established that H.O.S. techniques offer certain advantages over the traditional autocorrelation based methods, such as higher degree of noise reduction, ability to preserve non-minimum phase information, ability to identify nonlinearities and detect quadratic phase coupling. Thus, the investigation of H.O.S. for multichannel and multidimensional signals is expected to offer advantages over the existing approaches. The price to pay is high computational complexity.

In this paper, we use a Kronecker-product based representation and extend the definitions of H.O.S. to stochastic vector field processes. An approach for the identification of the parameters of non-causal and non-Gaussian MA-AR models for multichannel and multidimensional systems using output data only is proposed by extending ideas presented in [1], [2]. Contrary to autocorrelation based multichannel and multidimensional modeling methods, the proposed MA-AR model is allowed to be non-minimum phase, asymmetric non-causal or non-separable. The MA-AR parameter estimation problem is transformed to an equivalent problem of estimating the parameters of a pair of MA models. Then, an algorithm for identifying multichannel and multidimensional non-causal MA models is introduced.The algorithm is applicable to both the stochastic and the deterministic problems. The feasibility of the algorithm is examined with computer simulations in the bispectrum domain (third-order cumulants) and with two- dimensional tri-channel signals as is the case of colour images.

2 NON-CAUSAL MA-AR MODEL

Consider the following MA-AR modeling for a multidimensional and multichannel signal

$$y(n) = - \sum_{\substack{k=-p_1 \\ k \neq 0}}^{p_2} a(k)y(n-k) + \sum_{k=-q_1}^{q_2} b(k)w(n-k) = \sum_{k=-\infty}^{\infty} h(k)w(n-k) \quad (1)$$

where, $w(n)$ is a k-th order white, zero-mean, non-Gaussian ($s \times 1$) vector field process in the stochastic case, and a delta function in the deterministic case. Hence, its cumulants are multidimensional Kronecker delta function, i.e., $C_{kw}(t; \tau_1, \ldots, \tau_{k-1}) = \Gamma_{kw}(t)\delta(\tau_1)\ldots\delta(\tau_{k-1})$. where Γ_{kw} is an s^k-element vector. $y(n)$ is a ($s \times 1$) vector field output process and $a(k)$, $b(k)$ are coefficient matrices ($s \times s$) of AR and MA model parts respectively. The $h(n)$ is an impulse response matrix. We also assume that the $h(n)$ is absolutely summable so that the output cumulants are well defined. Then, the k-th order cumulants of the zero-mean linear space-invariant stationary vector field process $y(n)$ are given by [1], [2]

$$C_{ky}(\tau_1, \ldots, \tau_{k-1}) = \sum_{i=-\infty}^{\infty} [\bigotimes_{j=0}^{k-1} h(i+\tau_j)]\Gamma_{kw} \quad (2)$$

where $\otimes$ denote the Kronecker product operator. [3] Note that in the third-order cumulant domain $C_{3y}(\tau_1, \tau_2) = E\{y(m) \otimes y(m+\tau_1) \otimes y(m+\tau_2)\}$.

3 THE TWO-MA METHOD

Let $Z[]$ denote Z-transform of a random vector. Then, $A(z) = Z[a(i)], B(z) = Z[b(i)]$ and $H(z) = Z[h(i)] = [A(z)]^{-1}B(z)$. By taking the Z- transform of (2)

$$C_{ky}(z_1, \ldots, z_{k-1}) = [\bigotimes_{i=0}^{k-1} H(z_i)]C_{kw}(z_1, \ldots, z_{k-1}) = [\bigotimes_{i=0}^{k-1} H(z_i)]\Gamma_{kw} \quad (3)$$

By applying properties of the Kronecker product we obtain

$$[\bigotimes_{i=0}^{k-1} A(z_i)]C_{ky}(z_1, \ldots, z_{k-1}) = [\bigotimes_{i=0}^{k-1} B(z_i)]\Gamma_{kw} \quad (4)$$

Let us define $\mathcal{A}$ and $\mathcal{B}$ as follows ($i_0 = 0$)

$$\mathcal{A}(i_1, \ldots, i_{k-1}) = \sum_{i=-p_1}^{p_2} [\bigotimes_{j=0}^{k-1} a(i+i_j)], \quad \mathcal{B}(i_1, \ldots, i_{k-1}) = \sum_{i=-q_1}^{q_2} [\bigotimes_{j=0}^{k-1} b(i+i_j)]\Gamma_{kw} \quad (5)$$

By properly substituting $\mathcal{A}$ and $\mathcal{B}$ into (5) and by taking inverse Z-transform we get

$$\sum_{i_1=-p}^{p} \ldots \sum_{i_{k-1}=-p}^{p} \mathcal{A}(i_1, \ldots, i_{k-1})C_{ky}(n_1-i_1, \ldots, n_{k-1}-i_{k-1}) = \mathcal{B}(n_1, \ldots, n_{k-1}) \quad (6)$$

where $\mathbf{p} = \mathbf{p_1} + \mathbf{p_2}, \mathbf{q} = \mathbf{q_1} + \mathbf{q_2}$. Note that $\mathcal{B} = 0$ for $\mathbf{n_j}$ outside the interval $[-\mathbf{q}, \mathbf{q}]$. Thus, by properly choosing the $\mathbf{n_j}$'s in (6) we can retain only the left hand side and estimate the $\mathcal{A}$ s as the solution of an over-determined system of linear equations. Then, the $\mathcal{B}$ s can be estimated via (6). In other words by defining $\mathcal{A}'(\mathbf{i_1}, \ldots, \mathbf{i_{k-1}}) = \mathcal{A}\Gamma_{\mathbf{kc}}$ where, $\Gamma_{\mathbf{kc}}$ is any constant matrix ($s^k \times 1$), we obtain the following two MA models

$$\mathcal{A}'(\mathbf{i_1},\ldots,\mathbf{i_{k-1}}) = \sum_{i=-\mathbf{p_1}}^{\mathbf{p_2}} [\bigotimes_{j=0}^{k-1} \mathbf{a}(\mathbf{i}+\mathbf{i_j})]\Gamma_{\mathbf{kc}}, \quad \mathcal{B}(\mathbf{i_1},\ldots,\mathbf{i_{k-1}}) = \sum_{i=-\mathbf{q_1}}^{\mathbf{q_2}} [\bigotimes_{j=0}^{k-1} \mathbf{b}(\mathbf{i}+\mathbf{i_j})]\Gamma_{\mathbf{kw}}$$
(7)

The double MA method is applicable to non-causal as well as causal MA-AR models. The key point of this method is that it reduces the MA-AR parameter estimation problem to two MA parameter estimation problems.

4 C(q,τ) ALGORITHM FOR MA PARAMETER ESTIMATION

Consider the MA non-causal model ($\mathbf{q_1} \neq 0$): $\quad \mathbf{y}(\mathbf{n}) = \sum_{k=-\mathbf{q_1}}^{\mathbf{q_2}} \mathbf{b}(\mathbf{n}-\mathbf{k})\mathbf{w}(\mathbf{k})$.
Then, the output k-th order cumulants are obtained from (2), i.e.,

$$\mathbf{C_{ky}}(\tau_1,\ldots,\tau_{k-1}) = \sum_{i=-\mathbf{q_1}}^{\mathbf{q_2}} [\bigotimes_{j=0}^{k-1} \mathbf{b}(\mathbf{i}+\tau_j)]\Gamma_{\mathbf{kw}} \quad (8)$$

By defining (**unvec** denotes unity vector):

$$\overline{C}_{\mathbf{ky}}(\tau) = \mathbf{unvec}_{s \times s^{k-1}}(\mathbf{C_{ky}}(\mathbf{0},\ldots,\mathbf{0},\mathbf{q},\tau)) \quad (9)$$

and by making the assumptions that $\mathbf{b}(\mathbf{0}) = \mathbf{I}$, $\mathbf{b}(-\mathbf{q_1})$ and $\mathbf{b}(\mathbf{q_2})$ have full rank, and $\mathbf{unvec}_{s \times s^{k-1}}\Gamma_{\mathbf{kw}}$ has rank s, then $\overline{C}_{ky}(\tau)$ has full rank s. The, it can be shown that

$$\mathbf{b}(\tau - \mathbf{q_1}) = \overline{C}_{\mathbf{ky}}(\tau)\overline{C}_{\mathbf{ky}}^{\mathbf{T}}(\mathbf{q_1})[\overline{C}_{\mathbf{ky}}(\mathbf{q_1})\overline{C}_{\mathbf{ky}}^{\mathbf{T}}(\mathbf{q_1})]^{-1} \quad (10)$$

By applying this procedure to the two MA models $\mathcal{A}'$ and $\mathcal{B}$ of section 3 separately, we can estimate the known coefficients $\mathbf{a}(\mathbf{k})$ and $\mathbf{b}(\mathbf{k})$ of the MA-AR model.

5 SIMULATION RESULTS

Simulations were performed in the third-order cumulant domain domain to test the MA algorithm of section 4. Dimensions of test images were 76×76×3. The input signal $\mathbf{w}(\mathbf{n})$ generated was i.i.d., exponentially distributed, two-dimensional and tri-channel vector field with zero mean and unit variance. Then, two tri-channel two-dimensional textural type signals $\mathbf{y}(\mathbf{n})$ were created by passing $\mathbf{w}(\mathbf{n})$ through the two tri-channel systems

Sys. 1: $\mathbf{B_{0,0}} = \begin{pmatrix} 1 & 0 & 0 \\ 0 & 1 & 0 \\ 0 & 0 & 1 \end{pmatrix} \quad \mathbf{B_{0,1}} = \begin{pmatrix} 1.5 & 0 & 0 \\ 0 & 0.95 & 0 \\ 0 & 0 & -1.2 \end{pmatrix} \quad \mathbf{B_{1,1}} = \begin{pmatrix} 0.9 & 0 & 0 \\ 0 & -0.8 & 0 \\ 0 & 0 & -0.75 \end{pmatrix}$

Sys. 2: $\mathbf{B}_{0,0} = \begin{pmatrix} 1 & 0 & 0 \\ 0 & 1 & 0 \\ 0 & 0 & 1 \end{pmatrix}$ $\mathbf{B}_{0,1} = \begin{pmatrix} 1.5 & 0 & 0 \\ -0.89 & 0.95 & 0 \\ 1.1 & 0 & -1.2 \end{pmatrix}$ $\mathbf{B}_{1,1} = \begin{pmatrix} 0.9 & 0 & 1.3 \\ 0 & -0.8 & 0 \\ 0 & 1.15 & -0.75 \end{pmatrix}$

Thus, system 2 assumes correlation between channels as well. Third-order cumulant estimates of $\mathbf{y}(n)$ were obtained by replacing the expectation operation with sample averaging. The image was segmented into several overlapping (62.50% overlapping in our simulation) segments, sample estimates of cumulants were obtained for each segment (size of moving window is 8×8), and averaged to obtain the final sample cumulant estimates..

The average of the estimated coefficients from ten independent realizations of each experiment and the corresponding mean square error (MSE) are given in Tables 1 and 2. The estimated coefficients are biased and some of them have large errors due to the following reasons: i) The small size of the generated image which in turn affects the size of the moving window in the estimation of third order cumulants. It is well known that cumulants obtained from small data records are severely biased and exhibit large variance. ii) The deviation of the generated input signal $\mathbf{w}(n)$ from being truly i.i.d. for the small number of data samples generated. Both conditions are even harder to satisfy with multidimensional and multichannel type signals. Obviously better coefficients estimates will be obtained with more accurate cumulant estimates. This can be accomplished by either using more frames of color image or increase the size of moving window and the percentage of segment overlapping in different lags of estimated cumulants. An alternative approach for improving the cumulant estimates that is based on bootstrapping techniques is under investigation.

Table 1. Ideal and Estimated coefficients of matrices $\mathbf{B}_{0,1}$ and $\mathbf{B}_{1,1}$ for system 1

	$B(1,1)$	$B(1,2)$	$B(1,3)$	$B(2,1)$	$B(2,2)$	$B(2,3)$	$B(3,1)$	$B(3,2)$	$B(3,3)$	MSE
$B_{0,1}$	1.5000	0.0000	0.0000	0.0000	0.9500	0.0000	0.0000	0.0000	-1.2000	
$\hat{B}_{0,1}$	1.1519	0.1071	-0.6337	-0.0415	0.9663	0.0785	0.0415	0.1094	-1.0800	0.063
$B_{1,1}$	0.9000	0.0000	0.0000	0.0000	-0.8000	0.0000	0.0000	0.0000	-0.7500	
$\hat{B}_{1,1}$	0.6197	0.2421	-0.4838	-0.0335	-0.4056	0.0490	0.0616	0.1619	-0.9787	0.068

Table 2. Ideal and Estimated coefficients of matrices $\mathbf{B}_{0,1}$ and $\mathbf{B}_{1,1}$ for system 2

	$B(1,1)$	$B(1,2)$	$B(1,3)$	$B(2,1)$	$B(2,2)$	$B(2,3)$	$B(3,1)$	$B(3,2)$	$B(3,3)$	MSE
$B_{0,1}$	1.5000	0.0000	0.0000	-0.8900	0.9500	0.0000	1.1000	0.0000	-1.2000	
$\hat{B}_{0,1}$	1.1732	0.0684	0.4078	-0.3744	0.8846	0.2197	0.8517	0.1938	-0.8901	0.088
$B_{1,1}$	0.9000	0.0000	1.3000	0.0000	-0.8000	0.0000	0.0000	1.1500	-0.7500	
$\hat{B}_{1,1}$	0.4383	0.0385	0.8724	-0.1700	-0.5000	-0.1094	0.4583	0.8566	-0.6552	0.093

6 REFERENCES

1. G. B. Giannakis et al., IEEE Trans. Autom. Control, No. 34 (1989) 783
2. J.M. Mendel, Proc. IEEE, No. 79 (1991) 278
3. J. W. Brewer, IEEE Trans. Circuits and Systems, No. 25 (1978) 772

Application of multichannel two-dimensional AR modeling to color image processing

I. Pitas and P. Kilindris

Department of Electrical Engineering, University of Thessaloniki, Greece

Abstract
Multichannel AR models have been developed for color image analysis. It is proven by simulations that such models fit better to color images than their single channel counterparts because they take into account inter-channel correlation. The reported method has been successfully used for color texture segmentation as well as for color texture synthesis.

1. MULTICHANNEL AR MODELING

Color image processing is a hot research topic due to the variety of its applications (e.g. HDTV, graphics). The recent trend is to consider color images as being multichannel (vector) images, in order to take into advantage the strong correlation that exists among the color image channels (R,G,B) [1,2]. The extension of AR two-dimensional models to the multichannel case is of great importance in color image processing, because they can express inter-channel correlations. Thus, they are expected to be better models of color images than their independent single-channel counterparts. Such models have already been proposed for use in multichannel two-dimensional power spectrum estimation and in color image restoration [2]. The goal of this paper is to present applications of multichannel AR modeling in color image analysis/synthesis and segmentation.

Color image pixels $\mathbf{x}[m, n]$ are vectors of length 3. The three-channel two-dimensional prediction model is of the form:

$$\hat{\mathbf{x}}[i,j] = \sum_m \sum_n \mathbf{A}[m,n]\mathbf{x}[i-m, j-n] + \mathbf{w}[i,j] \tag{1}$$

in which the model coefficients $\mathbf{A}[m, n]$ are 3×3 arrays. The double summation is over any causal window (e.g. quarter plane or non-symmetric half plane window). In the following, we use quarter plane windows for notation simplicity. The two-dimensional *normal* equations are:

$$\sum_m \sum_n \mathbf{A}[m,n]\mathbf{R}_{xx}[i-m, j-n] = \begin{cases} \mathbf{P}_w, & \text{for } [m,n] = [0,0] \\ 0, & \text{otherwise} \end{cases} \tag{2}$$

Table 1: Prediction MSE for multichannel and single channel AR modeling.

Image	Window	Multichannel MSE	Single-channel MSE
LENNA	3×3	442	468
LENNA	5×5	401	442
BABOON	3×3	650	827
CAR	3×3	654	686

The biased 2-D three-channel autocorrelation estimated at lag $[k, l]$ is given by:

$$\hat{\mathbf{R}}_{xx}[k,l] = \begin{cases} \frac{1}{MN} \sum_{m=0}^{M-1-k} \sum_{n=0}^{N-1-l} \mathbf{x}[m+k, n+l]\mathbf{x}[m,n], & \text{for } k \geq 0, l \geq 0 \\ \frac{1}{MN} \sum_{m=0}^{M-1-k} \sum_{n=-l}^{N-1} \mathbf{x}[m+k, n+l]\mathbf{x}[m,n], & \text{for } k \geq 0, l < 0 \\ \hat{\mathbf{R}}^T[-k, -l], & \text{for } k < 0, \text{ every } l \end{cases} \quad (3)$$

where $\hat{\mathbf{R}}_{xx}[k, l]$ is a 3×3 array. Unbiased multichannel autocorrelation estimators have been used as well. In both cases, the signal mean has been removed before autocorrelation matrix estimation. Several simulation experiments have been performed on the well known color images LENNA, BABOON and CAR. Quarter plane support regions of size 3×3, 4×6, 5×5, 6×6 have been employed. The prediction error MSE=$\|\mathbf{x} - \hat{\mathbf{x}}\|$ is shown in Table 1. The multichannel prediction MSE is 5%-30% lower than single-channel prediction MSE. As expected, the biggest reduction has been observed for images having highly textured areas, e.g. in BABOON. The first conclusion from these experiments is that multichannel AR models describe better the color image than their single-channel counterparts. In the worst case, they have similar performance to the single-channel AR models. The price that is paid for the better performance is the computational complexity for the calculation of the autocorrelation matrices and for the the solution of the normal equations. Fast methods for their solution that are extensions of the ones of the single-channel case have been devised.

2. COLOR IMAGE SEGMENTATION AND SYNTHESIS

Two applications of multichannel color image modeling have been investigated. The first one is color image segmentation and identification of the boundaries of textured image regions. It is well known that classical segmentation techniques (e.g. the ones that use region mean and dispersion) fail in such cases. Let us suppose that we know seeds of such an image region. Their pixels can be used to estimate the autocorrelation matrices (3). The normal equations can be solved and the multichannel AR model can be applied to the entire image. The prediction error $\|\mathbf{x}[i,j] - \hat{\mathbf{x}}[i,j]\|$ at each image pixel can be used for region segmentation by thresholding. The image regions having similar AR models to the seed region have low prediction error. Regions having different texture produce high prediction error and can be distinguished easily by thresholding. Multiple texture image regions can be segmented by using more than one seed regions and AR models. An application of color texture segmentation based on this method is shown in Figure 1. Figure 1a consists of two image regions. They cannot be distinguished easily by the human eye. The prediction error image 1b is thresholded. Region segmentation is very good, as can be seen in Figure 1c.

The second application is color texture synthesis. This application is particularly important in computer graphics for texture mapping. The method operates as follows. A sample texture region is modeled, by using (2-3). The model coefficient matrices are used in (1) to construct a synthetic image. An important problem is to drive the multichannel AR model by the white Gaussian noise process $\mathbf{w}[i,j]$ having a known correlation matrix $\mathbf{P}_w$. This matrix has been derived during the solution of the normal equations. This problem is solved in the following way. Let us suppose that we have a unit Gaussian random number generator. It can be used to construct a three-channel white noise process $\mathbf{v}$ having unit autocorrelation matrix $\mathbf{I}$. This process is passed through a linear transformation matrix $\mathbf{T}$ in order to transform it to $\mathbf{w}$. This transformation matrix must satisfy the following nonlinear equation:

$$\mathbf{P}_w = \mathbf{T}\mathbf{T}^T \qquad (4)$$

This equation can be solved by any method (e.g. Newton Raphson) in order to obtain the transformation matrix elements. When this matrix $\mathbf{T}$ is calculated, the unit white Gaussian noise is passed through it and, subsequently, through the multichannel AR model in order to produce the desired color texture. The results obtained by this method are very good. Figure 2a show the image BABOON. Its hair texture has been used to produce the synthetic texture shown in Figure 2b. The color of the synthetic texture closely resembles the color of the baboon hair. Unfortunately, BW printing cannot show the resemblance.

In conclusion, we can state that we have developed an interesting multichannel AR modeling method. This model is superior to single channel independent color image modeling. This model has been successfully used in color texture image segmentation and in color texture synthesis.

References

[1] B.R.Hunt, "Karhunen-Loeve multispectral image restoration, part I: theory", *IEEE Transactions on Acoustics, Speech and Signal Processing*, vol. ASSP-32, No.3, pp.592-599, June 1984.

[2] G. Angelopoulos, I. Pitas, "Multichannel Wiener filters in color image restoration based on AR color image modeling", *Proc. of the IEEE International Conference on Acoustics, Speech and Signal Processing*, Toronto, 1991.

Figure 1: (a) Color image consisting of two different textured regions. (b) Multi-channel prediction error image (c) Result of thresholding image 1b.

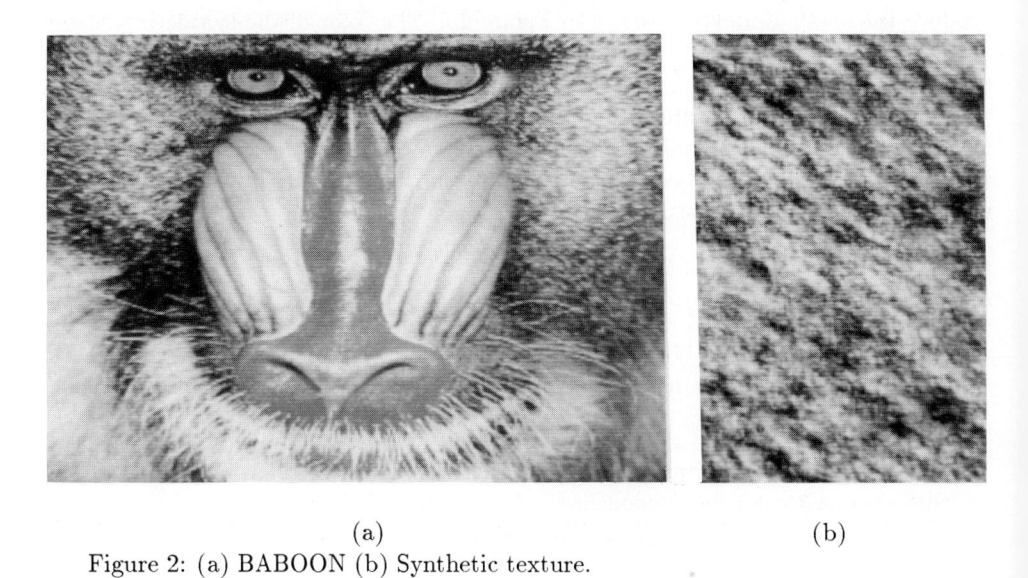

Figure 2: (a) BABOON (b) Synthetic texture.

Image Processing: Theory and Applications
G. Vernazza, A.N. Venetsanopoulos, C. Braccini (Editors)
© 1993 Elsevier Science Publishers B.V. All rights reserved.

Separable optimal filters for edge detection

M. Petrou and S. Yusof

Dept. of Electronic and Electrical Engineering, University of Surrey, Guildford GU2 5XH, United Kingdom.

Abstract

In this paper we examine the use of 1D optimal convolution filters for the detection of ramp edges. The idea is to replace one 2D convolution by two 1D ones performed in sequence. We investigate the way the parameters of the 1D filters have to be chosen so that the performance of the operation is not compromised. We present our theoretical results and apply them to some real images to show that the quality of the performance remains the same while the efficiency of the operation improves significantly.

1. INTRODUCTION

A lot of researchers have put much effort in the derivation of optimal convolution filters for the detection of edges (eg [1],[2],[3]). However, almost none of the "optimal filters" are actually used in practice. The majority of people who claim to use "the Canny edge detector", they actually use Gaussian filters. The most important reason for using Gaussian filters is their separability. While the optimal 2D convolution filters as developed by the workers mentioned above are not separable, the Gaussian filters are, thus enabling the user to replace a 2D cumbersome convolution by two 1D cascaded convolutions which are much faster.

The purpose of this paper is to extent the theory of optimal convolution filters and propose a set of separable 2D filters appropriate for the detection of ramp edges in an image.

2. THEORETICAL BACKGROUND AND PREVIOUS WORK

The filters derived using Canny's approach combine the effects of smoothing and differentiation in one go and for this reason they are antisymmetric. Spacek [3] suggested that the two proceedures can be separated by integrating the filter to derive the optimal smoothing filter which best preserves the features we want to detect. The differentiation for the location of the feature can be performed afterwards and need not be anything more sofisticated than a local differencing. The optimal smoothing filter being the integral of an antisymmetric function is symmetric and thus it can easily be extented to two dimensions by replacing the linear variable with the polar radius.

Petrou and Kittler assumed that the edges they wish to detect can be well modelled by the function:

$$c(x) = \begin{cases} 1 - e^{-sx}/2 & \text{for } x \geq 0 \\ e^{sx}/2 & \text{for } x \leq 0 \end{cases} \qquad (1)$$

where s is a parameter intrinsic to the imaging device, assumed responsible for the conversion of all step edges in the scene into ramps of this particular slope. Once fixed, it can be kept the same for all edges to be detected in all images captured by the same device. The optimal smoothing filter for this type of edges was derived to be:

$$h(x;s) = e^{Ax}[L_1\sin(Ax) + L_2\cos(Ax)] + e^{-Ax}[L_3\sin(Ax) + L_4\cos(Ax)] + L_5 x + L_6 e^{sx} + L_7 \tag{2}$$

where the dependance on the value of the parameter s has been explicitly stated in order to stress the point that for different values of s different filters are optimal. The parameters A and $L_1 - L_7$ were computed and tabulated for various values of s and filter size. It was found that no optimal filter size existed, but the size should be chosen by resolution considerations.

It is clear from the above expression that if the variable x is replaced by the polar radius r in order to extent the filter to 2D, the filter will not be separable.

3. THE SEPARABILITY REQUIREMENT

The basic idea of this work is that there is no need for the 2D filter either to be produced by circularizing $h(x)$ or to consist of two identical functions each being a function of one directional variable as is the case for the Gaussian filters. In what follows, we shall assume that we seak to define filters of the form: $\tilde{h}(x,y) = h(x;s_1)h(y;s_2)$ which can be used to smooth the image in a cascaded fashion before differentiating it to identify edges, and which are chosen optimally according to Canny's criteria so that the features to be detected are best preserved.

We start by considering the effect of using the filter given by equation 2 to convolve in the x direction only. Clearly, the edges presented to it at an angle $\phi = 90^0$ will be best preserved since that is how the filter was constructed. There is no reason, however, to expect that all the edges to be detected in the image will be aligned with the y axis. In fact, for an image with a lot of detail the edges will present themselves at various angles to the filter. As the filter smooths out the image it also affects the edges, increasing their slope not only in the direction of convolution but in the orthogonal direction as well. The profile of a random edge, therefore, will be affected in the y direction after convolving the image in the x direction. Thus the optimal filter for the convolution in the y direction should not be an identical function $h(y)$ but one with similar functional form but with different parameter values to reflect the change in the value of the slope s of the features that have to be optimally preserved.

Let us assume first that the edge we want to detect is orthogonal to the direction of convolution and it is best modelled by equation 1 with $s = 1$. The first question we have to ask is: "What is the parameter s_ϕ which would best model the edge along the x axis if it were at an angle ϕ with respect to the direction of convolution x?" It is very easy to show that $s_\phi = s\sin\phi$. The next question is: "Given that the edge is at an angle with respect to the direction of convolution, how does its slope change along the direction of convolution and perpendicularly to it after the convolution?" We express the answer to the last question by modelling the output edge in the two orthogonal directions and defining the best value of the parameter s. We use s_x and s_y to represent the best values of the s parameter along the respective orientations. The results for various values of the orientation angle ϕ are given in the table. The next question we have to answer, concerns the relationship between s_ϕ and s_y: "Is there a simple analytic relationship between the above mentioned parameters?" The answer is yes. If we plot s_y versus s_ϕ as seen on figure 1, it looks as if there must be such a relationship. The actual shape suggests a quadric. Indeed, after trial and error, it was found that the best fitting to the curve is given by:

$$s_\phi^n + s_y^n = s, \quad \text{with} \quad n = 1.6 \tag{3}$$

$\phi(^0)$	s_ϕ	s_x	s_y
10	0.1736	0.1673	0.9488
20	0.3420	0.3177	0.8729
30	0.5000	0.4485	0.7728
40	0.6428	0.5583	0.6654
45	0.7071	0.6026	0.6040
50	0.7660	0.6470	0.5429
60	0.8660	0.7149	0.4128
70	0.9397	0.7627	0.2776
80	0.9848	0.7911	0.1736
90	1.0000	0.8004	0.0000

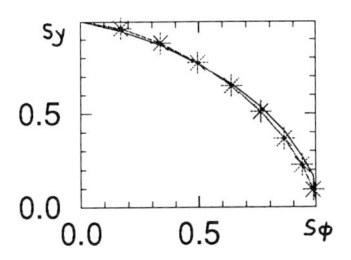

Figure 1: s_y versus s_ϕ. Line with stars: The fitted curve using the quadric.

The fitted curve is shown on figure 1 superimposed to the real curve.

The final question we have to consider is more difficult: Suppose we decided that the value of the parameter s that best models the edges of our image is $s = s_1$ and that we convolved the image in the x direction by $h(x; s_1)$. What is the value of s we should use for the filter with which we shall convolve the output image in the y direction so that the minimum damage will be done to the edges? Given that the edges will be detected by differencing along the x and y directions and examining for local maxima along the same directions, it is obvious that the limiting edges we expect to detect by differencing along the x direction are those at 45^0. (Edges at smaller angles than this will be detected as local maxima along the y axis.) The slope of these edges has been reduced most because they suffered by two effects: because of their orientation they will appear to the differentiator as shallow, and because they were convolved with a filter which had the "wrong" s for them (since their effective s along the x axis was 0.707 as seen from the table). It is clear therefore, that these are the edges we must try to damage least when we convolve in the y direction. We propose, therefore, to chose the filter parameter for the $h(y; s_2)$ function to be such that the function is optimally chosen for the preservation of the edges at orientations of 45^0. From equation 3 it is obvious that the parameter value s for this filter should be $s_2 = \left[s_1 - (s_1 \sin \phi)^{1.6} \right]^{1/1.6}$.

4. RESULTS AND CONCLUSIONS

Figures 2a and 2b show a real and a simulated image and figures 2c and 2d the results of the edge detector which uses two 1D convolutions for smoothing. Figures 2e and 2f show the results obtained by the same edge detection algorithm where the two 1D cascaded convolutions have been replaced by one 2D convolution. It is difficult to differentiate between the two results although figures 2c and 2d were produced at a fraction of the time needed for figures 2e and 2f.

We conclude that separability should no longer be an obstacle in taking advantage of the full benefit of the use of the optimal filters for the detection of edges, since good results can be obtained by using separable filters provided the parameters of the filters are properly chosen.

5. REFERENCES

1 J. Canny, A computational approach to edge detection, PAMI-8,679, 1986.
2 M. Petrou and J. Kittler, Optimal edge detectors for ramp edges, PAMI-13,483, 1991.
3 L. A. Spacek, Edge detection and motion detection, Image Vision Comput., 4,43, 1986.

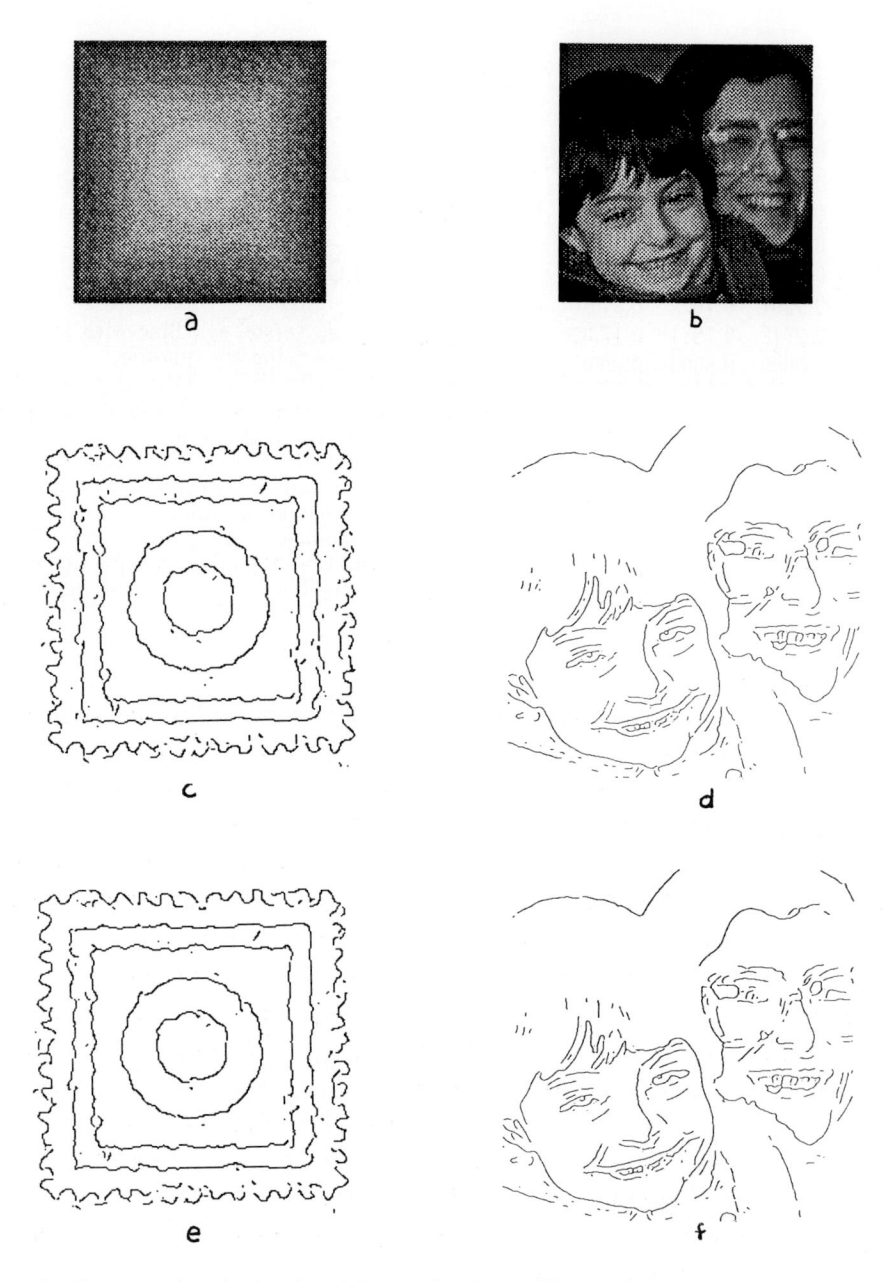

Figure 2: Top panels: A simulated image (with 100% added noise) and a real image. Central panels: The results of the edge detector that uses the separable filters. Bottom panels: Results of the edge detector that uses the nonseparable filters.

On the design of FIR filters for scanning rate conversion

P.Carrai, G.M. Cortelazzo, G.A. Mian

Dipartimento di Elettronica e Informatica, Università di Padova

via Gradenigo 6/A, 35131 Padova, Italy

Abstract

This work investigates the frequency domain characteristics of the 2-D minimax optimal linear phase FIR filters designed in the frequency-domain to meet diamond-shaped specs. A number of results and notions typical of one dimensional FIR filters minimax optimal are extended to this class of filters.

1. Introduction

The 2-D diamond-shape filters are defined on orthogonal lattices. The filter coefficients are assumed to form set $h(n_1, n_2)$, $n_1 = 0, \pm 1, ..., \pm M_1$; $n_2 = 0, \pm 1, ..., \pm M_2$. Phase linearity, quadrantal, octagonal or other symmetries, accordingly reduce the number of independent coefficients.

The specs of Fig. 1, can be considered the counterpart of the one-dimensional half-band specs (the internal region, denoted by A, is the pass-band, and the external region, denoted by B the stop-band; the undashed region is the don't care band, hosting the filter transition region). The scales are in terms of normalized (horizontal and vertical) frequencies. The geometry of the specs allows one to characterize the transition region by segment ΔF, obtained from the intersection of the transition region with a 45^0-slope line. Note that a 45^0-slope line is orthogonal to the pass-band region's border in the first quadrant. The length of ΔF will be referred to as normalized transition bandwidth. Other specs obtained from the specs of Fig. 1, by constraining the horizontal pass-band to half the sampling frequency and by varying the vertical pass-band and the stop-band region, were also considered. Such specs are perceptually very effective when the horizontal frequency is the temporal frequency. The normalized transition bandwith of such specs can be defined as the length of segment ΔF obtained from the intersection of the transition region with a line orthogonal to the pass-band border.

The filters were designed in minimax norm, by linear programming, a well-established technique for these application [1-3].

2. Order, bandwidth and error relationships

The 2-D half-band filters, i.e., the filters designed according to the specs of Fig. 1, exhibit a minimax error versus order characteristic, at various normalized transition region values ΔF, of the type of Fig. 2. The minimax error $\delta = \delta_p = \delta_s$ is the same in both bands. The filter coefficients are assumed to have octagonal symmetry. The filter order is expressed in terms of $N_1 = 2M_1 + 1$ and $N_2 = 2M_2 + 1$, with $N_1 = N_2$.

The data of Fig. 2 plot the value of $(N-1)\Delta F$ versus $20\log_{10}\delta$ (when $N = N_1 = N_2$). Fig. 2 shows the existence of a linear relationship between $(N-1)\Delta F$ and the minimax error (in dB). Such a relationship is best fitted by

$$N - 1 \simeq \frac{-20\log_{10}\delta - 0.53}{20.5\Delta F} \qquad (1)$$

Expression (1) is remarkably reminiscent of the one-dimensional order versus bandwidth and error relationship due to Kaiser [4].

The order versus bandwidth and minimax error characteristics, corresponding to the considered variation of the specs of Fig. 1 are best fitted by relationship

$$N - 1 \simeq \frac{-20\log_{10}\delta - 2.56}{18.3\Delta F} \qquad (2)$$

The relationships so far presented, refer to the case of identical pass-band and stop-band weights (and consequently errors). However, it can be found that if the pass-band and stop-band weights are different the relationships between order, bandwidth and errors are still of type (1)-(2), provided that quantity $20\log_{10}\delta$ of (1)-(2) is substituted with $20\log_{10}\sqrt{\delta_p\delta_s}$. In this case, similarly to the one-dimensional case, the quality of the linear fit is inferior to that of case $\delta = \delta_p = \delta_s$.

In video-applications it may be important to have transfer functions with a flat behaviour around the origin. Furthermore, in the most typical cases of video decimation and interpolation, the spectral repetitions to be eliminated are centered around normalized frequencies ($\pm 0.5, \pm 0.5$). In order to cope with both the above requirements one could use two approaches. The first consists in constraining the transfer function to be exactly one at the origin, and exatly zero at points ($\pm 0.5, \pm 0.5$). The second consists in weighting the error, near the origin and near points ($\pm 0.5, \pm 0.5$), heavier than the rest of the approximation set. Experimental results show that the second approach, based on the use of suitable parabolic weights in pass-band and stop-band, is superior to the first. This indicates that the reduction of degrees of freedom implied by the first approach represents an unnecessary sacrifice.

The similarity between the just seen results and those relative to the one dimensional case is worth noting.

3. Implications for multistage sampling structure conversion

In the one-dimensional case the advantages of multistage interpolators / decimators are well known [4]. Such advantages apply also to the 2-D case as the proposed work shows.

For vertico-temporal filtering the use of $N_1 \neq N_2$ is of great interest. As the number of used field memories coincides with $N_1 - 1$ (if the horizontal dimensions represents temporal direction), values of N_1 equal to three or five, currently are of great industrial interest. The performance of minimax optimal linear phase FIR filters designed according to the specs of Fig. 1 when N_1 is fixed (typically to a low-value) and N_2 varies will also be considered.

Fig. 3 shows the error versus order N_2 behaviour of the filters designed according to the specs of Fig. 1, with $N_1 = 3, 5, 7$ and $\Delta F_1 = .141$ (solid line), and for $\Delta F_2 = \Delta F_1/2$ (dotted line). The main indication of Fig. 3 is that the minimax error saturates, i.e., it does not decrease below limit values depending on N_1. Such a behaviour shows that there is no trade-off between the two-orders. This fact, obvious with separable filters, is not surprising as increasing the order only in one direction does not motivate an approximation improvement in both directions. The order N_2 threshold value at which the minimax error reaches saturation, for fixed N_1, typically corresponds to $N_2 \simeq 2N_1$. The minimax saturation error is found to decrease approximately 6 dB as N_1 goes to $N_1 + 2$. A detailed account of the presented results is given in [5].

This work was supported in part by C.N.R. - Progetto Finalizzato Telecomunicazioni, contract n. 92.00972.PF71

References

[1] J. Hu, L. Rabiner: "Design techniques for two–dimensional digital filters", IEEE Trans. on Audio Electroacoust. AU–20, 5, 249–257, Oct. 1972

[2] A. Biasiolo, G.M. Cortelazzo, G.A. Mian: "Design of multidimensional video FIR filters based on linear programming", Proc. SPIE's 33rd Int. Symp. on Optical & Optoelectronic applied science & Engineering, San Diego, CA, 1153–1160, Aug. 1989

[3] P.Siohan: "2-D FIR filter design for sampling structure conversion", IEEE Trans. Circuits and Systems for Video Technologies, VT–1, 4, 337–350, Dec. 1991

[4] A.V. Oppenheim, R.W. Schafer: "Discrete–time signal processing", Prentice–Hall, N.Y., 1989

[5] P.Carrai, G.M. Cortelazzo, G.A.Mian: "Characteristics of minimax FIR filters for video interpolation/decimation", submitted for publication

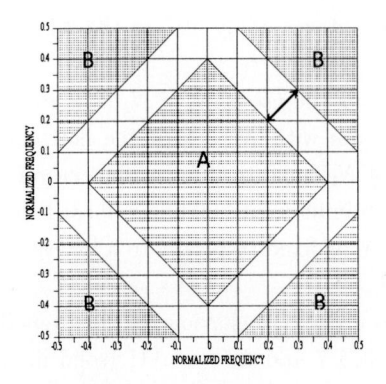

Figure 1: 2-D half–band specs

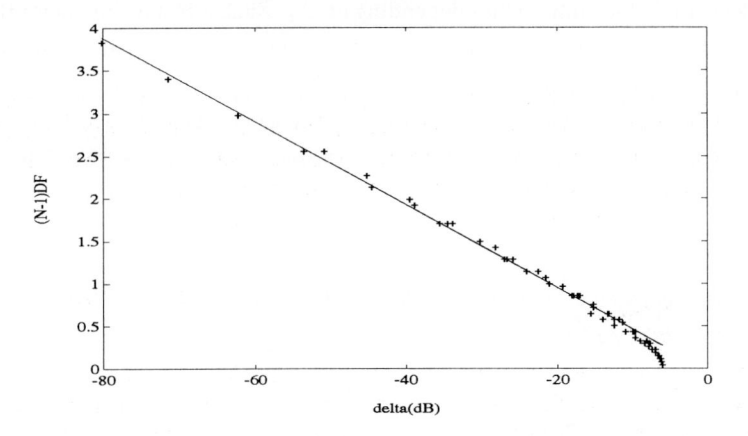

Figure 2: product $(N-1)\Delta F$ vs. order relationship;

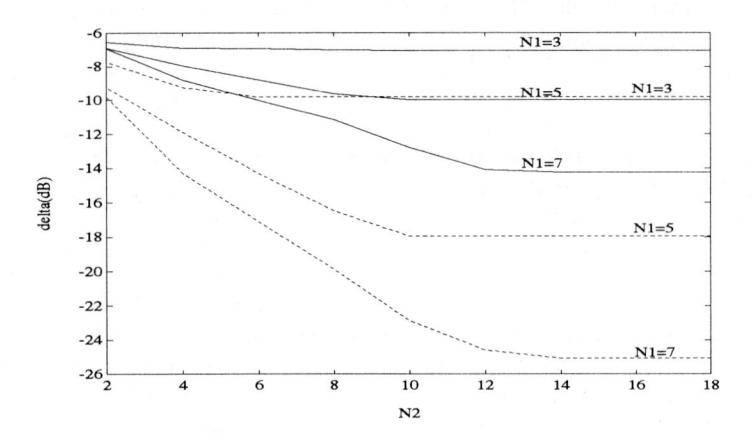

Figure 3: Error vs. N_2 plots for $\Delta F_1 = 0.141$ (dashed line) and $\Delta F_2 = \Delta F_1/2$ (solid line).

Contrast Enhancement through Multiresolution Decomposition

J.M. Jolion

LIGIA / LISPI, bat. 710, Université Claude BERNARD, Lyon I, F-69622 Villeurbanne Cedex, France

Abstract

We present a novel technique toward contrast enhancement using a multiresolution representation of an image. The contrast pyramid is extracted from this representation. The method enhances the image by means of modification of the contrast values at multiple scales, *i.e.* in the multiresolution domain. The enhanced image is then recovered by reversing the steps used in the construction of the multiresolution representation.

1. INTRODUCTION

Whenever an image is converted from one form to another, its visual quality may decrease. For instance, in image acquisition, the intensity obtained for a pixel is related to the image irradiance over its surface. Each pixel of a CCD camera integrates the light received on its surface. Moreover, especially in remote sensing applications, the intensity is related to a non local part of the observed scene. All these phenomena introduce contrast attenuation.

A classic approach toward visual quality enhancement of an image is, for instance, the histogram transformation technique [1, p.231-237]. Histogram "flattening" or "equalization" is an example of this technique. It enhances the image contrast by flattening the distribution of the most probable gray levels. As an alternative, the adaptive approach works on local contrast and makes use of values extracted from neighborhoods of different sizes or shapes of a given pixel.The pyramid based approach belongs to this second class of techniques [2].

Although the human visual system cannot accurately determine the absolute level of luminance, contrast differences can be detected quite consistently. Usually, this luminance contrast is defined as $C = (L/L_b) - 1$, where L denotes the luminance at a certain location in the image plane and L_b represents the luminance of the local background. More generally, L and L_b are computed from neighborhoods or receptive fields whose center P is the pixel to be processed, the neighborhood associated with L_b being greater than that of L. The value of the size of the neighborhood is an *a priori* information of such kind of techniques. It is clear that is has to be related to the size of the detail to be reinforced in the image. However rarely is this size unique for a given image. It is thus interesting to work simultaneously on several sizes for a given point.

2. BUILDING THE CONTRAST PYRAMID

The pyramid framework allows the manipulation of multiple neighborhood sizes. Let P be a node on level k ($0 \leq k \leq N$) in an intensity pyramid G. Its value $G_k(P)$ denotes the local luminance (*i.e.* in a local neighborhood which size is related to the size of the receptive field of P).

$$G_k(P) = \sum_{M \in Sons(P)} w(M) \cdot G_{k-1}(M) \quad \text{for } 1 \leq k \leq N$$

w is a normalized weight function which can be tuned to simulate the Gaussian pyramid [3]. The luminance of the local background is obtained from the luminances of the fathers of P. Thus, the background pyramid is built as follows.

$$B_k(P) = \sum_{Q \in Fathers(P)} W(Q) \cdot G_{k+1}(Q) \quad \text{for } 0 \leq k \leq N-1 \text{ and } B_N(P) = G_N(P)$$

W is a normalized weight function which takes into account the way P is used to build the luminance of its fathers. The contrast pyramid representation, ($C_0, ..., C_N$) is thus defined by

$$C_k(P) \equiv \frac{G_k(P)}{B_k(P)} \quad \text{for } 0 \leq k \leq N-1 \text{ and } C_N(P) = 1$$

Of course, the input image, G_0, can be recovered exactly from this representation by reversing the steps used in the construction.

$$G_k = C_k \cdot B_k \quad \text{for } 0 \leq k \leq N$$

3. HIERARCHICAL CONTRAST ENHANCEMENT

Different recombination rules yield different transformations of the input image. As an example, Toet [4] recently proposed a non linear scheme which, basically, changes the background based on the contrast value. This recombination algorithm starts at some level of the contrast pyramid and proceeds from coarse scales to finer scales. We proposed another approach to enhance the contrast of an image in which a new contrast value is built by

$\Gamma_N(P) = G_N(P)$ or a constant valued image and

$\Gamma_k(P) = f_k(|C_k(P) - 1|) \cdot B_k(P). \quad \text{for } 0 \leq k \leq N-1$

Γ_k denotes the contrast enhanced version of G_k, and f_k, a set of contrast modification functions satisfying

$$F(x) \begin{cases} < x & x < 1 \\ = 1 & x = 1 \\ > x & x > 1 \end{cases}$$

Power based functions are such modification functions. The pyramid decomposition allows different kinds of contrast modifications depending on the quality of the input image. The upper pyramid levels correspond to low spatial frequencies or coarse scale image representations while the lower pyramid levels correspond to high spatial frequencies or fine scale image representations. Consider a noise free image, in order to achieve better contrast enhancement, local contrasts (high frequencies) have to be more enhanced than global ones (low frequencies). This is very easy to do thanks to the pyramid decomposition. For instance, $f_k(x) = x^{2-k/N}$ gives more emphasis on the lower levels of the pyramid ($f_k(x)$ ranges from $f_N(x) = 1$ to $f_0(x) = x^2$).

When the image is noisy, local contrasts which are related to noise have to be attenuated. The visual quality of the image is improved by a band-limited contrast modification (attenuation or amplification). In the pyramid decomposition framework, it is possible to selectively enhance image details at certain spatial scales. Note that this method removes large scale luminance gradients and produces images which appear relatively independent of changes in the lighting conditions of the depicted scene.

The process can be iterated in order to satisfy various criteria. In our experiments, we used the entropy of the image histogram [1, p. 181], by taking as the final step the image resulting in the highest entropy. Another application of this technique is the simplification of an image by means of keeping only its contour related information. It is shown on some examples that the proposed scheme, if iterated, converges to quasi-binary images made of contours.

4. EXAMPLES

Figure 1 shows an example of this technique on night vision image (Figure 1a). For this application, the acquisition involves a light-amplification system resulting in noisy and low contrasted images. Figure 1b presents the enhanced image.

Figure 1d shows the result of applying the process in an iterative manner on the *Girl* image (see Figure 1c). The "fixing point" of this process is a quasi-binary image where emphasizes has been made on the transitions.

REFERENCES

1   A. Rosenfeld and A.K. Kak *Digital Image Processing*, Academic Press (1982).
2   J.M. Jolion and A. Rosenfeld *A Pyramid Framework for Early Vision*, Kluwer, to appear (1993).
3   P.J. Burt and E.H. Adelson "The Laplacian pyramid as a compact image code," *IEEE trans. on Comm*, 31(4) (1983) 532-540.
4   A. Toet "Adaptive multi-scale contrast enhancement through non-linear pyramid combinaison," *Pattern Recognition Letters*, 11 (1990) 735-742.

Figure 1 : (a) Night vision input image; (b) Enhanced image ; (c) *Girl* image; (d) Binary image obtained by iterative contrast enhancement

3-D RECONSTRUCTION OF LINE-LIKE OBJECTS IN TRINOCULAR VISION [1]

Ding Mingyue[2]
Institute of Pattern Recognition and Artificial Intelligence
Huazhong University of Science and Technology, Wuhan, Hubei 430074, P.R.China

F. M. Wahl
Institute for Robotics and Process Control, Technical University of Braunschweig
Hamburger Street 267, 3300 Braunschweig, Germany

Abstract In this paper first a masking technique to efficiently compute epipolar lines for stereo vision is described. The artificial point problem is analyzed and an algorithm based on space continuity constraint is proposed to remove artificial points. In order to ensure correct initial reconstructed points a correspondent test algorithm is introduced. Finally, some reconstruction experiments illustrate the feasibility of our approach.

1. INTRODUCTION

One basic task in robot vision is to acquire range data of objects. For example, in assembly of electrical products, it is needed to acquire the 3-D coordinates of flexible line-like objects such as wires. Most stereo vision algorithms are not suitable for this task because of two reasons: First, grey level information is of no further use in cases where the correspondence problem can not be solved uniquely if the surfaces of the objects have about an uniform grey level appearance; second, most of the algorithms nowadays are too complicated and too time-consuming.

In 1986, a 3-D measurement algorithm for terminals of line-like objects from shadow information had been proposed by Guo et al.[1-2]. They used the epipolar constraint to bring the terminals into correspondence with their shadows. In 1989, Stahs et al.[3] suggested a masking technique for the efficient computation of the epipolar constraint in the case of three point light sources; they utilized the intersection of three correspondent epipolar planes to yield the 3-D cooriantes directly, causing an additional speed-up. But these active approaches can only be used in indoor and close range applications. Furthermore, the artificial points generated by these approaches will produce false interpretations. Therefore it is necessary to further investigate the 3-D reconstruction of line-like objects.

2. MASKING PATTERN OF EPIPOLAR LINES

Suppose that $^{H}T_{C}$, $^{B^{(i)}}T_{H}$ (i=1,2,3 denote different camera positions) and $^{B}T_{W}$ are the transform matrices between hand and camera, robot base and hand as well as robot base and

---
[1]The work is supported by the Alexander von Humboldt Foundation in Germany
[2]Currently on leave at the Institute for Robotics and Process Control, Technical University of Braunschweig, Hamburger Street 267, 3300 Braunschweig, Germany

world coordinate system respectively (see Fig. 1). c, S_x, X_H, Y_H, a_0 and a_1 are the camera focal distance, the scale factor when digitizing a picture in x and y directions, the principal point coordinates in the sensor plane and the parameters of the radial symmetrical distortion model of the camera respectively [4]. The world coordinate C_{w_k} of the optical center of the camera k can be determined as follows:

$$T_k \times C_{w_k} = 0 \qquad (1)$$

where k = 1, 2, 3 and

$$T_k = (^{B^{(k)}}T_H \times {^H}T_C)^{-1} \times {^B}T_W$$

Let us determine the world coordinate of image point (i, j). First, compute the camera coordinate $X_{c_k} = (x_C, y_C, z_C)^T$ of (i, j):

$$x_C = (\frac{j - 256}{S_x} - X_H) \times (1 + a_0 \times r_{i,j}^2 + a_1 \times r_{i,j}^2) \qquad (2)$$

$$y_C = (i - 256 - Y_H) \times (1 + a_0 \times r_{i,j}^2 + a_1 \times r_{i,j}^2) \qquad (3)$$

$$z_C = c \qquad (4)$$

where

$$r_{i,j} = \sqrt{(\frac{j - 256}{S_x} - X_H)^2 + (i - 256 - Y_H)^2}$$

Then, its world coordinate X_{w_k} can be calculated as:

$$X_{w_k} = (^{B}T_W)^{-1} \times {^{B^{(k)}}}T_C \times {^H}T_C \times X_{c_k} \qquad (5)$$

So the epipolar plane through two camera optical centers and (i, j) can be determined. Scanning the whole image, the masking pattern of epipolar lines will be generated by quantizing the angel between the epipolar plane and a reference plane.

3. GENERATATION OF ARTIFICIAL POINTS

At point p_1 in image 1, we will find two epipolar lines $L_{12}(p_1)$ and $L_{13}(p_1)$. Suppose q_1, r_1 are the correspondent points of p_1 in images 2 and 3. According to the trinocular epipolar constraint, $L_{12}(p_1) = L_{21}(q_1) = m$, $L_{13}(p_1) = L_{31}(r_1) = n$ and $L_{23}(q_1) = L_{32}(r_1) = k$. m, n and k correspond to the three different epipolar planes. In an epipolar line net, they will form a triangle whose vertexes are p_1, q_1, r_1. If there are many points in space, the overlap between them will generate some space points which do not have real correspondence in object space; they are called *artificial points*. For example, suppose p_1, q_1, r_1, p_2, q_2, r_2 and p_3, q_3, r_3 are the correspondent points of space points P, Q, R in images 1, 2 and 3 respectively. They will form three triangles separately. At the same time, if P, Q, R are distributed such that their correspondent projection lines intersect the same space point, then p_1, q_2, r_3 will also form a triangle, generating an artificial point. Thus, artificial points are unavoidable in 3-D reconstruction if only the epipolar constraint is used. In human vision, man uniquely reconstructs the space point only from its two projections captured from the left and right eyes. The reason for this is that man uses other additional cues such as shading, texture, object simplicity, symmetry, continuity, meaningfulness etc.

4. REMOVAL OF ARTIFICIAL POINTS

In order to remove artificial points, a space continuity constraint suited to be combined with the above described masking technique of the epipolar constraint is proposed. Obviously, a continous line in an image produces a continous line in space. Therefore, for adjacent points in a line-like object, the differences between their 3-D reconstructed coordinates will not exceed a given threshold, — this is called *space continuity constraint*. This constraint is independent on camera geometry, but dependent on starting points. False starting points retain artificial points but reject object points. In order to ensure correct starting points, we introduced *a correspondence test algorithm*. The combination of projections from different space points is the main reason causing artificial points. If we remove this influence, the number of artificial points will be reduced. Suppose A_1, A_2 and A_3 are the three binary images of line-like objects computed from three views; assume all edges to be one pixel wide. First, we establish the epipolar line index arrays from A_2 and A_3. For a point q_1 in A_2, two epipolar lines m, k can be determined. Then, the point (m, k) in the epipolar line index array E_1 will increase by 1 (the initial value of the array is 0). After scanning A_2, the epipolar line index array E_1 is established. Similarly the epipolar line index array E_2 can be obtained from A_3. Secondly, let m, n be the two epipolar lines through the point p_1 in A_1. p_1, q_1, r_1 satisfy the epipolar constraint if we can find an epipolar line k such that $e_1(m, k)$ and $e_2(k, n)$ are not equal to 0. It can be seen, that E_1 and E_2 are kept constant in reconstruction. This increases the possibility of generating artificial points by accidental combination of projections from different space points. In order to decrease this possibility, *a variable epipolar line index array* has been used . If q_1, r_1 are the correspondent points of p_1, the point (m, k) in E_1 and (k, n) in E_2 will decrease by 1 so that their influence can be removed. In order to decrease the influence of the scanning order, *a second scanning* has been suggested. First, we scan A_1 from top left to bottom right. Subsequently we scan it in opposite direction. For a correct starting point, the result is the same in both scanning processes because it is generated by the projections of *a unique point corresponding to a real object point* . For an artificial point, it is generated by the combination of the projections from *different object points*. Therefore, the point which appears in the first scan will not appear in the second and vice versa. With this algorithm, correct initial points can be determinated .

5. EXPERIMENTS

Some experiments were conducted on a Micro VAX-II computer with KONTRON image processing system. The three images are captured by a camera, mounted on a robot, from three different positions. In order to increase the position accuracy, *a space averaging filtering algorithm* has been used. For one image point, we may find more than one reconstructed point. We substitute these by their space center, inceasing the position accuracy and decreasing the number of space points. Some results are shown in Fig. 2. Fig. 2(a) shows the three original views of line-like objects. Fig. 2(b) is the reconstruction of line-like objects without artificial point removal. We can see, that some artificial lines are so similar that it is impossible to remove them by postprocessing. Fig. 2(c) is the result of our approach. As can be seen, most of the artificial points are eliminated although there are some overlaps between the line-like objects. Our program is written in C langage and executed under VAX/VMS operating system. The whole run time for the 3-D reconstruction of a 512 × 512 image including the preprocessing is about 90-180 seconds, depending on the complexity of the images.

References:

[1] H.-L.Guo , M. Yachida and S. Tsuji, Three-dimensional measurements of terminals of lin-like objects from shadow information , Advanced Robotics, 1(1), 47-58, 1986

[2] —, Three-dimensional measurement of many line-like objects, Advanced Robotics, 1(2), 117-130, 1986

[3] T.Stahs F. Wahl, Three-dimensional Range Acquisition of Line-like objects, 3rd International Conference on Image Processing and its Applications, 1989, Univ. of Warwick, UK, 14-18

[4] —, Oberflaechenvermessung mit einem 3D-Robotersensor, ZPF-Zeitschrift fuer Photogrammetrie und Fernerkundung, No. 6, 1990, 190-202

Fig. 1. The coordinate systems

(a)

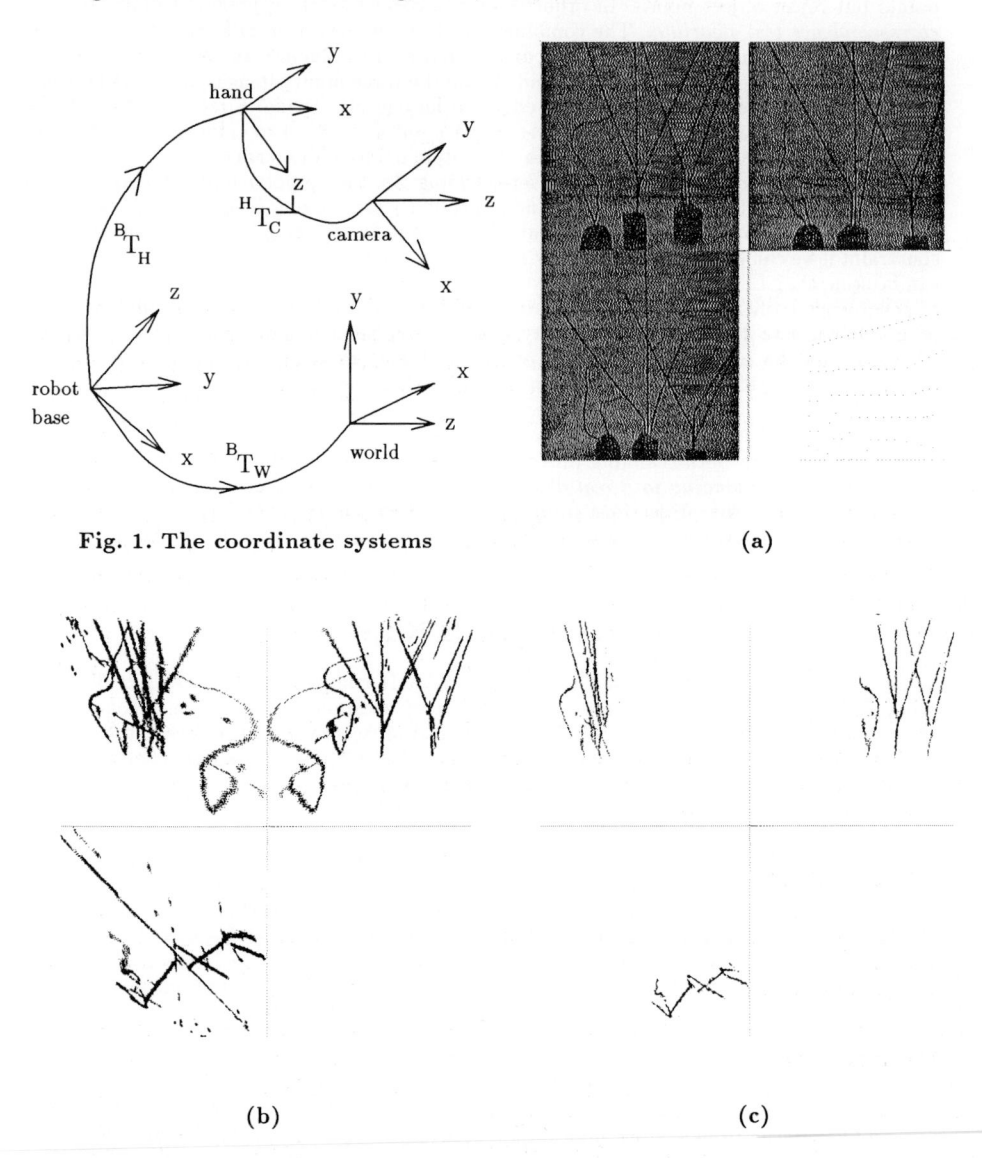

(b)　　　　　　　　　　　　　　(c)

Fig. 2 The comparsion of reconstruction

Pattern Wafer Image Segmentation Using Neural Networks

C.H. Chen*, G.H. You* and Pay-Shin King**

*Electrical and Computer Engineering Dept., University of Massachusetts Dartmouth
N. Dartmouth, MA 02747 USA. Email address: CCHEN @ UMASSD.EDU
**Inspex, Inc., 47 Manning Road, Billerica, MA 01865 USA

Abstract

In machine inspection of pattern wafer images, an important step is to perform reliably image segmentation at high speed to extract regions corresponding to potential defects. While model based methods such as the use of Markov random fields have been popular in image segmentation, the pattern wafer images are not particularly suitable for modeling. On the other hand, training samples are easily available, based on our knowledge about the defect and defect-free regions. Neural networks are found to meet the segmentation needs quite well.

We have examined the use of Probabilistic Neural Network (PNN), the popular back-propagation (BP) neural network, and a new Class-Sensitive Neural Network (CSNN) for pattern wafer image segmentation. Their performances and capabilities will be compared in this paper. In terms of classification (and thus segmentation) accuracy, CSNN is found to be the best and is typically 6% more in percentage correct classification than PNN and BP networks. The performances of PNN and BP networks are fairly similar. Computationally PNN takes less time as it is not an iterative operation. The inputs of neural networks are the 3 x 3 gray level values. Inputs of this size are quite suitable for small size neural networks. With both the computational and performance advantages, neural networks can become an important part of high speed pattern wafer image processing system to perform image segmentation for defect extraction.

I. INTRODUCTION

This project has been concerned with developing new and effective image processing algorithms to support Inspex's wafer inspection systems. An important part of the system operation is to perform reliably image segmentation at high speed to extract regions corresponding to potential defects. While model based methods such as the use of Markov random fields have been quite useful for many textured images, they are not particularly suitable for wafer images which are less structured. Neural networks which do not make any assumption on the image model, on the other hand, can be very suitable for the wafer inspection images. The segmentation is performed by pixel classification. Neural networks that have good classification capability are thus considered. In addition to the well known back-propagation trained neural networks [1] and the probabilistic neural network [2], the newly developed class-sensitive neural network (CSNN) [3] [4] is also used for image segmentation. Although image segmentation is a well studied topic in image processing, there are very few publications dealing with wafer image segmentation. It may be noted that a real-time image segmentation system using back-propagation network has been implemented for machine inspection [5].

2. OVERVIEW OF WAFER IMAGE PROCESSING

The Inspex's wafer image processing system [6] performs the image matching, image compression, area based operations, edge based operations, image comparison and defect finding, in addition to the usual image smoothing functions. The wafers are inspected under microscopes. At different light intensities two images of the same object will be different. Matching of the two images is performed by using light intensity normalization. It is found that the gray level difference between the two images versus the gray level is a linear function. The linear relation allows us to make simple linear increment in gray level to match the two images. Run length limited compression method is used for image compression. Area based operations convert the original image into a two tone or three tone image, by using histogram thresholding or pixel classification (image segmentation). The edge based operations compute the gradients and the directional gradients (see e.g. [7]). Image comparison is the stage to find out the difference between the inspected image and the model image. The resulting difference image is used to locate the defects. For images processed by area based operations, defect finding is carried out as follows. First a 3x3 window is used to filter out isolated points and single pixel curves. Then a 7x7 window is used to find minimum size defect and the boundary of the defect is searched. A rectangle is used to indicate the position of the defect.

3. NEURAL NETWORK IMAGE SEGMENTATION

A digitized pattern wafer image is very large and usually is much more than 512x512 in size. Neural network image segmentation certainly has the speed advantage which is quite essential in real-time operations. For neural network use, there are a large number of training pixels available based on our knowledge about the image. Fig. 1a and Fig. 1b are examples of typical pattern wafer images in gray scale. Fig. 2 is a small section of Fig. 1a for segmentation study, where training pixels are windowed. The inputs to neural networks are the 3x3 gray level values. Inputs of this size are quite suitable for small size neural networks. The features are the original gray scale values, while the output consists of three classes for three tones. Fig. 3 is the CSNN segmentation result for the image of Fig. 2. In terms of classification accuracy, CSNN achieves 99% correct, while the back-propagation network and probabilistic neural network are around 93% each. The probabilistic neural network is simpler computationally as it does not require iterations. The CSNN consists of several two-class neural network classifiers using back-propagation training but operated in parallel. Its computation time is thus slightly longer than the multi-class back-propagation neural networks. Based on the limited computer results presented, it can be concluded that neural networks are indeed effective for pattern wafer image segmentation for defect extraction in pattern wafer image processing.

Figure 1a. An example of pattern wafer image

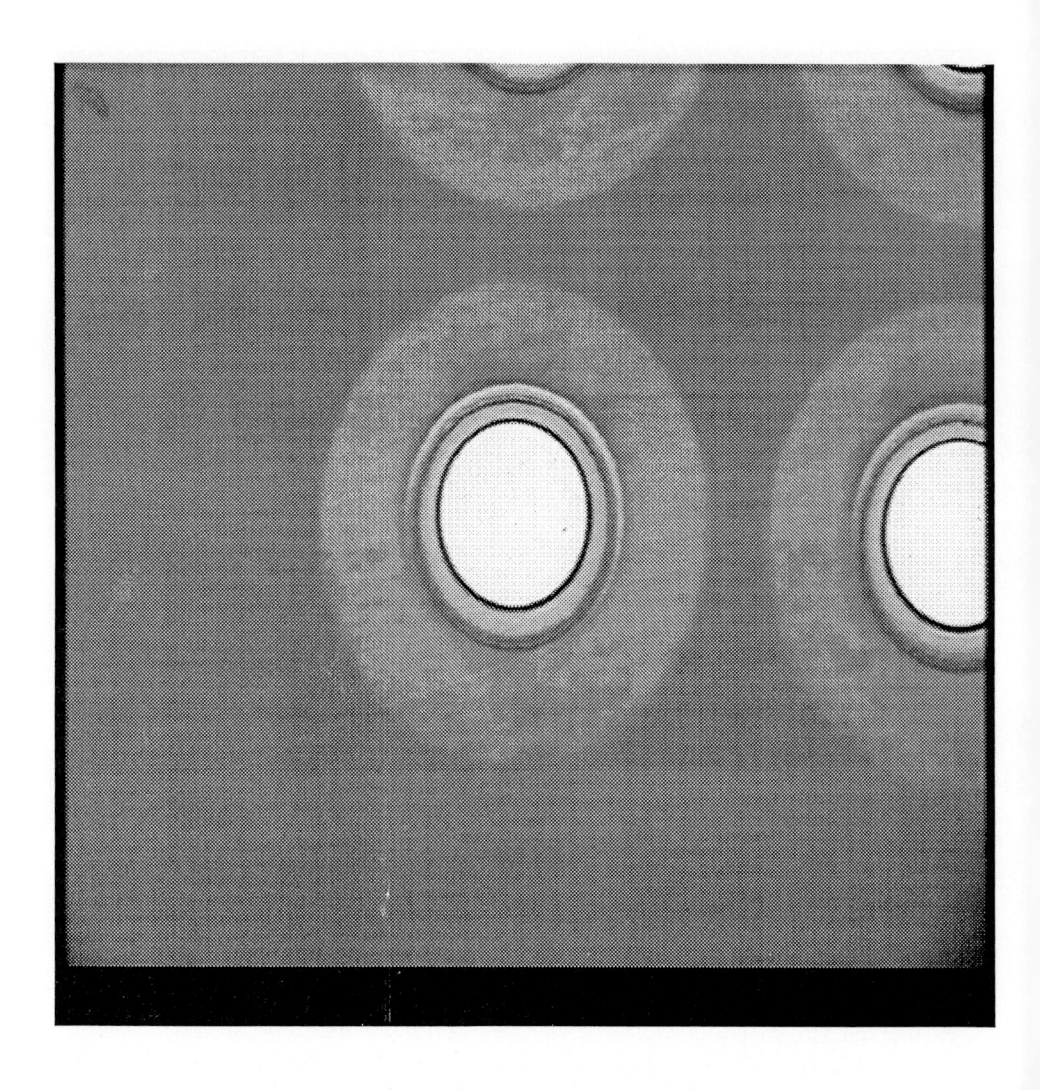

Figure 1b. Another example of pattern wafer image.

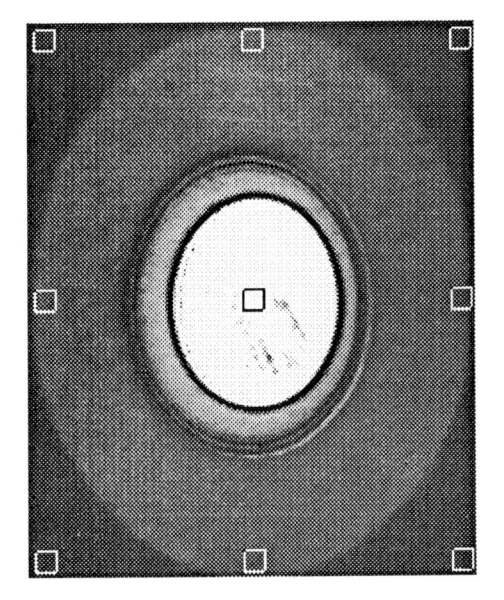

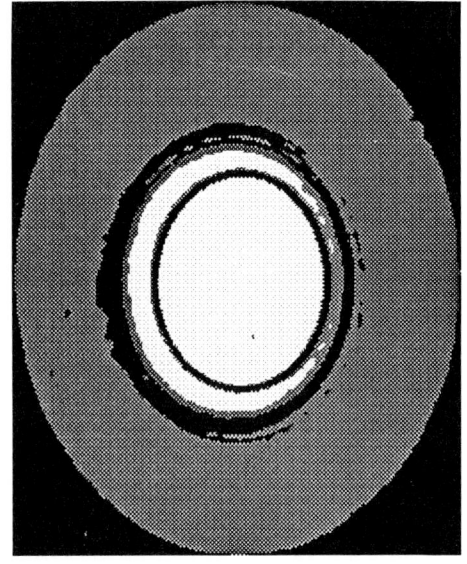

Figure 2.

The original image for segmentation study along with windowed regions for training.

Figure 3.

CSNN image segmentation result displayed in a three-tone image.

4. REFERENCES

1. D. Rumelhart, G.E. Hinton and R.J. Williams, Learning internal representations by error propagation, in D.E. Rumelhart and J.L. McClelland (eds.), Parallel Distributed Processing: Explorations in the Microstructure of Cognition, vol. 1: Foundations, MIT Press 1986.
2. D.F. Specht, Probabilistic neural network, Neural Networks, vol. 3, pp. 109-118, 1990.
3. C.H. Chen and G.H. You, Class-sensitive neural network, Neural, Parallel and Scientific Computing, vol. 1, 1993.
4. C. H. Chen, Artificial neural networks in pattern recognition and signal classification, this proceedings.
5. W.E. Blanz and S.L. Gish, A real-time segmentation system using a connectionist classifier architecture, in C.H. Chen (ed.), Neural Networks in Pattern Recognition and their Applications, World Scientific Publishing, Singapore and New Jersey, 1991.
6. G.H. You, Pattern wafer image processing research, Final Report prepared for Inspex, Inc., July 1992.
7. R.C. Gonzalez and R.E. Woods, Digital Image Processing, 3rd edition, Addison-Wesley, Reading, MA 1992.

Fast shape discrimination with a system of neural networks based on scaled normalized central moments

B.G. Mertzios and D. Mitzias

Automatic Control Systems Laboratory, Department of Electrical Engineering, Democritus University of Thrace,
67 100 Xanthi, GREECE, Fax : (30) 541-20275, 26947
e-mail: mertzios@xanthi.uucp

Abstract

A shape discrimination technique is presented, which uses as feature vectors a number of 2-D scaled normalized central statistical moments. A system of neural network is developed, which is comprised of more than one subnetworks and is very efficient for demanding shape discrimination tasks with high discriminating efficiency. Each neural subnetwork operates with different features corresponding to various levels of importance.

1. INTRODUCTION

2-D statistical moments of various quantities associated with planar curves have been used for shape discrimination [1],[2]. In this paper a recently presented set of scaled and normalized sets of 2-D central moments that are invariant under translation and magnification of the image are used [3]. These new moments appear to have better classification performance over the existing sets of moments and they are less sensitive to noise. A number of moments is selected as features, which carry the necessary information for pattern recognition. The discrimination of an unknown shape is achieved by testing a weighted least square cost function that measures the distance of its moments against the moments of the considered prototypes.

A system of two neural networks is developed, which is comprised of two subnetworks and is very efficient for demanding shape discrimination tasks with high discriminating efficiency. Each subnet operates with different features, corresponding to various levels of importance.

2. THE SCALED NORMALIZED CENTRAL SET OF MOMENTS

The considered 2-D Scaled Central (SC) moments of order (p,q) are defined as follows:

$$h_{pq} = \frac{\beta}{m_{00}} a(p,q) \sum_{i=1}^{M} \sum_{j=1}^{N} (x_i - \bar{x})^p (y_j - \bar{y})^q f(x_i, y_j), \quad p, q = 0, 1, 2, \ldots \quad (1)$$

where $a(p,q)$ is the scaling factor corresponding to the h_{pq} moment, and β/m_{00} is the normalization factor which sets the the zeroth-order moment m_0 to a predetermined normalization value β.

If the scaling factor a(p,q) is separable with respect to p and q, then it may be written in the form

$$a(p,q) = a_x^p a_y^p \qquad (2)$$

Then the horizontal and vertical scaling factors a_x, a_y are characterized by the properties:

$$\frac{h_{pq}}{h_{p,q-1}} = a_x \frac{\mu_{pq}}{\mu_{p,q-1}}, \text{ for } h_{p,q-1}, \mu_{p,q-1} \neq 0$$

$$\frac{h_{pq}}{h_{p-1,q}} = a_y \frac{\mu_{pq}}{\mu_{p-1,q}}, \text{ for } h_{p-1,q}, \mu_{p-1,q} \neq 0 \qquad (3)$$

Let K_x and K_y are defined by the relations
$K_x = \max\{|x_i - \bar{x}|, i = 1,2,...,MN\}$, $K_y = \max\{|y_j - \bar{y}|, j = 1,2,...,MN\}$

The 2-D central moments are increased exponentially as p and q increase, if E x_i, $j = 1,2,...,M$, such that $|x_i - \bar{x}| > 1$ and E y_j, $j = 1,2,...,N$, such that $|y_j - \bar{y}| > 1$. These latter conditions are equivalent to the conditions $K_x > 1$ and $K_y > 1$ respectively.
Then the scaling factors a_x and a_y are bounded by $0 < a_x < 1$ and $0 < a_y < 1$, and specifically they should be selected so that $a_x < 1/K_x < 1$, $a_y < 1/K_y < 1$.

As $p \to \infty$, $q \to \infty$, the factor $a_x^p a_y^p$ prevents h_{pq} to diverse to infinity. However, the proposed 2-D SC moments are not invariant under rotation, and therefore they carry information about the orientation of the shape in the scene. The 2-D SC moments may be used as recognition features after normalizing them appropriately with respect to rotation. The required number of the moments depends on: (i). the level of the existing noise in the considered application and (ii). the form of the considered shapes.

3. THE NEURAL BASED CLASSIFIER

The recognition procedure is achieved using a two layer classifier with two subnets in the first layer, as it is shown in Fig. 1. Each subnet operates with different inputs, which represent different features of the same pattern. It is pointed out that the dimension of the input vector can vary at each subnet, while the dimension of the output should be the same at each subnet. In this application each subnet is a three layer perceptron, which is trained with the backpropagation algorithm [4].

The second layer combines the outputs y_j^k, k=1,2, j=1,2,...,m of the subnets, in an efficient way, in order to achieve a high discrimination rate. The final output y(j) of the classifier is:

$$y_j = W_1 y_j^1 + W_2 y_j^2, \quad j = 1,2,...,m \qquad (4)$$

where W_k, k=1,2 are the weights of the comparator network (Fig. 1), which are determined by

$$W_k = \frac{1}{2}(W_k^r + W_k^e), \quad k=1,2 \tag{5}$$

In (6) the W_k^r, are called *recursion weights* and denote a measure of the discrimination efficiency of the kth subnet among the training patterns. Also in (6) the W_k^e are called *error weights* and denote the discrimination efficiency rate of the kth subnet, when partially destroyed prototypes are presented in subnets, after the learning phase. These weights are given by the formula:

$$W_k^r = 1 - \frac{R_k}{R_1 + R_2}, \quad W_k^e = 1 - \frac{E_k}{R_1 + R_2}, \quad k=1,2 \tag{6}$$

where R_k is the number of the recursions required to train the kth subnet for a specified error. Also E_k is the total square error between the desired and the original output of the kth subnet, which is ginen by:

$$E_k = \sum_{i=1}^{p} \sum_{j=1}^{m} e_{ij}^2 \tag{7}$$

where p is the number of training patterns and m the number of the outputs.
In the training phase of the subnets we use as prototypes the four shapes shown in Fig. 2. The training feature vectors for the subnet-1 are the first five SC moments $h_{i,i}^o$ of the original prototype. Moreover, the training feature vectors for the subnet-2 are the first five SC moments $h_{i,i}^o$ of the contour of the original prototype. In the sequel, the Subnets-1,2 learn to discriminate among the four prototypes, until the mean square error becomes less than a specified value for each subnet. Then the error weights W_k^r, k=1,2 are calculated.
After the training phase has been completed, the prototypes are partially destroyed and the first five SC-moments, from the destroyed prototypes shown in Fig. 3, are extracted. These moments form the inputs, which are applied to the subnets-1,2, in order to calculate the error weights W_k^e, k=1,2.

For the specific application it is found that $W_1^r = 0.41$ and $W_2^r = 0.59$. Note that the subnet-2, corresponding to the contours of the original shapes, is trained faster than the subnet-1. Equivalently, this means that the discrimination among the prototypes with subnet-1 is more difficult, than that with subnet-2. On the other hand $W_1^e = 0.62$ and $W_2^e = 0.38$, which means that subnet-2 is more sensitivity in the noise. This remark shows the efficiency of the proposed network, since it combines the outputs of the subnets, in order to result to a robust solution, when one of the subnets is unable to discriminate among one or more different complicated shapes.

4. REFFERENCES

1 K. Chen, "Efficient Parallel Algorithms for the computation of Two-Dimensional Moments," *Pattern Recognition* vol. 23, pp. 109-119, 1990.
2 K. Tsiricolias and B.G. Mertzios, "Statistical Pattern Recognition Using Efficient 2-D Moments with Applications to Character Recognition," *Pattern Recognition.* To appear.
3 B.G. Mertzios, "Shape Discrimination in Robotic Vision Usig Scaled Normalized Central Moments," *Proceedings of the IFAC Workshop of Computing Power and Control Theory*, Prague, Chechoslavakia, September 1-2, 1992.
4 D.E. Rumelhart and J.L. McClelland, (Eds.), *Parallel Distributed Processing*, Cambridge, MA: MIT Press, 1986.

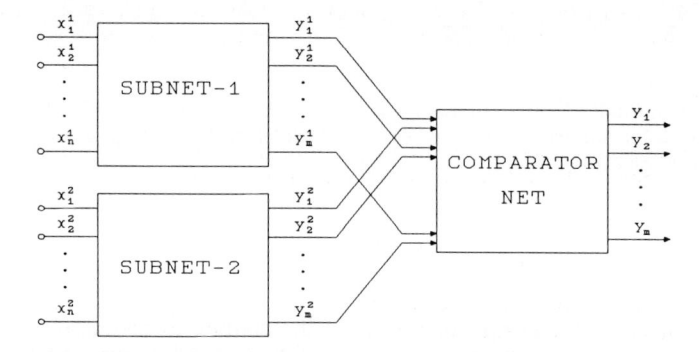

Figure 1. The neural based network classifier for the discrimination.

Figure 2. The four prototype shapes.

Figure 3. The destroyed prototype shapes.

A MULTISTAGE NEURAL NETWORK APPROACH TO IMAGE RESTORATION

O.K. Ersoy and R. Sundaram

School of Electrical Engineering, Purdue University, West Lafayette, Indiana, USA

Abstract

Neural networks have been proposed to perform image restoration. Some advantages are parallelism and ease of hardware implementation. We propose a multistage structure to error(energy) reduction using the discrete Hopfield-like neural network (HNN) in a "divide and conquer" scheme. We demonstrate the capability of the multistage method to both attain lower minimum on the error surface and to do so more rapidly than conventional optimization strategies. Two distinct multistage algorithms are proposed, one involving partial data at each stage and, the other using partial neuron information at each stage.

1. INTRODUCTION

The issue of recovering the desired signal from measurements, obtained after a linear transformation and the addition of corruption in the form of noise,

$$[Y] = [H][Z] + [N] \tag{1}$$

has been addressed in various contexts. Typically, the [H] matrix could represent degradation in signal/image restoration problems. The ill-conditioned nature of [H] renders any inverse filtering approach [2] useless in low signal-to-noise ratio (SNR) cases. Knowledge and use of $[H]^{-1}$ only serves to accentuate the noise which results in the poor quality of the reconstruction.

In order to circumvent this problem, a least-squares solution is sought, in which an error function including a regularization term is minimized. The error, representing a cost/energy function E is expressed vectorially as,

$$E = \frac{1}{2} \| [Y] - [H][\hat{Z}] \|^2 + \frac{1}{2} \lambda \| [D][\hat{Z}] \|^2 \tag{2}$$

where, $\| [U] \|$ represents the L_2 - norm of [U], λ is a predetermined constant, and [$\hat{Z}$] is the current estimate of [Z]. The second term in (3) provides regularization [1] by introducing smoothness in the estimate, [$\hat{Z}$]. The constant λ controls its relative influence on the first or least squares term thereby ensuring the proximity of [Y] to [$\hat{Z}$].

The matrix [D] is usually the second-order derivative operator. For example, the 2-D discrete problem could use the numerical approximation of the Laplacian operator [2]. A lexicographic notation is one of several ways to represent a 2-D signal sequence as a 1-D sequence. The estimation error can also be written as:

$$E([Z]) = -\frac{1}{2} [Z]^T [T][Z] - [\theta]^T [Z] \tag{3}$$

where, [T] is a reconfigured degradation matrix which will later represent the neuron interconnection weights and, θ is the bias vector. In this paper we proceed to use a discrete HNN and the simple sum of neuron state variables.

The form of the error function reveals non-zero self feedback components, i.e. $t_{jj} \neq 0$ which can result in divergence of the error function. To overcome this obstacle, a deterministic rule which checks the energy change for each neuron state change has been proposed and successfully implemented [2]. A stochastic rule, similar to simulated annealing [2], can be developed for convergence to a global minimum. In this paper, we consider a modified error energy function [4],

$$\hat{E} = E - \frac{1}{2} \sum_{j=1}^{N} \sum_{i=1}^{N} (h_{ij}^2 + d_{ij}^2) \sum_{l=1}^{M} x_{jl}(x_{jl} - 1) \qquad (4)$$

This serves to suppress the self feedback terms but entails a modification of the bias terms [2].

2. MULTISTAGE FRAMEWORK

In order to decompose the error expression in (3), we identify by N_1 all those signal values which we wish to leave untouched at any given stage and, by N_2, all those which we would like to modify during that stage. Then (3) can be expressed in two ways. In one case, we choose to alter all the neurons of certain data samples at each stage, and call this the "partial data" (PD) case. In the other case, we modify only certain neurons of all the data samples at each stage, and label this approach the "partial neuron" (PN) case. In the two cases, $\hat{E}$ can be expressed as follows:

PARTIAL DATA (PD) CASE:

$$\hat{E} = \frac{1}{2}\sum_{i=1}^{N}\left[\sum_{i\in N_2}\left\{\sum_{j\in N_2}\left[h_{ij}h_{ik}+\lambda d_{ij}d_{ik}\right]\right\}\sum_{l=1}^{M}\sum_{m=1}^{M}x_{jl}x_{km}\right] - \frac{1}{2}\sum_{i=1}^{N}\left[\sum_{j\in N_2}\left(h_{ij}^2+\lambda d_{ij}^2\right)\right]\sum_{l=1}^{M}x_{jl}^2$$

$$-\frac{1}{2}\sum_{i=1}^{N}\left[\sum_{j\in N_1}\left(h_{ij}^2+\lambda d_{ij}^2\right)\right]\sum_{l=1}^{M}x_{jl}(x_{jl}-1) + \frac{\lambda}{2}\sum_{i=1}^{N}L_i^2\frac{1}{2}\sum_{i=1}^{N}(y_i-K_i)^2$$

$$-\left[\frac{1}{2}\sum_{i=1}^{N}\left\{(y_i-K_i)(\sum_{j\in N_2}h_{ij})\right\} - \frac{1}{2}\sum_{j\in N_2}(h_{ij}^2+\lambda d_{ij}^2) - \left\{\lambda L_i \sum_{j\in N_2}d_{ij}\right\}\right]\sum_{l=1}^{M}x_{jl} \qquad (5)$$

$$K_i = \sum_{j\in N_1} h_{ij} \sum_{l=1}^{M} x_{jl} \text{ and, } L_i = \sum_{j\in N_2} d_{ij} \sum_{l=1}^{M} x_{jl}.$$

PARTIAL NEURON (PN) CASE:

$$\hat{E} = -\frac{1}{2}\sum_{j=1}^{N}\sum_{k=1}^{N}t_{jk}\left[\sum_{l\in N_2}\sum_{m\in N_2}x_{jl}x_{km}\right] - \frac{1}{2}\sum_{j=1}^{N}g_j\sum_{l\in N_2}x_{jl}^2 + \frac{1}{2}\sum_{i=1}^{N}[y_i-K_i]^2 + \frac{\lambda}{2}\sum_{i=1}^{N}L_i^2$$

$$-\frac{1}{2}\sum_{j=1}^{N}(h_{ij}^2+\lambda d_{ij}^2)\sum_{l\in N_1}x_{jl}(x_{jl}-1) \qquad (6)$$

$$K_i = \sum_{j=1}^{N} h_{ij} \sum_{l\in N_1} x_{jl}, L_i = \sum_{j=1}^{N} d_{ij} \sum_{l\in N_1} x_{jl}, \hat{\theta}_j = \sum_{i=1}^{N}[(y_i-K_i)h_{ij} - \lambda L_i d_{ij}] - g_j.$$

Our strategy of "divide and conquer" now focuses on the error in region N_2 while freezing the activity in region N_1. However after performing the suboptimal minimization of the error in region N_2 holding N_1 constant, we proceed to treat N_1 in a similar fashion, this time holding N_2 at its latest optimal condition. An iterative process is developed to converge to a desired stable state with minimum error. Obviously, the approach can be generalized to more number of stages.

3. OPERATING CONDITIONS

The stages in both approaches operate in cascade and do so by iteratively modifying the thresholded state variable. Otherwise, the Hopfield network equations are valid as follows:

$$\text{PN}: \begin{cases} u_i = \sum_{j=1}^{N} \sum_{l \in N_2} t_{jl} \, x_{jl} + \hat{\theta}_i & i = 1, 2, ..., N \\ z_i(k+1) = f(u_i) & k = 1, 2, 3, ... \end{cases} \quad (7)$$

$$\text{PD}: \begin{cases} u_i = \sum_{j \in N_2} \sum_{l=1}^{M} t_{jl} \, x_{jl} + \theta_i & i = 1, 2, ..., N \\ z_i(k+1) = f(u_i) & k = 1, 2, 3, ... \end{cases} \quad (8)$$

where k is the discrete time index and $z_j(k) = \sum_{l=1}^{M} x_{jl}(k)$; the values of t_{jl}, θ_i are identified by comparing the Hopfield energy function [4] to equations (5) and (6). Also,

$$f(u_i) = \begin{cases} 1 & u_i \geq 0 \\ -1 & u_i < 0 \end{cases} \quad (9)$$

is the bipolar nonlinear thresholding function.

4. EXPERIMENTS AND DISCUSSION

The multistage structure proposed above was adopted to perform image recovery. Two dimensional image data was blurred by a known impulse response (see (10) below) and corrupted by additive white Gaussian noise (AWGN) with zero mean, $\sigma^2 = 0.001$. The matrix [**H**] and, the 2D Laplacian operator (see (11) below) in block Toeplitz form, [**D**] were identified by a one-dimensional lexicographic ordering of the image data. In addition, selection of the neuron counts, $M = 64, 256$ per sample and $\lambda = 0.01$ preceded the computation.

$$h(i,j) = \begin{cases} 1/2 & (i,j) = (0,0) \\ 1/16 & (i,j) = (1,0), (0,1), (1,1) \\ 0 & \text{elsewhere} \end{cases} \quad (10)$$

$$\mathbf{L} = \frac{1}{6} \begin{bmatrix} 1 & 4 & 1 \\ 4 & -20 & 4 \\ 1 & 4 & 1 \end{bmatrix} \quad (11)$$

A comparison was made among single stage, two stage PN and PD, four stage PN structures. The pair of tables display the reduction of the normalized MSRE as a function of the number of iterations. The PD approach fails to make progress beyond a few iterations and will require a restart procedure to resume the search. The PN structures, on the other hand, proceed to deeper minima than single stage and PD systems.

The potential of multistage neural computation to solve unconstrained optimization problems has been revealed through this application to an inverse problem. The PN decomposition displays a probabilistic range of permissible data values at each stage. As the system descends to the deeper minima, this distribution shows less variability. The notion of temperature in annealing techniques has been translated to this stagewise reassignment of the range of data values.

REFERENCES

[1] O.K. Ersoy, *Signal/Image Processing and Understanding with Neural Networks*, Chapter in "Neural Networks: Concepts, Applications and Implementations," edited by P. Antognetti, V. Milutinovic, Prentice-Hall, 1989.

[2] Y-T. Zhou et. al., "Image Restoration Using A Neural Network," IEEE Trans. on ASSP, Vol.36, No.7, pp.1141-1151, July 1988.

[3] O.K. Ersoy et. al., "An Iterative Interlacing Approach for Synthesis of Computer-Generated Holograms," Applied Optics, Vol.31, No.32, pp. 6894-6901, November 1992.

[4] J.J. Hopfield and D.W. Tank, "Neural computation of Decisions in Optimization Problems," Biological Cybernetics, vol.52, pp. 141-152, 1985.

Table 1

MSRE ATTAINED AFTER PRESCRIBED ITERATIONS

(16x16 image, Neuron count = 64)

NN	5	20	35	50	75	100
1S HNN	0.722	0.650	0.584	0.524	-	-
2S PN HNN	0.727	0.659	0.594	0.523	0.445	0.374
4S PN HNN	0.727	0.661	0.599	0.543	0.509	-
2S PD HNN	0.773	-	-	-	-	-

Table 2

MSRE ATTAINED AFTER PRESCRIBED ITERATIONS

(16x16 image, Neuron count = 256)

NN	5	20	35	50	75	100
1S HNN	0.743	0.712	0.682	0.654	0.611	0.573
2S PN HNN	0.745	0.710	0.669	0.628	0.581	0.433
4S PN HNN	0.745	0.701	0.674	0.629	0.613	0.521
2S PD HNN	0.754	0.720	0.684	-	-	-

Image compression by Hadamard transform and neural network spectrum extrapolation

E. Costamagna[a], A. Podda[b], A. Turno, and A. Vargiu

[a]Istituto di Elettrotecnica, Università di Cagliari, Italy

[b]IBM SEMEA SUD, Scientific and Technical Solutions Centre, Cagliari, Italy

Abstract

Image data compression is performed by means of neural networks trained to extrapolate Hadamard high pass spectral components from supplied low pass components.

1. SPECTRUM EXTRAPOLATION

Neural networks have been utilized by varios authors to provide data compression in image coding and transmission, and significant results have been already obtained, as in [1],[2],[3]. However, to the knowledge of these authors, the relatively simple approach described in this paper was not investigated, although it seems to deserve some attention, and some first results appear to be encouraging.

The approach is based on the customary bidimensional Walsh-Hadamard transform [4], applied to image blocks of a limited number of pixels, typically 8x8. The derived low frequency spectral components (basis planes) are supplied to a trained neural network, which performs spectrum extrapolation. Spectral components are separated into low pass (LP) and high pass (HP) components according to a ratio of 1:3 (compression ratio 1:4), following some previous experience in spectrum extrapolation applied to very low resolution images [5].

During the learning phase, both the LP and HP spectral components derived from the sample images are supplied to the neural network, which is trained to associate the LP and the HP spectra, so that HP components can be guessed during the exploitation phase, when only the LP components are presented to the net. In our experience, the best results have been obtained by means of two layer associative memories trained by the error back propagation (EBP) algorithm, but similar behaviours have been observed in three-layer netwotks (EBP, 50 nodes in the hidden layer), although leading to larger errors or learning times. Both linear and non linear transfer functions (i.e., sine and sigmoid functions) have been utilized, and linear networks have been shown to be more effective in this application.

The spectral LP and HP components, obtained from the original images by linear correlators, are separately supplied to the 16 input and 48 output nodes of the net during the learning phase, and the connection weights are modified according to the delta rule (or dumped delta rule for three layers networks). Linear correlators have been somewhat implemented by two layer networks trained to correlate single basis planes: in fact, observing the very fast learning times, in which the network connection weights converge to the equivalent correlator coefficients, encouraged the authors to implement the present application of the Hadamard transform to neural net operations.

Spectrum extrapolation is performed by the so called generalization property of neural networks: when the LP components of a new image are supplied, the net attempts to associate to them HP components derived from the previous experience, and a complete spectrum is compounded from the true LP and the synthetized HP spectra.

The training phase is quickly accomplished: a very few iterations (less than 10) are sufficient to a linear network to learn one or several images up to a mean square error in the output of the order of 1%, which is only about 1/3 or 1/4 of the initial error, but nearly coincident with the asymptotic value attained after large numbers of iterations. Moreover, the generalization property is clearly displayed: when learning a new image, networks trained with similar but different images show initial errors very similar to those computed for the learned images, i.e., similar to the asymptotic values, and definitely less than errors from non trained networks.

Some results are illustrated in Figure 1. Top photographs show an original image (left side), the image obtained from its LP components only (right side), and the image obtained by means of adding to the LP components the HP spectrum supplied by the two layer net (16 gray levels). The net was trained to the same image 1 (batch procedure). Similarly, a network trained to image 2 gives for it the results shown in the following photographs (same order). Generalization and spectrum extrapolation effects are shown in the last two photographs: here, a net trained to one image (1 or 2) synthesizes the HP components of the other (2 or 1), leading to image improvements, with respect to the LP version, which seem to be well evident.

2. TRANSMISSION PROCEDURES

The described behaviour can be exploited in image transmission systems by means of transmitting only the LP spectral components: at the receiving end, spectrum extrapolation is performed by a net trained to images roughly similar to the transmitted ones. This seems to be well suitable at least for

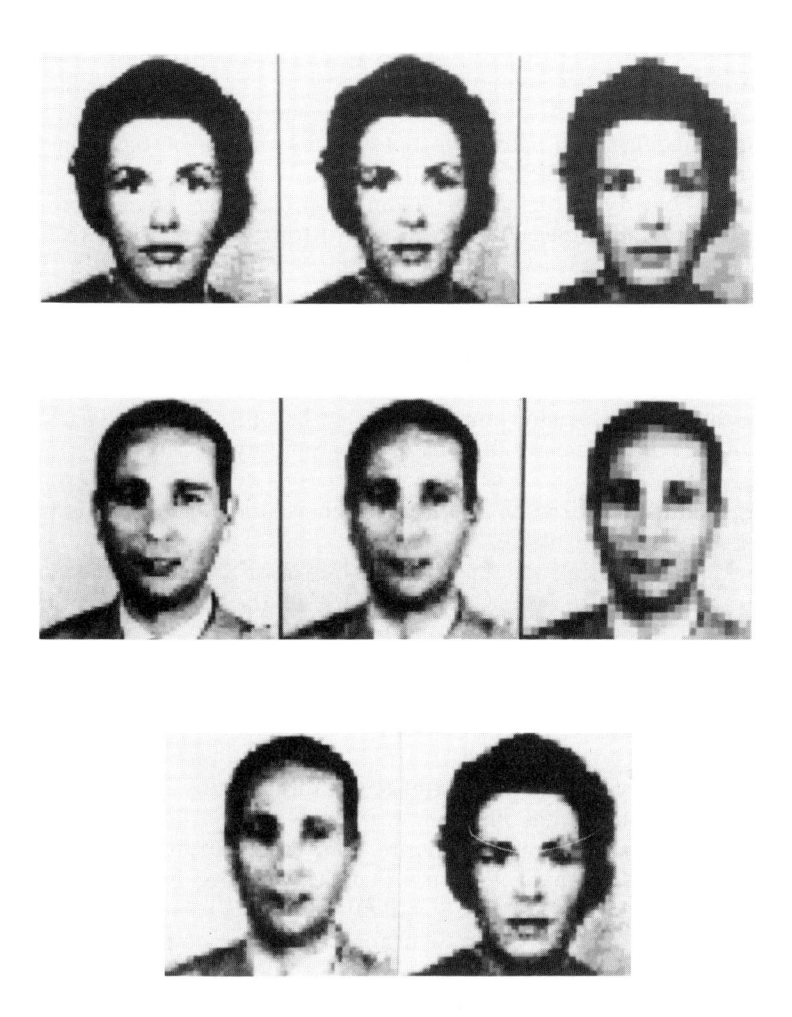

Figure 1. Top photographs: an original image (left side), the image obtained by its LP components (right side), and the image obtained by the spectrum extrapolation performed by a neural network trained to it. Following photographs: as above, with a similar image. Bottom photographs: images obtained by means of spectrum extrapolation performed by a network not trained to the subject image, but trained to the other.

transmission of images in which changes with time are slowly introduced, as in video conference or video telephone applications. Moreover, the small dimensions of the pixel blocks, compared to the complete image, and the small training times suggest to update the network by means of training phases distributed during transmission.

For instance, two types of on line training procedures have been tested. Following the first, a complete spectrum is transmitted from time to time, to update the receiving associative memory. The second is a further improvement, exploiting the image to image redundancy of television transmissions: full spectra are transmitted only when important image changes are observed, and spectrum extrapolation is performed during intervals. In practice, blind transmission of a complete spectrum for only one new pixel block at any new complete television image seems to be quite satisfactory. In any case, the overall transmission bit rates are not notably increased by updating information.

Similar procedures have to be applied to higher compression ratios to check the limits of suitable application.

This work was partially supported by Marconi S.p.a., Genova, Italy. Helpfull discussions with Prof. G.L. Sicuranza of the University of Trieste, Italy, and with Dr. G. Colusso are gratefully acknowledged.

3. REFERENCES

1 G.W. Cottrell, P. Munro, D. Zipser, Image Compression by Back Propagation: An Example of Extensional Programming, in N.E. Sharkey, Ed., Models of Cognition: A Review of Cognition Science. Norwood, NJ: Ablex, 1989.

2 G.L. Sicuranza, G. Ramponi, S. Marsi,"Artificial Neural Network for Image compression," Electronics Letters, 26 (1990) 477.

3 S. Marsi, G. Ramponi, G.L. Sicuranza, Image Compression Using a Perceptron, in E.R. Caianello, Ed., Third Italian Workshop on Parallel Architectures and Neural Networks, World Publishing Co., 1990.

4 W.K. Pratt, Digital Image Processing. New York: J.Wiley, 1978, parts 4, 5, 10 and 23.

5 E. Costamagna, "Application of spectrum extrapolation techniques to very low resolution image processing," Alta Frequenza, vol. LIII (1984) 50.

A SIMPLIFIED 2-D IIR DIGITAL FILTER AND ITS APPLICATION TO IMAGE PROCESSING

Takao HINAMOTO and Mitsuji MUNEYASU

Faculty of Engineering, Hiroshima University, Kagamiyama 1-4-1
Higashi-Hiroshima 724, Japan

Abstract
A simple 2-D IIR filter is considered that can be viewed as a special case of the Fornasini-Marchesini second model. If image processing is carried out using such a filter from four directions, we can effect smoothing, edge detection or edge enhancement without any distortion. The above filtering can flexibly be performed by choosing three parameters only. Some examples illustrate the utility of the proposed technique.

1. INTRODUCTION

Two-dimensional (2-D) digital filters are useful in image processing. An IIR filter requires a significantly smaller number of coefficients to meet a particular magnitude specification than does an FIR filter. However, since an IIR filter always has nonlinear phase, the image processed by applying an IIR filter is distorted in various ways, including blurring [1]. To avoid this distortion, zero phase can be achieved by applying an IIR filter from four directions.
This paper proposes a simple 2-D IIR filter which is viewed as a special case of the Fornasini-Marchesini second model [2]. This filter contains three parameters only. Applying this filter to image processing from four directions, it is possible to carry out smoothing, edge detection and edge emphasis without any distortion in the processed image. In addition, the filter stability is always guaranteed by choosing a parameter within the region specified in the paper.

2. 2-D IIR FILTERS FOR IMAGE PROCESSING

Consider the following 2-D IIR digital filter:

$$x_1(i,j) = ax_1(i-1,j) + ax_1(i,j-1) + \alpha u(i-1,j) + \alpha u(i,j-1)$$
$$y_1(i,j) = \beta x_1(i,j) + \gamma u(i,j) \tag{1}$$

where $a = (1-2\alpha)/2$ and $0 < \alpha < 1$. The original image and image processed by applying the filter, (1), are denoted by $u(i,j)$ and

$y_1(i,j)$, respectively. Fig.1 shows the ordering for computing the output of the filter, (1). The stability of the filter is ensured by choosing the parameter α as $0 < \alpha < 1$.

The transfer function of (1) is given by

$$H_1(z_1,z_2) = \frac{\alpha\beta(z_1^{-1}+z_2^{-2})}{1-a(z_1^{-1}+z_2^{-1})}+\gamma . \tag{2}$$

The filter (1) is represented using a unit sample response as

$$y_1(i,j) = \gamma u(i,j)+\alpha\beta\sum_{r\geq 1}\sum_{0\leq p\leq r}\frac{r!}{p!(r-p)!}a^{r-1}u(i-r+p,j-p) \tag{3}$$

By appropriately choosing the parameters α, β and γ we can freely design a 2-D filter with either lowpass, highpass, or bandstop characteristics. Table 1 indicates the relationship between the region of parameters and the filter function. Some examples of characteristics are shown in Figs. 2-4. The filter can be used for the smoothing, edge detection or edge emphasis of images.

3. 4-DIRECTIONAL FILTERING

To avoid distortion due to nonlinear phase, we use the other three filters:

$$\begin{aligned}x_2(i,j) &= ax_2(i-1,j)+ax_2(i,j+1)+\alpha u(i-1,j)+\alpha u(i,j+1)\\ y_2(i,j) &= \beta x_2(i,j)+\gamma u(i,j)\end{aligned} \tag{4}$$

$$\begin{aligned}x_3(i,j) &= ax_3(i+1,j)+ax_3(i,j+1)+\alpha u(i+1,j)+\alpha u(i,j+1)\\ y_3(i,j) &= \beta x_3(i,j)+\gamma u(i,j)\end{aligned} \tag{5}$$

$$\begin{aligned}x_4(i,j) &= ax_4(i+1,j)+ax_4(i,j-1)+\alpha u(i+1,j)+\alpha u(i,j-1)\\ y_4(i,j) &= \beta x_4(i,j)+\gamma u(i,j)\end{aligned} \tag{6}$$

These filters have the same coefficients as in (1). However, the ordering for computing the output of each filter is different from (1). If these filters are used from four directions, zero phase can be attained as a whole. The processed image is then derived from

$$\begin{aligned}y(i,j) &= \{y_1(i,j)+y_2(i,j)+y_3(i,j)+y_4(i,j)\}/4\\ &= \gamma u(i,j)+\sum_{r\geq 1}\sum_{0\leq p\leq r}\frac{r!}{p!(r-p)!}a^{r-1}\\ &\times\{u(i-r+p,j-p)+u(i-r+p,j+p)\\ &\quad+u(i+r-p,j+p)+u(i+r-p,j-p)\}\end{aligned} \tag{7}$$

where the impulse response of overall filter is spread out over four quadrants.

The transfer function of each filter is represented by

$$H_2(z_1,z_2)=H_1(z_1,z_2^{-1}), H_3(z_1,z_2)=H_1(z_1^{-1},z_2^{-1}), H_4(z_1,z_2)=H_1(z_1^{-1},z_2) . \tag{8}$$

The frequency response of the overall filter is obtained by

$$\frac{1}{4}\sum_{k=1}^{4}H_k(e^{j\omega_1},e^{j\omega_2}) = \alpha\beta\{A_1(\omega_1,\omega_2)/B_1(\omega_1,\omega_2) \qquad (9)$$
$$+A_2(\omega_1,\omega_2)/B_2(\omega_1,\omega_2)\}+\gamma$$

where
$$A_1(\omega_1,\omega_2) = \cos\omega_1+\cos\omega_2-2a\{1+\cos(\omega_1-\omega_2)\}$$
$$A_2(\omega_1,\omega_2) = \cos\omega_1+\cos\omega_2-2a\{1+\cos(\omega_1+\omega_2)\}$$
$$B_1(\omega_1,\omega_2) = 2(1-2a)(\cos\omega_1+\cos\omega_2)+4a^2\{1+\cos(\omega_1-\omega_2)\}$$
$$B_2(\omega_1,\omega_2) = 2(1-2a)(\cos\omega_1+\cos\omega_2)+4a^2\{1+\cos(\omega_1+\omega_2)\} .$$

4. IMAGE PROCESSING EXAMPLES

The original image of 50 x 50 pixels is shown in Fig.5 and the noisy image degraded by white noise is shown in Fig.6 where the magnitude of white noise is limited to 10% of the maximum of pixel values and the root mean square ratio of noise and signal is equal to 7.4%.
(1) Smoothing
   Applying a lowpass filter shown in Fig.2 to the noisy image in Fig.6, the image processed by four filters from four directions is shown in Fig.7.
(2) Edge Detection
   Applying a highpass filter shown in Fig.3 to the original image shown in Fig.5, the image processed by four filters from four directions is shown in Fig.8.
(3) Edge Emphasis
   Applying a bandstop filter shown in Fig.4 to the noisy image in Fig.6, the image processed by four filters from 4 directions is shown in Fig.9.

5. CONCLUSION

A simple 2-D IIR filter has been proposed that can be used for the smoothing, edge detection and edge emphasis of images by choosing three parameters only. The filter stability is always guaranteed by letting $0 < \alpha < 1$. When this filter is used so as to process an image from four directions, there is no distortion in the processed image due to zero-phase. Some examples have been given to illustrate the utility of proposed technique.

REFERENCES

[1] T.S.Huang,J.W.Burnett and A.G.Deczky:"The importance of phase in image processing filters",A Trans. Acoust.Speech. & Signal Process.**ASSP-23**,pp.530-542,Sept. 1975.
[2] E.Fornasini and G.Marchesini:"Doubly-indexed dynamical systems:State-space models and structural properties", Math. Syst. Theory,**12**,pp.59-72,1978.

Table 1. Relationship between parameters and filter function

filter function	parameters
smoothing	$0<\alpha<0.5, \beta>\gamma>0$
edge detection	$0.4\leq\alpha\leq0.6, \beta=-\gamma\ (\beta>0)$
edge emphasis	$0.8\leq\alpha<1.0, \beta>\gamma>0$

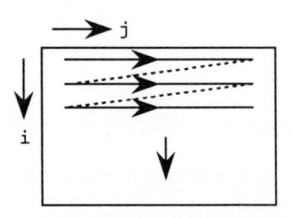

Fig. 1 Processing direction of a prototype filer.

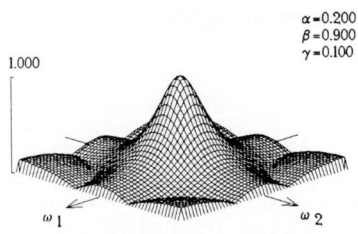

Fig. 2 Magnitude response of Eq.(2) with lowpass characteristics.

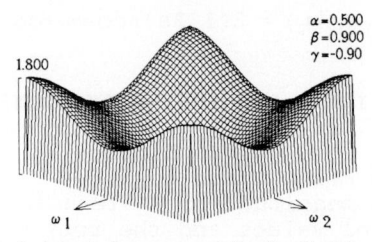

Fig. 3 Magnitude response of Eq.(2) with highpass characteristics.

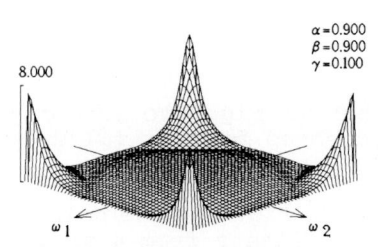

Fig. 4 Magnitude response of Eq.(2) with bandstop characteristics.

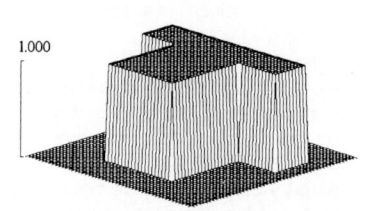

Fig. 5 Original 50×50 image.

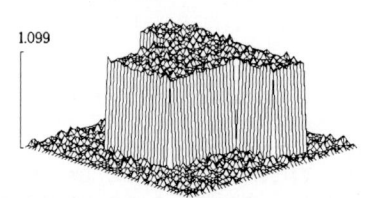

Fig. 6 Original 50×50 image with noise.

Fig. 7 Result of smoothing.

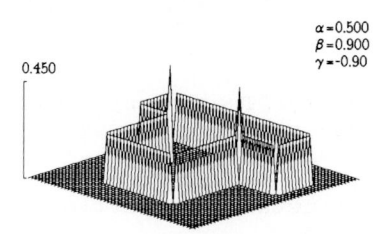

Fig. 8 Result of edge detection.

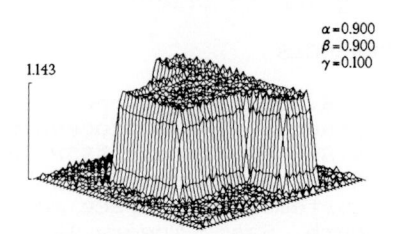

Fig. 9 Result of edge enhancement.

Nonlinear methods for the restoration and reconstruction of images with quasilinear solutions

B. Bundschuh

Universität–GH–Siegen, Hölderlinstraße 3, D–5900 Siegen, Germany

Abstract
If the statistical properties of an image are *not* stationary, linear shift *in*variant algorithms usually fail to provide good restorations. Discontinuities such as steps remain blurred while on the other hand there is considerable ripple where the image to be restored is smooth [1]. In this paper two ways out of this dilemma are shown and compared, linear shift variant restoration and nonlinear restoration.

1. LINEAR RESTORATION

Often a signal cannot be measured directly but only after having passed a low pass type system. The measured data d_{kl} are generated by twodimensional convolution of the original signal s_{ij} and the point spread function h_{mn}. The result of the convolution operation is corrupted by the measurement noise r_{kl}.

$$d_{kl} = \sum_i \sum_j h_{k-i,l-j} s_{ij} + r_{kl} \tag{1}$$

In order to obtain an estimate $\hat{s}_{ij}$ we can use restoration methods, which are based on a weighted linear shift invariant regularization function f_{mn}, usually a high pass filter.

$$E = E_d + E_r = \sum_k \sum_l \left(d_{kl} - \sum_i \sum_j \hat{s}_{ij} h_{k-i,l-j} \right)^2 + \alpha \sum_i \sum_j w_{ij} \left(\sum_m \sum_n f_{mn} \hat{s}_{i-m,j-n} \right)^2 \tag{2}$$

Partial derivation of E leads to a linear system of equations, which is solved iteratively. The weighting function w_{ij} is large where the image is smooth and small at discontinuities.

$$\sum_i \sum_j \hat{s}_{ij} \left(\sum_k \sum_l h_{k-i,l-j} h_{k-p,l-q} + \alpha \sum_m \sum_n w_{mn} f_{m-i,n-j} f_{m-p,n-q} \right) = \sum_k \sum_l d_{kl} h_{k-p,l-q} \tag{3}$$

It can be shown [1] that f_{mn} should be derived from the autocorrelation function of the original signal. The optimal weighting function was derived in [2].

$$w_{ij_{opt}} = 1 / \left\langle \left(\sum_m \sum_n f_{mn} s_{i-m,j-n} \right)^2 \right\rangle \qquad \alpha = \sigma_r^{-2} \tag{4a,b}$$

Eq. 4a shows that the optimal weighting function is matched to the regularization function in a remarkably simple way. According to [2] α should be equal to the reciprocal of the noise variance. $\langle ... \rangle$ is the ensemble average.

If w_{ij} is set to 1 the algorithm becomes shift invariant and also covers the shift invariant Wiener Filter [3]. If the original signal and the noise are Gaussian distributed it is equivalent to a Maximum a Posteriori Estimation [3]. If the image is smooth the quality of the shift invariant restoration can be very good. But if the assumption of stationarity is not valid, such as for most real world images, the restoration provides only an unsatisfactory compromise between resolution and smoothness. Discontinuities remain blurred and there is strong ripple in smooth regions. The special case $\alpha = 0$ covers the simple unconstrained Least Squares Estimation, the Pseudo Inverse and the Inverse Filter [3]. If the noise is Gaussian distributed it is equivalent to a Maximum Likelihood Estimation [3]. $\alpha = 0$ means that the noise is neglected. In most cases this leads to strong noise amplification [1].

2. NONLINEAR RESTORATION

Various restoration algorithms can be derived using the very general approach in eq. 5 where $F(|...|)$ is an arbitrary nonlinear transformation.

$$E = E_d + E_r = \sum_k \sum_l \left(d_{kl} - \sum_i \sum_j \hat{s}_{ij} h_{k-i, l-j} \right)^2 + \alpha \sum_i \sum_j F\left(\left| \sum_m \sum_n f_{mn} \hat{s}_{i-m, j-n} \right| \right) \quad (5)$$

Minimization of the error term E leads to the following nonlinear system of equations.

$$\sum_i \sum_j \hat{s}_{ij} \left(\sum_k \sum_l h_{k-i, l-j} h_{k-p, l-q} + \alpha \sum_m \sum_n f_{m-i, n-j} f_{m-p, n-q} \cdot \right.$$

$$\left. \cdot \sum_m \sum_n \dot{F}\left(\left| \sum_u \sum_v f_{uv} \hat{s}_{m-u, n-v} \right| \right) \cdot \left| \sum_u \sum_v f_{uv} \hat{s}_{m-u, n-v} \right|^{-1} \right) = \sum_k \sum_l d_{kl} h_{k-p, l-q} \quad (6)$$

This nonlinear approach introduces a new starting point for the development of image restoration algorithms. A comparison of eq. 3 and eq. 6 shows that the general structures of the systems of equations are similar if the definition in eq. 7 is used. Theoretically w_{mn} does not depend on the estimate in eq. 3 but it does in eq. 6.

$$w_{mn} \Rightarrow \dot{F}\left(\left| \sum_u \sum_v f_{uv} \hat{s}_{m-u, n-v} \right| \right) \cdot \left| \sum_u \sum_v f_{uv} \hat{s}_{m-u, n-v} \right|^{-1} \quad (7)$$

Since the term in eq. 4a is hardly known a priori, $s_{i-m, j-n}$ is replaced by $\hat{s}_{i-m, j-n}$ in practical applications. This means that the measured data, which provide the starting point of the iteration, respectively the intermediate results of the iteration are used to determine an estimate of the weighting function, which is updated after each step of the iteration [1]. The repeated convolutions and correlations during the iteration are calculated efficiently in the frequency domain. Because of the repeated update of w_{ij} a formal proof of convergence is extremely difficult. Therefore it was tested experimentally and fortunately no problems occurred. The same approach can also be used in the nonlinear case if the weighting function is replaced according to eq. 7. It is also updated after each step of the iteration. This leads to an iterative quasilinear solution. An example for the nonlinear transformation $F(|...|)$ is the L_q–norm. The special case $q = 2$ leads to linear shift invariant algorithms.

$$E_r = \alpha \sum_i \sum_j \left| \sum_m \sum_n f_{mn} \hat{s}_{i-m,j-n} \right|^q \Rightarrow w_{ij} \hat{=} q \cdot \left| \sum_u \sum_v f_{uv} \hat{s}_{i-u,j-v} \right|^{q-2} \quad (8a)$$

The Minimum Cross Entropy term according to Frieden [4] with $\hat{s}_{ij}$ strictly positive and $f_{mn} = \delta_{mn}$ is probably the most popular nonlinear restoration method.

$$E_r = \alpha \sum_i \sum_j \hat{s}_{ij} \cdot \ln(\hat{s}_{ij}) \Rightarrow w_{ij} \hat{=} \left(1 + \ln(\hat{s}_{ij})\right)/\hat{s}_{ij} \quad (8c)$$

The Maximum Entropy term according to Burg [5] with $\hat{s}_{ij}$ strictly positive and $f_{mn} = \delta_{mn}$ provides another well known solution, which is equivalent to eq. 4a.

$$E_r = \alpha \sum_i \sum_j \ln(\hat{s}_{ij}) \Rightarrow w_{ij} \hat{=} \hat{s}_{ij}^{-2} \quad (8b)$$

The problem of selecting a particular nonlinear transformation is most easily explained using a one dimensional example. The original signal is assumed to be constant on the two sides of a single step and we set $f_m = \delta_m - \delta_{m-1}$. If the L_2–norm is used according to eq. 8a, blurring of the step reduces the regularization error E_r. The algorithm becomes linear and shift invariant. If a modified Entropy term similar to eq. 8c is used, a large single step leads to a smaller regularization error than many small ones. This is optimal (see eq. 4a) if steps are the only type of discontinuity. If other types exist, this algorithm tends to create artefacts. The L_1–norm leads to a more versatile algorithm, which neither tends to blur steps too strongly neither to create too many artefacts [6].

3. RESULTS OF SIMULATION

Figure 1 shows the original signal, a random collection of blocks and wedges. Figure 2 shows a degraded version. The point spread function is the two dimensional Gaussian function $h_{mn} = c \cdot \exp(-\frac{1}{2}(m^2+n^2)/4)$. The noise is white and Gaussian distributed with a SNR of 20 dB. The regularization function is chosen as: $f_{mn} = 2 \cdot \delta_{mn} - \delta_{m-1,n} - \delta_{m,n-1}$.

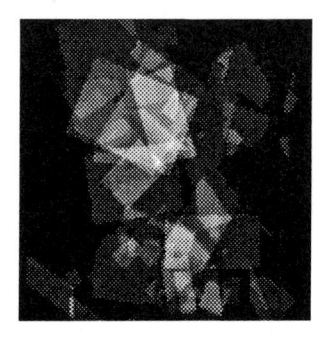

Figure 1. Simulated test signal

Figure 2. Blurred and noisy image

Figure 3 and 4 show the result of a linear shift variant restoration respectively the result of a nonlinear restoration using the L_1–norm.

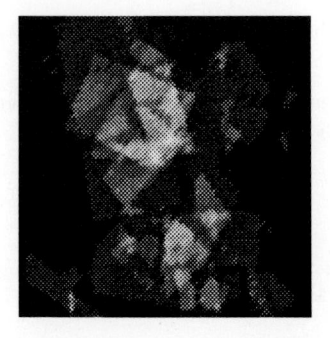

Figure 3. Result of linear restoration 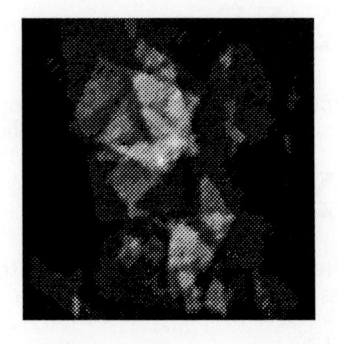   Figure 4. Result of nonlinear restoration

4. CONCLUSIONS

It was shown that linear shift variant and nonlinear image restoration can be regarded to be equivalent. A very general nonlinear error term, which is to be minimized, provides a base for various restoration methods. The minimum of the error term is determined as an iterative quasilinear solution of a nonlinear system of equations. The nonlinear approach allows the flexible design of algorithms for various types of inverse problems. Therefore the field of application is not restricted to image restoration.

5. ACKNOWLEDGEMENT

I would like to thank Olga Miliukova from the Institute of Information Transmission Problems in Moscow who provided the idea for image restoration using the L_1-norm.

6. REFERENCES

1. B. Bundschuh, Adaptive Image Restoration and Reconstruction, Third International Seminar on Digital Image Processing in Medicine, Remote Sensing and Visualization of Information, Riga, Latvia, April 1992
2. B. Bundschuh, A Linear Predictor as a Regularization Function in Adaptive Image Restoration and Reconstruction, 5th International Conference on Computer Analysis of Images and Patterns, Budapest, Sept. 1993, to be published
3. H.C. Andrews and B.R. Hunt, Digital Image Restoration, Prentice-Hall, Englewood Cliffs, N.J., 1977
4. B.R. Frieden, Restoring with Maximum Likelihood and Maximum Entropy, Journal of the Optical Society of America, Vol. 62, No. 4, 1972
5. J.P. Burg, Maximum Entropy Spectral Analysis, Ph.D. Dissertation, Stanford University, Stanford, California, 1975
6. O. Miliukova, On the Image as a Function of Bounded Variation, Third International Seminar on Digital Image Processing in Medicine, Remote Sensing and Visualization of Information, Riga, Latvia, April 1992

Image Processing: Theory and Applications
G. Vernazza, A.N. Venetsanopoulos, C. Braccini (Editors)
© 1993 Elsevier Science Publishers B.V. All rights reserved.

Detection and Correction of Speckle Degradation in Image Sequences using a 3D Markov Random Field

Robin D. Morris and W. J. Fitzgerald

Communications and Signal Processing Laboratory, Cambridge University Engineering Department, Trumpington Street, Cambridge, CB2 1PZ, England. Tel +44 223 33 2767

Abstract

We describe a method for restoring image sequences degraded by *speckle*. We model the sequence as a 3D Markov Random Field, using a motion compensated temporal neighbourhood. The conditional probability assigned to each pixel by the model is used to *detect* speckle. The flagged pixels are then *restored* by replacing them with the grey scale value with the highest conditional probability.

INTRODUCTION

In our experience of processing old motion pictures, in many cases the main degradation is not global white noise, but point degradations from dirt and scratches. We therefore do not wish to process the entire image, rather we want to detect the location of the point degradations and correct only those pixels. By not processing undegraded areas of the image we avoid blurring problems associated with global processing and long-range interaction problems often associated with Markov Random Field (MRF) models. A simple interaction model is also suitable, as we need to model only very small areas of the image.

An MRF model is used as the prior on our image. To perform Bayesian restoration we must also consider the likelihood. Considering the nature of speckle degradation, in effect we have two likelihood functions. For pixels degraded by speckle, the observed data tell us nothing about the true values at those points, the likelihood being uniformly distributed across the allowable states. For an un-degraded pixel, the likelihood is one for the pixel having the value observed, and zero for all other values. We must therefore decide which likelihood to use at each pixel. This is the rôle of the speckle detection part of the algorithm.

ALGORITHM

An MRF image model is specified by the conditional probability of a pixel having a particular value, given the values of the pixels in a finite neighbourhood [1]. That is we specify

$$p(X_i = x_i | X_j = x_j, i \neq j) = p(X_i = x_i | X_j = x_j, j \in \mathcal{C}_i) \qquad (1)$$

where C_i is the finite neighbourhood of X_i. To detect and correct defects in the current frame, we make use of the information within the current frame, and that in the previous and next frames. We choose the neighbourhood to be the eight-neighbours in the current frame, denoted C_i, together with one motion-compensated neighbour in each of the previous and next frames, denoted P_i and N_i respectively. The neighbourhood of each pixel is intersected with the lattice forming the images.

By the MRF-Gibbs distribution equivalence theorem (Hammersly-Clifford theorem) [2], the conditional probability for the grey scale value at X_i can be written as

$$p(X_i = x_i | X_{C_i} = x_{C_i}, P_i = p_i, N_i = n_i) =$$

$$\frac{1}{Z_i} \exp -\left(\alpha_1 \sum_{j \in C_i} \phi(x_i - x_j) + \alpha_2 \phi(x_i - p_i) + \alpha_2 \phi(x_i - n_i) \right) \quad (2)$$

where Z_i is known as the partition function and $\phi(\cdot)$ is the potential function. In this model we treat only interactions between pairs of pixels. We use

$$\phi(u) = -1/(1 + |u/\delta|) \quad (3)$$

as the potential function, where δ is a scale parameter. It has been noted that this potential function encourages local smoothness, but penalises very large changes in grey scale value only marginally more than modest changes. This allows different regions in the image to develop naturally without the need for a line process [3].

The detection and restoration algorithm proceeds as follows.
1. Using equation 2 calculate the conditional probability assigned by the model to each pixel.
2. Form a histogram of the conditional probabilities. This will split into two regions, corresponding to areas modelled well (ie undegraded areas), and areas modelled badly, that is speckle. See figure 1.
3. Choose the threshold at the minimum between the two peaks of the histogram.
4. At each pixel where the conditional probability is below the threshold, replace it by the grey scale value having the highest conditional probability.

This is equivalent to a single iteration of the ICM algorithm [4] *at those pixels detected as speckle*.

EXPERIMENTS

We present results from experiments comparing our algorithm with the multilevel median filter (MMF) proposed in [5]. The image sequence shows a member of our laboratory juggling. It has areas undergoing significant motion and so represents a difficult test for any restoration algorithm incorporating information from more than one frame, due to the difficulty of estimating motion vectors accurately.

Speckle was simulated by choosing a certain percentage of the pixels at random, and then replacing the original grey scale values by ones drawn independently from a uniform distribution over the allowed states.

Parameter values

The MRF image model depends on two parameters, α_1 and α_2, and the function $\phi(\cdot)$, which depends on δ. It is our experience that δ need change only with the number of grey levels; in our experiments a value of 6 was used with 6-bit images. α_1 was set to 1, and α_2, which controls how much weight is given to temporal neighbours, was set to 1.5. The threshold was chosen from figure 1 to be 0.01. The results discussed below show that the value of the threshold is not particularly critical.

Results

For the experiments, speckle was added to three sequential frames taken from the sequence. Three runs were performed using 1, 2 and 5% speckle. Motion estimation was performed using hierarchical block matching [6], with integer pixel accuracy which produced one motion vector per 7 by 7 pixel block in the image. This one motion vector was used for all the pixels in that block. Figure 2 shows the detector characteristics for this case. The algorithm is seen to be robust to differing levels of degradation. The false detection shows a cut-off, below which no false detections occured and the chosen threshold is seen to perform well. The true test is the visual quality of the restored frame. Figure 4 shows that the small amounts of speckle still visible are in areas of large motion, around the juggler's hands. Some artefacts are also visible along the boundary between the arms and the background. These are due to the motion estimation providing only one motion vector per block, and could be reduced by using smaller blocks in the estimation.

The MMF filter [5] was also applied to the degraded sequences. Whilst it did remove most of the speckle, being a *global* filter it also modified areas of the frame where no speckle had been added. Mainly because of this, in each of the three cases the MMF restoration had a mean squared error typically 40% larger than the restoration produced by our algorithm. (eg for 2% speckle, the mse for our algorithm was 0.9; for the MMF restoration it was 1.3. The degraded frame had an mse of 15.)

CONCLUSIONS

We have developed an algorithm for speckle removal from motion pictures. We have demonstrated its effectiveness in the presence of motion, and the uncritical nature of its parameters. We believe it to be a useful technique for restoring old films.

Acknowledgements

Funding was provided by the SERC. The authors would like to thank K. G. Lim for a number of conversations which initiated this work, and Anil Kokaram, for starring in the 'juggler' sequence, providing the motion estimation, and much useful advice.

REFERENCES

[1] S. Geman and D. Geman. Stochastic relaxation, Gibbs distributions and the Bayesian Restoration of Images. *IEEE Trans PAMI*, 6(6):721-41, November 1984.

[2] J. Besag. Spatial interaction and the statistical analysis of latice systems. *J. Royal Statistical Soc. B*, 36:192-326, 1974.

[3] S. Geman, D. E. McClure and D. Geman. A nonlinear filter for film restoration and

other problems in image restoration. *CVGIP: Graphical Models and Image Processing*, 54(4):281-289, July 1992.

[4] J. Besag. On the statistical analysis of dirty pictures. *J. Royal Statistical Soc. B*, 48(3):259-302, 1986.

[5] G. R. Arce. Multistage order statistic filters for image sequence processing. *IEEE Trans Signal Processing*, 39(5):1146-1163, May 1991.

[6] M. Bierling. Displacement estimation by hierarchical block matching. *SPIE: Visual Communications and Image Processing*, 1001:924-951, 1988.

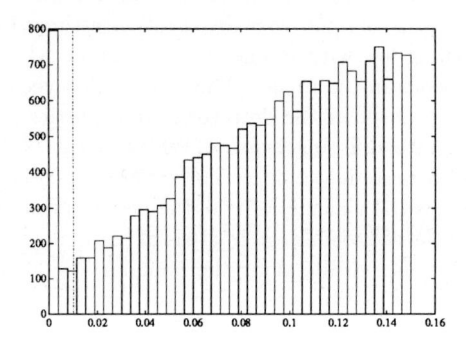

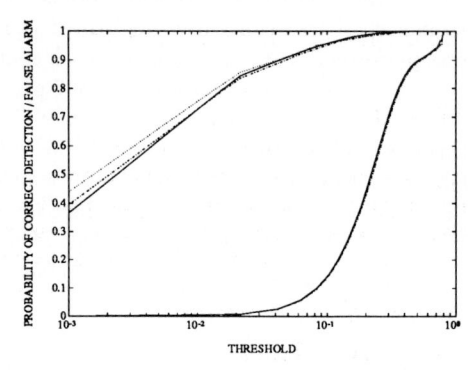

Figure 1: Partial Histogram of conditional probability values, 2% speckle. Vertical line is chosen threshold

Figure 2: Detector characteristic. solid 5%, dash-dot 2%, dotted 1%

Figure 3: Degraded Frame, 2% speckle

Figure 4: Restored Frame

Image Processing: Theory and Applications
G. Vernazza, A.N. Venetsanopoulos, C. Braccini (Editors)
© 1993 Elsevier Science Publishers B.V. All rights reserved.

EMBEDDING WAVELET TRANSFORMS ONTO PARALLEL ARCHITECTURES

Mounir Hamdi[a] and Richard W. Hall[b]

[a]Department of Computer Science, Hong Kong University of Science and Technology, Clear Water Bay, Kowloon, Hong Kong

[b]Department of Electrical Engineering, University of Pittsburgh, Pittsburgh, PA. 15261, U.S.A.

Abstract

Wavelet Transforms are very effective tools for signal analysis in many problems in image processing and computer vision for which Fourier based methods have been inapplicable, expensive for real-time applications, or can only be applied with difficulty. We present the parallel implementation of wavelet transforms onto parallel computer systems with emphasis on the hypercube. Two efficient embeddings of wavelet transforms onto the hypercube are studied along with the parallel algorithms associated with that. The time performance on basic wavelet transform algorithms are given in order to assess the different embeddings.

1. INTRODUCTION

The method by which an image is represented is very crucial in computer vision since the accuracy of the final result depends on the adequacy of these representations. They transfer the pixel-level data into more organized and goal dependent attributes so that the objects and events in a 3-D scene can be described and recognized from these descriptions. In particular, they should be complete and stable in the sense that necessary information for any particular task can be extracted and robust image processing algorithms in terms of noise sensitivity can be built upon these representations. The wavelet representations is such a representation and is based on the wavelet theory developed by mathematicians Grossmann, Morlet, Meyer, et al. [1, 2]. The wavelet representation enables us to understand and model the concepts of resolution and scale and to obtain a scale-independent interpretation of the image. Its application to multi-scale image processing has emerged as an exciting research problem during the past three years [1]. The multi-resolution analysis allows an image to be studied at different scales; a scaled image at the resolution 2^{j-1} can be decomposed into the sum of the approximation at the lower scale (resolution 2^j) and the difference image between two scales which can be represented by wavelet transform at the jth scale.

The abundance of applications of wavelet transforms and their computationally intensive nature has led to the consideration of embedding images represented in wavelet forms onto parallel architectures [9]. In this paper, we consider the fine-grain embedding of wavelet coefficients onto parallel architectures, that is, we assume that each processing element (PE) will store only one coefficient. These types of fine-grain (massive) parallel architectures are the most widely adopted parallel architectures for image processing and computer vision [8]. Thus, the main problem would be in the efficient placement of the wavelet coefficients onto the parallel architectures. We consider two different embeddings in this paper, and we evaluate their time performance on different parallel transform algorithms. Although these embeddings can be considered for different types of parallel architectures, we focus in this paper on their implementation on hypercube computers [6]. In section 2, we present the two different embeddings of wavelet coefficients onto hypercubes. In section 3, we present the implementation of fundamental parallel wavelet transform algorithms using the two embeddings. In section 4, we evaluate and compare the time performances of the two embeddings on these algorithms.

2. EMBEDDINGS OF WAVELET COEFFICIENTS ONTO HYPERCUBES

A hypercube architecture of size n^2 can be thought of as an array of PE's containing n rows and n columns, where the rows and columns are hypercube connected. For more details please refer to [3]. There are two different embbedings of wavelet coefficients that are considered for the hypercube as illustrated in Fig. 1. The original image is of size $n \times n$ and is labelled to be at level 0. The lower resolution images are at higher levels, $i = 1, 2, 3, ...$, which is characteristic of wavelet representations. The embedding illustrated in Fig. 1(a), which is similar to [4, 5, 9], the detail images D_i^1, D_i^2 and D_i^3 at level i of the multiscale transform and the lower resolution version of the image, A, are concentrated into sub-blocks within the hypercube. This embedding is denoted Block-Concentrated (**BC**). A second embedding has been developed, denoted Distributed (**Dist**), and is illustrated in Fig. 1(b) where the individual components of A, a, and the individual components of the detail images at level i, d_i^1, d_i^2 and d_i^3, are distributed over the parallel architecture. Specific detail image components at level i ($i > 0$) correspond to the same base image location are mapped to PE's in the hypercube which are 2^i distance apart for the **Dist** case and 2^{n-i+1} distance apart for the **BC** case. Thus, **BC** has higher communication cost at lower levels and lower communication cost at higher levels while the opposite holds for **Dist** especially for parallel architectures with low bandwidth.

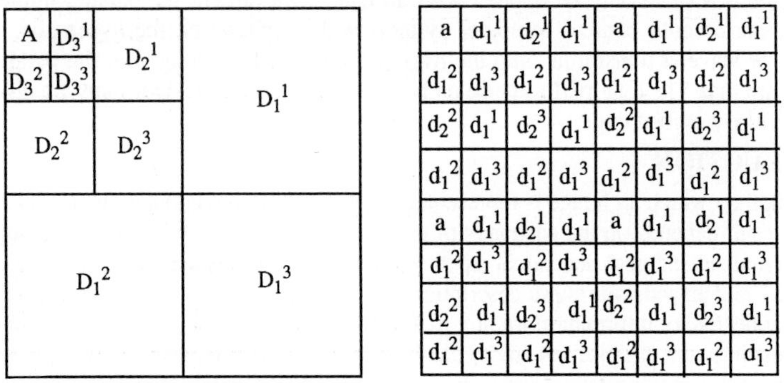

Fig. 1. Wavelet transform embedding in a hypercube architecture
(a) Block-Concentrated embedding (**BC**); (b) Distributed (**Dist**) embedding.

3. PARALLEL WAVELET TRANSFORM ALGORITHMS

Two fundamentals parallel wavelet algorithms are implemented and their time performances are analyzed with respect to the embeddings used. The first parallel algorithm decomposes a given image into its wavelet transform and the second algorithm reconstructs the original image from its wavelet transform. These two algorithms are essential in any parallel application of wavelet transforms [4, 5]. The decomposition process requires a set of linear convolutions along rows and columns of the hypercube with special data movements to place the results in the desired locations [4]. To perform the image decomposition using the emebedding of Fig. 1(a) on the hypercube, we need two data movement operations: convolution and shifting. We need to convolve, separately, the rows and the columns of the hypercube with one-dimensional filters, $\overline{H}$ or $\overline{G}$, as shown in [4]. The linear convolution algorithm used on the hypercube is based on the algorithm given in [7]. The shifting process is needed to move a lower resolution image or a detailed image spread on the whole hypercube or on a section of the hypercube of size $n \times n$. Thus, as can be seen from Fig. 1, we need shifting in four directions (up, down, right, left) to implement the parallel decomposition algorithm. To perform the image decomposition using the embedding of Fig. 1(b), the same data movement operations are used, and hence the same algorithm.

However, in this case the convolution operation would take longer than with the previous embedding since the coefficients are now spread all over the hypercube. On the other hand, shifting would take less communication steps because of the proximity of the pixels associated with this embedding that need to be shifted. The implementation of the parallel algorithms for the reconstruction of images from wavelet transforms is the inverse process of the decomposition algorithms. It also needs the same data movement operations previously explained. The reconstruction process requires the insertion of rows and columns of zeros in the hypercube, row and column linear convolutions, and data movements to restore image components to their original locations [4].

Now let us try to determine the number of communication steps needed by the decomposition and reconstruction algorithms using the two embeddings. Only the communication steps (e.g. movements of data from PE to neighboring PE) are evaluated in this analysis since both embeddings require the same number of computation steps (e.g. adds and multiplies within a given PE). The number of communication steps required for each algorithm over each embedding have been evaluated in closed form. Some typical results are illustrated in Table I for filter size of 8. The results are normalized by expressing the number of communication steps as a multiple of the linear dimension of the image, n. In most

Table I. Communication steps required by wavelet decomposition and reconstruction algorithms taken to/on 8×8 lowest resolution images (filter size is 8).

Image Size $n \times n$	Decomposition BC	Decomposition Dist	Reconstruction BC	Reconstruction Dist
1024×1024	$1.2\,n$	$1.5\,n$	$1.1\,n$	$3.5\,n$
512×512	$2.0\,n$	$2.2\,n$	$1.9\,n$	$3.6\,n$
256×256	$3.4\,n$	$3.5\,n$	$3.4\,n$	$3.9\,n$
128×128	$5.6\,n$	$5.6\,n$	$5.7\,n$	$6.1\,n$
64×64	$8.8\,n$	$8.1\,n$	$9.3\,n$	$11.3\,n$

practical applications, the number of resolution images is typically lower than that represented by Table I. Thus, in order to evaluate the two embeddings from a practical point of view, we assumed that the wavelet transforms are performed only to three levels of resolution for each image space size. Table II illustrates the number of communication steps needed for the construction algorithm and the decomposition algorithm under this assumption for both embeddings.

Table II. Communication steps required by wavelet decomposition and reconstruction algorithms working on images of only 3 levels of resolution (filter size is 8).

Image Size $n \times n$	Decomposition BC	Decomposition Dist	Reconstruction BC	Reconstruction Dist
1024×1024	$0.8\,n$	$.65\,n$	$.65\,n$	$.52\,n$
512×512	$1.45\,n$	$1.28\,n$	$.96\,n$	$1.13\,n$
256×256	$2.3\,n$	$1.89\,n$	$1.7\,n$	$1.95\,n$
128×128	$3.95\,n$	$2.45\,n$	$1.98\,n$	$2.37\,n$
64×64	$7.2\,n$	$6.6\,n$	$3.3\,n$	$4.23\,n$

4. COMPARISON OF THE TWO EMBEDDINGS

In Table I the wavelet transforms are computed to a level where the lowest resolution image is the size of the filter, i.e. 8 for these results. In table II the transforms are performed only to 3 levels of resolution for each image space size. In the former case more processing at higher levels of the multi-resolution structure is required than in the latter case. In multi-resolution processing where lower numbers of levels of resolution are used Table II data is more representative and where larger numbers of levels of resolution are required the Table I data is more representative.

We remark that in Table I data, where higher level processing is required, **Dist** typically requires more communication overhead than **BC**. These observations are somewhat reserved for Table II data where the processing is restricted to a fixed number of levels. Thus, where one can avoid building a multi-resolution representation with a large number of levels the **Dist** embedding appears to be more practical as a time efficient wavelet embedding for hypercube architectures. Thus, we can see that the performance of the two embeddings onto the hypercube, though a bit different from each other, are close in performance because of the higher bandwidth of the hypercube. These embeddings have also been considered for the embedding of wavelet coefficients on the mesh, however unlike the hypercube, their time performances on the reconstruction and the decomposition algorithms on wavelets are more diverse from each other due to the poor interconnectivity of the mesh network as opposed to the hypercube [9]. On the other hand, the results presented here are not that much superior to those for the mesh parallel computer. Thus, the extra hardware incurred by using a hypercube is hardly justified.

5. CONCLUSION

We have presented two different embeddings of wavelet coefficients onto the hypercube. These embeddings are essential for the parallel implementation of wavelet applications. The time performance of the two embeddings have been evaluated for the decomposition and the reconstruction of wavelet transforms. It is shown that each one of them slightly out-performs the other under the right circumstances. Hence, when a hypercube architecture is considered for wavelet implementation, the embedding method does seem to be that crucial even though **Dist** embedding seem to be more practical. However, one of the above embeddings should be used. Thus, although this investigation is not complete, it gives us an idea of where it would be preferable to use which embedding especially for parallel architectures where the connectivity of the network is not that rich. Finally, the embeddings and the analysis carried out on the hypercube can be easily extended to other parallel architectures, as we feel that the effectiveness of wavelets would be tightly coupled with the effectiveness of parallel architectures because of its computationally intensive nature.

References

1 C. K. Chui. *An Introduction to Wavelets*. Academic Press: Boston, 1992.
2 A. Grossmann and J. Morlet, "Decomposition of Hardy functions into square integrable wavelets of constant shapes," *SIAM J. Math Anal.*, 15, 1984, pp. 723-736.
3 M. Hamdi and R. W. Hall, "Compound graph networks for parallel image processing," *Proc. of the 1991 Workshop on Compt. Arch. for Machine Perception*, pp. 365-377, 1991.
4 S. G. Mallat, "A theory for multiresolution signal decomposition: The wavelet representation," *IEEE Trans. Patt. Anal. Mach. Intell.*, 11, 1989, pp. 674-693.
5 S. G. Mallat, "Multifrequency channel decompositions of images and wavelet models," *IEEE Trans. Acoustics, Speech Sign. Proc.*, 37, 1989, pp. 2091-2210.
6 R. Miller and Q. F. Stout. *Parallel Algorithms for Parallel Architectures*. MIT Press, 1991.
7 S. Ranka and S. Sahni, "Convolution on mesh connected multicomputers," *IEEE Trans. Patt. Anal. Mach. Intell.*, 12, 1990, 315-318.
8 V. Cantoni and S. Levialdi, "Multiprocessor computing for images," *Proceedings of the IEEE*, pp. 959-969, 1988.
9 R. W. Hall, Senol Kucuk, and M. Hamdi, "Wavelet transform embeddings in mesh architectures," submitted for publications.

Image Processing: Theory and Applications
G. Vernazza, A.N. Venetsanopoulos, C. Braccini (Editors)
© 1993 Elsevier Science Publishers B.V. All rights reserved.

Analysis of the influence of sampling, quantization and noise on the performance of the second directional derivative edge detector

Johan De Vriendt[1]

Laboratory for Communication Engineering, University of Ghent, Sint-Pietersnieuwstraat 41, B-9000 Gent, Belgium

Abstract
  In this paper we study the influence of sampling, quantization and noise on the performance of the second directional derivative edge detector. We considered two sampling models and different implementations of the smoothing filter and derivative operator. Furthermore, we studied the influence of the spread of the smoothing filter for different edge profiles. We compared these results with those obtained for the Laplacian edge detector.

1 INTRODUCTION

Many high-level visual processes make use of the edges detected in an image or in a sequence of images. These edges should have a physical significance to be of any value for the higher-level process. A large number of edge detectors have already been published in literature. A commonly used method for localizing edges is to associate edges with the zero crossings of a suitable second order differential operator of a smoothed version of the image. Marr and Hildreth [1] proposed the Laplacian of a Gaussian as a zero crossing edge detector, while Haralick [2] proposed the use of the second directional derivative instead of the Laplacian. Haralick also compared both edge detectors for a set of natural images. The effect of several processes that influence these edge detectors have been studied by Berzins [3] and De Vriendt [4]. Both authors do not discuss in detail the influence of sampling, quantization and noise on the accuracy of edge localization.

2 SAMPLING PROCESS AND FILTERS

We have considered two different sampling models in our analysis. In sampling model I, the continuous signal is smoothed by a Gaussian of spread σ_s and the result is sampled by a Dirac distribution. In model II, the intensity value of a pixel is found by integrating the continuous signal around the pixel's location. The smoothing is introduced to take into account the blurring effect of the imaging system's optics. Also the Gaussian smoothing filter and the derivative operators have to be implemented by digital filters. Both IIR

---
[1]The author is supported by the Belgian National Fund for Scientific Research(NFWO)

and FIR filters could be used. However, IIR filters are not considered here because of the numerous spurious zero crossings that are introduced by applying these filters. In our implementation, the filtering and the derivative operator (size $(2W+1)*(2W+1)$) are separated. Though, the derivative operator introduces some additional smoothing dependent on the width of the filter.

3 INFLUENCE OF SAMPLING, QUANTIZATION AND NOISE

3.1 Sampling

First, we only consider the sampling effects. The edges are located with subpixel accuracy by locally approximating the second directional derivative by a low order polynomial and searching for the zero crossings of this polynomial. The errors introduced by the sampling are highly dependent of the spread of the filter (smoothing and derivative filter). Increasing the spread of the filter increases the accuracy of edge localization for the ideal step edge (see figure 1), but reduces the accuracy of double step edge localization. This last result is not due to the sampling, but a result of the performance of the edge detector for two edges close to each other. This effect was studied in [4]. This shows that a compromise should be made: the spread (σ) of the smoothing filter should be small in order to decrease the influence of nearby edges and the spread should be large to reduce the influence of sampling. More smoothing prior to the sampling (larger σ_s) also results in a reduction of the error.

Due to the sampling, also the orientation of the edge influences the accuracy of edge localization. In figure 1, the error is given as a function of the position of the edge (the ideal step edge is given by U(x-l), with $|l| <1/2$ and U(x)= if x>0 then 1 else 0) and this for different orientations ($\sigma=1$, W=2). From this figure we see that $\theta = 0$ (edge parallel to one of the coordinate axes) results in the largest displacements. The displacements decrease monotonically as the angle ($\theta \leq \pi/4$) between the edge and one of the coordinate axis increases.

3.2 Quantization

The influence of the quantization of the intensity values also results in a displacement of the detected edge location. Due to the quantization, a change in the location of the true edge will not be reflected in the sampled intensity values if this change is below a certain threshold. This threshold depends on the sampling process and on the quantization step. Suppose the height is m (positive integer) times the quantization step, the maximum uncertainty of the localization for sample process I is given by

$$1 - \sigma_s \left[\Phi^{-1}\left(1 - \frac{1}{m}\right) - \Phi^{-1}\left(\frac{1}{m}\right) \right] \qquad (1)$$

with $\Phi^{-1}(x)$ the inverse function of the cumulative Gaussian distribution with variance 1. Equation (1) is valid for all σ_s for which the uncertainty is positive. In case of no smoothing prior to the Dirac sampling ($\sigma_s=0$) the uncertainty is 1 pixel, independent of

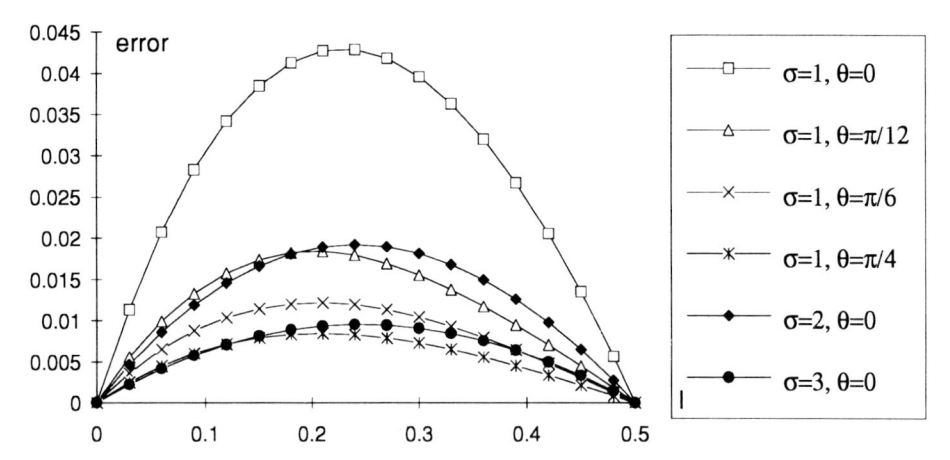

Figure 1: error as a function of the position of the edge, for different values of σ and θ.

the value of m. In case of sampling process II, the maximum uncertainty is given by $\frac{1}{m-1}$ pixels.

3.3 Noise

The influence of noise will be considered with and without sampling for the ideal step edge. We will also compare the results for the second directional derivative with those for the Laplacian. In case there is no noise, the Laplacian and the second directional derivative are equal for the ideal step edge. This is no longer the case if there is noise. Indeed, one can easily argument that the Laplacian is more sensitive to noise than the second directional derivative. The variance of the edge location $E[l_c^2]$ for the Laplacian (in the continuous domain) due to noise can be easily computed and results in:

$$E[l_c^2] = \frac{N_0}{2I^2} \frac{(\sigma^2 + \sigma_s^2)^3}{\sigma^6} \qquad (2)$$

where $N_0/2$ is the power spectral density of the noise and I the step height. So, we can conclude that the variance of the edge location decreases slowly by increasing the spread of the Gaussian filter. We expect that the influence of σ for the second directional derivative edge detector will be smaller, and will have a similar behavior.

The study of the influence of noise for the second directional derivative is more complicated due to the non-linear character of the edge detector. Assume additive, independent, identically distributed Gaussian noise is added to the pixel values. The probability distribution function (pdf) of the Laplacian of the discrete intensity function is Gaussian, but the pdf of the second directional derivative is certainly not Gaussian. Though, under certain restrictions we can prove that this pdf is approximately Gaussian. Furthermore, the approximation becomes better for higher SNR and larger σ. In table 1, the results are given for the second directional derivative (approximation and simulation) and the Laplacian (simulation), and this for several σ and SNR.

Table 1: Variance of the edge location due to noise for the second directional derivative and the Laplacian edge detector, and for different values of σ and SNR.

	sec. dir. der. (approx.)			sec. dir. der. (simul.)			Laplacian (simul.)		
σ	10 dB	20 dB	30 dB	10 dB	20 dB	30 dB	10 dB	20 dB	30 dB
1	.2476	.0789	.0251	.2285	.0755	.0248	.3406	.1136	.0369
2	.2114	.0671	.0213	.2021	.0653	.0210	.3197	.1040	.0333
3	.2018	.0640	.0205	.2002	.0639	.0204	.3174	.1029	.0327
4	.1983	.0628	.0199	.1955	.0623	.0198	.3147	.1011	.0321

From table 1, we conclude that the results correspond with our expectations. The second directional derivative results in a lower variance of the edge location (factor of about $\sqrt{2}$) than the Laplacian. The approximation of the pdf of the second directional derivative becomes better for higher SNR and for larger σ. Furthermore, the variance of the edge location decreases slowly for increasing σ, which is in accordance with equation(2).

The orientation of the edge also influences the variance of the edge location. The smallest variances are obtained for small θ. As θ increases, the variance increases. Though, the influence of the orientation on the variance is rather limited. The behavior of the Laplacian edge detector is very similar, though the influence of the orientation is larger.

4 CONCLUSIONS

We showed that an increase of the spread of the smoothing filter reduces the error due to sampling and noise for the ideal step edge. Though, a larger σ results in larger errors for non-ideal step edges (double step edge, non-linearities, etc...). These effects limit the value of σ. We also showed that the quantization leads to an uncertainty interval for the edge location. The length of this interval can be limited by a smaller quantization step. Also the blurring of the imaging system's optics reduces the length of the uncertainty interval, and reduces the errors due to sampling. Furthermore, we proved that the Laplacian edge detector is more sensitive to noise than the second directional derivative edge detector.

References

[1] D. Marr and E. Hildreth, "Theory of edge detection", Proc. Roy. Soc. London Ser. B 207, pp. 187-217, 1980.

[2] R.A. Haralick, "Digital step edges from zero crossings of second directional derivatives", IEEE Trans. Pattern Anal. Mach. Intell., vol. PAMI-6, pp. 58-68, 1984.

[3] V. Berzins, "Accuracy of laplacian edge detectors", Comput. Vision, Graphics, Image Processing, vol. 27, pp. 195-210, 1984.

[4] J. De Vriendt, "Accuracy of the zero crossings of the second directional derivative as an edge detector", will be published in Multidimensional Systems and Signal Processing.

Image Processing: Theory and Applications
G. Vernazza, A.N. Venetsanopoulos, C. Braccini (Editors)
© 1993 Elsevier Science Publishers B.V. All rights reserved.

Edge extraction and enhancement using coordinate logic filters

K. Tsirikolias and B.G. Mertzios

Automatic Control Systems Laboratory, Department of Electrical Engineering, Democritus University of Thrace, 67 100 Xanthi, Greece

Abstract
In this paper we present a number of Coordinate Logic (CL) Filters for edge extraction and enhancement. Also a comparison of the results with those resulting with morphological and convolution filters is done. The CL filters are nonlinear filters that execute a coordinate logic operation among the pixels of the image. They may be seen as multi-binary morphological type filters. The CL filters are very fast and can be easily implemented using logic circuits or cellular automata.

1. INTRODUCTION

This paper refers to edge extraction and enhancement using Coordinate Logic (CL) Filters [1],[2],[4]. The CL filters are nonlinear filters that execute a coordinate logic operation [3], which will be denoted by the symbol @, among the pixels of the image. The operator (@) represents one of the CL operators CAND, COR, CXOR or CNOT. If the signal is not binary, then the operations are executed among their corresponding binary values. The CL filters are very efficient in various image processing applications, such as lowpass and highpass filtering, noise cleaning, opening, closing, skeletonization, region filling, coding, shape smoothing, image magnification, as well as edge detection and feature extraction.
The CL filters are related to morphological filters, but constitute a separate new class of nonlinear filters. Indeed, the morphological filters may be seen as a class of rank order filters, which by definition involve some kind of sorting, while the CL filters can not. The CL filters can execute easily and fast the four basic morphological operations (erosion, dilation, opening and closing). Therefore, CL filters are expected to be suitable for all the variety of tasks that are executed by morphological filters. Moreover, CL filters operate in a different simpler and faster way than morphological filters do. Specifically, CL filters succeed the desired processing by executing only direct logic operations among the pixels of the given image. On the contrary, in general the morphological filters when applied to gray- level images, require additional operations.

2. DERIVATION OF THE COORDINATE LOGIC OPERATIONS

Definition : Coordinate logic operation
The coordinate logic operation @ of two numbers A and B in decimal

system, in their binary sequence is given by

$$C = [c_1 \ c_2 \ \ldots \ c_n] = A \ @ \ B \qquad (1)$$

where the ith bit c_i is the logic operation **o** of the corresponding a_i and b_i bits of the operands, i.e.

$$c_i = a_i \ o \ b_i, \quad i = 1, 2, \ldots, n \qquad (2)$$

The operation **o** may be the logical AND, OR or XOR. The most elementary Boolean functions among them are the AND and OR functions.

3. PROPERTIES OF COORDINATE LOGIC OPERATIONS

Let two integer positive numbers A and B in decimal system. Let also the CL AND and OR operations (CAND and COR respectively) of the numbers A and B be

$$D = A \ CAND \ B \qquad (3)$$
$$E = A \ COR \ B \qquad (4)$$

The following Theorem provides useful relations among A, B and D, E in decimal system.

Theorem

Let D and E be defined by (3.1) and (3.2) respectively. Then

$$0 \leq D \leq \min\{A, B\} \qquad (5)$$
$$\max\{A, B\} \leq E \leq (2^n - 1) \qquad (6)$$

where n is the word length [4]. From Theorem 1, it results that the COR and CAND of A and B represent a measure of the maximum and of the minimum functions of A and B respectively.

Corollary 1

Given A_m, $m = 1, 2, \ldots, M$ and D, E defined by (3.1) and (3.2), it holds

$$0 \leq D \leq \min\{A_i\}, \quad i = 1, 2, \ldots, M \qquad (8)$$
$$\max\{A_i\} \leq E \leq (2^n - 1), \quad i = 1, 2, \ldots, M \qquad (9)$$

4. DERIVATION OF THE COORDINATE LOGIC FILTERS

Let a 2-D digital signal will be denoted by a 2-D set

$$S = \{s(i,j), \ i = 1, 2, 3, \ldots, M, \ j = 1, 2, \ldots, N\} \qquad (10)$$

where M and N denote the finite dimensions of the signal in the horizontal and the vertical directions respectively. Then the dilation S_B^D and erosion

S_B^E of the image S by the filter structure B are defined using the COR and CAND operations. Various square structures may be obtained, depending on the size of the edge, on the place of the origin and on whether or not the origin is taken into account. Indicatively, one characteristic rhombus structure is described in the sequel.

$$\begin{matrix} & \circ & \\ \circ & [\circ] & \circ \\ & \circ & \end{matrix}$$

The output of the 2-D CL filter with the above filter structure is

$$f(i,j) = s(i-1,j) @ s(i,j-1) @ s(i,j) @ s(i+1,j) @ s(i,j+1) \qquad (11)$$

where the symbol @ denotes a CL Operation,(COR, CAND, etc).Note that all our applications are executed using the above filter structure. It is pointed out that the CAND and COR operations may result to values that are not included in the initial range of the values of the pixels. However, finally the combinations of CAND and COR operations, (i.e. closing and opening), take values that are included in the initial range.

5. APPLICATION TO EDGE EXTRACTION

CL filters are very efficient and exhibit excellent performance in edge extraction applications. Fig. 2 shows the result of the application of the CL filter
$$f(i,j) = [s(i,j-1) \text{ CXOR } s(i,j+1)] \text{ COR } [s(i+1,j) \text{ CXOR } s(i-1,j)] \qquad (12)$$
to the the original image of Fig. 1. The logic operation CXOR detects the existence of a difference in a specific direction, while the logic operation COR detects the existence of a difference in the active neighbor. The use of bigger structures produces bolder edges.

The edge detection in an image S with CL filters may be also achieved by extracting at first the erosion S of the image S and then by subtracting the eroded image from the original. The above operation is faster using CL filters than topological min/max operations. Various combinations may be used in order to obtain the edges of an image, such as $S_B^D - S$, $S - S_B^E$, $S_B^E - S_B^D$, e.t.c.

Fig. 3 shows the result of the application of $S_B^D - S$, where S is the original image and S_B^D is the dilation of the image S, using CL filters and the structure element B. Fig. 4 shows the the result of the application of $S_B^D - S$ using morphological filters. Fig. 5 shows the result of the application of the convolution mask

$$\begin{matrix} & -1 & \\ -1 & 4 & -1 \\ & -1 & \end{matrix}$$

Finally in Fig. 6 we can see the result of an extremely fast CL filter, since it requires only n Boolean operators, where n is the number of pixels that belongs to the structure element. Here we considered the filter structure and mask of (11), hence in this case n = 5. This filter is

$$f(i,j) = s(i,j) \text{ CAND CNOT}[s(i-1,j) \text{ CAND } s(i+1,j) \text{ CAND } s(i,j+1) \text{ CAND } s(i,j-1)]. \tag{13}$$

6. CONCLUSIONS

We presented some edge detection and enhancement applications using CL filters. The CL filters execute coordinate logic operations among the pixels of the image and they are very efficient for various image processing applications. They may be seen as multi-binary morphological kind of filters. The CL filters are very fast and can be easily implemented using logic circuits or cellular automata.

7. REFFERENCES

1. K. Tsirikolias and B.G. Mertzios, "Logic Filters in Image Processing," *Proceedings of the International Conference on Digital Signal Processing*, pp. 285-287, Florence, Italy, September 4-6, 1991.
2. B.G. Mertzios and K. Tsirilolias, "Coordinate Logic Filters in Image Processing Applications," *Proceedings IEEE Winter Workshop for Nonlinear Digital Signal Processing*, Tampere, Finland, Jan. 17-20, 1993
3. D. L. Dietmeyer, *Logic Design of Digital Systems*, Allyn and Bacon Inc.
4. B. Mertzios and K. Tsirikolias, "Coordinate Logic Filters in Image Processing". Submitted for publication.

Fig. 1. The original image.

Fig. 2. The result of the application of the CXOR filter.

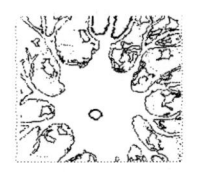

Fig. 3 The result of the application of the $S_B^D - S$ operation using CL filters.

Fig. 4. The result of the application of the $S_B^D - S$ operation using morphological filters.

Fig. 5. The result of the application of a convolution filter.

Fig. 6. The result of the application of the filter (13).

A Structured Regularized Approach in Multichannel Image Restoration

Michael E. Zervakis

Department of Computer Engineering, University of Minnesota, Duluth,
MN 55812, USA

Abstract

In this paper we develop the framework for the extension of the constrained mean-square-error (CMSE) estimation scheme to the multichannel domain and we study its performance in image restoration. The CMSE approach is interpreted as a regularized optimization scheme that utilizes the prototype Wiener structure in the smoothing process of the estimate. Through its structured regularizing functional, it accounts for the simultaneous suppression of the noise process and the preservation of sharp detailed structure. In particular, the paper focuses on the efficient implementation of the multichannel CMSE algorithm and on the selection of its regularization parameter.

1. INTRODUCTION

With the advent of digital electronics and computers, multichannel representation and multichannel processing become increasingly important in a wide variety of research areas. Due to the multiple encoding of information, multichannel images convey much more information than single-channel images. The redundancy and the complementary nature of information within channels makes multichannel image processing significantly different than single-channel processing, not only because of the increased dimensionality, but also due to the need for identification and exchange of information among all different channels. Several extensions of single-channel approaches to the multichannel domain show that the use of multichannel information can effectively improve the result of the restoration process. Existing multichannel approaches include the Wiener filtering approach [1], the Kalman filtering approach [2], and the constrained least-squares (CLS) estimation scheme [3]. Despite their simplicity, however, these approaches do not favor the reconstruction of sharp edges and often produce noise and signal artifacts.

A regularized approach that can overcome several disadvantages of other techniques is based on the constrained mean-square-error (CMSE) estimation scheme [4]. In Section 2, we develop the framework for the extension of the CMSE approach to the multichannel case and we study the factors that affect its performance. Moreover, we derive a simplified estimator for low signal-to-noise ratios (SNR). The selection of the regularization parameter is addressed in Section 3. The CMSE approach is demonstrated through a color restoration example.

2. THE STRUCTURED CMSE APPROACH

Consider the additive-noise model of a multichannel image-formation system $\{g = Hf + n\}$, where f and g denote the original and the degraded image, n denotes the noise vector, and H denotes the matrix induced by the point-spread function (PSF) of the system. The vector notation results from the multichannel representation by arranging rows above columns within each channel, and then arranging channels on top of each other. For K channels, the overall degradation matrix is written in block form as $\{H = [H_{ij}], \ i,j = 1,\ldots,K\}$. The off-diagonal block-elements H_{ij}, $i \neq j$, enable the consideration of channel interference in the image formation process, representing channel leakage in multispectral imagery, or registration errors in time-varying sequences, for instance. In addition, these elements can be utilized for the simultaneous registration and restoration of time-varying images.

The CMSE restoration approach is interpreted as a structured regularized scheme. The smoothing functional in this approach involves the multichannel minimum-mean-square-error (MMSE) estimate, which is essentially utilized as a prototype constraint-image. The corresponding objective function is written as:

$$\min_{\hat{f}} E \left\{ \mu ||f - \hat{f}||^2 + ||C(g - H\hat{f})||^2 \right\}, \quad (1)$$

where μ and C denote the regularization parameter and the regularizing operator, respectively. The CMSE approach combines the MMSE criterion and the least-mean-squares (LMS) criterion, which is weighted through the de-correlating operator C. The matrix C has the same structure as the degradation matrix H, but it reflects the operation of a high-pass multichannel filter. An appropriate structure for C is that of the 3-D Laplacian operator. Within the CMSE criterion, the MMSE component is directly interpreted as regularizing the pseudo-inverse solution, which forms the solution of the LMS criterion. With the high-pass structure of the operator C, the restoration of low frequencies is left to the MMSE estimation scheme. Alternatively, at high frequencies, the CMSE approach utilizes a combination of the MMSE and the LMS scheme, with emphasis to the LMS term, in order to account for both efficient preservation of the detailed structure and partial suppression of the noise process. The CMSE approach always derives a meaningful estimate, which is conceptually located between the Wiener estimate and the pseudo-inverse solution.

The quadratic optimization problem in (1) derives the estimator:

$$\hat{f} = [\mu I + H^t C^t C H]^{-1} [\mu R_{fg} + H^t C^t C R_{gg}] R_{gg}^{-1} g, \quad (2)$$

where R_{gg} and R_{fg} represent the correlation matrices between $\{g, g\}$ and $\{f, g\}$ respectively. Similar to other restoration approaches in the multichannel domain, the computational complexity implied by the CMSE estimator is enormous. In order to reduce this complexity, we utilize the assumptions of wide-sense stationarity and space-invariance between two specific channels. This channel-stationary characteristic imposes a partially block-circulant matrix structure, so that each matrix component in (2), at each pair of channels, has a block-circulant form. In this case, the operations involved in the computation of the CMSE estimate can be decomposed into operations among composite block-circulant blocks performed in the 2-D discrete Fourier-transform (DFT) domain.

As the SNR in the multichannel model decreases, it is expected that the optimal value of the regularization parameter μ increases to weaken the effect of the pseudo-inverse (LMS) solution in the CMSE estimate. Thus, at low SNR, the algorithm operates with a relatively large value of the regularization parameter μ. In this case, the first matrix inversion in the structure (2) of $\hat{f}$ can be expanded into Taylor series. By eliminating higher order terms, the first-order approximation of this inversion is written as:

$$[\mu I + H^t C^t C H]^{-1} \cong \frac{1}{\mu} I - \frac{1}{\mu^2} H^t C^t C H. \quad (3)$$

With this approximation, the implementation of the CMSE algorithm involves the inversion of a single matrix, the auto-correlation matrix of the data. In a number of experiments tested with different original images and degradation models, the simplification in (3) efficiently characterizes the CMSE estimate for SNR at least smaller than 35dB.

3. SELECTION OF THE REGULARIZATION PARAMETER

The regularization parameter μ affects the quality of the estimate in regularized restoration approaches. In this section we present a set-theoretic approach for the selection of this parameter, which is motivated by the Tikhonov-Miller formulation to the regularized inversion of ill-posed problems. According to this formulation, the CMSE estimate can be interpreted as a vector in the intersection of the *mean-square-error* set:

$$C_f = \left\{ \hat{f} : E \{ ||f-\hat{f}||^2 \} \le \delta_f^2 \right\} , \quad (4a)$$

and the *constrained-residual* set:

$$C_n = \left\{ \hat{f} : E \{ ||C(g-H\hat{f})||^2 \} \le \delta_n^2 \right\} . \quad (4b)$$

Moreover, the regularization parameter is set to the ratio $\{ \mu_{st} = \delta_n^2/\delta_f^2 \}$.

The limit on the constrained power of the residual can be estimated based on the noise statistics. In the case of white noise in each channel, this limit is efficiently computed:

$$\delta_n^2 = E \{ ||C(g-Hf)||^2 \} = trace \{ CR_{nn}C^t \} = \sum_{l=1}^{K} \sum_{k=1}^{K} \sigma_k^2 \, trace \{ C_{lk} C_{lk}^t \} , \quad (5)$$

where R_{nn} denotes the correlation matrix of the noise process and σ_k^2 denotes the variance of noise in the k-th channel. Since the individual matrices C_{lk} assume a block-circulant form, this operation can be performed in the 2-D DFT domain.

A meaningful bound δ_f of the mean-square-error set can be derived based on the expected variance of the estimator $\hat{f}$ and the confidence (or the probability) with which $\hat{f}$ is known to belong in the set C_f [5]. The expected variance of any estimator of the random vector f is bounded by the total information matrix J_τ, which is derived for the multichannel case as [5]:

$$V_f = E \{ ||f-\hat{f}||^2 \} \ge trace \{ J_\tau^{-1} \} = trace \{ \Sigma_{ff} [I + H^t R_{nn}^{-1} H \Sigma_{ff}]^{-1} \} , \quad (6)$$

where Σ_{ff} denotes the covariance matrix of the original image. This approximate bound is derived under the assumption that the original image is normally distributed.

The second quantity that determines the bound δ_f, namely the confidence characterizing the membership of the CMSE estimate $\hat{f}$ in the set C_f decreases as the variance of the noise process increases. In accordance with these considerations, we propose the following form of the bound δ_f:

$$\delta_f^2 = trace \{ [I + R_{nn}]^{-1} \} \, trace \{ J_\tau^{-1} \} , \quad (7)$$

with J_τ^{-1} given in (6). The computation of this bound is performed in the multichannel DFT domain. Overall, the set-theoretic formulation provides a fast and efficient method to select the regularization parameter based on the bounds δ_f and δ_n. Moreover, the set-theoretic technique can be readily adjusted to the selection of a parameter-vector that reflects different characteristics in each channel and imposes channel-dependent performance in the restoration algorithm [5]. Another selection technique based on the optimization of the weighted-least-squares measure is presented in [5].

The CMSE approach is briefly demonstrated through the restoration of the Lena image. The dimensionality of this color image is 128 x 128, with 8 b/pixel per color channel. The red and the blue channels are distorted by a PSF that represents a defocused lens of diameter 13 pixels, whereas the green channel is blurred by a similar degradation of diameter 9 pixels. The linearly degraded frames are further corrupted by white Gaussian noise resulting in SNR equal to 20dB in each channel. Figures 1(a) and 1(b) depict the intensity of the red channel along one horizontal line within the spatial support of the image. The degraded image is restored by the multichannel CLS and CMSE approaches. The CLS approach employs three channel-dependent regularization parameters, which are selected through the set-theoretic formulation using the corresponding estimate-sets [3]. The CMSE approach implements the simplified estimate-structure in (3) and utilizes a single parameter for all channels, which is selected through the proposed set-theoretic technique. Despite its incomplete consideration, the CMSE approach expresses superior performance in the reconstruction of both the smooth structure and the sharp edges of the image. The CMSE approach preserves similar performance in a variety of applications tested and provides a flexible, yet powerful tool in multichannel signal processing.

4. REFERENCES

1 N.P. Galatsanos, and R.T. Chin, "Digital Restoration of Multichannel Images", *IEEE Trans. on Acoustics, Speech, and Signal Proc.*, vol. ASSP-37, no. 3, March 1989.

2 N.P. Galatsanos, and R.T. Chin, "Restoration of Color Images by Multichannel Kalman Filtering", *IEEE Trans. on Signal Processing*, vol. ASSP-39, no. 10, Oct. 1991.

3 N.P. Galatsanos, A.K. Katsaggelos, R.T. Chin, and A.D. Hillery, "Least Squares Restoration of Multichannel Images", *IEEE Trans. on Signal Processing*, vol. ASSP-39, no. 10, Oct. 1991.

4 M.E. Zervakis and A.N. Venetsanopoulos, "Design of a New Restoration Algorithm Based on the Constrained Mean-Square-Error Criterion", *Multidimensional Systems and Signal Processing*, vol. 3, pp. 381-408, 1992.

5 M.E. Zervakis, "Optimal Restoration of Multichannel Images Based on Constrained Mean-Square Estimation," *Journal of Visual Communications and Image Representation*, vol. 3, no. 4, Dec. 1992.

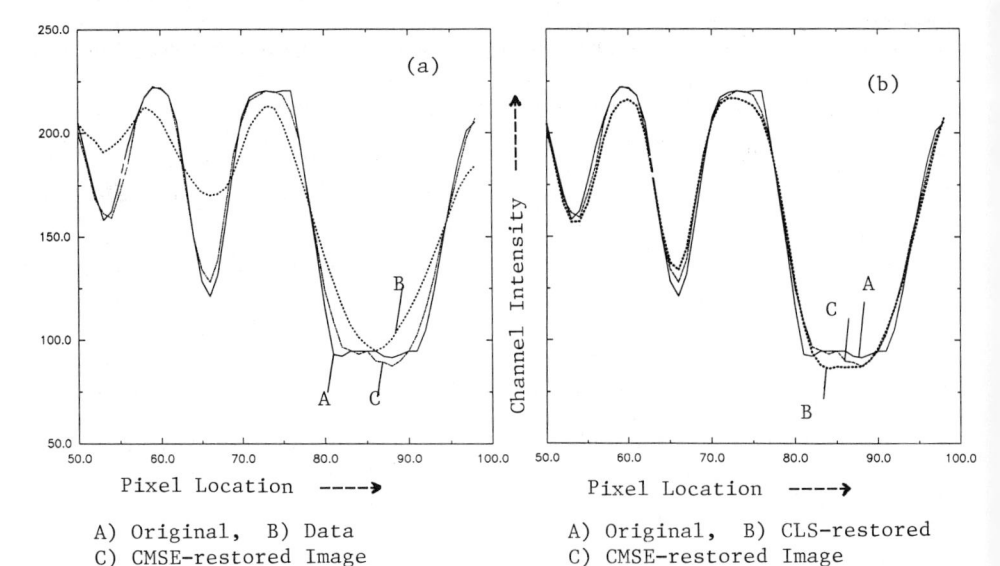

Figure 1: Multichannel Restoration of Lena; Intensity Line on Red Channel.

Design of VLSI Array Processors for Very High Speed 2-D Digital Filtering of High Definition Images

Yasushi Iwata, Masayuki Kawamata, and Tatsuo Higuchi

Department of Electronic Engineering, Faculty of Engineering Tohoku University, Aoba, Aramaki, Sendai 980 Japan

Abstract— This paper designs and evaluates VLSI array processors for very high speed 2-D state-space digital filters. The architecture of the VLSI array processors is a linear systolic array, of which processing elements (PEs) are simple and homogeneous 1-D state-space digital filters. We adopt a hierarchical behavioral description language and synthesizer for the design and evaluation of the VLSI array processors. Eight PEs can be integrated into one 14.70×14.98 mm^2 VLSI chip using $1\mu m$ CMOS technology. The processing system which is composed of 128 designed VLSI array processors at 25 MHz clock can process a $1,024 \times 1,024$ image in 1.47 msec.

1 Introduction

The use of high speed two-dimensional (2-D) digital filters has found important applications in many areas, including biomedical image processing, image transformation and video signal processing, intelligent robots, and so on [1], [2]. Because 2-D digital filter algorithms possess a large amount of inherent concurrency, the computational concurrency of 2-D digital filters render them well suited for multiprocessor VLSI implementation.

This paper designs and evaluates VLSI array processors for very high speed 2-D digital filters of high definition images. The overall architecture of the signal processing system based on multiprocessor environment is a linear systolic array, of which processing elements (PEs) are simple and homogeneous 1-D digital filters. The design concept of the signal processing system is to assign each PE to a single pixel in an image scan line, which is concurrently computable. The number of PEs is equal to the number of rows of the processing images. Several PEs can be integrated into a VLSI array processor as a single chip and the overall processing system consists of many VLSI array processors.

For the design and evaluation of the VLSI array processors, we use a hierarchical behavioral description language and synthesizer, because they have the following advantages: (1) The description can be changed easily, since the VLSI array processors are designed by the software description of their behavior. (2)Technology independent design is possible. Thus we can easily adopt the progress of new technologies. (3)The turnaround time for the design of the VLSI array processors can be reduced in comparison with using the conventional computer-aided engineering tools.

2 Architecture

Consider a 2-D digital filter described by the following 2-D state equations [3]:

$$\begin{pmatrix} \boldsymbol{x}^h(m+1,n) \\ \boldsymbol{x}^v(m,n+1) \end{pmatrix} = \begin{pmatrix} \boldsymbol{A}_{11} & \boldsymbol{A}_{12} \\ \boldsymbol{A}_{21} & \boldsymbol{A}_{22} \end{pmatrix} \begin{pmatrix} \boldsymbol{x}^h(m,n) \\ \boldsymbol{x}^v(m,n) \end{pmatrix} + \begin{pmatrix} \boldsymbol{b}_1 \\ \boldsymbol{b}_2 \end{pmatrix} u(m,n) \qquad (1)$$

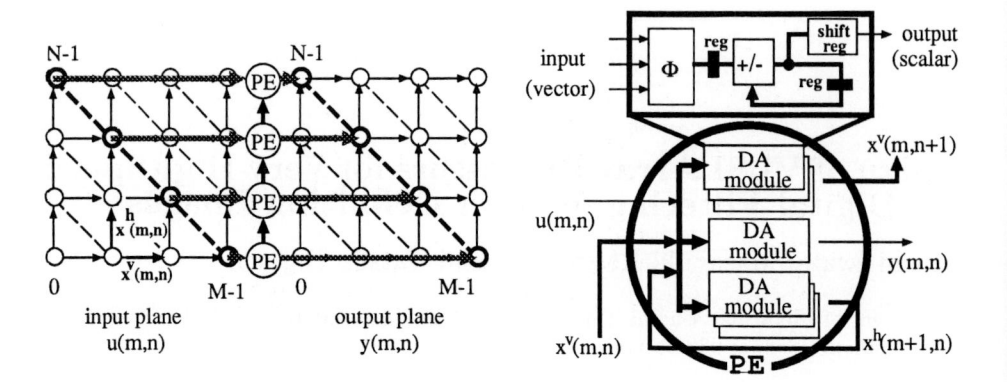

Figure 1: System architecture and its I/O sequence of the parallel processing system.

Figure 2: Block diagram of PE.

$$y(m, n) = \begin{pmatrix} c_1 & c_2 \end{pmatrix} \begin{pmatrix} x^h(m, n) \\ x^v(m, n) \end{pmatrix} + d \cdot u(m, n) \quad (2)$$

where $x^h(m, n)$ is the horizontal state vector of order $P \times 1$, and $x^v(m, n)$ is the vertical state vector of order $Q \times 1$. The input $u(m, n)$ and output $y(m, n)$ are scalars, and matrices A_{ij}, b_i, c_i, and d are coefficient matrices. These 2-D state equations are very powerful equations to represent variety of 2-D filter structures.

Because we have designed the system architecture in [4] based on graph-based design methodology [5], this section briefly discusses the architecture of the processing system.

According to the Eqs. (1) and (2), several outputs can be computed in parallel, since these computations are not mutually constrained by any precedence relation. For example, the output $y(4, 0), y(3, 1), y(2, 2), y(1, 3)$, and $y(0, 4)$ are mutually independent and can be computed in parallel. Because the locus of the sample locations, which can be computed in parallel, is a straight line with slope "−1" in the input plane, these concurrently computable pixels are diagonally processed by PEs. The resulting architecture of the signal processing system is a linear systolic array and each PE updates the state vectors and computes the output according to the Eqs. (1) and (2). Fig. 1 shows the system architecture and its I/O sequence of the VLSI array processors for a 2-D state-space digital filters. In this architecture, The communications between PEs and computations in PE can be done at the same time so that the data flow will not be interrupted. As a result, we can make good use of pipelining between PEs. The structure of the VLSI array processors is characterized by high inherent parallelism, modularity, regularity, local interconnections, and high throutput.

The computations of Eqs. (1) and (2) consist of $(P+Q+1)$ inner products of $(P+Q+1)$ dimensional vectors, one of which vector is constant and the other of which is variable. We adopt distributed arithmetic[6] (DA) for the computation of the inner products. All the inner products can be executed in parallel since each inner product is independent. One PE executes $(P+Q+1)$ inner products in parallel by $(P+Q+1)$ DA modules. The main concept of the DA algorithm is to look up the function Φ, which stores the partial product, and to accumulate the shifted partial product. Fig. 2 shows the block diagram of PE and the DA module.

For high definition real-time images, we consider the following target specifications:

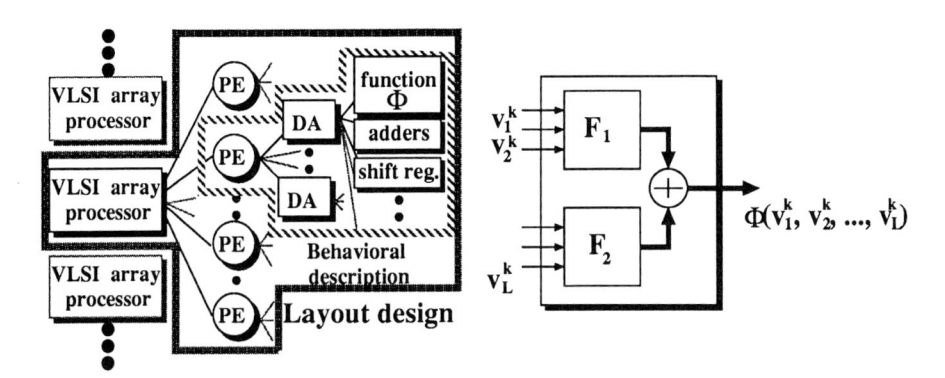

Figure 3: Hierarchical structure of the VLSI array processors.

Figure 4: Small scale implementation of the function Φ.

(1) 1024×1024 processing image with the scan rate of 60 frames/sec, (2)The input and final output signals are represented in 16 bits, (3) the filter order is (4,4)-th. According to the system architecuter, the processing system consists of 1024 homogeneous PEs and each PE consists of nine DA modules.

3 Design of the VLSI array processors

Since one VLSI array processor consists of a cascade connected several PEs, we can divide the design process of the VLSI array processor into two steps. The first step is to design one PE by behavioral description language and synthesizer. The second step is to integrate several PEs into one VLSI chip by layout CAD tool. We use "PARTHENON" system as the behavioral description language and synthesizer[7]. PARTHENON is the integrated VLSI design tool, which consists of the behavioral description language, behavioral simulator, logic synthesizer, and technology mapping optimizer. The hierarchical module structure of the VLSI array processor is shown in Fig. 3.

One DA module consists of a few simple parts such as the function Φ, adders, and registers. Although the conventional way to realize the function Φ is by ROM, the ROM implementation occupies much area on VLSI chip. To reduce the chip size, we implement the function Φ by logic gates. This is accomplished by address partition method as shown in Fig. 4. The functions F_1 and F_2 are enough small scale functions to be implemented by logic circuits. As a result, the number of gates of the function Φ implemented by address partition method is approximately 10% of that implemented by ROM.

Other components of DA module (i.e., adders, shift registers, registers, and so on) can be easily described. The total behavior of each DA module and PE is independently simulated. Then, all the input and output data at each node of PE can be traced and verified.

4 Performance evaluation

To synthesize the real logic circuits of PE, the behavioral description of PE is transferred to the logic synthesizer and optimizer. The chip performance including the interconnection delay is estimated by layout design. After layout design of one PE, the layout of the VLSI array processor is obtained by a cascade connection of several PEs. Eight PEs can be integrated into a single VLSI chip under the constraint of chip area and power dissipation based on a $1\mu m$ CMOS technology. Fig. 5 shows the chip layout of the VLSI array processors.

Table 1: Performance of the VLSI array processors.

Technology	1.0 μm CMOS
System clock	25MHz
Inner product execution	720 nsec
Processing time for (1,024 × 1,024) image	1.47msec/frame[†]
Number of gates	129,138
Number of transistors	516,552
Chip core size	14.70×14.98mm^2
Power dissipation	4.0 W@25MHz
Power supply	5V ± 10%
Package	223 pin CPGA

†performed by 128 chips (1,024 PEs)

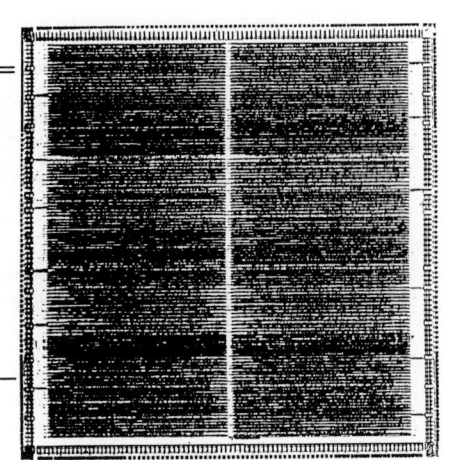

Figure 5: Layout of the VLSI array processor.

The VLSI array processor is driven at 25 MHz clock and executes the computation of one pixel in 720 nsec. The parallel processing system which is composed of 128 designed VLSI array processors can execute the 2-D state-space digital filtering of (4,4)-th order for a 1,024 × 1,024 image in 1.47 msec/frame, where the interconnection delay between the VLSI array processors is considered. The performance of the parallel processing system is over the target specifications. Table 5 summarizes the performance of the VLSI array processor.

5 Conclusion

The VLSI array processors for very high speed 2-D state-space digital filters of high definition images have been designed and evaluated using a hierarchical behavioral description language and synthesizer. The structure of the parallel processing system is characterized by high inherent parallelism, modularity, regularity, local interconnections, and high throughput. The parallel processing system, which is composed of 128 designed VLSI array processors at 25 MHz clock, can execute a 1,024×1,024 image in 1.47 msec.

References

[1] Spyros G. Tzafestas, *Multidimensional Systems*, Marcel Dekker, 1985.

[2] Jae S. Lim, *Two-Dimensional Signal and Image Processing*, Prentice-Hall, 1990.

[3] R. P. Roesser, "A discrete state-space model for linear image processing," *IEEE Trans. Automat. Contr.*, vol. 20, pp. 1-10, Jan. 1975.

[4] T. Yamakage and M. Kawamata, " VLSI architecture for 2-D state-space digital filters," *IEICE Trans.*, vol. J72-A, pp. 916-922, June 1989.

[5] S. Y. Kung, *VLSI Array Processors*, Prentice-Hall, 1988.

[6] A. Peled and B. Liu, "A new hardware realization of digital filters," *IEEE Trans. Acoust., Speech, Signal Processing*, vol. 22, pp. 456-462, June 1974.

[7] Y. Nakamura, "An integrated logic design environment based on behavioral description," *IEEE Trans. Computer-Aided Design*, vol. 6, pp. 322-336, Mar. 1987.

Non-Convex Optimization for Image Reconstruction with Implicitly Referred Discontinuities*

L. Bedini[a], M.G. Pepe[b], E. Salerno[a], A. Tonazzini[a]

[a]Consiglio Nazionale delle Ricerche, Istituto di Elaborazione della Informazione, Via S. Maria, 46, I-56126 Pisa, Italy

[b]Scuola Normale Superiore, Centro Elaborazione Informazione Beni Culturali, Via Della Faggiola, 19, I-56126 Pisa, Italy

Abstract

Edge-preserving image reconstruction can be performed by minimizing appropriate cost functionals in which image intensity and discontinuities are explicitly referred. Equivalent and less expensive reconstructions can be obtained by using a class of functionals that only depend on the image intensity and have been shown to implicitly refer to an underlying discontinuity process. We analyze the performances of two of these functionals and show that Graduated Non-Convexity (GNC) algorithms permit a further reduction of the computational costs if compared with Simulated Annealing algorithms, commonly used in non-convex optimization.

1. INTRODUCTION

Correct edge location has always been an important issue for image processing applications. In approaches based on Markov Random Field models and the Maximum A Posteriori estimation, this problem is solved by explicitly introducing a *line process*, located in the inter-pixel sites of the image. This type of approach is highly flexible in exploiting prior knowledge on both image intensity and discontinuities [1-3], but, on the other hand, it implies the minimization of generally non-convex cost functionals and the use of computationally demanding stochastic relaxation algorithms. Recently proposed methods exploit cost functionals in which the discontinuities are referred implicitly rather than explicitly. These functionals are still non-convex, however they only depend on the intensity process and their minimization is thus less expensive [4-6]. Moreover, the *duality theory* [7] demonstrates that, for a particular class of such functionals, the solutions obtained via these methods are equivalent to those deriving from the explicit introduction of a line process. With reference to image reconstruction from sparse samples, we consider two particular cost functionals of this class in order to compare their performances. The minimization is accomplished using a version of the Simulated Annealing algorithm [7] which reduces the computational cost. A deterministic GNC algorithm is also derived which helps to further reduce the computation times [6,8].

2. THEORY

The general form of the cost functional to be minimized is the following:

---

* This work has been partially supported by the Italian National Research Council (CNR), Special Program *Information Systems and Parallel Computing*.

$$F(\mathbf{f}) = P(\mathbf{g},\mathbf{f}) + \sum_{c \in \mathcal{C}} \phi\left(D^k(\mathbf{f}; c)\right) \qquad (1)$$

where the *data term*, P, takes into account the consistency between the estimated image **f** (*intensity process*) and the available data **g**. The second term is the *regularization term* and contains all the prior constraints on the intensity and, implicitly, on the discontinuities of the image. The function $D^k(\mathbf{f}; c)$ is defined over the clique system $\mathcal{C}$ associated with an assigned neighbourhood system, and represents the k-th partial derivative (k=1,2,3) of **f** relative to the clique c. The *neighbour interacting function* ϕ is an increasing function of D^k and thus enforces a k-th order smoothness constraint on **f**; its form, however, can be chosen so that this constraint is weakened where it is more likely to have a discontinuity. In this way, local interactions between the intensity process and its discontinuities can be accounted for.

As a first approach, we considered the data generation model $\mathbf{g} = H\mathbf{f} + \mathbf{n}$, where H is the degradation matrix and each component of **n** (the *noise vector*) is a zero mean Gaussian variable with variance σ^2; we adopted the first order neighbourhood system and placed our constraints over the first partial derivatives of the image. Two cost functionals having the properties described above are the following [7]:

$$F_1(\mathbf{f}) = \frac{\|\mathbf{g} - H\mathbf{f}\|^2}{2\sigma^2} + \sum_{i,j} \frac{\lambda^2 |f_{i,j} - f_{i,j-1}|}{(\lambda^2/\alpha)|f_{i,j} - f_{i,j-1}| + 1} + \frac{\lambda^2 |f_{i,j} - f_{i-1,j}|}{(\lambda^2/\alpha)|f_{i,j} - f_{i-1,j}| + 1} \qquad (2)$$

$$F_2(\mathbf{f}) = \frac{\|\mathbf{g} - H\mathbf{f}\|^2}{2\sigma^2} + \sum_{i,j} \frac{\lambda^2 (f_{i,j} - f_{i,j-1})^2}{(\lambda^2/\alpha)(f_{i,j} - f_{i,j-1})^2 + 1} + \frac{\lambda^2 (f_{i,j} - f_{i-1,j})^2}{(\lambda^2/\alpha)(f_{i,j} - f_{i-1,j})^2 + 1} \qquad (3)$$

The positive parameter λ^2 determines a compromise between data consistency and smoothness, while the quantity α/λ^2 can be interpreted as a sort of threshold for the difference between two adjacent pixels, above which the smoothness constraint becomes increasingly weak thus permitting a discontinuity to be created.

For the minimization of (2) and (3), we adopted a Simulated Annealing algorithm similar to that reported in [7], employing a Metropolis algorithm to update the pixels [8]. In order to obtain acceptable computation times, we adopted a linear decay schedule for the temperature, and the value proposed to update each pixel is randomly selected not over the entire image range but only from the union of small intervals around the current value of the pixel itself, its neighbours and the corresponding observed data. The algorithm terminates when the cost difference between the images obtained with different temperature values falls below a pre-determined threshold for a number of consecutive iterations. We also studied the possibility of using the GNC algorithm [6], which is based on deterministic minimizations of a sequence of successive approximations of the cost functional. The first approximation of this sequence must be convex and the last must coincide with the original cost functional. The starting point of each minimization is the image resulting from the minimization of the previous functional. We derived a sequence of approximations for the functional F_1 [8]; the derivation of a sequence for F_2 should be straightforward.

3. EXPERIMENTS

The test images used for our experiments were real and synthetic piecewise smooth images. The degraded images were obtained by randomly selecting a percentage of the pixels and adding noise. The intensity process is quantized in 256 gray levels. We carried out experiments with different percentages of missing samples and different values for the noise variance and for the regularization parameters λ and α, evaluating the quality of the reconstructed images by means of the root mean squared errors ε with respect to the original images.

As an example, we report in Figure 1 the reconstructions of a real 200×200 image (a portrait of the painter Caravaggio) obtained minimizing the functional $F_1(f)$ via Simulated Annealing and GNC. For display purposes, in Figure 1 b), the missing samples have been set at 255 (white pixels). Another example is shown in Figure 2, relative to a 128×128 real image of printed characters ("Letters"). The reconstructions are now obtained via Simulated Annealing minimizing both F_1 and F_2. In Figure 3, similar results are shown for a 128×128 synthetic image ("Line").

a)   b)   c)   d)

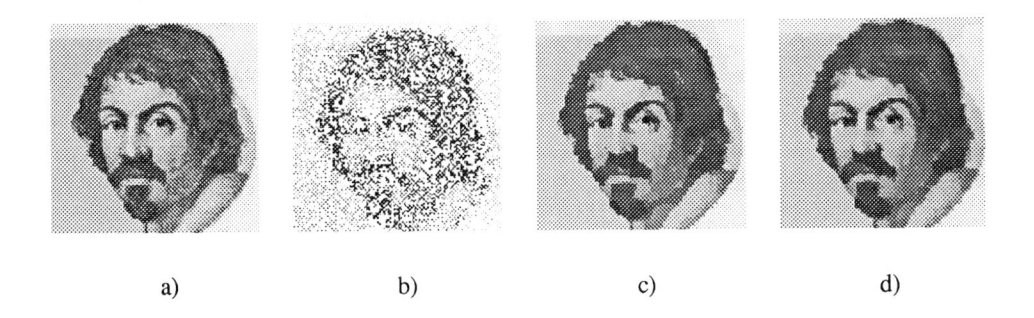

Figure 1. a) original "Caravaggio" image; b) data image with 50% randomly selected pixels and Gaussian noise with $\sigma = 12$; reconstructions with $\lambda = 10$ and $\alpha = 1000$, cost functional F_1: c) Simulated Annealing ($\varepsilon = 12.9$); d) GNC ($\varepsilon = 11.5$).

a)   b)   c)   d)

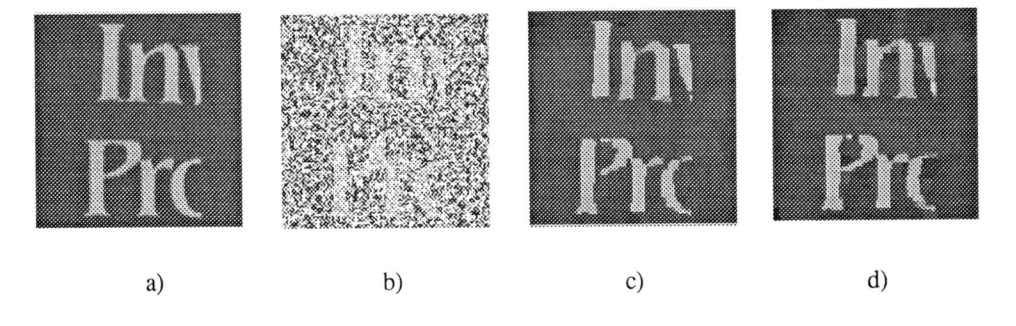

Figure 2. a) original "Letters" image; b) data image (50% pixels, $\sigma = 25$); Simulated Annealing reconstructions with $\lambda = 25$ and $\alpha = 4000$: c) cost functional F_1 ($\varepsilon = 11.1$); d) cost functional F_2 ($\varepsilon = 11.8$).

4. DISCUSSION

The feasibility of applying methods that implicitly consider the discontinuities to solve the problem of image reconstruction from sparse samples has been confirmed by our experiments, and some comparisons are now possible between the performances of the optimization algorithms and between those of the cost functionals considered.

From a qualitative evaluation of the reconstructed images it is found that, whatever the algorithm and the cost functional employed, the results are nearly equivalent, as an efficient recovery of the image discontinuities and a noise reduction is always achieved. Considering the root mean squared error as a quantitative evaluation index, we found that the performance of the functional F_1 is slightly better than that of F_2.

a) b) c) d)

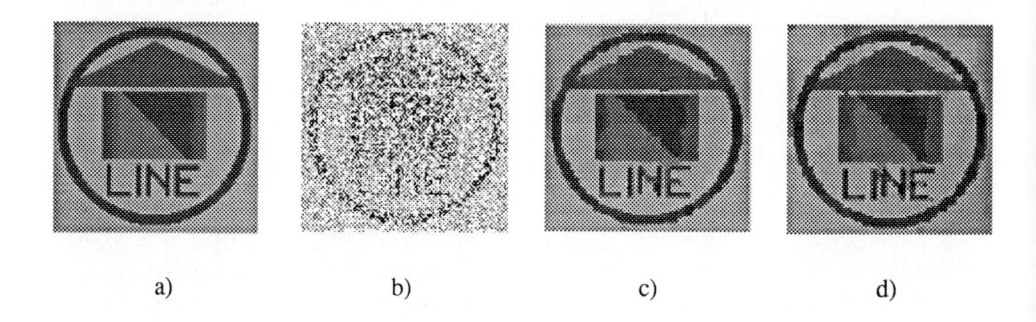

Figure 3. a) original "Line" image; b) data image (50% pixels, $\sigma = 25$); Simulated Annealing reconstructions with $\lambda = 15$ and $\alpha = 2200$: c) cost functional F_1 ($\varepsilon = 13.5$); d) cost functional F_2 ($\varepsilon = 15.1$).

For GNC, complete reconstruction procedure took about one half of the computation time needed by Simulated Annealing. We found that the stochastic algorithm gives results which are closer to the global minimum than GNC. This does not always correspond to a lower value of the mean squared error because the image model may not exactly fit the problem at hand. In fact, the original image does not in general correspond to the global cost minimum, and this is confirmed in our case. The investigation should thus include a search for suitable models for the different classes of images. With this purpose in mind, we are planning to extend our study to more complicated cost functionals, including different forms of neighbour interacting functions and first or higher order derivatives.

5. REFERENCES

1 S.Geman and D.Geman, IEEE Trans., PAMI-6 (1984) 721.
2 J.L.Marroquin et al., J.Am.Stat.Ass., 82 (1987) 76.
3 L.Bedini and A.Tonazzini, Image and Vision Comp., 10 (1992) 108.
4 A.Blake and A.Zisserman, Visual Reconstruction, MIT Press, Cambridge, Mass., 1987.
5 D.Geiger and F.Girosi, IEEE Trans., PAMI-13 (1991) 401.
6 A.Blake and A.Zisserman, Patt.Rec.Lett., 6 (1987) 51.
7 D.Geman and G.Reynolds, IEEE Trans., PAMI-14 (1992) 367.
8 L.Bedini et al., CNR, P.F. Sist. Inf. & Calc. Par., Int. Rep. R/2/85, Rome, Italy, 1992.

Asynchronous Pyramids

S. Bataouche, J.M. Jolion

LIGIA / LISPI, bat. 710, Université Claude BERNARD, Lyon I, F-69622 Villeurbanne Cedex, France

Abstract
In this paper we study the hierarchical reasoning and especially the pyramidal processing of image data. First of all, we are concerned with the way information will be transferred from one pyramidal cell to another when these cells work asynchronously. As example, we will consider the asynchronous hierarchical extraction of connected components in binary images.

1. INTRODUCTION

In general computer vision systems, the goal of the early vision tasks is to extract useful information by means of a search of image components. These elements are characterized separately (for instance by feature's values) and as groups by neighborhood graph or relationships like locative inclusion, part-whole, and class inclusion. In this paper, we address the problem of adding one more characteristic to the description : the time it took to extract the element.

The context of our study is the general framework of hierarchical reasoning for early vision [1]. It is well known that humans are able to recognize at a glance a known but unexpected object. This task is carried out in only a small number of steps in neural hardware [2]. An analysis of the complexity of the visual mechanism shows that hierarchical internal representations and hierarchical reasoning are plausible approaches [3,4].

2. ASYNCHRONOUS HIERARCHICAL REASONING

Consider a classical hierarchical architecture, the so-called pyramid. Up to now, the pyramid processing mode has been considered as a serial-parallel mode. The input information takes place on the first level of the pyramid (each cell of this level, or base, is allocated a pixel or a window of the image). Then, the information is recursively fused up to the apex of the pyramid yielding a transformation of a local and distributed set of data into a global and centralized set of data.

Inside a level, the processing mode is massively parallel, and the process on this level do not start until the process on the level below have finished their work. The elements of the pyramid have then three successive states : wait, work, end. Thus, processing mode is very well suited for classical computations. However, it is not so good for symbolic processing where decisions take place. The execution time for a given level is related to the slower process on this level. This implies (especially for symbolic manipulation) that some information are available before we decide that the "level processing" is over.

We introduce a new framework which makes explicit this phenomenon. We say that an information computed on level k is "complete" if and only if it will not be modified on higher level of the pyramid when taking into account bigger part of the image. So, for any level above the level k, this complete information has just to be transfered to the next level up to the apex. We argue that, as it will no longer change, this information can pop out faster than uncomplete information. In order to achieve this phenomenon, we introduce the asynchronous pyramid. In such a structure, the state of a cell is given by Figure 1. A received information is analyzed and processed or only sent to the parents, depending on its "completeness". As shown in Figure 1, a complete information is available faster than an information which needs to be processed. Thus, each extracted information is characterized by its computation time. This characteristic depends on the information and its context (the nature of the other information contained in the image).

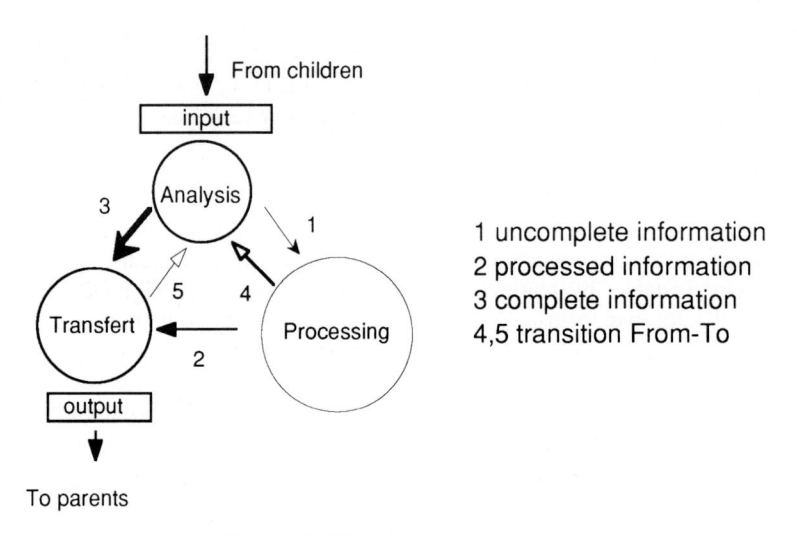

Figure 1. Processing cycle

1 uncomplete information
2 processed information
3 complete information
4,5 transition From-To

When working with a pyramid, one had to be very careful about the bottleneck effect and then about the message passing management. The bottleneck effect is due to the architecture. In a non-overlapped quad-pyramid, one cell receives information from his four children. From bottom to top, the number of cells decreases and the amount of messages grows as the complexity of the scene. Moreover, to have an asynchronous output, the message passing management must be well designed. For instance, the "complete" messages mustn't be delayed by the "uncomplete" ones.

3. EXAMPLE: EXTRACTION OF CONNECTED COMPONENTS

We applied this framework for hierarchical extraction of connected components in binary images. In this particular case, the completeness of an information is easy to understand as shown in Figure 2. The cell sees the three components A, B, and C. When arriving, the connected component A is already known as complete. So, it has only to be transfered, without any processing, up in the pyramid. On the contrary, B is uncomplete. However, when processed, the cell knows that it couldn't be completed by other connected components coming from the other sons. So, B is transfered as uncomplete. The C connected component is made of four sub-components coming from the four sons of the cell. When arriving, the three first parts are known as uncomplete. The processing step results in a bigger and bigger connected component still uncomplete but "completable". When the fourth part has been processed, the C connected component becomes complete and can then be transfered up in the pyramid. Thus the three connected components have different execution time.

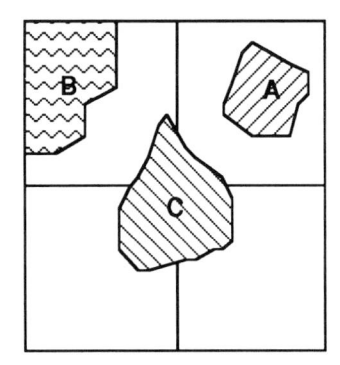

Figure 2.

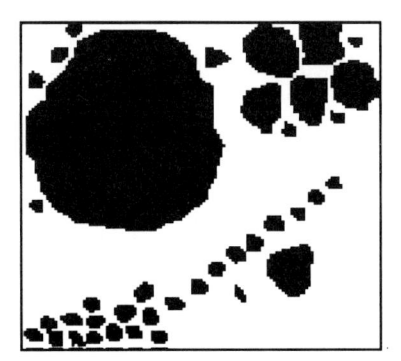

Figure 3.

Let us consider the binary image given by the Figure 3. It contains 37 blobs of different sizes. We have compared the synchronous and asynchronous processes of connected components extraction. We notice that due to the delay the maximum number of messages accumulated at one cell is reduced in the asynchronous process. The messages are processed when they arrive, so the bottleneck effect is reduced.

In the asynchronous mode, a cell has a smaller number of messages to process simultaneously. So, the global process is faster and the connected components are extracted more quickly than in the synchronous process. Figure 4 shows the extraction time of the 37 blobs (the time is function of the number of basic operations being used for the extraction).

So, the time it took to be extracted is one more characteristic of a blob. The asynchronous process extracts the components in the order of their size category.

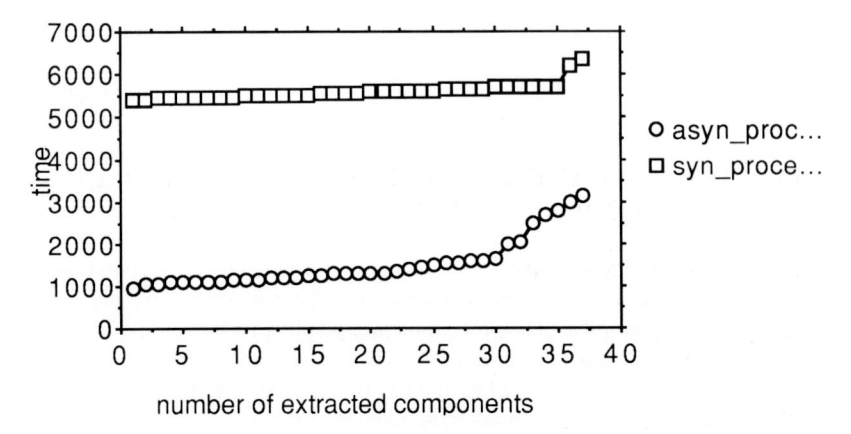

Figure 4. time vs number of extracted connected components

However, it is not always easy to define the completeness of an information. For instance, in the particular case of feature extraction by hierarchical clustering of the feature space [5,6], the completeness of a cluster is related to its size and compactness. We thus have to use a fuzzy definition.

REFERENCES

1. J.M. Jolion and A. Rosenfeld *A Pyramid Framework for Early Vision*, Kluwer, to appear (1993).
2. S.Thorpe "Image Processing by the Human Visual System," *Eurographics'90*, Technical report series, EG 90 TN 4, 1-34 (1992).
3. J.K. Tsotsos "How Does Human Vision Beat the Computational Complexity of Visual Perception ?," *Computational Processes in Human Vision : An Interdisciplinary Perspective*, edited by Pylyshyn, Z.W., 286-338 (1990).
4. A.Rosenfeld "Recognizing Unexpected Objects: A Proposed Approach," *Int.Journal of Pattern Recognition and Artificial Intelligence*, 1(1), 71-84 (1987).
5. J.M. Jolion, P. Meer, P. and S. Bataouche"Robust Clustering with Applications in Computer Vision," *IEEE Trans. on Pattern Anal. and Machine Intel.*, 13(8), 791-802 (1991).
6. S. Bataouche and J.M. Jolion "A Hierarchical and Robust Process for Information Retrieval," *Progress in Image Analysis and Processing*, edited by Cantoni, V. *et al.*, World Scientific, 510-517,(1992).

Image Processing: Theory and Applications
G. Vernazza, A.N. Venetsanopoulos, C. Braccini (Editors)
© 1993 Elsevier Science Publishers B.V. All rights reserved.

A NONLINEAR ADAPTIVE FILTER FOR EDGE-PRESERVING REDUCTION OF ADDITIVE OR MULTIPLICATIVE IMAGE NOISE

L. Alparone[a], S. Baronti[b], R. Carlà[b]

[a]Dipartimento di Ingegneria Elettronica, University of Florence,
via di S. Marta, 3, 50139 Firenze, ITALY

[b]Istituto di Ricerca sulle Onde Elettromagnetiche IROE - CNR,
via Panciatichi, 64, 50127 Firenze, ITALY

Abstract

In this paper a nonlinear filter is suggested that is based on first order local statistics, and is suitable for adaptive reduction of either additive or multiplicative image noise. A selective smoothing of noise is obtained by replacing a pixel when its value belongs to either of the tails of the local histogram. The replaced value is the inner boundary (percentile) of the tail itself, whose area is adaptively defined through a gaussian function of a local feature. The filter fits the actual noise level of images and preserves structures and textures, still maintaining the local mean like a linear filter, without introducing any blur on edges.

1. INTRODUCTION

A general approach to noise reduction considers mathematical models to describe signal, noise, and also how signal is affected by noise; additive or multiplicative models are widely used. Stationarity hypotheses are usually made on noise, but are seldom satisfied by signals thus leading to unsatisfactory results. For this reason, algorithms that are not based on specific models can be effective for wide classes of problems. This approach is typical of nonlinear filters, whose theory is generally difficult, but whose performances are often far better than those of linear ones.

Many specific algorithms have been proposed for reduction of multiplicative noise, such as the Lee filter [1], generalized by Kuan et al.[2]. According to that scheme, the optimum estimate ($\hat{y}$) of the actual radiometric value (y) is

$$\hat{y} = E[y] + (z - E[z])(1 - \sigma_u^2/LVC^2)/(1 + \sigma_u^2)$$

where (z) is the observed pixel value, E[] denotes the expectation operator, σ_u^2 the noise variance, and LVC is the local variation coefficient, defined as $\sigma_z/E[z]$, σ_z being the local standard deviation of (z). Such an expression stems from the minimization of $E[(y-\hat{y})^2]$, when a multiplicative model (z = y·u) is assumed to describe the influence of the unitary-mean noise (u). For an additive noise model the above expression is modified by simply replacing LVC with the Local Standard Deviation (LSD) and suppressing σ_u^2 at denominator.

The algorithm presented in this paper is also based on local statistics and is a generalization of the schemes reported in [3-4].

2. THE ALGORITHM FOR ADAPTIVE NOISE REDUCTION

Local histogram Hist(l), l=0,L_{MAX} (where L_{MAX} is the full-scale value, namely 255 for 8 bit images), and the local feature LF(i,j) are computed on a sliding window of predefined odd sizes (m,n) centered on each pixel of coordinates (i,j). Two values $A_l(i,j)$ and $A_r(i,j)$, complementary in the range (0÷1) and defining the areas of the tails of the normalized local histogram, are derived from LF(i,j). Each input value f(i,j) is compared with levels $P_l(i,j)$ and $P_r(i,j)$, defined as the A_l, A_r percentiles respectively, by the following rule:

$$CDF[P_l(i,j)-1] \leq A_l(i,j) < CDF[P_l(i,j)] \quad (1)$$
$$CDF[P_r(i,j)-1] < A_r(i,j) \leq CDF[P_r(i,j)]$$

with CDF the local Cumulative Distribution Function. The output value g(i,j) is defined by:

$$g(i,j) = \text{Median}\{P_l(i,j), f(i,j), P_r(i,j)\} \quad (2)$$

Adaptivity is achieved by deriving $A_l(i,j)$ and $A_r(i,j)$ dependently on LF(i,j). In homogeneous areas (low LF) the filter should strongly smooth the image, whereas on edges (high LF) the image should be left almost unchanged; elsewhere the filter should steadily decrease its smoothing effect for increasing LF. Smoothing is the heaviest when $A_l(i,j)=A_r(i,j)=0.5$, causing $P_l(i,j)=P_r(i,j)$. From (2) it is apparent that the filter becomes a median. Conversely, the strongest degree of preservation is obtained when $A_l(i,j)=0$ and $A_r(i,j)=1$, hence both $P_l(i,j)$ equals the minimum value inside the local window and $P_r(i,j)$ the maximum; in that case, pixels are not changed.

The above considerations suggest defining an adaptivity function of the local feature LF, $A_l(i,j)=F(LF(i,j))$, having its maximum for LF=0, slowly varying for low LF and then rapidly decreasing, like a gaussian function $A \cdot \exp[-LF(i,j)^2/\sigma^2]$, where A is a constant, chosen dependently on the desired adaptivity range, and σ is a shaping factor. Since σ^2 defines the strength of smoothing, it may be related to the variance of noise σ_u^2; taking $A=0.5 \cdot \exp(1)$ and clipping the result to 0.5 yields

$$A_l(LF(i,j)) = 0.5 \cdot \min\{1, \exp[-(LF(i,j)^2/\sigma^2 - 1)]\} \quad (3)$$

If σ is chosen equal to σ_u, the algorithm behaves like a median filter on uniform areas, where either LVC approximates the standard deviation of multiplicative noise or LSD measures the standard deviation of additive noise. In any case, the smoothing effect may be increased and decreased along with σ as in a linear gaussian filter, making the algorithm easily fit more or less noisy images, same as in [1] and [2]. After that, $A_r(i,j)$ may be obtained symmetrically as $1-A_l(i,j)$; $P_l(i,j)$ and $P_r(i,j)$ are found as percentiles of the local histogram according to (1). The average level is easily kept also when the level changes are unbalanced, i.e. if the local histogram is not symmetric around its mean, by correcting $A_l(i,j)$ and $A_r(i,j)$ dependently on the asymmetry of the local histogram:

$CT(i,j) = [MN(i,j)-MD(i,j)]/L_{MAX}$

in this way, the correct values of the percentile areas are

$A_{lc}(i,j) = A_l(i,j)[1+CT(i,j)]$
$A_{rc}(i,j) = [1-A_l(i,j)][1-CT(i,j)]$

where $A_l(i,j)$ is given by (3). $A_{lc}(i,j)$ and $A_{rc}(i,j)$ are actually used in (1) to retrieve P_l and P_r.

3. EVALUATION CRITERIA, RESULTS AND COMPARISONS

The proposed algorithm, referred from here as Adaptive Percentile Clipping (APC) filter, has been applied to images affected by additive or multiplicative noise. Evaluations have been carried out according to SNR improvement (only SNR results for multiplicative noise are reported here), visual interpretation, and computation cost.

For multiplicative noise model an SNR measure may be defined as

$$SNR = \sigma_y^2/E[y]^2\sigma_u^2 \qquad (4)$$

where σ_y^2 is the variance of the radiometric signal (y), E[y] is the mean of (y) and σ_u^2 is the variance of unitary-mean noise (u). Particularly significant is the SNR recovery (ΔSNR, SNR variation between the original and filtered images), because it accounts for both smoothing uniform areas and preserving high-variance areas.

The generalized Lee and the median filter have been compared with ours. Filtering parameter have been chosen in order to obtain the best performances. Table 1 reports ΔSNRs for two 16 look, C-band HH-polarization, L-band VV-polarization, amplitude SAR images: the APC filter is the best followed by the generalized Lee. Visual analysis on these images confirmed that the APC and generalized Lee filters are the most promising, as noise is strongly smoothed in uniform areas, whereas structured areas, edges and point targets are preserved.

FILTER	ΔSNR (dB)	
	C-HH	L-VV
median	3.20	2.94
gen. Lee	6.90	4.09
APC	7.95	4.71

Table 1 ΔSNR improvement for two C-HH and L-VV SAR images.

Additive (zero-mean, SNR=1, Gaussian) and multiplicative (unitary-mean, σ_u^2=0.125, Chi-square) noise has been superimposed on the test image girl, figure 1 and 2 respectively (top-left). Median (3x3 2 iterations), APC (5x5 1 iteration), and Generalized Lee (5x5 1 iteration) filters are shown for comparison purpose (clockwise disposition). APC and generalized Lee filters exhibit similar performances and are promising on the whole, while the median is somewhat inferior. The APC filter confirms its high selectivity, and its ability to strongly smooth low contrast areas while effectively preserving high detail areas as edges or structures. This properties are also maintained when the filtering window is increased in size or by acting on the noise level parameter; in this way slowly varying gradients are preserved as well.

All the schemes exploit fast algorithms, whose computational cost is approximately proportional to the length of the vertical side of the window. The median is the fastest, followed by the generalized Lee.

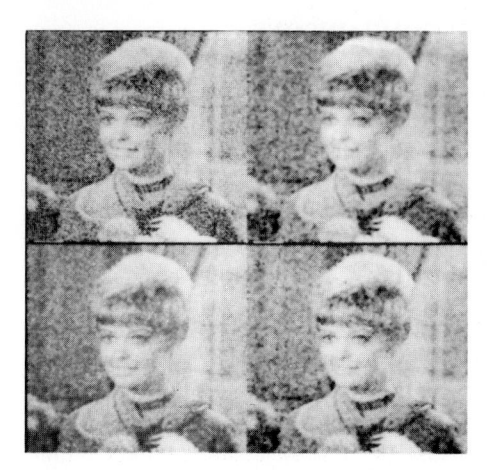

Figures 1 (additive noise, left) and 2 (multiplicative noise, right). The noisy test images (top-left) have been filtered by median (top-right), Gen. Lee (bottom-left) and APC (bottom-right) algorithms.

4. REFERENCES

1 J.S. Lee, "Digital Image Enhancement and Noise Filtering by Use of Local Statistics", *IEEE Trans. Pattern Anal. Machine Intel.*, vol. PAMI-2, no. 2, pp. 165-168, Mar 1980.
2 D.T. Kuan, A.A. Sawchuck, T.C. Strand and P. Chavel, "Adaptive Noise Smoothing Filter for Images with Signal-Dependent Noise", *IEEE Trans. Pattern Anal. Machine Intel.*, vol. PAMI-7, no. 2, pp. 165-177, Mar. 1985.
3 R. Carlà, V.M. Sacco and S. Baronti, "Digital Techniques for Noise Reduction in APT NOAA Satellite Images", in *Proc. of IGARSS '86*, Zurich, pp. 995-1000, 1986.
4 L. Alparone, S. Baronti, R. Carlà and C. Puglisi, "A New Adaptive Digital Filter for SAR Images: Test of Performance for land and crop Classification on Montespertoli Area", in *Proc. of IGARSS '92*, Houston, Texas, pp. 899-901, 1992.

Prior Distributions for Tomographic Reconstruction

N. Pendock

Department of Computational and Applied Mathematics,
University of the Witwatersrand, Johannesburg, South Africa

Abstract
Tomographic image reconstruction may be regarded as an ill-posed inverse problem for which two types of information are usually available : data collected by some remote sensing experiment and prior information on the expected form of the reconstructed image. Ignoring prior information is equivalent to choosing a uniform prior which allows all solutions, including physically non-realizable reconstructions. Positivity and an assumption of independence of image pixels to be reconstructed implies an entropic prior which may be easily modified to incorporate spatial correlations in the reconstructed image. We examine various priors for seismic tomography and compare their performances on a synthetic cross-hole seismic tomography example.

1. INTRODUCTION

Tomography comes from the Greek word *tomos* meaning slice and the tomographic method consists of collecting projections of some object via a remote sensing process and then inferring some properties of the object. The technique has had much success in medical imaging where various waves including x-rays and ultrasound are passed through the body and the intervening tissue is thereby imaged. In the geophysical application acoustic, radar, radio and seismic waves are used to image rock at scales from *in situ* geological mapping in a mine to global models of attenuation and velocity. Lines [3] identifies three phases of the geophysical tomographic method : collection of data; formulation of a forward model either assuming straight ray paths or using ray tracing techniques and inversion of the forward model to produce an image of velocities or attenuations. The inversion process is usually *ill-posed* for several reasons :
in the interests of tractability, the forward model is usually a simplification of the problem and may be a poor approximation to the underlying physical reality;
the data are often contaminated with noise which renders the forward model inconsistent;
the data are usually a finite set of observations of the continuous natural reality and thus the tomographic reconstruction is a discrete realization of a continuous phenomenon.
From the multiple images which fit the data, we would like to choose a single one as the "answer". A Bayesian approach to reconstruction allows us to incorporate prior information into the inversion process and in so doing provides us with the ability to choose sensible images.

2. BAYESIAN INVERSIONS

A Bayesian tomographic reconstruction strategy consists of deciding what is known *a priori* about the image to be estimated, representing this by means of a probability distribution and then updating this using any data available to achieve a posterior distribution. The relationship between the posterior, prior and data likelihood is known as Bayes' theorem. Probability theory is now silent on which particular image to choose from the posterior distribution. Historically, the distribution maximum was chosen corresponding to the most likely reconstruction, but another choice could be made, for example the distribution mean. An alternate approach to a single choice would be to sample the posterior distribution many times to get a "feel" for the type of reconstructions indicated by the data.

How should we represent any prior information we may have? Jaynes [2] proposes a basic desideratum for the assignment of prior distributions : *if two problems have the same prior information we should assign the same prior probabilities.* In addition, we would like the priors to be as uninformative as possible so that the data may tell their side of the story.

2.1 Prior Distributions

We can usually be sure there exist real numbers a and b such that all reconstructed pixel values fall in the interval $[a, b]$. If we assume no other prior information, then the most reasonable prior assignment is the uniform density over $[a, b]$. This choice of prior does not mean that each value in the interval is equally likely, just that we do not have any reason to assign one value a higher probability that another.

In seismic tomography, we typically wish to reconstruct an image $S = \{s_i\}$ of rock slowness. To assume maximum ignorance about $p(S)$ we choose a uniform prior distribution. But what if instead of estimating S we were to estimate the velocity distribution $V = \frac{1}{S}$? Now $dS = \frac{1}{V^2}dV$ and so assuming a uniform distribution for S implies a prior distribution for V proportional to $\frac{1}{V^2}$. But this is unreasonable, since we could make the same argument for V to have a uniform prior implying $p(S) \propto \frac{1}{S^2}$. With Jaynes' desideratum in mind, we require that S and V should have the same prior distributions. $p(V) = \frac{1}{V}$ is the only form the prior may have for V and S to have the same prior distribution. If we ignore any spatial correlations between slowness values and assume $\{s_i\}$ to be independent, the corresponding Jeffreys' prior is $p(S) \propto \prod_i \frac{1}{s_i}$. The most likely slowness image is achieved by maximizing $p(S)$ which is equivalent to maximizing $log\, p(S) = -\sum log\, s_i$, the Burg entropy.

Skilling [5] makes an argument for an alternate entropic prior $p(S) = -\sum s_i\, log\, s_i$ by considering an image as having been generated by a team of monkeys throwing photons onto a photographic plate. As the number of photons increases without limit the images generated by the monkeys cluster around the entropy maximum. This form of the entropic prior is the same as the function derived by Shannon [4] as the only consistent information measure for a discrete probability distribution. Jaynes [1] extended Shannon's results to continuous distributions by introducing the invariance measure $m(t)$ which makes the entropy $\int s(t) - m(t) - s(t) \, log \, \frac{s(t)}{m(t)} \, dt$ invariant under change of variable. The inclusion of m caters for situations where the state of maximum uncertainty is characterised by non-equal pixel values and may be used to incorporate prior spatial information into the reconstruction in a natural way. Returning to our geophsyical problem, if we knew that the region to be reconstructed consisted of horizontal layers of similar slowness, one possible m distribution is $m_i = \sqrt{s_{i-1}s_{i+1}}$.

3. A SYNTHETIC EXAMPLE

In order illustrate the effect of different prior information on tomographic reconstructions, we used the synthetic seismic example of Squires and Cambois [6]. We consider a model of constant slowness (one) and parameterize it into 8×8 square cells. Eight transmitters and eight receivers are positioned with equal spacing on the opposite vertical sides of the model. Sixty four data points T are then calculated, one from each of the sixty four ray paths. Our forward model is the even determined linear system $A_{64 \times 64} S_{64 \times 1} = T_{64 \times 1}$. A is singular with condition number $\approx 10^{33}$. Inspection of the singular values led us to *guess* that fifty are useful in estimating S. Setting the other fourteen values to zero, we may compute the pseudo-inverse of A and this minimum norm least squares reconstruction is correct.

We next added zero-mean Gaussian noise with variance 10% of the mean data value to produce a data set used in all the subsequent reconstructions. Reconstructions were made using various priors and are shown in Figures 1 to 4. The damped least-squares solution $S = (A^T A + \sigma I)^{-1} A^T T$ (Figure 1) was obtained by damping the minimum norm least-squares solution with a damping factor σ equal to 1% of the maximum eigenvalue of $A^T A$. This solution is physically not realizable as negative slowness values have been estimated. The positive least-squares solution (Figure 2) corresponds to a prior $p(S) \propto \frac{1}{k}$ where k is some arbitrarily large positive constant and the resulting reconstruction is very spiky. The role of entropy as a regularizer is clear in the MaxEnt (Figure 3) reconstruction and the result is similar to the damped least squares image although it has the advantage of being a well-defined answer as the entropy is a convex function and so has a unique maximum. The choice of the damping parameter σ is somewhat *ad-hoc* and each damping value will generate a different slowness distribution. Spatial continuity was encouraged in the final reconstruction (Figure 4) by choosing a sampling prior $m_i = \sqrt{s_{i-1}s_{i+1}}$ for S and produces the best reconstruction.

4. CONCLUSIONS

The choice of prior distribution for tomographic reconstructions may have a large effect on the form of the reconstructed image. Some powerful priors are available even if the *a priori* information is sparse : positivity of pixel values and the fact that the image to be reconstructed is a positive additive distribution indicate an entropic form for the prior. In addition, a philosophical case may be made for entropic priors since maximum entropy reconstructions are the most *honest* solutions consistent with the measured data in that they have the least amount of structure not indicated by the data.

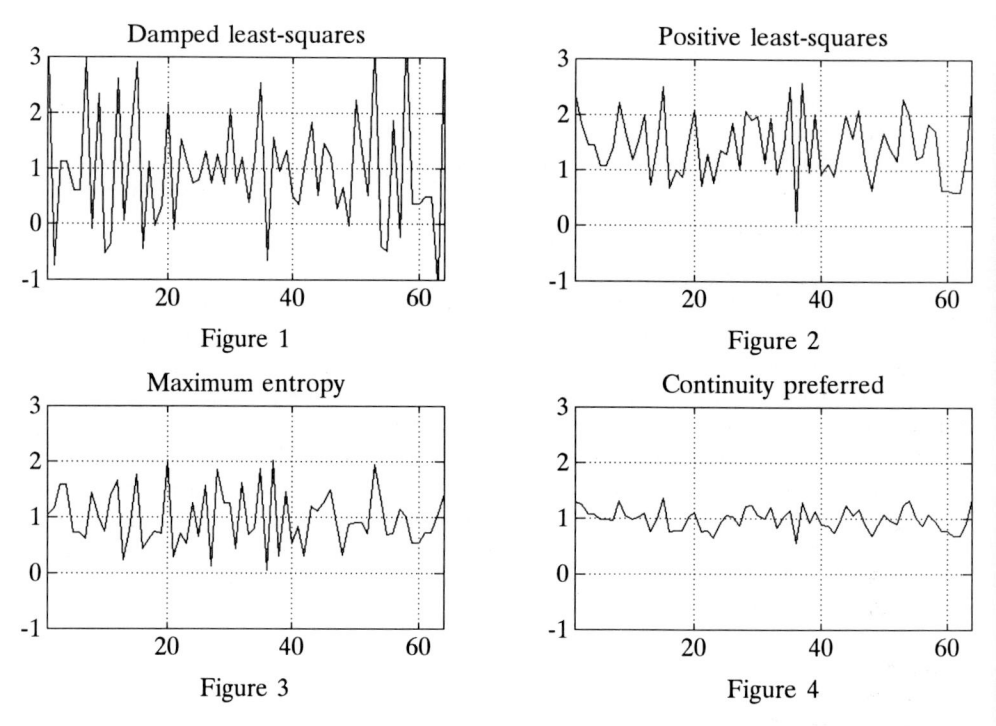

Tomographic reconstructions for various priors. The true solution is the constant slowness value 1.

5. REFERENCES

1 E. T. Jaynes, Information theory and statistical mechanics, in *Statistical Physics* **3**, K. W. Ford (ed.), New York, W.A. Benjamin Inc, 1963.
2 E. T. Jaynes, Prior probabilities, IEEE Trans. Systems Science and Cybernetics, **SSC4**, 1968.
3 L. Lines, Applications of tomography to borehole and reflection seismology, Geophysics : The Leading Edge of Exploration, 1991.
4 C. Shannon & W. Weaver, The mathematical theory of communication, Univ. Illinois Press, 1949.
5 J. Skilling, Theory of maximum entropy image reconstruction, in *Maximum Entropy and Bayesian Methods in Applied Statistics*, J H Justice (ed.), Cambridge University Press, 1986.
6 L. J. Squires & G. Cambois, A linear filter approach to designing off-diagonal damping matrices for least-squares inverse problems, Geophysics, **57**, 7, 1992.

Multivariate Estimation Based on Gradient Search

Kaijun Tang, Jaakko Astola and Yrjö Neuvo

Signal Processing Laboratory, Tampere University of Technology, P.O. Box 527, SF-33101 Tampere, Finland

Abstract: In this paper, a multivariate estimator based on gradient search is proposed to handle noisy multivariate signals. It is another extension of the univariate median to multivariate case in the sense of a minimum sum of distances. The estimator results in better noise attenuation than the vector median filter and even outperforms the componentwise processing for short/medium and long-tailed noises. In addition, the result of a noisy bivariate edge filtering indicates that it has a rather good edge-preserving ability.

1. INTRODUCTION

The successful application of the univariate median and related nonlinear filters in real-valued signal processing has motivated a new investigation of the multivariate counterparts. The reason is because we, sometimes, involve handling multivariate signals. Although multivariate signals can be processed by operating on each component independently with univariate techniques, individual application of these techniques to each component of multivariate signals may result in significant loss of information. The vector median filter is one of nonlinear multivariate filters [1, 2], where multivariate signals are treated using vector technique instead of componentwise one.

In the univariate case, the sample having a minimum sum of distances to input samples is one of the input samples. However, the same conclusion does not apply to the multivariate case. The vector median can not achieve in general a minimum sum of distances. In this paper, we find such a new multivariate minimizing a sum of distances to input samples. Since the analytical solution does not exist, we have to use the gradient search algorithm. This estimator has been applied to parameter estimation in so-called matched median filter [3]. However, it has more efficient noise attenuation compared with the componentwise counterpart, in the presence of the medium/short and long-tailed distribution noise. In addition, we test the edge response of the estimator by use of a bivariate edge corrupted by Gaussian noise. It shows a fairly good edge preserving ability. As the estimate is not necessarily one of the input samples, the artifacts of streaking and blotching are reduced greatly when it is applied for color image filtering. In the following, a multivariate estimator based on gradient search is introduced. Then, the performance of several multivariate filters is compared with that of the multivariate estimator. Conclusions follow.

2. MULTIVARIATE ESTIMATOR

Given a set of N p-variate samples $\mathbf{X}(i)$ for $i = 1, 2, \cdots, N$ with $\mathbf{X}(i)=[x^{(1)}(i), x^{(2)}(i), \cdots, x^{(p)}(i)]$, the output of the multivariate estimator is given by

$$\mathbf{Y} = \arg\min_{\mathbf{X}} \sum_{i=1}^{N} \|\mathbf{X} - \mathbf{X}(i)\| \tag{1}$$

where $\|\cdot\|$ denotes L_2 norm.

If $\mathbf{X}$ is restricted to one of the N multivariate samples, $\mathbf{X}$ is defined as the vector median. As the analytic solution does not exist, a desired approximate estimate can be obtained by computing the gradient of the sum of the distances, As a general case, we consider a weighted version of the multivariate estimator. A function of the weighted sum of the distances is written as

$$f(\mathbf{X}) = \sum_{i=1}^{N} W_i \|\mathbf{X} - \mathbf{X}(i)\| \tag{2}$$

An iterative gradient search algorithm is represented algebraically as

$$\mathbf{X}_{k+1} = \mathbf{X}_k - \mu \nabla_k f \quad k = 0, 1, \cdots \tag{3}$$

where k is the step or iteration number and μ is step size. The gradient function is computed as follows,

$$\nabla_k f = \begin{bmatrix} \sum_{i=1}^{N} W_i \frac{x_k^{(1)} - x^{(1)}(i)}{\|\mathbf{X}_k - \mathbf{X}(i)\|} \\ \sum_{i=1}^{N} W_i \frac{x_k^{(2)} - x^{(2)}(i)}{\|\mathbf{X}_k - \mathbf{X}(i)\|} \\ \vdots \\ \sum_{i=1}^{N} W_i \frac{x_k^{(p)} - x^{(p)}(i)}{\|\mathbf{X}_k - \mathbf{X}(i)\|} \end{bmatrix} \tag{4}$$

where $x_k^{(j)}$ and $x^{(j)}(i)$ denote the j-th component of $\mathbf{X}_k$ and $\mathbf{X}(i)$, respectively.

In order to guarantee the existence of the gradient function, $\mathbf{X}_k - \mathbf{X}(i) \neq 0$ should hold for every k and i. If $\mathbf{X}_k$ is equal to $\mathbf{X}(i)$, $\|\mathbf{X}_k - \mathbf{X}(i)\|$ will be removed from the weighted sum terms. In the case that the solution is not unique, the multivariate sample closest to the center of the window is preferred to others. Therefore, the center sample within the window is selected as an initial value of the gradient search. The choice of the step size is determined empirically. The small step size leads to relatively slow convergence rate. The large step size implies the fast convergence rate but leads to instability of the gradient search. Usually, it is difficult to evaluate the bound of the step size for the stability of the gradient search algorithm.

3. PERFORMANCE ANALYSES

In order to evaluate the performance of the multivariate estimator quantitatively. Two sets of computer simulations were performed. In the first example, we test the noise attenuation of the estimator for different exponential-type noises. In the second example, we examine the edge response of the estimator at a noisy bivariate step edge. Unless otherwise stated, all of the multivariate noises are zero mean unit variance i.i.d. noise in each component and independent between the components. The length of the tested multivariate filters is 7 and all the weights are one.

3.1 Noise Attenuation

Three random sequences with 10000 samples were generated independently and consist of a 3-variate noise sequence. We have produced three 3-variate noise sequences with different distribution functions. Then, the three sequences were filtered with the filters, respectively. The variance of each of the filtered sequences is computed by taking average over its three components. We compare the multivariate estimator with other multivariate filters. The results show that the multivariate estimator has the excellent noise attenuation ability as shown in Table I. It is even better than the componentwise counterpart (the marginal median filter) in different noisy conditions and approaches the mean filter in Gaussian noise. Although the mean filter has a small difference in the noise attenuation, it is poor in outliers removal and edge preservation [4]. This fact suggests that the simultaneous multivariate processing can outperform the componentwise processing.

3.2 Edge Response

In order to illustrate the edge preserving property of the multivariate estimator, a bivariate step edge with a transition from a constant value (0,0) to (2,5) is generated, each side having 50 samples. It is corrupted by Gaussian noise as plotted in Fig. 1(a), where the time order is not implied. The edge responses of marginal median filtering, vector median filtering and multivariate estimating are shown in Fig. 1(b)-(d), respectively. The results indicate that the multivariate estimator has better noise smoothing than the marginal median filter and the vector median filter without edge blur.

4. CONCLUSIONS

In this paper, we have analyzed the noise attenuation and edge preservation properties of the multivariate estimation based on gradient search. As the estimate is a centralized sample in a vector space spanned by certain number of multivariate samples, the multivariate estimator shows a good robustness.

References

[1] C. A. Pomalaza-Raez and Y. Fong, *Estimation of the location parameter of a multispectral distribution by a median operation*, in Proc. 11th Int. Symp. on Machine Processing of Remotely Sensed Data, West Lafayette, IN. 1985, pp. 41-48.

[2] J. Astola, P. Haavisto and Y. Neuvo, *Vector median filters*, Proc. IEEE, vol. 78, April

Table I. OUTPUT MEAN SQUARE ERRORS OF MULTIVARIATE FILTERS

Multivariate Filter	output mean square errors		
	Biexponential	Gaussian	Uniform
Mean Filter	0.1411	0.1409	0.1495
Marginal Median Filter	0.1160	0.2057	0.3427
Vector Median Filter	0.2194	0.3393	0.5129
Extended Vector Median Filter	0.2028	0.2990	0.4083
Multivariate Estimator	0.1040	0.1464	0.1920

1990, pp. 678-689.

[3] Qin Liu, Jaakko Astola and Yrjö Neuvo, *Matched median filter as detector for QAM signals*, The Intern. Symposium on Circuits and Systems, Singapore, June, 11-14 1991, pp. 97-100.

[4] I. Pitas, A. N. Venetsanopoulos, *Nonlinear digital filters: principles and applications*, Kluwer Academic, 1990.

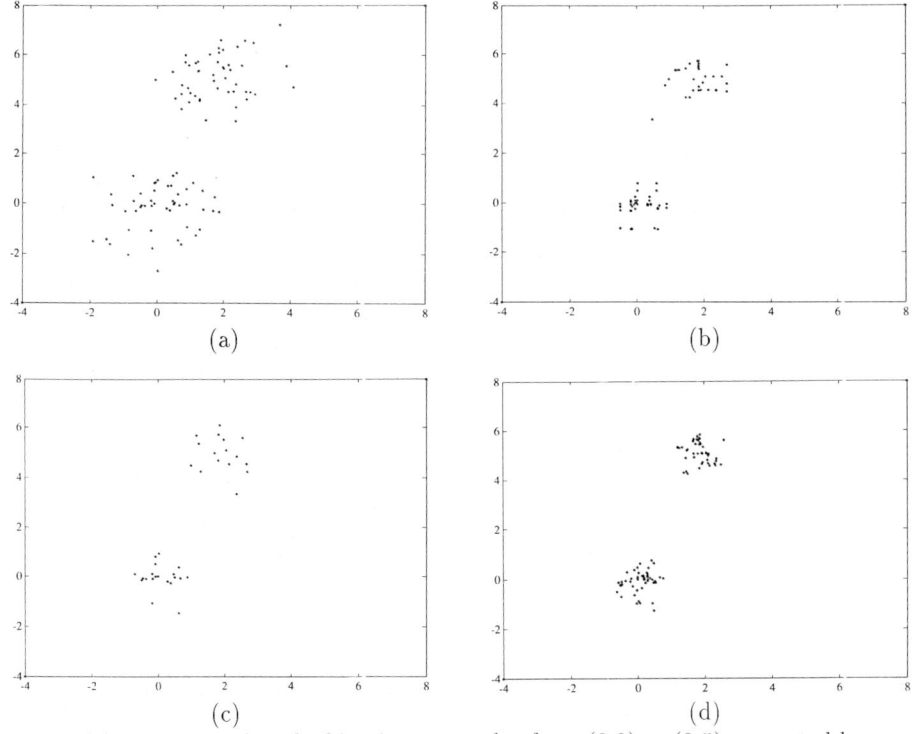

Fig. 1 (a) Cartesian plot of a bivariate step edge from (0,0) to (2,5) corrupted by zero mean unit variance Gaussian noise . (b) Output of marginal median filter. (c) Output of vector median filter. (d) Output of multivariate estimator.

Optimal Weighted Median Filtering for Image Restoration

Ruikang Yang, Lin Yin, Moncef Gabbouj, and Yrjö Neuvo

Signal Processing Laboratory, Tampere University of Technology,
P.O. Box 553, SF-33101, Tampere, Finland

Abstract

In this paper, an algorithm using nonlinear programming is developed for designing optimal weighted median (WM) filters which minimize noise subject to a pre-determined set of structural constraints on the filter's behavior. Optimal weighted median filters under several structural constraints will be designed based on the algorithm. Simulations are carried out to demonstrate the performance of these optimal weighted median filters.

1. INTRODUCTION

Weighted Median (WM) filters have proven to be particularly effective in image recovery [2]. This is because WM filters can preserve or reconstruct edges, lines and other image details while also removing noise or other processes which have altered the image.

WM filters were first developed by Justusson [1]. One and multi-dimensional weighted medians have been studied and applied in various areas by several researchers, see [2].

Weighted median filters can preserve image details, i.e. lines, edges, by selecting appropriate weights. On the other hand, for a given set of image details, there may be many weighted median filters which can preserve the given details. The question is how to select one which suppresses noise the best. This is referred to as optimal filtering under structural constraints [2]. In this paper, an algorithm is developed based on the statistical analysis of weighted median filters [3]. It is shown that the optimal WM filter can be designed by use of nonlinear programming.

In Section 2 we first introduce the basic concept of WM filters, and then, the optimal problem is formulated. It is shown that the optimal problem can be solved by nonlinear programming. In Section 3, based on the algorithm, we design optimal WM filters under several structural constraints. Simulations are carried out to demonstrate the performance of these optimal WM filters. Conclusions are drawn in Section 4.

2. WEIGHTED MEDIAN FILTERS

Definition 1: Let $\{X(\cdot,\cdot)\}$ and $\{Y(\cdot,\cdot)\}$ be the input image and the output image, respectively, of a WM filter with a $(2L+1) \times (2M+1)$ window $W = \{(s,t)|-L \leq s \leq L, -M \leq t \leq M\}$. Then the output of the WM filter with weight vector $\underline{W} = \{W_{s,t}|(s,t) \in W\}$, where $N = (2L+1)(2M+1)$, is given by

$$Y(i,j) = \text{MED}\{W_{s,t} \diamond X(i-s, j-t)|(s,t) \in W\},$$

where MED[·] denotes the median operation and $\diamond$ denotes duplication, i.e. $n \diamond X = \underbrace{X, ..., X}_{n\ times}$.

The threshold T of a WM filter is defined as $T = \dfrac{1}{2}\sum_{i=1}^{N} W_i$

In image restoration, it is important to preserve many image details. However, giving a set of detail-preserving requirements, e.g. lines, sharp corners, there may exist a number of WM filters preserving these structures. The designer's task is, then, to pick an WM filter among them in order to achieve the maximum noise attenuation. The following theorem formulates the optimal problem as a nonlinear programming problem.

Theorem: For WM filters with window size $N = 2K + 1$, designing optimal WM filters under structural constraints is equivalent to minimizing a nonlinear function under detail-preserving requirements, i.e.

$$\text{Minimize} \sum_{i=1}^{K}\sum_{b \in S_i} L_i U(\underline{W}^t b - T) \qquad (1)$$
$$\text{subject to}$$
$$\mathbf{A}\,\underline{W} \geq \underline{0}, \text{ (structural constraints)}$$

where $\mathbf{A}$ is an $I \times N$ matrix. $\mathbf{A}\underline{W} \geq B$ denotes I-inequalities which represent the structural constraints, and $S_i = \{b | b \in \{0,1\}^N, \omega(b) = i\}$, where $\omega(\cdot)$ denotes the Hamming weight.

One may notice that the objective function is not differentiable since it involves the unit-step function $U(\cdot)$. Such optimization problems can be quite difficult to solve. The unit-step function can, however, be approximated by a sigmoidal function $U_s(\cdot)$.

3. APPLICATIONS

The success of median and weighted median filters in image processing and other application areas has been reported in the literature [2]. Weighted medians, when properly designed, can preserve edges, sharp corners and at the same time remove noise effectively. In this section, we will investigate optimal WM filters with structural constraints.

Consider WM filters with 3×5 window,

$$\underline{W} = \begin{pmatrix} W_1 & W_2 & W_3 & W_4 & W_5 \\ W_6 & W_7 & W_8 & W_9 & W_{10} \\ W_{11} & W_{12} & W_{13} & W_{14} & W_{15} \end{pmatrix}$$

Suppose that we want to preserve horizontal lines and vertical lines, according to the properties of WM filters, the weights must satisfy the following inequalities,

$W_6 + W_7 + W_8 + W_9 + W_{10} \geq T$
$W_3 + W_8 + W_{13} \geq T$

Taking these inequalities as the structural constraints in (1), using nonlinear programming, we found the optimal WM filter which preserve horizontal and vertical lines. It is listed in Table 1 and denoted by WM1. Similarily, we designed the optimal WM filters under several other structural constraints. They are listed in Table 1 and denoted by WM2, WM3 and WM4, respectively.

Table 1: Several optimal WM filters

	structural constraints	weight vector
WM1	$W_3 + W_8 + W_{13} \geq T$ $W_6 + W_7 + W_8 + W_9 + W_{10} \geq T$	$\begin{pmatrix} 2.111 \ 2.111 \ 14.33 \ 2.111 \ 2.111 \\ 6.667 \ 6.667 \ 17.89 \ 6.667 \ 6.667 \\ 2.111 \ 2.111 \ 12.33 \ 2.111 \ 2.111 \end{pmatrix}$
WM2	$W_1 + W_2 + W_3 + W_6 + W_7 + W_8 \geq T$ $W_3 + W_4 + W_5 + W_8 + W_9 + W_{10} \geq T$ $W_6 + W_7 + W_8 + W_{11} + W_{12} + W_{13} \geq T$ $W_8 + W_9 + W_{10} + W_{13} + W_{14} + W_{15} \geq T$ $W_3 - W_{13} > 0$	$\begin{pmatrix} 4.208 \ 4.208 \ 10.00 \ 4.208 \ 4.208 \\ 8.750 \ 12.50 \ 20.33 \ 11.25 \ 7.500 \\ 4.208 \ 4.208 \ 10.00 \ 5.458 \ 5.458 \end{pmatrix}$
WM3	$W_1 + W_2 + W_3 + W_6 + W_7 + W_8 \geq T$ $W_3 + W_4 + W_5 + W_8 + W_9 + W_{10} \geq T$ $W_6 + W_7 + W_8 + W_{11} + W_{12} + W_{13} \geq T$ $W_8 + W_9 + W_{10} + W_{13} + W_{14} + W_{15} \geq T$ $W_7 - W_9 > 0$	$\begin{pmatrix} 4.208 \ 4.208 \ 10.00 \ 4.208 \ 4.208 \\ 8.750 \ 12.50 \ 20.33 \ 11.25 \ 7.500 \\ 4.208 \ 4.208 \ 10.00 \ 5.458 \ 5.458 \end{pmatrix}$
WM4	$W_3 + W_8 + W_{13} \geq T$ $W_6 + W_7 + W_8 + W_9 + W_{10} \geq T$ $W_2 + W_8 + W_{14} \geq T$ $W_4 + W_8 + W_{12} \geq T$	$\begin{pmatrix} 0.000 \ 5.043 \ 6.543 \ 5.043 \ 0.000 \\ 2.522 \ 2.522 \ 21.17 \ 2.522 \ 2.522 \\ 0.000 \ 5.043 \ 3.543 \ 5.043 \ 0.000 \end{pmatrix}$

In order to demonstrate the performance of these optimal WM filters, we applied them to filter two images corrupted by impulsive noise and Gaussian noise, respectively. The probability of impulses was 0.12 and the height was set to ± 200. The variance of Gaussian noise was 1000. The mean square error (MSE) and mean absolute error (MAE) are listed in Table 2. For comparison, the results by the standard median (SM) filter are also listed.

Table 2: Results of the filtered images

	Noisy stream bridge by impulsive noise Prob= 0.12, size= 200		Noisy stream bridge by Gaussian noise variance=1000	
	MSE	MAE	MSE	MAE
SM	214.088104	9.051422	318.218811	13.849895
WM1	132.540680	5.814171	314.257874	13.481792
WM2	140.791313	6.365288	290.944122	13.187550
WM3	136.344513	6.286636	285.415558	13.070602
WM4	173.845810	4.264816	421.810455	16.041459

Fig.1(a) and (b) shows an original image and the image corrupted with impulsive noise. Fig.1 also shows the noisy image filtered with (c) WM1, (d) WM2, (e) WM3 and (f) WM4. Note that there are still some white dots (impulsive noise) left in the images filtered by WM1 and WM4. WM2 and WM3 produce quite good results.

(a). (b). (c).

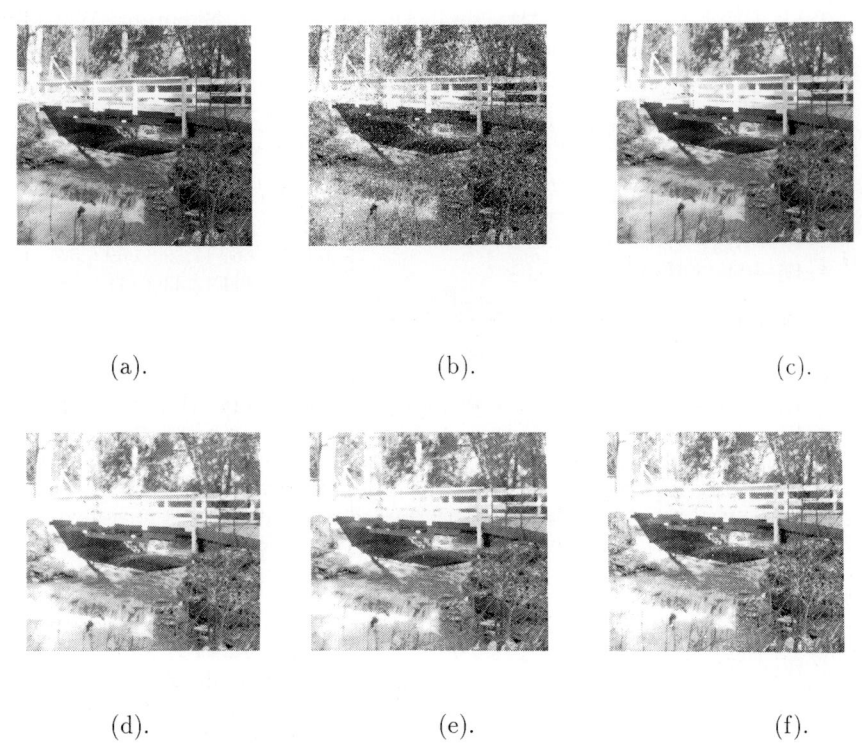

(d). (e). (f).

Figure 1. (a) Original image, (b) noisy image (impulsive noise), (c) WM1, (d) WM2, (e) WM3 and (f) WM4.

4. CONCLUSIONS

We have developed an algorithm for finding optimal WM filters under structural constraints for image restoration. It was shown that the optimal WM filter can be obtained by solving a nonlinear programming problem. optimal WM filters in image processing were discussed. This method can also be used in one dimensional processing, where under some structural constraints, e.g. pulses of a given length, it is desirable to design optimal 1-D WM filters.

5. REFERENCES

1. B.J. Justusson, "Median filtering: statistical properties, in *Two Dimensional Digital Signal Processing II*, T.S. Huang, ED., Springer-Verlag, Berlin, 1981
2. E. Coyle, J.-H. Lin, and M. Gabbouj, "Optimal Stack Filtering and the Estimation and Structural Approachs to Image Processing", *IEEE Trans. On Acoust., Speech and Signal Processing*, vol.37, no.12, Dec. 1989.
3. R. Yang, L. Yin, M. Gabbouj, and Y. Neuvo, "2-D optimal weighted median filters," *IEEE Workshop on visual signal processing and communications*, Sept.2-3, 1992, North Carolina, USA, pp.31-35.

Hierarchical Fitting of Constrained Composite Curves

Alexander M. Geurtz[a] and Riccardo Leonardi[b] and Murat Kunt[a]

[a]Signal Processing Laboratory, Swiss Federal Institute of Technology, EPFL-Ecublens, CH-1015 Lausanne, Switzerland

[b]Dept. of Electronics for Automation, University of Brescia, Italy

Abstract

We propose a method for recursively estimating the parameters of a complex object, consisting of several connected components, from features detected in a grey level image. Each of the individual components are rigid simple (*primitive*) objects that undergo rotation, translation, and scaling. Physical relations between the objects that form the complex shape as well as other *a priori* knowledge are translated into constraints on the solution space. No attempt is made to establish a closed-form solution. Instead, upper and lower bounds both on the object parameters and on the constraints add a degree of freedom that aids in handling the uncertainty in the data and improving the convergence of the algorithm. A simultaneous fuzzy segmentation allows to smoothly solve the important segmentation problem. The hierarchical method proposed in this paper reduces the problem of getting trapped in one of the numerous local minima of the resulting optimization problem. The particular primitive model used here is a circle (which becomes an ellipse after the transformation) in two-dimensional space, and the feature points used are sparse contour points.

1. INTRODUCTION

The work is performed in the context of a project on the analysis of human movement from image sequences. The body is modelled as a set of jointed objects that are considered representative for the overall shape and the task is to estimate the parameters of these objects from features detected on the image (sequence). One of the earliest and most complete results using image sequences were obtained by O'Rourke and Badler [1]. They introduced a constraint network where possible solutions were validated by comparing them with the set of constraints. Recently, physically based systems have gained significant interest. Terzopoulos and Metaxas [2] and Pentland and coworkers [3, 4] were the first to introduce Finite Element Methods for shape and motion estimation and to apply them successfully to range and image data.

This paper differs from the above-cited literature in that it completely integrates segmentation of the feature data with the (constrained) estimation of the articulated object parameters in a hierarchical way. The introduction of bounds on both the parameters and the constraints allows to control the amount of uncertainty and to eventually be able to include an assessment of the accuracy.

The organization of the paper is as follows. After an initial introduction of the (simplified) model, we discuss the parameter estimation applied at one level. Then, we will illustrate how global information is diffused from a coarse to a finer object level by expliciting the rules to do so. Finally, we will show some results on noisy real world data.

2. MODEL

Each part of the body is modelled by an implicit curve $f(\underline{x},\underline{p}) = 0$ of points $\underline{x}$. The parameter vector $\underline{p}$ contains the orientation, size, and position parameters of this simple (rigid body) transformation, coined RST-transform. In order to keep the subsequent method as simple as possible, but without affecting too much its generality, the circle has been chosen as the primitive implicit curve. This closed curve becomes an ellipse after the rigid RST-transformation. In case of a complex object consisting of multiple transformed primitive objects, each individual part of the total object has its own parameter vector and the composite vector $\underline{\mathbf{p}} = (\underline{p}_1, \ldots, \underline{p}_{N_0})$ consists of all the parameter vectors associated with each of the N_0 primitive objects.

Physical limitations between the primitive objects are expressed by lower and upper bounds on the (non)linear constraint functions $\underline{C}(\mathbf{p})$. A typical constraint function would be that two neighboring parts meet at one of their endpoints, or that certain size parameters are equal. In principle, the lower and upper bounds should be zero, indicating that constraints must be satisfied exactly. An important feature of the method presented here is that the constraints are allowed to be loosened a little bit if necessary. This added flexibility greatly improves the convergence and therefore the final result. It adds of course new variables to be set (the bounds). The total model at a given level of detail consists of the set of parameters plus their physical constraints and bounds.

3. CONSTRAINED COMPOSITE CURVE FITTING

At a given level, the parameters of the composite object are estimated in the following way. As an error measure the In/Out measure is taken here, as it is widely used in the literature, despite some of its drawbacks [5, 6]. This measure quantifies the deviation from zero of the implicit function $f(\underline{x},\underline{p})$ (it is not, however, a Euclidean distance). In case of a composite object, a major problem is the segmentation of the extracted data into regions that can be associated with the objects that form the global structure. A fuzzy segmentation approach is chosen in this work: With each data point $\underline{x}_i$ a normalized vector $\underline{u}_i = (u_i^{(1)}, u_i^{(2)}, \ldots, u_i^{(N_0)})$, with N_0 the number of primitive objects, is associated indicating the strength of the correspondence of the point to each of the individual objects. This representation has the advantage of being able to handle in a smooth way the inherent segmentation uncertainty in areas where individual parts meet. A disadvantage is that the dimensionality of the problem increases significantly.

The general error functional for fitting N_0 objects with their spatial constraints described by the spatial constraint vector $\underline{C}(\mathbf{p})$, given N data points, is

$$J(\underline{\mathbf{p}}, \underline{u}_i, i = 1 \ldots N) = \sum_{i=1}^{N} \sum_{j=1}^{N_0} u_i^{(j)} \|z_i(\underline{p}_j)\|^2 \qquad (1)$$

subject to the bounds on the spatial constraints $\underline{C}(\mathbf{p})$ as before and to the N (linear) constraints derived from the normalization of the fuzzy membership vectors $\underline{u}_i$. The error $z_i(\underline{p}_j)$ is the deviation of feature point $\underline{x}_i$ evaluated with the parameters of curve j. The above minimization problem subject to bounds on the parameters and on the nonlinear constraints is solved by a sequential programming procedure [7].

4. THE HIERARCHICAL SCHEME

In a very general context, computer vision problems can be viewed as search problems, a search of the most consistent representation among all of the possible ones. A powerful way of attacking such problems is by performing the search in a hierarchical way. Hierarchical methods have been applied successfully to such areas as grey level and feature segmentation, motion estimation, and signal analysis. The advantages of such a multi-level approach are manifold, of which two are of particular interest here. The first advantage is that, as the search space is limited when going from one level to the next, the total complexity is greatly reduced. The second is that it may guide the segmentation process: After a global positioning identifying roughly the area of interest, a further subdivision is made using the previously obtained result plus the necessary spatial constraints as an initial estimate.

The performance of any hierarchical scheme will depend largely on the way the data and their properties are projected from one level to the next. We have adopted the following simple rules for downprojecting the properties (the parameter vector **p**) and the segmentation from one level to the next. Given one object, the radial density distribution is estimated by *histogramming* the orientations of the lines going from the center of the object found through each of the data points. The two or three distinct peaks are taken as initial estimates for the orientations of the underlying two- or three-body structure. The sizes are downprojected by initializing them at half the size of the coarse simple object. The new, initial, centers are derived from their orientation and width estimates. The fuzzy segmentation is initialized by determining the relative distances of each of the data points to each of the new centers. We do not generate the data in a hierarchical way (bottom-up), only the analysis is done in such a way (top-down).

5. RESULTS AND CONCLUSIONS

The algorithm was tested on noisy real world feature data, taken from the grey level image of a person walking. After edge detection, thresholding, thinning, and some manual cleaning a set of contour points was obtained. A 25 % subset of the contour points was taken as the feature data, of which again only 25 % is shown in Figure 1a. First, a single implicit curve was imposed on these data yielding the result shown in Figure 1b. The initial curve was the tilted ellipse that can be seen in the center. The radial projection of the data points, shown in Figure 1c, exhibits three significant peaks, leading to the final result as shown in Figure 1d.

In conclusion, we have presented a hierarchical method for estimating composite curves from contour location data. The physical limitations on the curves that constitute the global object are introduced as nonlinear constraints on the solution space of the resulting optimization problem. The problem of segmenting the data into regions is handled by an integrated fuzzy representation and included in the overall search problem. Weak bounds on both parameters and constraints allow handling the uncertainty in the data and improving the convergence of the algorithm. The representation makes it easy to add constraints if desired, because no closed-form solutions are derived. Current research topics are a more rigorous way of projecting bounds and parameters and a quantitative assessment of the resulting accuracy.

6. REFERENCES

[1] J. O'Rourke and N.I. Badler. Model-based image analysis of human motion using

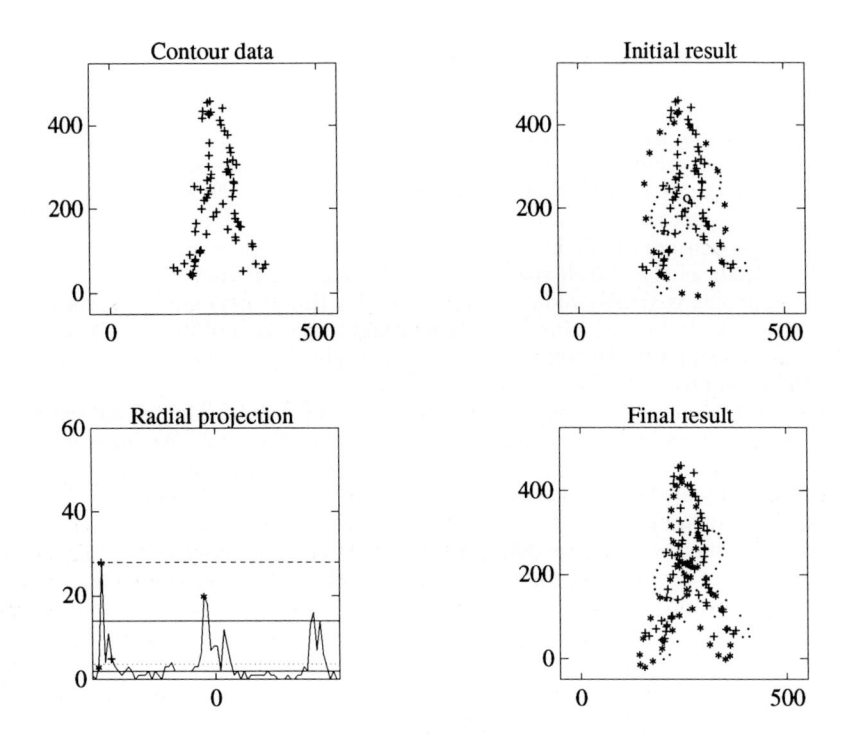

Figure 1: From left to right and top to bottom: a) 25 % of the retained contour points (+++), b) fitting one curve (initial estimate ..., final estimate ★★★), c) radial projections from $-\pi$ to π, d) fitting three curves (initial estimate ..., final estimate ★★★)

---

constraint propagation. *IEEE Trans. Pattern Analysis and Machine Intelligence*, 2(6):522–536, November 1980.
[2] D. Terzopoulos and D. Metaxas. Dynamic 3D models with local and global deformations: Deformable superquadrics. *IEEE Trans. Pattern Analysis and Machine Intelligence*, 13(7):703–714, 1991.
[3] A. Pentland and B. Horowitz. Recovery of nonrigid motion and structure. *IEEE Trans. Pattern Analysis and Machine Intelligence*, 13(7):730–742, 1991.
[4] A. Pentland and S. Sclaroff. Closed-form solutions for physically based shape modelling and recognition. *IEEE Trans. Pattern Analysis and Machine Intelligence*, 13(7):715–729, 1991.
[5] F. Solina and R. Bajcsy. Recovery of parametric models from range images: The case for superquadrics with global deformations. *IEEE Trans. Pattern Analysis and Machine Intelligence*, 12(2):131–147, February 1990.
[6] Alok Gupta, Luca Bogoni, and Ruzena Bajcsy. Quantitative and qualitative measures for the evaluation of the superquadric models. In *Proc. Workshop on Interpretation of 3-D Scenes*, pages 162–169, Austin, TX, USA, Nov 27-29 1989.
[7] Numerical Algorithms Group. *NAG Fortran Library Documentation, Release 15*. NAG Ltd, Oxford, UK, 1991.

2-D Subset Averaged Median Filters

Risto Suoranta and Kari-Pekka Estola*
Machine Automation Laboratory, Technical Research Centre of Finland,
P.O. Box 192 SF-33101 Tampere, Finland. Tel. +358 31 163 636, Fax. +358 31 163 494,
Internet: Risto.Suoranta@kau.vtt.fi.
Electronics Laboratory, Technical Research Centre of Finland,
P.O. Box 200 SF-90571 Oulu, Finland. Tel. +358 81 551 2300, Fax. +358 81 551 2320,
Internet: Kari-Pekka.Estola@ele.vtt.fi

Abstract - In this presentation we consider the properties of a subset averaged median estimate ($SAME$) in case of 2-dimensional filtering. The subset averaged median estimator based filters belong to the class of order statistic filters (OSF). $SAME$-filters are controlled by two parameters: window size L and subset size q. By changing the subset size q we are able to control both the noise attenuation and robustness properties of the proposed filter. With subset size q being near the window size L we obtain a filter behaving like a median filter. On the other hand with small q $SAME$-filter shares many properties with a mean filter. Using numerical studies it is shown that by applying proposed filter to image restoration a good noise attenuation is achieved together with robust edge preserving behavior. Moreover, comparisons are also made with some well-known OS-filters like α-trimmed mean filter.

1. INTRODUCTION

Nonlinear filters have many attractive properties considering image processing applications. Especially the good noise suppression of impulsive noise while preserving edge like changes is a very desired feature. Since Tukey proposed the median filter in 1971 it has been most widely used nonlinear filtering method in image processing. The standard median filter defined by Tukey performs simple operation of sorting samples in filter window and selecting the output sample from the midposition of sorted set of input samples. This Standard Median Filter (SMF) rejects impulsive noise efficiently and does not blur the edges when the amount of additive noise is moderate. However, the suppression of additive wide band noise is more poor than with corresponding linear filters and the edge preserving property is also reduced when the filtered signal is contaminated with high variance additive noise.

To improve the performance of median filters many different approaches are considered. To enhance the edge preservation of the filter several different nonlinear filter structures are developed where the time domain information is exploited. For example weighted median filter or hybrid filters which have both linear and nonlinear substructures. Another approach, mainly to achieve better noise rejection, is to employ more than one sample after ranking operation by using the linear combination of sorted data set. These filters are called Order Statistic Filters (OSF) or L-filters. L-filters contain a broad class of different filters including median, mean, α-trimmed mean filters and many others. In most cases there is clear trade-off between the noise suppression and edge preservation properties. One obvious solution is adaptive filter structure which is proposed by several researchers [1, 4].

Our approach to cope with the trade-off between robustness and noise attenuation and to enhance the performance of SMF is to utilize a robust estimator called Subset Averaged Median Estimator (*SAME*) in 2-D filtering. *SAME* is a low variance estimator of median and it is shown to be quite robust against the changes of the underlying distribution [6, 7]. Furthermore *SAME*-filter can be easily realized as an ordinary OS filter or by using the approximation of *SAME* which applies multistage structure called direct form *SAME* filter (*DSAME*) [8].

2. SUBSET AVERAGED MEDIAN ESTIMATE

The basic assumption in *SAME* is that by applying several sample median estimates of subsets z_i and by averaging them we obtain an estimate with lower variance than by using one sample median of the whole sample set. Due to the fact that the order of samples (pixels) does not affect the result of the median operation we can select several subsets z_i of size q from the original sample set of size L. Applying the sample median operator to the subsets and by averaging results we obtain a new median estimate *SAME*.

Proposed *SAME* estimator is controlled by two parameters: L describing the support area from which the estimate is computed and the subset size q defining the number of samples in each subset z_i. Since there is no requirement that samples in subsets should be successive in time or space there are K_{tot} different subsets where K_{tot} can be determined by applying the combinatorial theory. That is

$$K_{tot} = \binom{L}{q} = \frac{L!}{q!(L-q)!}. \quad (1)$$

The Subset Average Median Estimate *SAME(L,q)* and corresponding operator is defined as

$$\overline{med_q}(X(k)) = E\{med(z_i)\}, \quad (2)$$

$$z_i = \{x_{i_1}, x_{i_2}, ..., x_{i_q}\} ; z_i \subseteq X(k) ; i_j \subseteq Z_L ; i = 1, ..., K_{tot}$$

$$Z_L = \{k-M, ..., k, ..., k+M\}$$

$$X(k) = \{x_{k-M}, ..., x_k, ..., x_{k+M}\}$$

$$L = 2M+1 \quad ; q = 2m+1 \quad ; 1 \le q \le L$$

where $E\{\}$ denotes the expectation operator and $med()$ is the operator of sample median. In 2-D case $X(k_{ij})$ is the set all samples defined by two dimensional mask which is positioned so that the center of the mask is in the position of the pixel (i,j). Based on the definition we see that when $q=L$ the *SAME* equals the sample median of length L and when $q=1$ *SAME* equals sample mean of length L.

3. OS-FILTER REALIZATION

OS-filter is defined as a linear combination of sorted data set. The output of OS-filter of length L for data set $X(k)$ is defined as

$$y(k) = a^T X_{(k)}, \quad (3)$$

where $X_{(k)}$ is the sorted data set of $X(k)$ so that $x_{(k)}(1) \le x_{(k)}(2) \le ... \le x_{(k)}(L)$, and weights for each order statistic are given in vector $a = [a_1, a_2, ..., a_L]^T$.

By applying the *set theory* [3] we have derived an exact relation for *SAME(L,q)* and OS-filter representation. In other words it is possible to generate a set of OS-filter weights which corresponds to *SAME* with appropriate parameters of L and q. That is,

$$\overline{med_q}(X(k)) = a_{L,q}^T X_{(k)}. \qquad (4)$$

An equation for weight vector $a_{L,q}$ is derived in [6]. The elements of $a_{L,q}$ are,

$$a_i = \begin{cases} \binom{L-i}{m}\binom{i-1}{m} / \binom{L}{q} & ;i = m+1,\ldots,L-m \\ 0 & ;i = 1,\ldots,m, L-m+1,\ldots,L \end{cases}, \qquad (5)$$

where $q=2m+1$. For given $a_{L,q}$ it holds

$$\sum_{i=1}^{L} a_i = 1 \qquad (6)$$

$$a_i = a_{L-i+1} \qquad ;i = 1,\ldots,L$$

which are necessary and sufficient conditions for *SAME(L,q)* to be unbiased estimate for location of constant signal with noise having symmetric noise probability density function.

4. NUMERICAL COMPARISONS

We have compared SAME-filter to set of well-known OS-Filters. Compared filters are *mean* and *median* (SMF) filters, *alpha trimmed mean* (α-TM) with three different parameter values ($\alpha=3/9, 2/9, 1/9$), *Midpoint filter, LMS optimized OS-filter* for Laplacian noise (LMS-OS) proposed by Bovik et al. [2], *Subset Averaged Median Estimate (SAME)* filter with three different parameter values ($q=3, 5, 7$). Implicitly also filters α-TM(0/9), α-TM(9/4), *SAME(9,1)* and *SAME(9,9)* are included in the comparison due to the fact that these filters correspond *mean* and *median* respectively.

Six different test signals are used. Stationary noise with thee different characteristics: Uniform, Laplacian and Gaussian noise. Two checker boards with different sized black and white (0-1) squares ($L-1 \times L-1$ and $4L \times 4L$) added with $N(0,0.1)$ noise. Square sizes are chosen so that in the first case filter is all the time in nonstationary area and in second case filter is about 40% of its time in nonstationary area. Last two test signals are well-known images of Lena. The image is first filtered without any noise to find out how much bias error each filter introduce to the noise free image. In the last test signal the image is contaminated by 10% salt and pepper noise together with Gaussian noise $N(0,15)$.

Based on the results shown in Table 1 we see that *SAME* filter behaves much like α-trimmed mean filter. However, if we compare *SAME* and α-trimmed mean in pairs where each filter have the same amount of non-zero coefficients we see that *SAME* filter is more robust against changes of distribution of input data. Standard Median Filter outperforms *SAME* if the image contains lot of sharp edges like it is case in small square sized checker board image. In Figure 1. we see the noise corrupted image of Lena (480x480 pixels) together with three filtered images.

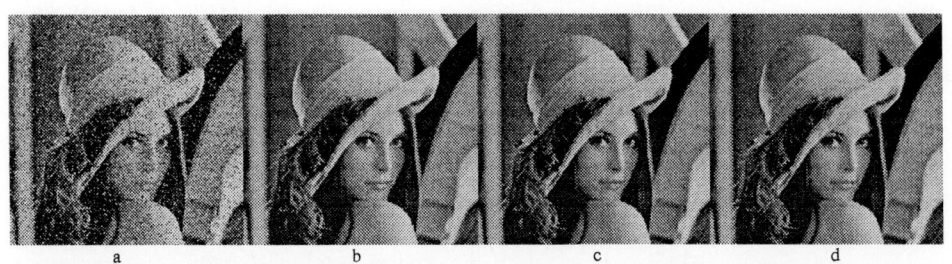

Figure 1. Image of Lena (480x480 pixels): a) corrupted by 10% salt and pepper noise together with additive Gaussian noise $N(0,15)$, b) filtered with *Subset Averaged Median Estimate* filter [$SAME(9,7)$], c) filtered with α-trimmed mean filter [α-TM(3/9)], d) filtered with Standard Median Filter [median(9)]. Applied mask in all test cases were 3x3 square mask.

Table 1: Mean Square Error attenuation in the case of ten different filtering methods. Attnuations are in dB's.

Test signal filter	Gaussian noise	Uniform noise	Laplacian noise	Checker board (L-1 x L-1) + $N(0,0.1)$	Checker board (4Lx4L) + $N(0.0.1)$	Image of Lena with no noise	Lena with 10% s&p + $N(0,15)$
mean(9)	-9.4	-9.5	-9.5	13.0	7.1	15.7	-8.1
median(9)	-7.5	-5.6	-10.8	2.5	-3.5	14.5	-12.0
α-TM(3/9)	-8.2	-6.5	-11.1	11.0	0.2	14.3	-12.4
α-TM(2/9)	-8.7	-7.3	-11.0	12.2	4.5	14.6	-12.0
α-TM(1/9)	-9.1	-8.3	-10.6	12.7	6.2	15.0	-10.4
MidPoint(9)	-6.9	-12.5	-3.7	14.0	9.9	18.8	-0.3
LMS-OS(9)	-8.4	-6.7	-11.1	10.8	1.5	14.3	-12.3
SAME(9,3)	-9.0	-7.8	-10.9	12.3	5.0	14.7	-11.3
SAME(9,5)	-8.6	-7.0	-11.1	11.5	2.6	14.4	-12.2
SAME(9,7)	-8.2	-6.4	-11.1	10.0	-0.5	14.3	-12.4

5. CONCLUSIONS

In this paper we have considered the properties of a subset averaged median estimate (*SAME*) in case of 2-dimensional filtering. Using numerical studies it is shown that by applying proposed filter to image restoration a good noise attenuation is achieved together with robust edge preserving behavior. The robust performance of the proposed filter is caused in many sense by the logical multistage structure of the filter in which the first stage performs as robust nonlinear operator and the second stage can be regarded as a linear averaging operator which efficiently attenuates noise.

References

[1] I. Pitas, A. N. Venetsanopoulos: *Nonlinear Digital Filters: Principles and Applications*, Kluwer Academic Publishers, Boston, MA, USA, 1990.

[2] A. C. Bovik, T. S. Huang, D. C. Munson: "A Generalization of Median Filtering Using Linear Combinations of Order Statistics", IEEE Trans. Acoust., Speech and Signal Processing, Vol. ASSP-31, pp.1342-1350, No. 6, Dec, 1983.

[3] A. Papoulis: *Probability and Statistics*, Prentice-Hall Inc. Englewood Cliffs, NJ, USA, 1990.

[4] C. Kotropoulos, I. Pitas: "Constrained Adaptive LMS L-filters", Signal Processing, Vol. 26, No. 3, March 1992.

[5] S. R. Peterson, Y. H. Lee, S. A. Kassam: "Some Statistical Properties of Alpha-Trimmed Mean and Standard Type M Filters", IEEE Trans., Acoustics, Speech and Signal Processing, Vol, ASSP-36, pp 707-713, No. 5, May, 1988.

[6] R. Suoranta, K-P. Estola: "New Class of Order Statistic Filter for Running Median Estimation", Proc. ICASSP-93, IEEE, Minneapolis, MN, USA, 1993.

[7] R. Suoranta, K-P. Estola: "New Low Variance L-Estimator for Percentile Estimation", Proc. IEEE Winter Workshop on Nonlinear Digital Signal Processing, pp 1.2-6.1 - 1.2-6.6, Tampere, Finland, 1993.

[8] R. Suoranta, K-P. Estola: "Subset Averaged Median Estimators", Proc, IEEE Int. Conf. on Circuits and Systems, Chicago, Illinois, USA, 1993

Face Identification and Feature Extraction Using Hidden Markov Models

F. Samaria[a,b] and F. Fallside[a]

[a]Cambridge University Engineering Department, Trumpington Street, Cambridge CB2 1PZ, United Kingdom, tel.+44 223 332752

[b]Olivetti Research Lab, Old Addenbrookes Site, 24a Trumpington Street, Cambridge CB2 1QA, United Kingdom, tel.+44 223 343000

Abstract

This paper details work done on automatic face identification. A new approach to the problem is proposed involving the use of Hidden Markov Models. We illustrate how these models allow the automatic extraction of facial features and the classification of face images. Some experiments are presented to support the plausibility of this approach. Successful results were obtained under the constraints of homogeneous lighting and constant background.

1. INTRODUCTION

Substantial research effort has gone into trying to understand how to build a successful model for face identification, both by psychologists [1] and information scientists. The importance of face identification, other than for the intrinsic challenge of the task, stems from its numerous potential applications: workstation security [2], criminal identification, video-mail retrieval, building security and home video actor-based indexing. We present a new architecture to classify face images that uses Hidden Markov Models (HMMs). Faces are treated as two dimensional objects and the model automatically extracts *statistical* facial features. However, it can at the same time make use of *structural* information, yielding an *hybrid* model.

2. PROPOSED APPROACH

We propose a method that automatically extracts facial features from a set of training data and successively uses them to identify test images. We achieve this using a particular context-dependent classifier [7] based on Hidden Markov Models (HMMs). Rabiner [4] provides a comprehensive tutorial on HMMs. They have been extensively and successfully used for speech recognition applications [3]. The one dimensional (1D) nature of speech signals (along the time axis) is suited to analysis by HMM. A 1D observation sequence can be obtained from an image by sampling it using a sliding window, where each element of the sequence is a vector of pixel intensities. The way in which the sliding window passes over the image determines the ordering of the sequence. The windowed data generates a spatial sequence [5] that can be used to train an HMM, which will then partition the sequence into a number of feature states. The term *feature state* here refers to a block of pixels (of the same size as the sliding window) which has sufficient statistical significance within the image to result in the HMM assigning a state to represent it.

An HMM is primarily characterised by a transition probability matrix **A** and an output probability matrix **B** [4]. After training an HMM on a sequence generated as above, each matrix will capture a different aspect of the face image. Matrix **A** will record the occurrences of transitions from one feature state to another across the face and the thickness of the various features. The latter can be seen by considering the terms representing the probability of going from a state to itself, *i.e.* the probability of staying in the same feature state. Matrix **B** will record what the feature states look like for each subject. Both matrices will provide strong discrimination for the various subjects.

3. INITIAL EXPERIMENTAL RESULTS

Two sets of simple spatial HMMs were built to test some of the ideas presented in the previous section. In both cases, a separate HMM was trained for each subject in the database and the resulting models were used to classify unknown images. The training and testing were carried out using the *HTK: Hidden Markov Model Toolkit V1.3* [8].

3.1. Ergodic Models

A total of 10 training images (of size 256 × 256, 8-bit grey levels) for each subject were used to train an HMM. Each image was sampled into an observation sequence using a 64 × 64 window. The window initially moved left to right stepping 48 pixels at a time, until it hit the right margin of the picture. It then moved 48 pixels down and started sampling right to left, and so on. This implies an overlap of 16 × 64 pixels when sliding horizontally and 64 × 16 pixels when sliding vertically.

Figure 1: Training data and magnified states for ergodic model

An 8-state ergodic HMM was built. The reason for choosing 8 states was that, by inspection, approximately 8 distinct regions seem to appear in the face image (eyes, mouth, forehead, hair, background, shoulders and two extra states for boundary regions). Figure 1 shows some training data, together with the 8 states found by the HMM. Some states, such as the background, the eye and the neck, are easily recognisable. Others have been generated by superposition of features and are less distinguishable. They nonetheless characterise the subject, providing strong discrimination.

3.2. Left-to-right Models

A total of 5 training images (of size 184×224, 8-bit grey levels) per subject were used to train each HMM. Each image was sampled into an observation sequence using a block of 16 lines, moving vertically 4 lines at a time (*i.e.* with a 12-line overlap). Each still image was effectively converted into a sequence of overlapping line blocks spatially ordered in a top-to-bottom sense. The overlapping was present to allow the model to capture features in a manner independent of their vertical position (a disjoint partitioning of the image could result in features occurring across block boundaries being truncated).

This sequence was then partitioned into 5 states, each representing a *feature band*. We wish to establish a correspondence between statistical feature bands and facial bands as understood by humans. Assuming that each face is in an upright, frontal position, we expect facial bands to occur in a predictable order, *i.e.* forehead, then eyes, then nose, and so on. This form of ordering suggests the use of a left-to-right HMM where only transitions between adjacent states in a left-to-right manner will be allowed. Figure 2 shows the training data for one subject in the database and the means for the 5 states found by the HMM for that subject. It is interesting to notice that we can identify some

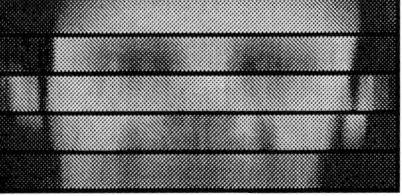

Figure 2: Training data and magnified states for left-to-right model

of the states as physical facial features. We can, for instance, identify the second state as being the eye band. This means that the eyes have sufficient statistical significance to induce the HMM to assign a state to represent them. Experiments [6] have shown evidence that one of the salient feature areas of the face for identification used by humans is the eye region. The remaining states extracted by the HMM can also be identified with other facial regions. Some states "leak" into their next one: it is possible, for example, to see part of the eye state lightly appearing in the next band. This is due to those observation vectors (or line blocks) which have, as far as state allocation is concerned, ambiguous statistical properties and actually give a better statistical match if placed in the next state. They will therefore contribute to the overall mean for that state vector thus causing the leaking effect.

4. CONCLUSIONS

Successful identification results were obtained with a database of 20 subjects when the test image was extracted from the same video sequence as the training images. This demonstrates that, provided lighting and background are kept the same, the model can cope with variations in facial features due to small orientation changes. This has been a considerable problem in all template-based approaches to face identification.

This method also provides a way of extracting statistical facial features. It is interesting to note that such statistical features may correspond to ordinary features (*e.g.* the eyes). Therefore this approach, if appropriately tuned, could be used to extract facial features automatically.

Acknowledgements

This work is supported by a Trinity College Internal Graduate Studentship and an Olivetti Research Lab CASE award. Their support is gratefully acknowledged. We wish to thank many of our colleagues in Cambridge: Andy Hopper, Leslie French, Phil Elwell and Martin Brown of Olivetti Research, for useful discussions and for the image capture software; Steve Young and the Speech Group at CUED for the HMM software; Owen Jones and Gabor Megezy of the Pure Mathematics Department, Barney Pell of the Computer Lab, and Steve Hodges, Gopal Chand and Charlie Allen of CUED for the many ideas discussed together.

5. REFERENCES

[1] V. Bruce. *Recognising Faces*. Laurence Erlbaum Associates, 1988.

[2] R. Gallery and T.I.P. Trew. An architecture for face classification. *IEE Colloquium on 'Machine Storage and Recognition of Faces', Digest No: 1992/017*, 2:1–5, 1992.

[3] T.W. Parsons. *Voice and speech processing*. McGraw-Hill, 1986.

[4] L.R. Rabiner. A tutorial on hidden markov models and selected applications in speech recognition. *Proceedings of the IEEE*, 77, 2:257–286, January 1989.

[5] F. Samaria. *Face Identification using Hidden Markov Models*. 1st Year Report, Cambridge University Engineering Department, 1992.

[6] J. Shepherd, G. Davies, and H. Ellis. Studies of cue saliency. In G. Davies, H. Ellis, and J. Shepherd, editors, *Perceiving and remembering faces*, pages 105–131. Academic Press, 1981.

[7] C.W. Therrien. *Decision estimation and classification*. Wiley, 1989.

[8] S.J. Young. *HTK: Hidden Markov Model Toolkit V1.3,Reference Manual*. Cambridge University Engineering Department, 1992.

From face sideview to identification

Laurent Najman, Régis Vaillant, and Étienne Pernot

L.C.R, Thomson-CSF, Domaine de Corbeville, 91404 ORSAY-Cedex, France.

1 Introduction

This paper presents a method and its results for the automatic identification of persons from face sideviews observation. We find this application characteristic of a problem class which is encountered in pattern recognition. To solve this type of problem, i.e. outline classification between very similar classes, we selected what we think to be a general strategy well adapted to this kind of problems: *preprocessing* to extract outlines and to work out characteristic points which will be used to perform a *geometrical* data normalization; *construction of an outline representation* which brings to the fore the differences between classes, and *selection* of a classifier which gives the best results on the database.

2 Preprocessing

Outline extraction: The outline extraction problem is not very difficult, for the face sideview is observed by placing a white background behind the subjects. For speed of execution and robustness, we decided to use a mathematical morphology procedure. Our process divides in several steps:
1. We derive a thresholded image from the acquired one. We compute an erosion of this thresholded image. The result is an image with one big blob which is located in the hair.
2. Computation of the "morphological gradient". It is well adapted to our case as the shape is highly contrasted with the background.
3. The exact outline is obtained by a watershed [6], using the big blob of step 1. as the marker which identifies the face and an arbitrary point in the background as a second marker.

Characteristic points extraction: We extract a set of characteristic points and we use them for a *geometrical* data normalization. For this goal, we have to define two frames on the face sideview outline, one for the forehead-eyes-nose part, and the other for the chin-neck part. Due to the bone structure, each of this set is a rigid part on the face, even if they are able to move relatively from each other.

We worked out a method which proved to be robust. The key points of our method are the use of the global face structure to find each point position, and various steps of computation in view of refining the exact point position. More details can be found in [3]

Position of the neck: we first take into account the whole profile, and we calculate the global face orientation in the image. It is given by the direction of the least square line of the global face sideview profile set. The neck is the farthest point from this line, on the bottom of the profile (see fig. 1.a).

Position of the nose: the neck provides us an essential information: what is the exact extent of the face sideview profile. We know the uppermost point of the profile which is related in a way to the "forehead" A and the lowermost point is the neck B. These notations and the following ones are shown on figure 1-a. We consider the line $\mathcal{D}_1$ which is the least square fitting line of the points of the profile between A and B. We consider a strip whose axis is perpendicular to the line $\mathcal{D}_1$ and whose center is the middle C of A^P and B^P. A^P and B^P are the orthogonal projection of A and B on the line $\mathcal{D}_1$. The width of the strip is half the distance from A to B. We consider that the nose is the point D contained in this strip which is the farthest on the left from the oriented segment A^P to B^P.

Position of the chin, the forehead and the eye: the following point that we determine is the chin E. It is the farthest point on the left of the oriented segment from D to B. Using the same kind of procedure, we determine the forehead as point F. The last characteristic point is the eye G. It is the farthest point on the left to the oriented segment between D' and F.

Using this procedure, we were able to extract the characteristic points for more than a thousand face sideviews. We obtain success in almost every case (less than 2% of error, easily identifiable, *e.g.* a big moustache can be confused for the nose). An example of application is shown on figure 1-b.

Normalization: The next problem is to normalize our data. We have the following degrees of freedom: translation, rotation and scaling in the image plane. These "effects" can be corrected using an appropriate procedure which leaves invariant the ratio of distances and the angles. There is another effect: deformation if the face sideview does not belong to a plane which is parallel to the image plane. It is important to notice that every profile has the same size for us. The corresponding information is definitely lost in the projection operation associated with the formation of the image.

We clearly have two parts in a face: around the chin and between forehead and bottom lip. So we use two normalizing referentials, one containing points around the nose, the other containing points around the chin. The first one has as origin the eye, the nose being on the ordinate axe. The second one has as origin the chin, the neck being on the ordinate axe. With this two frames, we have two rigid parts.

We then take points equally spread in these two referentials, using for both the distance eye-nose as a characteristic distance. With this process, we always have the same number of profile representative points, which allows us to give these data to a classifier.

3 Classification

In this section, we present the results we obtained on our problem of automatic recognition of individuals from face sideview. We focus our attention on two topics : *study of the sampling* which is necessary to represent the curve, and *study of an adequate representation* of the measurement points.

The database was constructed with 10 individuals, 9 males and one female. We took 41 snapshots of each, in various occasions. Individuals could slightly move the head, in any direction but the ones in which the face sideview profile disappeared.

For each profile, we construct a vector which is the representation of the set of sample points. We consider the parametrization by arclength of the nose and the chin curves. Each of these parametrizations is seen as producing two curves, the first one being the abscissa deviation in function of the parameter, and the second one being the ordinate deviation in function of the parameter. For each of these four sets of points, we calculate their least square line, and we construct our data vector with the difference between a curve point and the corresponding point on the least square line.

Classification method: we then try, with the help of $NeuroClasse^{TM}$, different classification techniques [4]. $NeuroClasse^{TM}$ is a software environment devoted to supervised classification, integrating many conventional and neural classification algorithms. Arbitrarily, we choose 80 points as a first choice of number of sampling points. We then have to look for a classifier with a 160 dimensional input and a 10 dimensional output.

We first try various classification methods, and with 75% of the database as the training set and 25% as the test set. We tuned the parameters of each methods.

The best results we obtain are around 90% for generalization, with a learning rate of 100%. We select the three best methods: Principal Components Analysis followed by a Quadratic Discrimination, k-Nearest Neighbors, and Gradient Back Propogation. We then compute an average generalization rate and try to assess the results robustness, using a statistical mode which allows us to compute an average generalization rate and the standard deviation of this generalization rate. The results are shown on table 1.

One can notice that the standard deviation is relatively high, between 2% and 4%. We might decrease the standard deviation by enlarging the database.

Number of sampling points: we have done several experiments where the number of points was changed. We always kept the same ratio of sampling points: 35% of the points are taken around the chin, and 65% of the points are taken around the nose. The total number of points varied from 8 to 100. We obtain the curve of the generalization percentage. The generalization rate increases up to 30 sampling points. It is afterwards more or less stable. Best results obtained are shown on the table 1. Our method appears to be not very sensitive to the number of sampling points.

4 Conclusion

In this paper, we have presented a statistical approach to the problem of outline classification in several classes. Our method relies on the addition of a careful preprocessing of the data and the use of classification algorithms.

We have experimented a very efficient technique for the presentation of our data to the statistical classifier. Rather than collecting directly in a vector the set of sampling points, we put in this vector the deviation between these points and the least square fitting line. This improve notably our results.

We have experimented our ideas on the problem of identification of persons from face sideview observation. We have obtained results: 90% of good recognition of a basis of 10 individuals, with 41 images for each one, 31 of the images being in the training set and 10 of the images being in the test set. This result is encouraging, but has to be validated on a larger database. On the other hand, this percentage is probably insufficient for a real application. The following ideas will surely improve the generalization rate: one can use more efficiently the information provided by the image itself by giving to the classifier data like grey levels or other features [5]. Another way to enhance the generalization rate is combining informations which could be obtained from a sequence of face sideview profiles rather than an unique image.

We have not developed these ideas for the moment, as the primary goal of this paper was to study a classification problem when the input is image outlines.

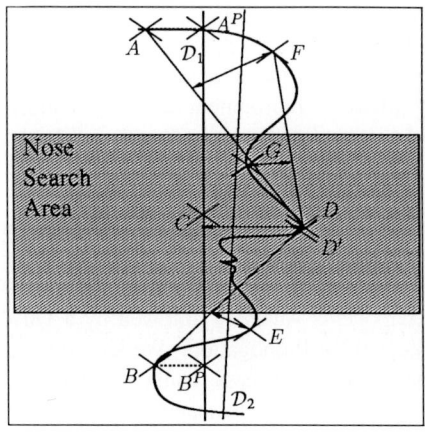

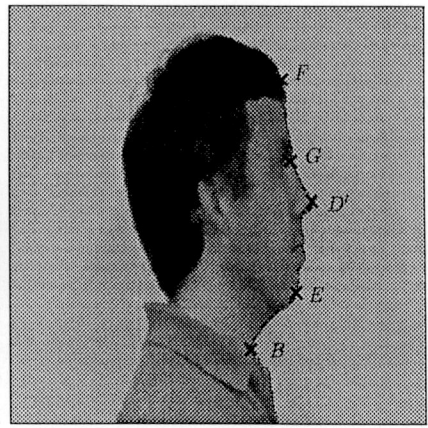

(a) Points and lines notation   (b) Example of application

Figure 1: Characteristic points extraction

Method	Parameter	Num. of sampl. pts	Gen. rate	Num. of sampl. pts	Gen. rate
PCA + Quad. Discr.	keeping 80% of the variance	80	88.4% (± 3.2)	100	90.2% (± 3.2)
k-Nearest Neighbors	one neighbor	80	90.8% (± 2.3)	80	90.8% (± 2.3)
Gradient Back Propagation	21 neurons in hidden layer	80	88.4% (± 4)	32	90.2% (± 2)

Table 1: Results of the best methods

References

[1] Akimoto, T. and Wallace, R. and Suenaga, Y. Feature extraction from front and side views of faces for 3D facial model creation. In *IAPR Workshop on Machine Vision Applications*, pages 291–294, Tokyo, 1990.
[2] L. D. Harmon, M. K. Khan, R. Lasch, and P.F. Ramig. Machine Identification of Human Face. *Pattern Recognition*, 13 No 2:97–110, 1981.
[3] L. Najman, R. Vaillant, and É. Pernot. From Face Sideviews to Identification. *Revue Technique Thomson*, 1992.
[4] É. Pernot and F. Vallet. NeuroClasse^{TM}: A software environment for classification of signals by neural and conventional methods. In *International Conference on Artificial Neural Networks (Helsinki)*, June 1991. p. 557-568.
[5] Mathew A. Turk and Alex P. Pentland. Face Recognition Using Eigenfaces. In *Computer Vision and Pattern Recognition*, 1991.
[6] L. Vincent. Algorithmes morphologiques à base de files d'attente et de lacets : Extension aux graphes. Thèse Ecole des Mines de Paris, May 1990.

A new algorithm for efficient subgraph matching

H.Bunke and B.T.Messmer

Institut für Informatik und angewandte Mathematik Universität Bern, Längassstr. 51, CH-3012 Bern, Switzerland; bunke@iam.unibe.ch, messmer@iam.unibe.ch

Abstract

A new procedure for subgraph (SG) isomorphism detection is described. The proposed method is efficient if the number of prototypes is large and if there are common substructures occurring in different prototypes. It can be shown that in the limit the time complexity of the proposed method becomes independent of the number of prototype graphs to be detected.

1 Introduction

Graph matching is a fundamental class of methods in computer vision and image analysis, and a large number of different techniques as well as various applications have been proposed [1]. However, a serious problem with graph matching is its high computational complexity. SG isomorphism detection and more complicated problems like error tolerant graph matching are known to be NP-complete and require algorithms of exponential time complexity. In the application of graph matching to the area of computer vision, often a graph repesenting an unknown input scene is matched to a number of prototype graphs, which represent model objects. The problem of computational efficiency becomes even more serious if a large number of prototypes is involved. In order to avoid the naive sequential testing of each prototype against the input graph, a number of solutions have been proposed in the literature, like model clustering and binary search trees [2], symbolic and attribute differences [3], or efficient nearest-neighbor search in an abstract metric space [4].

In this paper, we consider the problem of finding all occurrences of a number of prototype graphs in an input graph. We propose a new method for the efficient representation and use of a set of prototype graphs for SG isomorphism detection. The method can be understood as a particular type of indexing where structures of varying complexity (ranging from edges to complete prototypes) index all protoype graphs in which they occur. The indexing method has some similarities with the RETE-algorithm that was proposed in the context of forward chaining rule-based systems [5]. An application of a similar method to the matching of dynamically changing graphs was described in [6]. In the worst case, our new method is not slower than conventional sequential SG isomorphism detection based on tree search (TS). However, in the best case the computational complexity of the new method is much better. In the limit, it becomes independent of the number of different prototype graphs.

2 The matching procedure

In this section, we give an informal introduction to the proposed SG matching algorithm by means of an example. Fig. 1 shows two prototype graphs, g_1 and g_2, and Fig. 2 a test graph, g. The problem considered in this example is to find all occurrences of g_1 and g_2 in g.

In an off-line step, g_1 and g_2 are compiled into a network, the so-called *network of protoype graphs* (NPG). This network is a compact representation of the prototypes in the sense that common parts, i.e., SGs that are contained in more than one prototype occur only once in the

NPG. An NPG representing g_1 and g_2 is shown in Fig. 3. Generally, an NPG comprises four different types of nodes. First, there is one and only one *input-node* to the network. Next, there are *edge-checkers* in the NPG. Each edge-checker has exactly one incoming n-edge [1] that originates at the input-node. An edge-checker has a number of outgoing n-egdes, each leading to a SG checker or a prototype node (see below). The task of an edge-checker is three-fold. It tests any edge $e = (x, y)$ that is sent from the input node whether (a) the edge e has a certain edge label λ; (b) the source node x of e has a certain node label α ; (c) the sink node y of e has a certain node label β. Any edge e that fulfills all three conditions is forwarded from the actual edge-checker along all outgoing n-edges to other n-nodes.

The third type of n-nodes contained in the NPG is *SG-checker*. A SG-checker has exactly two incoming n-edges via which it receives either edges or SGs from its two predecessor n-nodes. These incoming n-edges are called the *left* and the *right input* n-edge, respectively. Associated with the left and right input edge of a SG checker is a local memory in which any edge or SG is stored that is forwarded to the SG checker. These memories are called *left* and *right memory* of the SG checker, respectively. The task of a SG checker is to test whether an edge or a SG stored in the left memory can be combined with an edge or a SG from the right memory to form a larger SG. Such a test consists of checking if certain nodes in the left memory are identical with certain nodes in the right memory. A SG checker has $n \geq 1$ outgoing n-edges each leading to another SG checker or a protoype node (see below) . If the test of a SG checker is successful, then the SG resulting from the combination of the parts from the left and the right memory is forwarded via these n-edges to other n-nodes. Finally, an NPG contains a number of *prototype nodes*, each representing a prototype. That is, each SG that is sent to a protoype n-node is a SG of the graph to be tested.

In order to find all occurrences of g_1 and g_2 in g (see Figs. 1 and 2), we input one edge of g after the other into the NPG shown in Fig. 3 a). That is, each edge (x_i, x_j) of g is sent via the input node to each of the edge checker nodes. Next, the SG checkers in the NPG become active. First, each SG checker tries to combine the contents of its left and right memories such that the node identity tests are satisfied. Each such resulting partial SG is sent along all outgoing n-edges to the next SG checkers, which will try to combine it with their left or right memories. These actions continue until all SG checkers have finished processing. Each instance of g_1 and g_2 contained in g will be stored in the corresponding prototype node. Fig 3 b) illustrates the left and right memory contents of the network nodes and all instances of g_1 and g_2 after the input g has been completely processed. We conclude that there is one SG occurrence of both g_1 and g_2 in g.

3 Computational Complexity

We consider the situation where there is exactly one occurrence of one of the prototypes in the input graph. Let
N = number of different prototype graphs,
I = number of edges in the input graph,
M = maximum number of edges in one prototype graph,
M_1 = number of edges that occur in all prototype graphs,
M_2 = number of edges that are unique to each prototype, where $M_1 + M_2 = M$

---

[1] It must be distinguished between the nodes and the edges of the graphs to be tested for SG isomorphism and the nodes and egdes of the NPG. For this reason we refer to the nodes and edges of the NPG as "network-nodes" and "network-edges", or "n-nodes" and "n-edges", for short, while the terms "nodes" and "edges" refer to the graphs to be tested for SG isomorphism.

The computational complexity of a conventional SG isomorphism detection procedure that is based on TS, treating each prototype individually, is

$$O(NM^3I^M) \text{ and } O(NM^3I) \qquad (1)$$

in the worst and best case, respectively (For more details see [7]). By contrast, our proposed method has a computational time complexity of

$$O(M_1^3 I^M + NM_2 M^2 I^M) \text{ and } O(M_1^3 + NM^2 M_2) \qquad (2)$$

in the worst case and best case, respectively. We notice that the two expressions in (2) become equal to $O(NM^3I^M)$ and $O(NM^3)$ for $M_1 = 0$, i.e., $M = M_2$. This corresponds to the one extreme case where there are no common parts in the prototype graphs. Notice that in this case the worst case is equal to (1) while the best case is better that (1) by a factor of I. In the other extreme case, we have $M_2 = 0$, i.e., $M = M_1$. This means that all the prototypes are identical, or, in other words, the common part that is shared by all prototypes is maximum. In this case, the two expressions in (2) become equal to $O(M^3I^M)$ and $O(M^3)$, respectively. Comparing with (1) we notice that now the time complexity is no longer dependent on the number of prototypes, neither in the worst nor the best case.

4 Experimental Results

The proposed graph matching procedure has been implemented in C++ and runs on SUN workstations. A number of experiments were conducted to verify the results of the theoretical complexity analysis given in section 3. In these experiments prototype and input graphs were randomly generated. The number of nodes of each prototype was a constant in the graph generation process, but the edges and the labels of each graph were randomly generated. An alphabet of ten node labels were used. In the experiments there were N different prototypes with K nodes each. All prototypes shared a common part consisting of L nodes. The input graph for each run of the matching procedure was the aggregation of all the prototypes, i.e. it was a graph consisting of $K * N$ nodes which contained exactly one instance of each prototype. Table 1 gives a comparison between the computation time needed by the new procedure and the conventional method which is based on sequential prototype testing and TS. All times reported in Table 1 are in seconds. While the size of the common SG increases, the overall size of the input graph remains constant. As can be expected from the complexity analysis, the performance of the TS becomes worse with increasing similarity, while the new method performs faster, due to its ability of matching common parts in models only once. For a small number of prototypes (typically less than ten) and a small size of the common SG (typically less than ten nodes) the conventional procedure is faster than the new method. This is due to more complicated data structures and some implementational overhead inherent to the new method. But for larger graphs, larger common parts (greater similarity among the models) and a larger number of prototypes, the new method becomes more and more superior to conventional TS.

5 Conclusions

A new procedure for SG isomorphism detection has been described in this paper. The proposed method is efficient if the number of prototypes is large and if there are common substructures occurring in different prototypes. It can be shown that in the limit the time complexity of

K	N	L	Tree-Search	New Method	Difference
40	10	5	13.09	13.89	-0.80
		10	11.81	12.92	-1.11
		15	11.57	11.66	-0.09
		20	13.88	11.73	1.15
		25	15.74	11.66	4.08
		30	17.55	8.75	8.80
		35	19.56	7.70	11.86
30	15	5	18.14	16.71	1.43
		10	21.12	17.21	3.91
		15	19.94	12.35	7.59
		20	19.84	10.84	9.00
		25	25.76	7.41	18.35
35	15	5	25.51	19.23	6.28
		10	21.61	17.50	4.11
		15	25.77	16.08	9.69
		20	23.85	16.78	7.07
		25	27.91	15.03	12.88
		30	39.32	11.72	27.60

Table 1: Comparison of the proposed method with conventional SG isomorphism detection based on TS. All computation times are in seconds. K =number of nodes in each prototype; N =number of prototypes; L =number of nodes in common SG shared by all prototypes.

the proposed method becomes independent of the number of prototype graphs to be detected. Working with only one prototype graph, the best case time complexity of the new procedure is better than that of TS based SG isomorphism detection while in the worst case it has the same time complexity. A number of computer experiments have been performed confirming the results of the computational complexity analysis. In order to cope with noisy data and segmentation errors, a generalization of the proposed method to error tolerant matching will be required. This topic is currently under investigation.

References

[1] Bunke, H. : *Structural and syntactic pattern recognition*, in Chen, C.-H., Pau , L.F., and Wang, P.: Handbook of Pattern Recognition and Computer Vision, World Scientific Publ. Co., Singapore, 1993

[2] Shapiro, L.G., Haralick, R.M.: *Organization of relational models for scene analysis*, IEEE Transactions PAMI-4, 595–602, 1982

[3] Shapiro, L.G.: *The use of numerical relational distance and symbolic differences for organizing models and for matching*, in Rosenfeld, A. (ed.): Techniques for 3-D machine perception, North Holland, 255–270, 1986

[4] Feustel, C.D., Shapiro, L.G.: *The nearest neighbor problem in an abstract metric space*, Pattern Recognition Letters, Vol. 1, 125–128, 1982

[5] Forgy, C.L.: *RETE, a fast algorithm for the many pattern / many object pattern match problem*, Artificial Intelligence Vol. 19, 17–37, 1982

[6] Bunke, H., Glauser, T., Tran, T.-H.: *Efficient matching of dynamically changing graphs*, in Johansen, P., Olsen, J. (eds.): Theory and Applications of Image Analysis, World Scientific Publ. Co., Singapore, 110-124, 1992

[7] Bunke, H., Messmer, B. : *Subgraph isomorphism detection using a RETE-type network*, under preparation

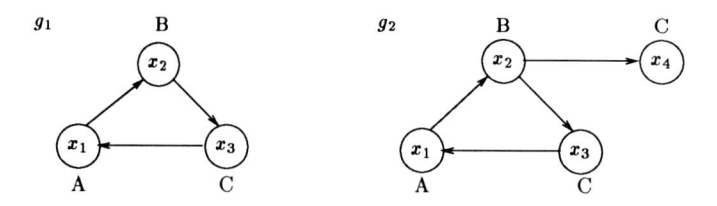

Figure 1: Two prototype graphs g_1 and g_2; $x_1, x_2, \ldots$ are node identifiers and A,B,C are node labels. There are no edge labels in this particular example.

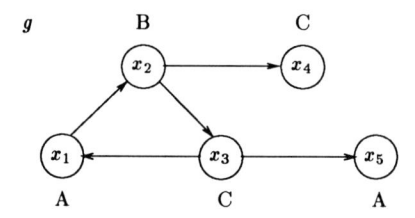

Figure 2: A test graph g. The problem is to find all occurrences of g_1 and g_2 shown in Fig 1 in this graph.

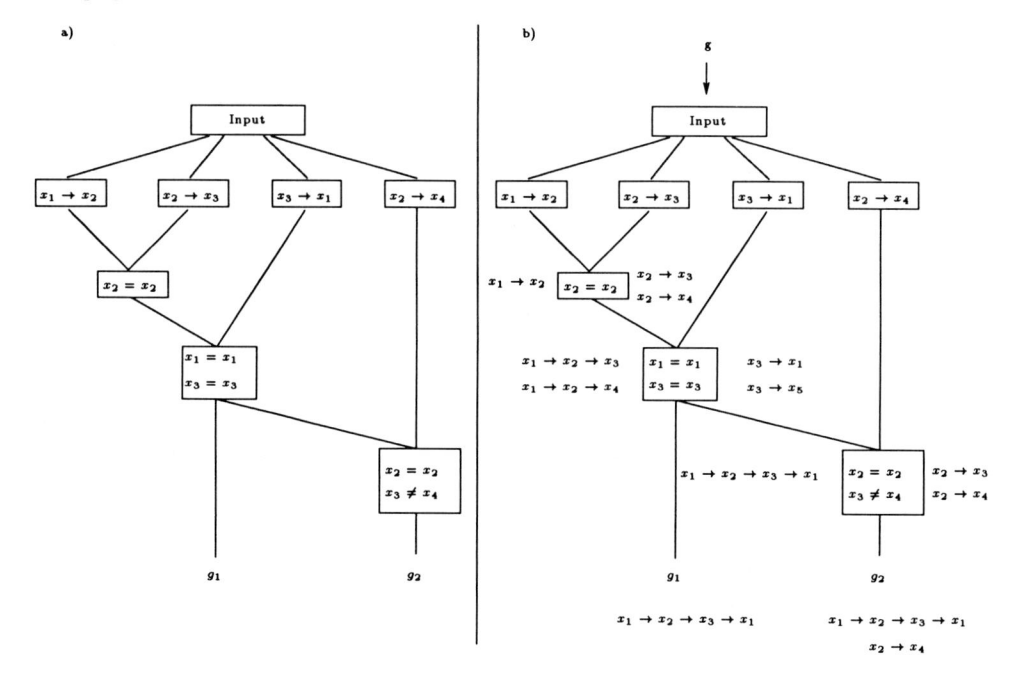

Figure 3: a) The NPG representing g_1 and g_2 shown in Figs 1 and 2. For each SG checker, the corresponding node identity tests are shown. b) A more detailed representation of the NPG. The contents of the local memories after termination of the SG isomorphism detection procedure are displayed.

2D pattern recognition using multiscale median and morphological filters

C.Jeremy Pye, J.Andrew Bangham, Stephen J. Impey and Richard V.Aldridge

School of Information Sciences, University of East Anglia, Norwich NR4 7TJ, UK

Abstract As defined here, the granularity is a complete characterisation of a signal. It is obtained by subtracting the output of each stage of a cascade of median or morphological filters from its input. By matching the granularity of an image with that of a target pattern, a pattern selective system can be implemented. In some ways it is a nonlinear analogue of a linear matched filter that is particularly appropriate for features with sharp edges.

1. Introduction

Mathematical morphological methods are often applied to shape characterisation and recognition problems. The concept of umbrae that underlies the approach is very similar to that of stacks that has emerged from the development of another group of rank filters, namely median filters. But, until recently, the main motivation behind the development of the latter has been different, namely to reduce random uncorrelated noise in signals.

Both filter types are nonlinear and furthermore the outputs from each are non differentiable and this distinguishes them from more conventional systems for which a mathematical understanding is well developed. In addition, the filters are neither associative nor commutative and so it is not easy to predict the properties of different arrangements of filters. Consequently the properties must be documented empirically.

One particular structure is now proving useful namely a cascade of filters (datasieve) of increasing scale[1, 2, 3, 4]. This scale increasing property can be exploited as the first step in an image pattern recognition system. Early reports suggest that it is useful in practical image analysis [2].

The datasieve is used to define the granularity of a signal. Granules are the set of 'pulses' obtained by subtracting the output of each filter from its input. In the case of one dimensional (1D) signals the granules are indeed pulses of characteristic length, but in 2D, erosion of corners makes for more complicated granules. The granularity of an object is a new representation of its spatial shape and size and this paper shows how it can be used as a basis for pattern recognition.

2. Methods

2.1 Morphological filters. This study uses morphological filters with a flat structuring element, B, operating on a sequence of discrete samples indexed with i. If B is a window of w samples ($w=m+1$) and the signal under study is A, then 'erosion' and 'dilation' operations are given by the following respectively:

$$\delta_m(A_{i,j}) = Min\{A_{i,j+k}, k \in B_w\}$$

$$\varepsilon_m(A_{i,j}) = Max\{A_{i,j+k}, k \in B_w\}$$

and morphological opening and closing are

$$\gamma_m = \delta_m \varepsilon_m \qquad \psi_m = \varepsilon_m \delta_m$$

The asymmetry introduced by these filters with respect to positive and negative going pulses is reduced by a combination of the two operations. Multiscale decomposition can be achieved by using cascades of opening and closing operations performed in a sequence first with $m=1$ and then $m=2$ and so forth. Two basic sieves are M_m and N_n defined with m so that a comparison can be made with median based filters, as follows.

$$M_m = \gamma_m \psi_m, \dots, \gamma_2 \psi_2, \gamma_1 \psi_1$$

$$N_m = \psi_m \gamma_m, ..., \psi_2 \gamma_2, \psi_1 \gamma_1$$

The result of each step in the sieve is to delete excursions of the signal that are m samples wide; so there is a scale related ablation of the signal. The granularity image of the signal at each scale is calculated by (likewise $I^M{}_m$)

$$I_m^N(A) = N_{m-1}(A) - N_m(A)$$

2.2 Median filters To be comparable the median based filter should also remove pulses that are m samples wide at each step consequently, at each stage (scale) of the decomposition, the signal is median filtered using a window of z samples ($z = 2m + 1$).

$$\mu_m(A_{i,j}) = Med\{A_{i,j+k}, k \in B_w\}$$

Multiscale decomposition is achieved using a cascade of median filters starting with $m=1$, followed by $m=2$ and so forth.

$$S_m = \mu_m, ... \mu_2, \mu_1$$

Unlike M and N the sieve is self-dual. The granularity image at each scale is

$$I_m^S(A) = S_{m-1}(A) - S_m(A)$$

In 1D, provided each I_m in the sieve only removes pulses of scale m, each pulse in I_m can be characterised by the triple, $g(x,a,m)$ (x is offset, a pulse amplitude and m mesh). This is the case with a root S, M and N and approximately true for S. [1]. Thus I_m can be thought of in terms of the granularity at each mesh m, $G_m = \{ g(x,a,m) \}$.

The total granularity of a signal is defined by $\mathbf{G} = \{G_m, m=1...n\}$, where n is number of data samples. The $A \to \mathbf{G}$ transformation is invertable in the sense that A can be recovered from $\mathbf{G}$ by expanding the granules and adding the resulting vectors.

In 2D the granules are more complex so in this paper the 2D G_m will refer to the set of non-zero pixels in each granularity image I_m, this makes computation slower but simpler.

2.3 Pattern recognition Let the granularity's of the target pattern be $\mathbf{G}^t$ and the unknown signal be $\mathbf{G}^u$. At its simplest in 1D the percentage of granules that match at position i as $\mathbf{G}^t$ is translated through $\mathbf{G}^u$ is P_i. In 2D $\mathbf{G}^t$ is passed through in 2 dimensions, i and j to produce P_{ij}. The matched sieve output signal is obtained by summing all $\mathbf{G}^u$ that match granules in $\mathbf{G}^t$ (at their respective positions) provided $P_i > T$ the threshold percentage.

Further refinements include; Fig.3, a tolerance is introduced on alignment of granules(i), Fig.4, T is set to 50% together with a further requirement, that at least 12.5% of $G^u{}_m$ must match $G^t{}_m$, ie. there must be a degree of matching at every m. In Fig.5 the percentage of $G^u{}_m$ that must match $G^t{}_m$ is equal to T, namely 50%, in addition only the Q most significant (in amplitude) granules in each $G^t{}_m$ were included.

3. Results

Fig.1 shows the decomposition by S of two 1D scan lines from an image. The granularity image associated with the first eye is indicated by the ↔. It is slightly different from the granularity of the isolated eye signal, Fig.2 S, because the two eyes are in different surroundings. Fig. 2 also shows the granularity images obtained by M, N and S. Fig. 3 shows the result of matching the target granularities to the signal and the non-eye signal components are attenuated.

It is less easy to show $\mathbf{G}$ in 2D so attention is focused on $\{P_{ij}\}$ and the output image. Fig.4 A, shows a test image that contains 4 copies of a target pattern. The arrowed copy is blurred by convolution. The C panels in Fig.4C show $\{P_{ij}\}$ for both filters. A good match is associated with white highspots. The D panels show the associated output image after thresholding. All four targets can just be seen. The bottom panels show the results in the presence of 30% salt and pepper noise. Clearly the 2D patterns can be recognised even in the presence of noise. It is also noticeable that the blurred target produces a large signal whereas impulses do not register. This seems to be the dual of linear filters in which impulses spread to all frequencies.

Fig.5, shows the result of using the centremost eye as a target. White spots at the centre of Fig.5C and D show the target perfectly matching itself and by setting a suitable threshold, just those regions that achieve a reasonable match can be seen in Fig. 5E and F. For comparison the results of cross-correlating the target with Fig.5A using a conventional linear filter are shown in Fig.6.

Conclusions

The granularity of 1 and 2D images can be used for pattern recognition. The differences between granularities produced by S, N and M (not shown) filters are small. The granularity representation of signals can be better for recognition purposes than the frequency representation. Future study should concentrate on quantifying the recognition success rate and developing methods for generalising the target mask and threshold parameters by using many training examples.

1 Bangham, J.A. Properties of a series of nested median filters, namely the datasieve. (1993) IEEE Trans. Signal Processing SP 41:31-42
2 Bangham, J.A. and Campbell, T.G. Sieves and wavelets: multiscale transforms for pattern recognition. (1993) IEEE Workshop on nonlinear signal processing, Proc. IEEE Workshop on Nonlinear Signal Processing pp1.1-4.1 to 1.1-4.6
3 Salembier, P. and Kunt, M. (1992) Size sensitive multiresolution decomposition of images with rank order based filters. Signal Proc. 27:205-241
4 Marshall, S. and Matsopoulos, G.K (1993) Morphological Data Fusion in Medical Imaging. Proc. IEEE Workshop on Nonlinear Signal Processing pp6.1-5.1 to 6.1-5.6

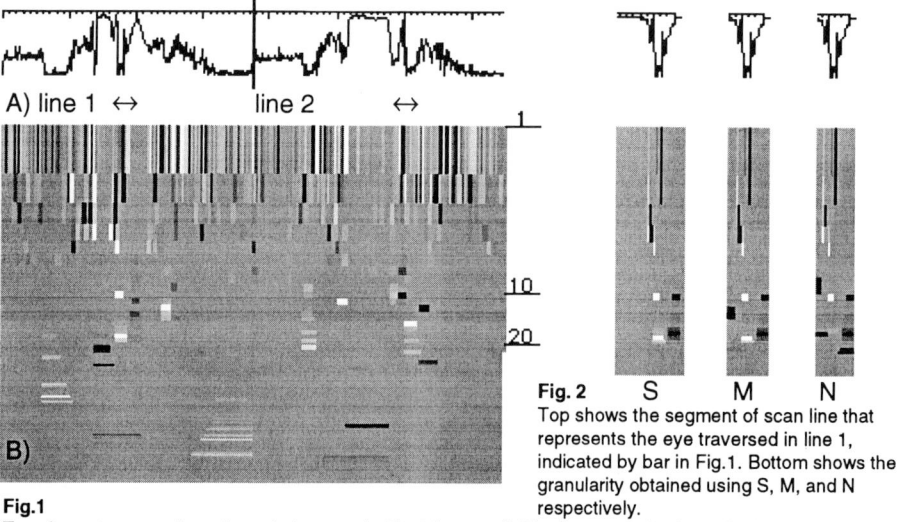

Fig. 2 S M N
Top shows the segment of scan line that represents the eye traversed in line 1, indicated by bar in Fig.1. Bottom shows the granularity obtained using S, M, and N respectively.

Fig.1
Top shows two scan lines through the eyes in Fig.5 (arrowed). The lines pass horizontally through one eye each (the head is slightly tilted). The eye locations are indicated with heavy black lines. Large tick marks indicate 100 pixels plotted across the abscissa. Bottom shows the granularity of the scan lines obtained using a root median sieve. m is plotted logarithmically down the ordinate and amplitude is plotted as intensity.

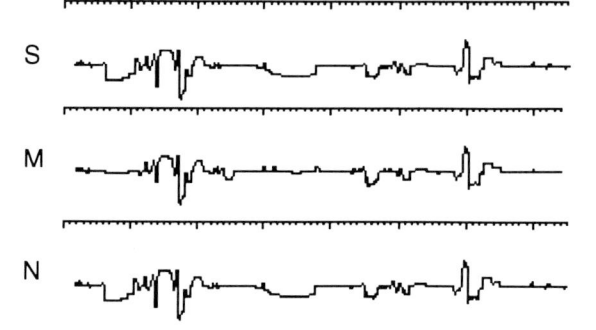

Fig. 3
The result of matching the granularites shown in Fig.2 to the granularities of the original data (top of Fig.1). The abscissae are aligned. For example, the set of granules in Fig.2 S are stored as a set of lists, one for each m. These are then compared to the lists of of granules shown in Fig.1 bottom. If >80% of granules match within $x \pm 2$ units then the matching granules are added to the output at the appropriate position along x.. Trace S shows the result using S for the decomposition, likewise M and N.

Fig.4. Recognition of test pattern:

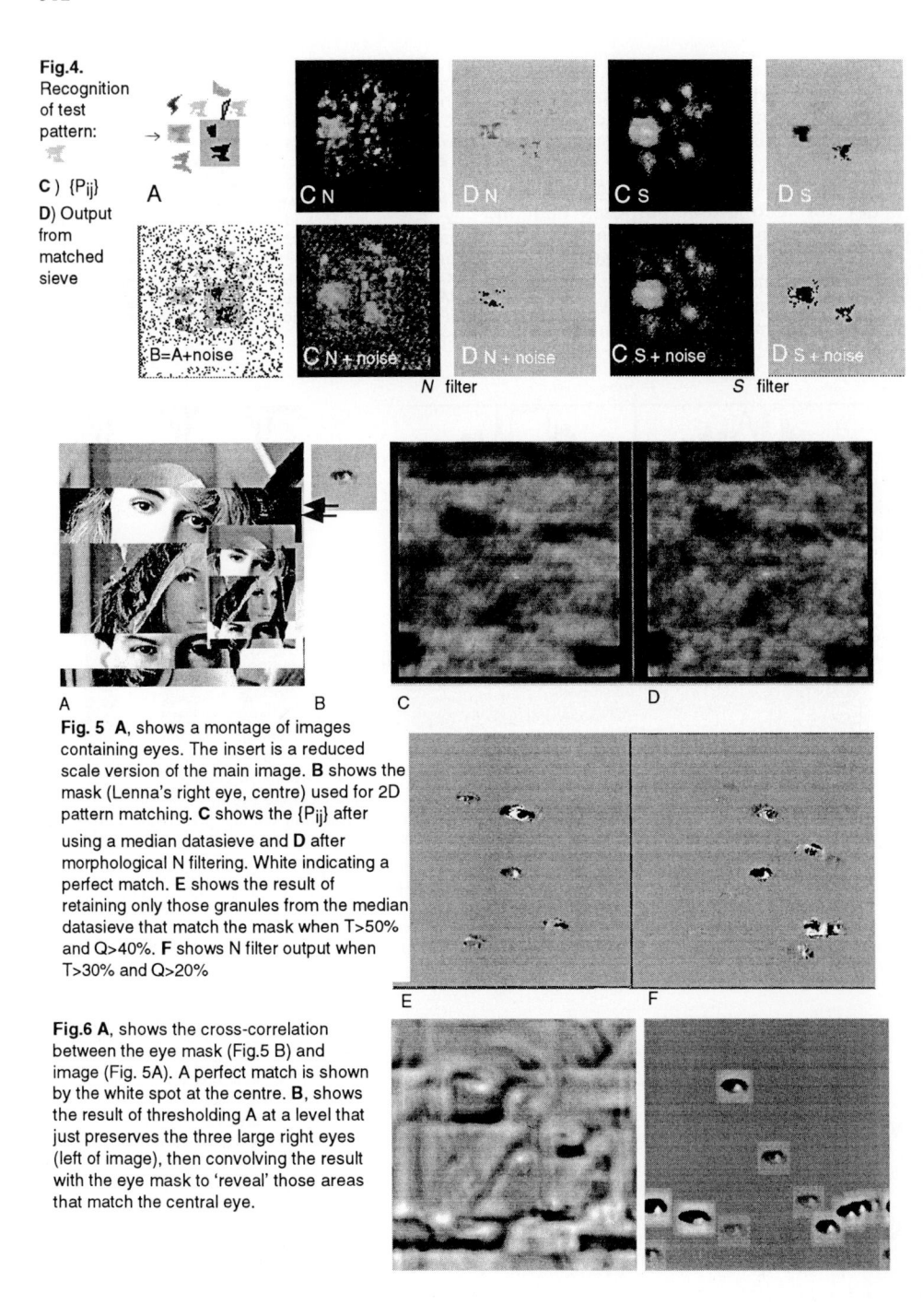

C) {P_{ij}}
D) Output from matched sieve

Fig. 5 A, shows a montage of images containing eyes. The insert is a reduced scale version of the main image. **B** shows the mask (Lenna's right eye, centre) used for 2D pattern matching. **C** shows the {P_{ij}} after using a median datasieve and **D** after morphological N filtering. White indicating a perfect match. **E** shows the result of retaining only those granules from the median datasieve that match the mask when T>50% and Q>40%. **F** shows N filter output when T>30% and Q>20%

Fig.6 A, shows the cross-correlation between the eye mask (Fig.5 B) and image (Fig. 5A). A perfect match is shown by the white spot at the centre. **B**, shows the result of thresholding A at a level that just preserves the three large right eyes (left of image), then convolving the result with the eye mask to 'reveal' those areas that match the central eye.

IMAGING OF VEHICLES WITH LINEAR ARRAYS

G. Jacovitti [a], A. Minniti [b], A. Neri [c], T. Tasselli [b]

[a] INFOCOM Dpt., University of Rome "La Sapienza", Rome, Italy

[b] CITEC S.p.A., Divisione Ricerca e Sviluppo, Rome, Italy

[c] University of Rome III, Rome, Italy

Abstract
   In this work, we directly resort to the concepts of the optimum estimation theory to develop an optimal strategy (based on the Maximum Likelihood criterion) for estimating the velocity of vehicles from time/space scans obtained with linear arrays.
   As a central result of our contribution, we show that, when the observation noise is a Gaussian random field, the likelihood function reduces to a simple function of a filtered Radon transform of the scan.

I. INTRODUCTION

   Application of artificial vision in vehicle traffic control constitutes a challenging field in image processing. Most usual imaging system employed in such applications are conventional video cameras. However, these devices, best suited for TV production, are often not adequate for traffic control, due to the insufficient time sampling, interlacing effects, etc.
   An imaging system particularly intended for computer vision of vehicles in railroad and parking access areas has been proposed by the authors [1]. Basically it consists of a CCD vertical linear array placed into a focal plane, performing an auto scan of the vehicles during their passage (see Fig.1). The main advantage of such system is that it provides a continuous image of the vehicles, unaffected by horizontal perspective distortions due to the proximity of the sensors.
   On the other hand, the spacing between the vertical sampled lines is non uniform due to the variable speed of the observed vehicle. This causes an unpredictable scale distortion along the horizontal axis (see Fig. 2).
   Correction of these distortions requires an accurate estimation of the instantaneous velocity of the vehicles. This can be done by means of a second CCD array, aligned with the line of motion. In fact, the collection of the horizontal slices forms an image where the motion shift of the slice results in parallel striate patterns (see Fig. 3). The slope of the patterns is inversely proportional to the vehicle speed.
   Referring to small portions of the image where the structures are nearly rectilinear, the reference for the horizontal scale is extracted by estimating their slope.
   Several methods are currently available for detection and direction estimation of striate patterns. In particular the straight line Hough transform is a well-known technique [2].
   Directly resorting to the concepts of the optimum estimation theory, we have developed an optimal slope estimation strategy, based on the Maximum Likelihood (ML) criterion.
   In this contribution, after a summary of such an approach, we describe the associate processing scheme and give its theoretical performance bounds.

II. VELOCITY ML ESTIMATION

Let $r(\mathbf{x})$ be the recorded horizontal scan, defined on the interval I of the Euclidean plane with points $\mathbf{x} = (x_1, x_2)^T$. Specifically, x_1 is the time and x_2 is the horizontal coordinate along the array. Let us model $r(\mathbf{x})$ as a striate pattern with cross section signal $g_0(t)$, first translated of an unknown quantity s and then rotated by the angle θ to be estimated, plus an observation noise $n(\mathbf{x})$. Let $g(\mathbf{x}; \theta, s)$ be the translated and rotated version of the striate pattern. Then, the observed image can be expressed as:

$$r(\mathbf{x}) = g(\mathbf{x}; \theta, s) + n(\mathbf{x}). \tag{1}$$

Without loss of generality, let us represent $r(\mathbf{x})$ with its series expansion coefficients $\{r_{ij}\}$ w.r.t. to a given orthogonal basis. Then, rearranging the representation coefficients into a one dimensional vector denoted by $\mathbf{r}$, we write by definition the ML estimate of θ as the value $\hat{\theta}$ maximizing the conditional probability of $\mathbf{r}$ given θ, i.e.:

$$\hat{\theta} = \text{Arg} \max_{\theta \in \Theta} p_{\mathbf{R}/\theta}(\mathbf{r}/\theta). \tag{2}$$

Proceeding as in [3], the ML estimation can be reduced to the problem of maximizing the ML functional $\Lambda(\mathbf{r}, \theta)$ given by the conditional probability of $r(\mathbf{x})$ divided by any arbitrary function which does not depend on θ and s. Thus, under the hypothesis of statistical independence of θ and s, the likelihood functional can be written as:

$$\Lambda(\mathbf{r}, \theta) = \int_S \Lambda(\mathbf{r}, \theta / s) p_S(s) ds \tag{3}$$

where the conditional likelihood function $\Lambda(\mathbf{r}, \theta / s)$ is now proportional to the probability of $\mathbf{r}$ conditioned to both θ and s.

In particular for a stationary, white, zero mean, gaussian observation noise, with variance $(N_0 / 2)$, the logarithm of $\Lambda(\mathbf{r}, \theta / s)$ is (see [3]):

$$\ln \Lambda(\mathbf{r}, \theta / s) = \frac{2}{N_0} \iint r(\mathbf{x}) g(\mathbf{x}; \theta, s) dx_1 dx_2 - \frac{1}{N_0} \iint [g(\mathbf{x}; \theta, s)]^2 dx_1 dx_2. \tag{4}$$

When the observed image is large enough with respect to the support Γ, and to the range of s, the last integral in Eq. (4) represents the energy of $g(\mathbf{x}; \theta, s)$, and therefore is shift and rotation invariant and can be dropped out. In addition, since the reference signal is a striate pattern of length L, the likelihood function $\Lambda(\mathbf{r}, \theta / s)$ can be written in terms of the Radon transform $\Phi_\theta(t)$ of the observed image, defined by the line integral:

$$\Phi_\theta(t) = \int_{x_1 \cos\theta + x_2 \sin\theta = t} \mathbf{r}[x_1(\tau), x_2(\tau)] \, d\tau. \tag{5}$$

In fact, it can be shown that the conditional log likelihood function given by Eq.(4) can be written in terms of the Radon transform, passed through a filter matched to the pattern cross section $g_0(t)$, as the following convolution:

$$\ln \Lambda(\mathbf{r}, \theta / s) = \frac{2}{N_0} \Phi_\theta(s) * g_0(-s). \tag{6}$$

As a consequence, for uniformly distributed offset s, the functional to be maximized is:

$$\Lambda(\mathbf{r}, \theta) = \int_S \exp\left[\frac{2}{N_0} \Phi_\theta(s) * g_0(-s)\right] ds. \tag{7}$$

In other words, the ML estimate is obtained by finding the value of θ maximizing a generalized tomographic projection of the filtered Radon transform along the offset s.

Therefore, an optimal processing scheme for estimating the velocity of the vehicles from the horizontal scan is depicted in Fig. 5.

The expected performance of the ML estimator can be analyzed through the Rao-Cramer bound. In particular, using the Fourier series expansion of the polar representation of the striate pattern:

$$g(\rho, \alpha) = g_0(\rho \cos \alpha) = \sum_{n=-\infty}^{+\infty} C_n(\rho) e^{jn\alpha} \tag{8}$$

we have obtained the following significant lower bound:

$$Var\left[\hat{\theta}(\mathbf{r})\right] \geq \left[\left(\frac{8\pi^2}{N_0}\right) \int \rho \, \overline{f^2}(\rho) E(\rho) d\rho\right]^{-1}, \tag{9}$$

where $E(\rho)$ and $\overline{f^2}(\rho)$ are the energy density and the rms angular frequency of the reference signal at radius ρ:

$$E(\rho) = \sum_{n=-\infty}^{+\infty} |C_n(\rho)|^2, \tag{10} \qquad \overline{f^2}(\rho) = \frac{\sum_{n=-\infty}^{+\infty}\left|\frac{n}{2\pi}C_n(\rho)\right|^2}{E(\rho)}. \tag{11}$$

An example of application of the method is provided in of Fig.4 where the reconstructed image, obtained from the vertical scan using the horizontal scan based ML velocity estimate, is reported. The scans where taken with an experimental equipment operating in real time at a rate of 500 lines per second. The equipment makes use of an array of four DSPs, to process up to one vehicle per second, running up to 100 Km/h.

REFERENCES
[1]   S. Del Lungo, G. Jacovitti, A. Neri, T. Tasselli, "Vehicle Cassification by Passive Optical Linear Arrays", *Digital Signal Processing 91*, pp.708-713, C. Cappellini and G. Constantinides (eds.), Elsevier Science Publisher B. V., 1991.
[2]   B. JAHNE, *Digital Image Processing*, pp. 249, Springer and Verlag, 1991.
[3]   H.L. Van Trees, *Detection, Estimation, and Modulation Theory*, John Wiley and Sons, Inc., 1968.

316

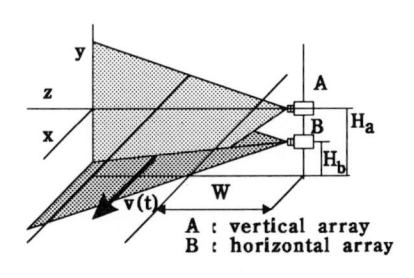

Fig.1 Sensors arrangement

Fig.2: Vertical scan

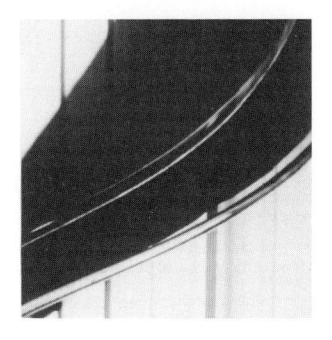

Fig. 3: Horizontal scan

Fig.4: Reconstructed image

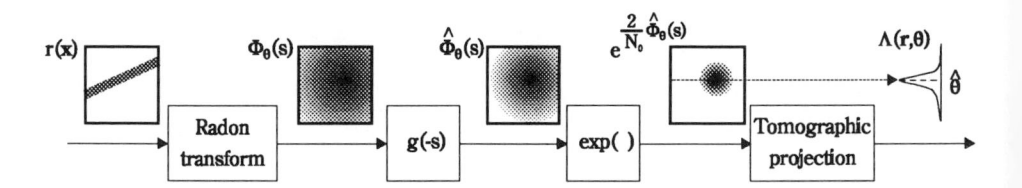

Fig. 5: Velocity estimation processing scheme

Optical image processing technique speeds up algorithms of computer graphics

Y. B. Karasik

Computer Science Department, Tel Aviv University, 69978 Tel Aviv, Israel

Abstract
The technique of optical computing is applied to develop new algorithms of computer graphics whose time complexity is $O(1)$ whereas algorithms of conventional computer graphics have polynomial time complexity. This technique explicitly demostrated on example of developing new filling algorithm which fills the interior of a polygon in constant time whereas conventional filling algorithms have at least $\Omega(n \log N)$ time complexity, where n is the number of boundary pixels of a polygon, and N is the diameter of the polygon expressed in terms of pixels.

1. INTRODUCTION

The filling of the interior of a polygon is an important problem in image processing, pattern recognition and computer graphics. This problem has received much attention from computer scientists who have proposed many solutions intended for ordinary display devices, the best of which [1], as far as we know, has $O(n \log N)$ time complexity, where n is the number of pixels contained in the boundary and N is the number of pixels contained in the diameter of a polygon.

The same problem also emerges in optical image processing where images are represented on special optical devices called spatial light modulators (SLM) in the form of distributions of transmittance (or reflection) coefficients of their pixels.

The main advantage of SLM, in comparison with usual display devices, is that we can control the transmittance coefficient of light both optically and electronically. The first way is preferable because optically we can control all pixels in parallel.

Optical control of the transmittance coefficient proceeds as follows. Let $A(i,j)$ be the amplitude of an incident plane lightwave at pixel (i,j). If the SLM is in control mode, then the transmittance coefficient of the pixel becomes $A(i,j)$ (in working mode SLM simply transmits incident light through each pixel relaxing its amplitude proportionally to the optical density of the pixel).

Obviously, two plane waves $A(n,m) \cdot e^{i(\omega t + const)}$ and $B(n,m) \cdot e^{i(\omega t + const)}$ can be added to yield a new wave $(A(n,m) + B(n,m)) \cdot e^{i(\omega t + const)}$. Thus, optically, we can perform
- **addition:** $C(n,m) = A(n,m) + B(n,m)$ of two images in parallel for every n,m.

Analogously, in parallel for each pixel we can also perform the following operations (the reader is reffered to recent comprehensive survey [3] for more explanation):
- **subtraction :** $C(n,m) = A(n,m) - B(n,m)$;
- **scalar multiplication:** $C(n,m) = k \cdot B(n,m)$ which is obtained as a result of the passage of light through both the SLM containing $B(n,m)$ and the SLM whose pixels have optical density

k. We assume that photodetectors cannot distinguish between two electromagnetic waves if their amplitudes differ by less than unit under appropriate scaling. Thus, elements of the matrices we deal with are positive integers and scalar multiplication of an image is, in fact, the operation: $C(n,m) = \lfloor k \cdot B(n,m) \rfloor$.
• **filtering of an image at a given level** L :

$$C(n,m) = Filter_L(B(n,m)) = \begin{cases} 1, & \text{if } B(n,m) \geq L; \\ 0, & \text{otherwise}; \end{cases}$$

• **convolution**:

$$C(n,m) = \sum_{i=1}^{N}\sum_{j=1}^{M} A(n-i, m-j) \cdot B(i,j) = A(n,m) * B(n,m);$$

These primitives can serve as the basis for the technique of symbolic substitution (SS)[2] which allows one to extract/remove all occurences of image B in image A optically in constant time.

We see that optical computing provides a variety of very complicated operations on images which can be performed in constant time. Therefore, based on these operations, we can try to reduce time-complexity of the filling. This paper aims to propose a new filling algorithm having $O(1)$ time-complexity.

2. AN ALGORITHM FOR RECONSTRUCTION OF ANY PLANE IMAGE FROM ITS REFINED BOUNDARY

To obtain the idea of the algorithm turn from a discrete plane consisting of pixels to a continuous plane consisting of points and consider any horizontal line $y = a$ intersecting the plane figure.

What segments of this line lie in A? To answer the question we should assign a weight to each intersection point $(x_i(a), a)$ by the following recurrence formulae:

$$weight(x_i(a), a) = \sum_{k=1}^{i-1} weight(x_k(a), a) + \begin{cases} 1, & \text{if the } i^{th} \text{ intersection is of} \\ & \text{the first type (see fig. 1 a, b);} \\ \frac{1}{2}, & \text{if the } i^{th} \text{ intersection is of} \\ & \text{the second type (see fig. 1 c, d);} \\ -\frac{1}{2}, & \text{if the } i^{th} \text{ intersection is of} \\ & \text{the third type (see fig. 1 e, f);} \\ 0, & \text{if the } i^{th} \text{ intersection is of} \\ & \text{the fourth type (see fig. 1 g, h),} \end{cases}$$

where i is the number of intersection in order from left to right.

It is easy to see that each segment of the line whose left endpoint has an odd weight, lies in A.

We can also attribute a weight to other points (x, y) by the following formulae:

$$weight(x, y) = \sum_{x_i(y) < x} weight(x_i(y), y).$$

Obviously, only points with an odd weight lie in A.

Now attribute a type to each point of ∂A as follows: the type of a boundary point is the type of the intersection between ∂A and the horizontal line passing through this point. Since interior points of horizontal boundary segments cannot be intersection points, we attribute to them the fourth type. Thus, we have:

$$\partial A = \bigcup_{i=1}^{4} S_i,$$

where S_i is the set of boundary points of the ith type.

It is easily seen that

$$weight(x,y) = S_1 * ray(x,y) + S_2 * \tfrac{1}{2} \cdot ray(x,y) + S_3 * -(\tfrac{1}{2}) \cdot ray(x,y),$$

where

$$ray(x,y) = \begin{cases} 1, & \text{if } x \geq 0, y = 0; \\ 0, & \text{otherwise,} \end{cases}$$

and convolution between set S and any function $f(x,y)$ is defined as follows:

$$S * f(x,y) = \sum_{(p,q) \in S} f(x-p, y-q).$$

Let us now turn from the plane consisting of points to the plane consisting of pixels and generalize the approach stated above. Instead of the function $ray(x,y)$ consider the image

$$RAY(n,m) = \begin{cases} 1, & \text{if } m = 1, n \geq 1; \\ 0, & \text{otherwise.} \end{cases}$$

Instead of boundary points of the ith type consider boundary pixels of the ith type (see fig. 2). We assume here that each boundary pixel is adjacent to exactly two others because those boundary pixels which do not satisfy this property can be removed from the image without breaking its topology (we omit the proof of this assertion). The removal is performed in constant time using technique of SS (we omit details in this version).

Let S_i be now a set of boundary pixels of the ith type and

$$WEIGHT(n,m) = S_1 * 2 \cdot RAY(n,m) + S_2 * RAY(n,m) - S_3 * RAY(n,m).$$

It is easy to see that the interior of A consists only of those pixels (n,m) which have $WEIGHT(n,m)$ not proportional to 4. Hence, we can propose the following filling algorithm.

Step 1. Remove redundant boundary pixels which are not of type 1, 2, 3 or 4.

Step 2. For each $i = \overline{1,3}$ recognize boundary pixels of the ith type . For this aim correlate the boundary with masks depicted on fig. 2 respectively and filter the resulting images at level 3. As a result we obtain images S_1, S_2 and S_3.

Step 3. Compute

$$WEIGHT(n,m) = S_1 * 2 \cdot RAY(n,m) + S_2 * RAY(n,m) - S_3 * RAY(n,m).$$

Step 4. Compute

$W(n,m) = \lfloor WEIGHT(n,m)/4 \rfloor.$

Step 5. Compute

$INTERIOR(n,m) = WEIGHT(n,m) - 4 \cdot W(n,m) = \begin{cases} 2, & \text{if the pixel } (n,m) \text{ is interior;} \\ 0, & \text{otherwise.} \end{cases}$

Since each step of the algorithm requires constant time for its performance, we obtain: *The interior of a plane image can be filled optically in constant time.*

References

[1] H. H. Atkinson, I. Gargantini, and T. R. S. Walsh, Filling by quadrants or octants, *Computer vision, graphics, and image processing*, vol. 33, pp. 138–155 (1986).

[2] S. D. Goodman and W. T. Rhodes, Symbolic substitution applications to image processing, *Applied Optics*, 27, pp. 1708–1714 (1988).

[3] D. G. Feitelson, *Optical Computing: A survey for computer scientists*, MIT Press, Cambridge, Massachusetts, London, England, 1988.

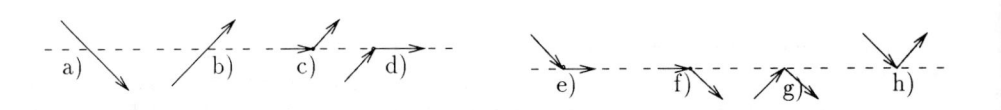

Figure 1: four types of intersections; directions of boundary segments are shown by arrows, points of intersections are indicated as rings.

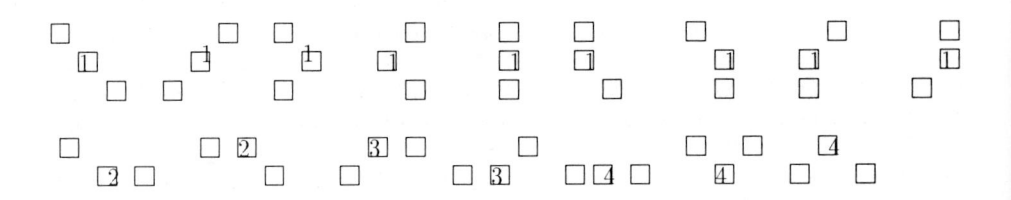

Figure 2: boundary pixels of the first, second, third and fourth types marked as 1, 2, 3 and 4 respectively.

Neural Networks for Remote Sensing Data Fusion

Vito Cappellini[a], Alessandra Chiuderi[b], Stefano Fini[a]

[a]Fondazione Scienza per l'Ambiente, Viale Galileo 32, 50125 Florence, Italy.

[b]Università di Firenze, Facoltà di Ingegneria, Via di S. Marta 3, 50139 Florence, Italy.

Abstract

This paper focuses on the possibilities offered by neural networks applied to multisensor image data processing.

Neural Networks (NNs) are employed to perform fusion of visible and SAR data in order to obtain a landcover classification on an agricultural area in the surroundings of Florence (Italy). The results obtained by using neural techniques are reported and discussed.

1. INTRODUCTION

The development of remote sensing activities related to Earth monitoring generate large amount of data. It is thus useful to develop techniques to fuse together data which are sensitive to different physical parameters collected with different sensors. Microwave, infrared and visible channels provide information spreading all over the electromagnetic spectrum; the combination of these data thus synthetizes dielectric, thermal and chemical information.

The term *data fusion* denotes the process by which it is possible to extract more information from the combination or integration of data collected by different channels than from each separate channel [1].

Two are the main issues proposed in the literature for merged data: photointerpretation by human expert, and automatic image processing, such as classification. In the former case, techniques such as HIS or PC transformation are employed [2], while in the latter, each source of information is considered as a different channel, and the set of numerical values obtained by composing the available information is dealt with as a multidimensional variable.

The main problem arising with this approach is due to the fact that some kind of assumptions on the statistical distribution of the data themselves must be made. When dealing with just one source of data, it is possible to assume that the multidimensional variable associated with the data has a Gaussian distribution, still obtaining good results [3]; unfortunately no theoretical reason supports us while making the same kind of assumption when dealing with data provided by different sensors.

In the last few years, on the other hand, neural networks [4,5] are becoming an interesting processing tool in every field of applied science. Processing of remote sensing data can take great advantages from this new methodology [6] which has been proved to be particularly suitable for processing data collected by different sensors, and this is especially due to the fact that neural networks require no kind of hypothesis on data distribution.

In particular, among other issues [e.g.7], neural networks find one of their most interesting applications in the landcover classification task, as in [e.g.8,9].

In these papers the authors employ a Multilayer Perceptron network, trained by the Back Propagation algorithm [4], a totally supervised method; this means that the task cannot be accomplished, not even a part of it, without the ground truth data. This entails a great disadvantage since gathering the ground data can be very expensive or sometimes impossible and thus, the use of supervised techniques should be avoided whenever it is possible [10].

The network we propose in this paper, on the contrary, is composed of two separate parts, an unsupervised one and a supervised one, which can be trained in a second time allowing us to obtain some preliminary, but still acceptable, results even without the ground data [11], and, successively, to improve these results by adding a supervised phase, if and when the training data will be available.

2. MATERIALS AND METHODS

In the following sections a brief description of both data and tools employed for the landcover classification task is given; in the first section we report the technical characteristics of the instruments employed for data acquisition together with the characteristics of the test site, while in the second section a brief description of the neural network is given.

2.1 Data Collection

The test site is situated in Montespertoli, an agricultural zone in the surroundings of Florence (Italy).

The sensors employed in this study are a multifrequency (band P,L,C) multipolarization SAR and a multispectral scanner (TMS) operating with 10 channels in the visible region and 2 channels in the thermal band. The two data sets have different spatial resolution: 25 meter by 25 meter for the TMS data while 6 meter in range and 12 meter in azimuth for the SAR image. These data were collected during the MAC-Europe '91 campaign [12].

The combination of data relative to different regions of the electromagnetic spectrum, exploits complementary information on the area: information about structure, texture, roughness is provided by SAR data, while chemical and thermal properties can be inferred from the TMS data by taking into account the visible and IR channels respectively.

SAR data have been first re-sampled to 24 meter by 24 meter of spatial resolution and, successively, have been co-registered on the TMS data set by means of a 5th order polynomial leading to residuals of 0.284 and 0.330 in the x and y directions respectively. Resampling was performed by means of cubic interpolation.

2.2 The Network

The neural network employed is a Counterpropagation type one [5]. It consists of two layers, the first one being unsupervised and the second one supervised.

The input of the network is given by the multidimensional vector obtained after the coregistration of the data set.

The first layer is a monodimensional self-organizing Kohonen map [13], the metric being the scalar product.

At the end of the first phase, the input vectors are subdivided into groups according to their similarities, or equivalently, according to their position in the feature space (up to now, no kind of supervision was performed). The output of the first layer can however be considered as a landcover classification when the ground truth data are not available.

If, as it is the case, a training set is available, a second layer is added to the network and a supervised training of the network is performed, refining the results obtained so far.

The second layer is given by as many units as the classes to be discriminated on the scene, every unit being connected with all the units of the unsupervised layer. Every pattern of the training set is associated with the class it belongs to.

At every step of the training algorithm, an input vector is presented to the network, the unit of the first layer which achieves the maximum scalar product is identified, and only the weights connecting this unit to the output layer are updated.

3. RESULTS AND DISCUSSION

In [11] the performances of the NN Kohonen map were compared with a more classical method such as the Maximum Likelihood: the comparison showed that the unsupervised Kohonen map outperformed the ML classifier. In table 1 the results of the Counterpropagation NN classification on the training set are reported by means of the confusion matrix. In the experiment, the input of the network was given by the six TMS channels corresponding to the visible channels of the TM, and the three polarizations of the L band of SAR data. The overall accuracy is 91.6% for the training set, while for the test set it reduces to 78.32%.

The accuracies reported show that the network performs well for the classification of the first three classes, while the classification rates for vineyard, uncultivated soil and grassland are lower. This can be ascribed to the choice of the training set: wheat, wood and alfaalfa classes were much more numerous than the remaining three, due to the colture distribution on the area.

The choice of training areas having more or less the same spatial extension, in order to present to the network the same number of training examples for each class, should overcome this difficulty and thus allow us to achieve better results.

Table 1
NN classification accurancy on the training set (in percentage)

Class	Wheat	Wood	Alfaalfa	Vineyard	Uncult.	Grassland
Wheat	97.07			2.47	0.23	0.23
Wood		98.47	1.53			
Alfaalfa			95.50			4.50
Vineyard	7.27			86.36	6.37	
Uncultiv.	3.50	1.72		30.43	64.35	
Grassland			17.54			82.46

The results obtained so far indicate NNs as a suitable tool for landcover classification purposes, and more generally for remote sensed data processing. In particular the proposed network showed interesting potentialities, due to the twofold unsupervised-supervised structure. The data presented to the network were not preprocessed; in the future the classification rates should be improved by performing filtering and decorrelation operations, in particular on SAR data, before the neural computation.

4. ACKNOWLEDGEMENTS

The present work has been supported by the Italian Research Council (C.N.R.), in the framework of the research "Fusion of SAR and Remotely Sensed Visible and Infrared Data"; partial support was also provided by the Italian Space Agency (A.S.I.).

The authors wish to thank Dr. Leandro Chiarantini for his helpful cooperation during the preparation of the present work.

5. REFERENCES

1. E. Waltz, J.Llinas: Multisensor Data Fusion. Artech House, Boston London (1990)
2. P.S. Chavez, S.C. Sides, J.A. Anderson: Comparison of Three Different Methods to Merge Multiresolution and Multispectral Data: Landstat TM and SPOT Panchromatic. PE&RS, vol. LVII, n. 3 (1991)
3. I.L. Thomas et al.: Classification of Remotely Sensed Images. Adam Hilger, Bristol England (1987)
4. D.E. Rumelhart, J.L. McClelland: Parallel Distributed Processing. MIT Press, Cambridge MA (1986)
5. P.D. Wasserman: Neural Computing, Theory and Practice. Van Nostrand Reinhold, New York (1989)
6. J.A. Benediktsson, P.H. Swain, O.K. Ersoy: Neural Network Approaches Versus Statistical Methods in Classification of Multisource Remote Sensing Data, IEEE Geoscience and Remote Sensing, vol 28, N. 4 (1990)
7. L.E. Pierce, K. Sarabandi, F.T. Ulaby: Application of an Artificial Neural Network in Canopy Scattering Inversion, Proceedings of IGARSS'92, Vol II, Houston Texas.
8. H. Bishof, W. Schneider, A. Pinz: Multispectral Classification of Landstat-Images Using Neural Networks, IEEE Geoscience and Remote Sensing, vol 30, N. 3 (1992)
9. J. Key, J.A. Maslanik, A.J. Schweiger: Classification of Merged AVHRR and SMMR Artic Data with Neural Networks, PE&RS Vol. 55, N. 9 (1989)
10. E.J.M. Ringot, R. Kowk, J.C. Curlander. S.S. Pang: Automated Multisensor Registration: Requirements and Techniques, PE&RS, Vol. 57, N. 8 (1991)
11. V. Cappellini, F. Butini, A. Chiuderi, S. Fini: Reti Neurali per la Classificazione del Territorio: uno Strumento per la Fusione dei Dati Telerilevati, Proc. of A.I.T. (1992)
12. P. Canuti, G. D'Auria, P. Pampaloni, D. Solimini: MAC-91 on Montespertoli: an experiment for agrohydrology, Proceedings of IGARSS'92, Houston, Texas.
13. T. Kohonen: The self-Organizing map, Proc. IEEE, vol 78, N. 9 (1990)

Tissue characterization by MRI: data segmentation using 3D feature map

S. Vinitski[a], S. Seshagiri[a], F. B. Mohamed[a], G. Frazer[b], A.E. Flanders[a], C.F. Gonzalez[a], D.G. Mitchell[a]

Departments of [a]Radiology and [b]Neurosurgery, Thomas Jefferson University Hospital, 132 South Tenth Street, Philadelphia, Pennsylvania, 19107, U.S.A.

ABSTRACT

The purpose of our study was to design software for tissue segmentation based upon 3D feature maps and to test this software using biological tissues.

This work is based on the algorithm developed by Cline et al. which calculates the probability distribution of each tissue of interest. Two image inputs (e.g., proton density and T2-weighted) are required. We hypothesized that adding a third input image would make the separation of clusters in 3D feature space greater, with resulting improvement in tissue segmentation.

The algorithm describing probability distribution was modified to accommodate the third dimension and the multiple display routine was created. Data sets were obtained from MR images. A 1.5T GE Signa scanner and SUN computer were used.

Results have shown that 2D segmentation images showed variable degrees of tissue segmentation, and that prior to experiments, it was not possible to know which MR images would produce the best results. However, the segmentation image based on 3D feature maps was always either better than or equal to the best of three 2D segmented images. Images appeared crisper and sharper, and the random color points (i.e., statistical noise) were greatly reduced. Further, when the third image was of deliberately low contrast, the resulting 3D segmentation image effectively "filtered out" inferior input. This methodology was successfully applied to tissue characterization in neuro imaging, MRI of atherosclerotic plaque and in the imaging of pulmonary embolism with the use of contrast agents.

In conclusion, the proposed technique produces accurate tissue segmentation and, thus, merits further evaluation and development.

INTRODUCTION

It has been hoped, almost from the birth of MRI, that different magnetic relaxation properties of normal and diseased tissues, namely proton density, T1 spin-lattice relaxation time and T2 spin-spin relaxation times, would enable characterization of these tissues. This hope was based on the fact that contrast in MRI (i.e., difference in signals) is, indeed, derived from the variation in relaxation properties. Numerous studies have investigated the relationship between tissue characterization and magnetic relaxation properties. Unfortunately, the general finding was that although differentiation does exist, the relaxation times of normal and diseased tissues are overlapping (1). This overlapping increases with a field strength. The high field strength MRI at which most clinical scanners operate provides a much better signal-to-noise ratio. Consequently, the ability to discriminate lesions is usually better at high field (2).

Still, work in tissue characterization continues. There are a number of approaches to pursue: the use of different contrast agents; sophisticated pulse sequences, spectroscopy and artificial intelligence. The spectroscopic approach is based on a) multinuclear spectral differences between different tissues, b) rate of chemical exchange, and c) relaxation time of different "isolated" spectral peaks.

The presented work deals with the application of artificial intelligence to tissue characterization in magnetic resonance imaging. Segmentation and three-dimensional (3D) display of magnetic resonance (MR) images of the head are prerequisites for recognition of the multiple tissue surfaces needed for surgical planning and 3D display of normal anatomy and of pathological lesions. Medical diagnosis with MR involves visual classification of tissues with separate groups using both medical experiment and anatomic knowledge. However, a quantitative classification is required before automatic segmentation and the construction of a 3D surface of the anatomical features. The gray level thresholding that is commonly used to extract bone surfaces in CT images does not work well in the more complex MR images of the head because different tissues have overlapping intensity ranges.

The purpose of our study was to design software for tissue segmentation based upon 3D feature maps and to test this software using biological tissues.

METHODS AND MATERIALS

This work was based on the algorithm developed by Cline et al. (3). The underlying assumption is that each tissue in question can be described by the normal Gaussian distribution. Cline et al. used two input images (proton density and T2-weighted) in their work. Unfortunately, many tissues (defined probabalistically) have overlapping MR signal intensities. We hypothesized that adding a third input image would make the separation of the clusters in a 3D feature space greater, with a resulting improvement in tissue segmentation.

If set of n images with different contrasts is used, then multivariate distribution is utilized. This distribution is determined by the samples obtained by the operator, who must identify at least some part of the tissue of interest. We used three-image input, thus, seeded pixel intensities can be seen as clusters in a 3D feature space. After completion of tissue identification by operator, probability distribution was calculated for each tissue (4). Following this, we calculated feature maps of a) three pairs of two images, and b) one set of three images. The resulting map(s) is an array of the most probable specific tissue classification in each point in the sample space. Next, 2D segmentation images were created in which each pixel is described by a color unique to this tissue. Currently, up to twelve colors are used. The resulting set of 2D segmented images (derived from any 2D or 3D feature map) is smoothed by the filter, approximating 3D diffusion equation (3). A connectivity algorithm extracts surfaces of interest (5) and a dividing cube algorithm (6) constructs a surface of selected tissue. Finally, these surfaces can be rendered as desired. The display routine was created to allow simultaneous display of three 2D segmentation images based on a data set containing three or more imaging series. The SUN SparcStation2 was utilized. Data sets were obtained from MR spin-echo, gradient and inversion recovery imaging. A dedicated data acquisition software program was also created and used in some experiments (7). In-vivo studies involved humans and animals. A 1.5T GE Signa scanner was used.

RESULTS

Results have shown that when using three input MR images, the three 2D segmentation images showed variable degrees of tissue segmentation, and prior to experiments, it was not possible to know which pair of images would produce the best results. The 3D feature map creation required intensive computations: 2.5 hours using SparcStation2 with 32MBytes memory and 90MBytes swapping space. However, the segmentation image based on 3D feature maps was always either better than or equal to the best of three 2D segmented images. Images appeared crisper and sharper, and the random color points (i.e., statistical noise) were greatly reduced. Further, when the third image was of deliberately low contrast, the resulting 3D segmentation image effectively "filtered out" inferior input.

Additionally, we applied the algorithm for tissue segmentation based on 3D feature maps for volumetric measurements. To minimize the effect of RF inhomogeneity and to increase the accuracy of measurements, a manually defined volume of interest throughout a stack of slices was created. A number of color voxels, with known dimensions, pertaining to the specific tissue was then integrated within the chosen volume of interest. It is assumed that partial volume effect was taken into account in the previous step, 3D filtering (3). The calculated volume was compared with that measured prior to imaging experiments and post-processing. The volumes in question ranged between 20 and 700 ml and up to six "tissues" were classified simultaneously. The results showed the range of error to be between 6% and 11%. These results indicated a good correlation between the proposed technique for tissue characterization, including its quantification, and measured distribution of tissues within the volume of interest.

This methodology was successfully applied in-vivo to tissue characterization in patients (neuro imaging and MRI of atherosclerotic plaque in carotids) as well as in animal models of brain infarct and pulmonary embolism, PE. Using image post-processing, we were able to identify brain tumor, edema and sclerotic lesions in neuro imaging; artery wall, cap and filling of plaque in patients examined by MRI prior to endarterectomy. In animal experiments, tissue characterization software also clearly identified pulmonary embolism in rabbit as well as experimental infarct areas in cat brain.

DISCUSSION

There are many image segmentation techniques. For example, Katz et al. used a 3D extension of Marr-Hildreth operator for segmentation of the brain from 3D Fourier MRI (9). The neural network algorithm constraint satisation) was used to segment multimodality imaging by Lin et al. (10). Fauros et al. (11) used fuzzy partition for segmentation of scintigraphic images. Hall et a. (12) compared neural network (cascade correlation) with unsupervised fuzzy clustering techniques in segmentation of MR images of the brain.

This work is based on multispectral methods used initially in the analysis of satellite images. While we did not directly compare this approach with that described above, we extended the work of Cline et al. into the 3D feature map. This approach is strongly operator-dependent who must, at least, partially identify tissues of interest. However, we did not observe inconsistencies in the results when we asked observers to "seed" tissues a number of times spread over the long periods. It has been shown by Venturi et al. (13) that a target orient approach gives better results. MRI can be characterized by the inhomogeneity of RF signal at high (>1T) field strength within imaging volumes. Thus, by using volumes of interest, we could provide a series of more accurate sub-images.

In conclusion, the proposed technique provides relatively good tissue segmentation and, thus, merits further evaluation and development. Work in tissue segmentation using a four-dimensional feature map is in progress. This computation-intensive approach should improve tissue separation by post-processing. Misregistration artifacts are addressed by faster data acquisition.

REFERENCES

1 D.G. Mitchell, D.L. Burk Jr, S. Vinitski, et al., AJR, 149 (1987) 831.
2 S. Vinitski, R.H. Griffey, JMRI, 1 (1991) 451.
3 H.E. Cline, W.E. Lorensen, R. Kikinis, F. Jolesz, JCAT, 14 (1990) 1037.
4 M. James. Classification Algorithms. New York: Wiley, 1985, p. 209.
5 H.E. Cline, C.L. Dumoulin, et al., Magn. Reson. Imaging, 5 (1987) 345.
6 H.E. Cline, W.E. Lorensen, S. Ludke, et al., Med. Phys., 15 (1988) 320.
7 S. Vinitski, D. Mitchell, et al., Proc. IEEE Eng. Biol. Med., 12 (1990) 82.
8 S. Vinitski, D.G. Mitchell, S.G. Einstein, et al., JMRI, 1993 (in press).
9 W.T. Katz, M. Merickel, Proc. IEEE Biol. Med. Eng., 14 (1992) 1920.
10 Lin W, Tsao E, Chen C, Feng Y. Neural networks for medical image segmentation. Northwestern Univ., Electrical Eng. and Computer Sci.
11 P. Faurous, J. Fillard, Proc. IEEE Biol. Med. Eng., 14 (1992) 1922.
12 L.O. Hall, A. Bensaid, IEEE Trans. Neural Network, 3 (1992) 672.
13 G. Venturi, P. Capitani, Proc. IEEE Biol. Med. Eng., 14 (1992) 1928.

Image Processing: Theory and Applications
G. Vernazza, A.N. Venetsanopoulos, C. Braccini (Editors)
© 1993 Elsevier Science Publishers B.V. All rights reserved.

Mammographic Image Analysis with Tree Structured Order Statistic and Wavelet Filters

Laurance P.Clarke, Wei Qian, Huaidong Li, Robert Clark, Edward Saff and Brad Lucier

University of South Florida, Center for Engineering and Medical Image Analysis(CEMIA), College of Medicine, Engineering, Art and Science, and H.Lee Moffitt Cancer Center and Research Institute, MDC Box 17, Tampa, Florida,33612

Abstract

This paper proposes a new class of hybrid filters called tree structured order statistic and wavelet filters and discusses the applications to microcalcification segmentation in mammography. This method was evaluated using fifteen representative digitized mammograms where similar sensitivity (true positive (TP) detection rate 100%) and specificity (0.1 average false positive (FP) MCC's/image) was observed but with varying degrees of detail preservation important for characterization of MCC's.

1. INTRODUCTION

The papers [1-9] reported on the development of several novel automatic methods for both image enhancement and feature extraction with the initial application directed at MCC's detection. The methods proposed, as outlined in Table 1, include the use of: (a) tree structured central weighted median filters (TSF's) that preserve image details with improved noise suppression, compared to single filter design [1]; the parameters fixed based on initial training image set and (b) the use of adaptive order statistic filters (AOSF) designed for better detail preservation across a larger image set [8]. The tree-structured order statistic filters are the generalization of tree structured central weighted median filters. The central weighted median filter is a special case of smoothing order statistic filter. The tree structured filter proposed in this paper for image enhancement that uses order statistic filters instead of central weighted median filters used in [4, 5] as basic sub-filtering blocks. The feature extraction methods include the use of: (a) two channel tree structured wavelet transform (TSWT) where the spatial frequencies and related orientations in the image can be determined automatically in the transform processes [3] and (b) quadrature mirror filter (QMF) bank as an alternative tool for feature extraction where the parameter coefficients allow three channel approaches for improved detail preservation [7]. Both feature extraction methods include multiresolution *decomposition* and *reconstruction* algorithms reported for the first time, that can be specifically tailored to extract specific features within a specific frequency range, such as those observed for MCC's [3]. A preliminary evaluation of the method proposed in this paper and the methods, as outlined in Table 1, was applied to (a) simulated phantom images containing linear and circular structures with varying noise level and (b) fifteen digitized mammograms of varying breast density containing biopsy proven malignant MCC's to estimate the relative sensitivity and specificity of detection for each method with specific emphasis on detail preservation for accurate characterization of MCC's. This paper proposes a class of hybrid filters called tree structured order statistic and wavelet filters for microcalcification segmentation in mammography. The class of order statistic filters is very efficient for noise suppression with adaptive flexibilities because (a) it offers a compromise in performance between linear filters and nonlinear filters and (b) it is possible to design an optimal (among OS filters) mean-square-error (MSE) filter for estimating the signal immersed in mixed noise.

	Filter	Feature Extraction
Method 1	TSF	TSWT 2 channel
Method 2	TSF	QMF 3 channel
Method 3	AOSF	TSWT 2 channel

Table 1: Summary of three methods used for MCC detection using multiresolution decomposition and reconstruction.

2. FILTER DESIGN

A filter is called wavelet filter if the filter operation consists of computing the correlation of the input signals with each of the rows of a wavelet matrix. Harr matrices, finite Fourier transform matrix, Hadamard matrix and Hadamard-Wash transform matrix are examples of wavelet matrices. A wavelet filter split the input image into four subimages. One of the four subimages can be thought of having a coarser scale corresponding to lower frequency part of the image than the input image. The other three can be thought of having finer scale. One gives the vertical high frequencies and horizontal low frequencies, the second the horizontal high frequencies and vertical low frequencies, and the third the high frequencies in both horizontal and vertical directions.

The standard wavelet transform decomposes subimages with a pyramidal algorithm in the low frequency channels[3,4]. The transform is suitable for the signals consisting primarily of smooth components so their information is concentrated in the low frequency regions. Since the most significant information of the microcalcifications often appears in the high frequency and middle frequency channels, the standard wavelet transform does not apply properly. Thus, an appropriate way to perform the wavelet transform for microcalcification analysis is to detect the significant frequency channels and then to decompose them further. Continuation of the procedure yields a tree of wavelet transform called tree-structure wavelet filter, which can perform the microcalcification segmentation and classification.

In the practical applications of digital mammography, some microcalcifications are faint, background structures are complex and noise is mixed. So, an efficient enhancement filter for noise suppression, feature preservation is incorporated to the tree structured wavelet filter. The hybrid combination is similar to the hybrid filters published in [9]. The requirements for the enhancement in digital mammography include the improvement of image contrast, enhancement of the fine details of parenchymal tissue structures, suspicious areas and microcalcification clusters, while optimally suppressing noise. In this paper, we proposed a new class of tree-structured order statistic filters that has more robust characteristics for noise suppression and detail preservation for mammographic image processing. The central weighted median filter is a special case of smoothing order statistic filter. The tree structured filter proposed in this paper for image enhancement that uses order statistic filter instead of central weighted median filters used in [4, 5] as basic sub-filtering blocks. The use of linear and curved filter windows was also explored for the tree structured filter to enhance the features.

The properties of this filter was compared to a single filters that include a median filter, a central weighted median filter and a central weighted median filter with adaptive windows using both visual criteria and simulated noise criteria. In all instances the tree structured filter demonstrated better detail preservation and noise suppression.

3. EXPERIMENTAL RESULTS
3.1. Simulated calcifications

A simulated image was generated containing three sets of specks similar in dimensions and geometry to those present in the American College of Radiology (ACR) phantom [1-3] Three sets of six circular specks arranged in the vertices and center of a pentagon are generated with diameters of 540 µm, 320 µm, and 200 µm. Two noise levels (level 1 and 2) were added to the simulated data corresponding to average (1.5%) and maximum (3.5%) optical density fluctuations measured on mammographic films. A Gaussian multiplicative probability distribution function was used to model the added noise; this is a good approximation of the actual Poison, signal dependent noise processes encountered in mammography. A 10% signal-to-background level was considered as determined from existing digitized films. A detailed description of this simulated image is given elsewhere [1-3].

A comparison of the performance of the method (Table 1) was compared with and without the image enhancement preprocessing step, to illustrate the importance of noise suppression. For the noiseless case a TP detection rate of 100% (18/18) was observed in each case with FP detection of zero. As noise was added the TP detection remained at 100% but a significant FP detection was observed when preprocessing step was removed due to the presence of extensive noise throughout the image. The cascaded use of the image enhancement and multiresolution decomposition and reconstruction resulted in 100% sensitivity for all specks with no FP detection. The method in this paper and the methods in Table 1 varied slightly in relation to the shape of the individual microcalcification for each size.

The simulation studies clearly demonstrate that the use of wavelet approaches alone are not successful for mammographic images with varying noise content

3.2. Digital Mammogram Analysis

A total of fifteen mammograms were digitized at a resolution of 105 µm/pixel, 12 bits deep (4096 gray levels) by a DuPont NDT Scan II digitizer (DuPont, Wilmington, Delaware). The mammograms were selected to contain at least one cluster of microcalcifications located in regions of relatively high parenchymal density. From the 15 images, 8 cases had a parenchymal density greater than 50% and 5 had density of less than 25%. The malignancy of the MCCs was biopsy proven in all cases and the location was identified on the digitized images and film by expert radiologists. Evaluation of the microcalcifications was done visually and included a comparison of the segmentation results with the original film and corresponding digitized mammograms in terms of the number, shape and size of TP MCCs as well as number of FPs. For the fifteen images analyzed, the sensitivity of detection was 100% (TP detection rate), with a similar average false positive (FP) detection rate only of the order of 0.1 MCC's/image However, the methods demonstrated varying levels of detail preservation by visual observation.

4. CONCLUSION
The sensitivity and specificity of detection for MCC's using this preliminary image data set

was significantly improved compared to traditional filtering and feature extraction methods, with the added advantage of operator independence in the CAD procedures. The success of the proposed methods are primarily due to the combined use of both advanced filtering methods for noise suppression and multiresolution approaches for feature extraction. These improvements are critically required for the practical realization of CAD methods in a realistic cancer screening setting where important features need to be fully characterized to reduce the unacceptable high false positive (FP) detection rates currently reported for MCC or suspicious tumor areas [1].

REFERENCES

[1]. CLARKE L.P., QIAN W., KALLERGI M. AND *et al*, Nonlinear filtering techniques for improved classification of mammographic parenchymal patterns in cancer screening. *Proc. of the 34th Annual Meeting of AAPM*,Calgary, Alberta, Canada; August 23-27, 1992.

[2]. CLARKE L.P., QIAN W., KALLERGI M., and CLARK R.A.,Computer assisted diagnosis (CAD) in mammography. *Proc. of the Symposium for Computer Assisted Radiology (SCAR)*, pp. 116-121, Baltimore, Maryland; June 14-17 1992.

[3]. QIAN W.,, CLARKE L.P., LI H.D., KALLERGI M. CLARK R.A., and SILBIGER M.L., Tree-structured nonlinear filters and Wavelet Transform for Microcalcification Segmentation in Mammography. Proceedings of the SPIE/IS&T Symposium on Electronic Imaging Science and Technology, San Jose , CA, Jan. 31-Feb. 5, 1993.

[4]. QIAN W., CLARKE L.P., KALLERGI M., and CLARK R.A., Digital mammographic screening: detail preserving tree structured nonlinear filters. *IEEE Trans. Med. Imag.* (submitted for publication) 1992.

[5]. QIAN W.,KALLERGI M.,CLARKE L.P.,LI H.D.,CLARK R.A.,and SILBIGER M.L.,Tree-structured filter and wavelet tranform for microcalcification segmentation in digital mammography. *Med. Phys.* (submitted for publication).

[6]. QIAN W., KALLERGI M., CLARKE L.P., LI H.D., CLARK R.A., and SILBIGER M.L., Digital Mammography: Tree-Structured Wavelet Decomposition and Reconstruction for Feature Extraction. *International Journal of Pattern Recognition Techniques and Artificial Intelligence* (Submitted for publication)

[7]. QIAN W.,, CLARKE L.P., LI H.D., KALLERGI M. CLARK R.A., and SILBIGER M.L., Application of M-Channel Quadrature Mirror Filters for Microcalcification Extraction in Digital Mammography *Computerized Medical Imaging and Graphics* (Submitted for publication)

[8]. QIAN W., CLARKE L.P., LI H.D.,KALLERGI M., CLARK R.A., and SILBIGER M.L., Digital Mammography: Adaptive Order-statistic Filtering and Wavelet Transform for Image Analysis. *IEEE Trans. Med. Imag.* (submitted for publication) 1993.

[9]. QIAN W.,KALLERGI M. and CLARKE L.P. Order Statistic-Neural Network Hybrid Filters for Gamma Camera-Bremsstrahlung Image Restoration. *IEEE Trans. Med. Imag.* March, 1993

Bistatic radar imaging in laboratory

J.Bertrand*, P.Bertrand** and M.Roisin**

(*) LPTM, University Paris VII, 75251 Paris Cedex 05, France

(**) ONERA/DES, BP 72, 92322 Chatillon, France

Abstract

A continuous wavelet formulation of bistatic radar imaging is proposed. The merit of this approach is to put the uncertainty relations under control. This point is of special importance when studying the dependence of the images on the frequency and on the direction of observation. Indications are given for the implementation of the technique by using the fast Mellin transform.

1. INTRODUCTION

The object of radar imaging is to get information on the position of reflecting points of a target by processing the values of its complex backscattering coefficient. In the present case the basic observations are performed in an anechoic chamber, as shown on Fig.1, for various positions of the transmitting and receiving antennas and for various frequencies. Problems of calibration of the apparatus are supposed solved so that the backscattering coefficient is available. From a theoretical standpoint, this coefficient is precisely defined by:

$$H(f, \theta_i, \theta_r) = \lim_{R \to \infty} \sqrt{4\pi R^2} \, \frac{E_r(R, \theta_r)}{E_i(R, \theta_i)} \, e^{4i\pi Rf/c} \tag{1}$$

where E_i is a plane wave incident field, E_r is a spherical wave reflected field and the phase origin has been chosen at the origin of coordinates.

The radar image is defined as a theoretical object computed from the backscattering coefficient. Axiomatically, it can be introduced as a distribution of small independent reflectors with definite positions, working frequencies and orientations. The definition will be satisfactory if the correspondence between the backscattering coefficient and the image is covariant with respect to a change of reference frame, allowing different observers to get a consistent idea of the target.

A solution to this problem has been proposed in the monostatic case [1] using a continuous wavelet analysis. Here we will show how the same type of approach can work in bistatic situations. The procedure will be sketched in section 2 for the two special cases of constant bistatic angle and constant incident angle. Section 3 is devoted to the presentation of an original method of image computation using a fast Mellin transform.

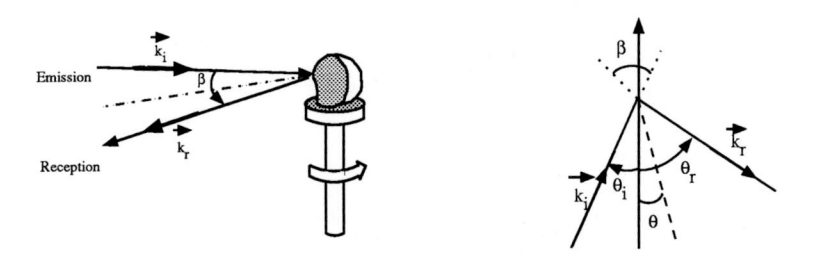

Figure 1. Experimental setting

2. GENERATING IMAGES BY WAVELET ANALYSIS

The transformation law of the backscattering coefficient H in a change of reference frame characterized by a dilation a, a translation $\mathbf{b}$ and a rotation R_ϕ) can be derived. The introduction of the bistatic angle β and the mean angle θ (cf. Fig.1) allows to write this law as:

$$H(k,\theta;\beta) \longrightarrow H'(k,\theta;\beta) = a\, e^{-2i\pi k|\cos(\beta/2)|\mathbf{u}.\mathbf{b}}\, H(ak,\theta - \phi;\beta) \qquad (2)$$

where $\mathbf{k} = k\mathbf{u}$ and $\mathbf{u}$ is the unit vector $\mathbf{u} = (-\sin\theta, \cos\theta)$. The factor a in front of (2) ensures that the radar cross-section $\mid H \mid^2$ has the dimension of a surface. Two different cases will now be distinguished depending upon whether the bistatic or the incidence angle is constant.

Constant bistatic angle

In that case, the experimental setting is invariant by the whole similarity group and the image function $I(\mathbf{x}, k, \theta; \beta)$ is required to transform pointwise according to:

$$I(\mathbf{x},k,\theta;\beta) \longrightarrow I' = I(a^{-1}R_{-\phi}(\mathbf{x}-\mathbf{b}), ak, \theta - \phi;\beta) \qquad (3)$$

when H undergoes the change defined by (2).

The consistency of the two transformations (2) and (3) introduces a constraint on the expression of I in terms of H. As in the monostatic case [1], this constraint suggests the use of a wavelet analysis. In the present case, we choose a basic wavelet $\Phi_0(\mathbf{k};\beta)$ that is supposed to represent the backscattering coefficient of a target located around $\mathbf{x} = 0$, reflecting mainly at a frequency corresponding to $k \equiv 2f/c = 1$, in direction $\theta_r = \beta/2$ when illuminated at the same frequency in a direction $\theta_i = -\beta/2$ and we associate this wavelet with the image point $(\mathbf{x} = 0, k = 1, \theta = 0)$. In a change of reference system labelled by $(a, \mathbf{b}, \phi)$, function Φ_0 is transformed according to (2) into a wavelet $\Phi_{\mathbf{x},k}$ attached to the point $(\mathbf{x} = \mathbf{b}, k = a^{-1}, \theta = \phi)$. The set $(\Phi_{\mathbf{x},k})$ thus defined is an overcomplete family of functions that is the same for all observers. The wavelet coefficient is then defined by:

$$C(\mathbf{x},\mathbf{k};\beta) = \int_{R^2} H(k',\theta';\beta)\, e^{2i\pi|\cos(\beta/2)|\mathbf{k}'\cdot\mathbf{x}}\, \Phi_0^\star(k'/k, \theta'-\theta;\beta)\frac{1}{k}\, k'dk'd\theta' \qquad (4)$$

It satisfies the so-called isometry relation which reads:

$$\int_{R^2\times R^2} |C(\mathbf{x},\mathbf{k};\beta)|^2\, d\mathbf{k}d\mathbf{x} = \frac{\kappa}{\cos^2(\beta/2)}\int_{R^2} |H(\mathbf{k};\beta)|^2\, d\mathbf{k} \qquad (5)$$

where

$$\kappa(\beta) = 2\int_{R^2} |\Phi_0(\mathbf{k};\beta)|^2\, \frac{d\mathbf{k}}{k^2} \qquad (6)$$

If Φ is such that the admissibility condition $\kappa < \infty$ is satisfied, a reconstruction formula holds. In that case, the image I is defined as:

$$I(\mathbf{x},\mathbf{k};\beta) = \kappa^{-1}\cos^2(\beta/2)\,|C(\mathbf{x},\mathbf{k};\beta)|^2 \qquad (7)$$

According to (5) it can be interpreted as a probability density on space $(\mathbf{x},\mathbf{k})$.

Constant incident angle

If an analysis of the data is performed at a given angle of illumination θ_i but at different bistatic angles β, the invariance group is restricted to dilations a and translations $\mathbf{b}$. Its action on the backscattering coefficient $H(k,\beta;\theta_i)$ can be inferred from relation (2). The image function $I(\mathbf{x},k,\beta;\theta_i)$ is again required to transform pointwise. In the present case, the group is too small to motivate a conventional wavelet analysis. Nevertheless, for each bistatic angle $\beta = \beta_0$, it is possible to construct a family of wavelets labelled by the points $\mathbf{x},k$. To this end a function $\Phi(k,\beta;\beta_O)$ associated with the image point $(\mathbf{x}=0,k=1)$ has to be chosen. A transformation $(a,\mathbf{b})$ then moves that point to $(\mathbf{x}=\mathbf{b}, k=a^{-1})$ and the function Φ to $\Phi_{\mathbf{x},k}$. Next we put a relation between the basic wavelets for different values of the bistatic angle β_0 by assuming that:

$$\Phi(k,\beta;0) \equiv \Phi_0(k,\beta) \qquad \Phi(k,\beta;\beta_0) = \Phi_0(k,\beta-\beta_0) \qquad (8)$$

A modified wavelet coefficient is then defined by:

$$C(\mathbf{x},k,\beta) = \int_{R^2} H(k',\beta')\frac{1}{k}e^{2i\pi k'|\cos(\beta'/2)|\mathbf{v}\cdot\mathbf{x}}\Phi_0\left(\frac{k'}{k},\beta'-\beta\right)|\cos(\beta'/2)|\,k'dk'd\beta' \qquad (9)$$

where $\mathbf{v}$ is the unit vector $\mathbf{v} = (-\sin(\theta_i+\frac{\beta}{2}), \cos(\theta_i+\frac{\beta}{2}))$. The extra cosine factor has been included in order to obtain an isometry formula given by:

$$\int_{R^2\times R^2} |C(\mathbf{x},k,\beta)|^2\, d\mathbf{x}dkd\beta = \kappa\int_{R^2} |H(k,\beta)|^2\, kdkd\beta \qquad (10)$$

where $\kappa = 2\int_{R^2} |\Phi_0(k,\beta)|^2\,\frac{dk}{k} < \infty$. With this definition a reconstruction formula for H can be written in terms of the coefficients C and it is reasonable to define the image $I(\mathbf{x},k,\beta)$ as $1/\kappa$ times the squared modulus of the wavelet coefficient (9). An interesting property of this definition is that it reduces to a known formula [2] when the wavelet Φ_0 is chosen equal to the characteristic function of the interval $([1-\Delta k/2, 1+\Delta k/2], [-\Delta\beta/2, \Delta\beta/2])$ thus yielding an $\mathbf{x}$-image for $k=1, \beta=0$.

3. PRACTICAL ASPECTS OF THE PROCEDURE

For the practical application of the above scheme an efficient algorithm will now be sketched and the choice of the generic wavelet will be briefly discussed. Both cases will be treated simultaneously and the backscattering coefficient will be simply denoted by $H(k, \psi)$ where ψ stands for either θ (case β =constant) or β (case θ_i = constant).

The acquisition of the data $H(k, \psi)$ is usually performed in polar coordinates and the computation of the wavelet coefficients using formulas (4) and (9) involve dilations on function Φ. To avoid polar reformatting and resampling, we will work directly in polar coordinates and use a Mellin transform to process the dilations. More precisely, a Mellin transform with respect to k is performed on H and yields:

$$\mathcal{M}[H](\gamma, \psi) = \int_0^\infty H(k, \psi) \, k^{2i\pi\gamma} \, dk \tag{11}$$

This formula can be inverted and a Parseval relation holds. Furthermore the operation $\mathcal{M}$ transforms any dilation on the argument k of H into multiplication by a phase. Such properties lead to new expressions of the wavelet coefficients in terms of $\mathcal{M}[H](\gamma, \psi)$ that can be computed by FFT's. The main tool is the discretized version of (11) [3]. It expresses the relation between N geometrical samples of $H(k, \psi)$ on the interval $[k_{min}, k_{max}]$ and N regularly spaced samples of its Mellin transform according to:

$$\mathcal{M}[H](m/\ln Q, \psi) = \sum_{n=P}^{P+N-1} q^{n(r+1)} H(k_{min} q^n, \psi) \, e^{2i\pi nm/N} \tag{12}$$

where $q = (k_{max}/k_{min})^{1/N}$ and $Q = q^n$.

The condition to avoid aliasing is found to be:

$$N > \frac{2L}{c} k_{max} \ln(k_{max}/k_{min}) \tag{13}$$

where L is the maximal dimension of the target. The samples in ψ are chosen regularly spaced and the treatment of the ψ- dependence does not introduce any new feature.

A good wavelet is highly concentrated both in **x** and **k**-space so as to make the image as sharp as possible. We choose the product of a minimal Klauder wavelet in k-space by a gaussian in ψ:

$$\Phi_0(k, \psi) = k^{2\pi\lambda} \, e^{-2\pi\lambda k} \, e^{-\psi^2/2\sigma^2} \tag{14}$$

where the parameters λ and σ control the spread.

The above algorithm uses only FFT's and can be seen to be very efficient. Moreover it allows a scanning of the image frequency and directivity parameters by performing a simultaneous computation of images corresponding to various values of these parameters. This is in sharp contrast with usual treatments which compute one image at a time.

REFERENCES

[1] J.Bertrand, P.Bertrand and J.P.Ovarlez, "Dimensionalized wavelet transform with application to radar imaging", Proc. IEEE ICASSP-1991.

[2] H.-J Li, F.-L Lin, Y.Shen and N.H.Farhat, "A generalized interpretation and prediction in microwave imaging involving frequency and angular diversity", J. Electromagnetic Waves Appl. **4**,pp.415-430, 1990.

[3] J.Bertrand, P.Bertrand and J.P.Ovarlez, " Discrete Mellin transform for signal analysis" Proc. IEEE ICASSP-90, pp.1603-1606, 1990.

Knowledge-Map Based Automatic Inspection and Change Detection in Complex Heterogeneous Environments

M.Barni[1], V.Cappellini[1], M.Mattioli[1], A.Mecocci[2], S.Zipoli[1]

[1] Department of Electronic Engineering, University of Florence, via S.Marta n.3, 50139 FIRENZE

[2] Department of Electronic Engineering, University of Pavia, via Abbiategrasso n.209, 27100 PAVIA

Abstract

This work deals with change detection in non-structured environments. A robot, always covering the same route, at fixed places takes a shoot of the scene it sees. The aim is to detect changes between two subsequent shoots of the same scene. A knowledge-map based architecture is presented which is capable of inspecting complex real environments by changing its processing tools and detection strategy in a space dependent way, in order to maximize the overall change detection capability.

1. INTRODUCTION

Automatic inspection of general environments to detect changes due to anomalies, intruders, or other undesirable causes, is an important task in many practical applications. In general the real world scenes that must be inspected are complex and comprise heterogeneous objects and entities. Typically the inspection task consists in verifying the presence of some entities, their position in the scene, their status, and the eventual appearance of new entities. In general each entity requires specific tools and processing methodologies in order to be reliably analyzed and validated.

The system described in this paper is capable of inspecting complex real environments by changing its processing tools in a space dependent way. The system has been designed to work with a roving robot which inspects its environments by following some predetermined paths. During its functioning the robot stops at a certain number of fixed locations which must be monitored and takes a shoot of the current scene, every time more or less from the same viewpoint and viewing direction. The scene so obtained is analyzed to establish if all the interesting objects and regions are in their *safe conditions*; such conditions are defined during a preliminary training phase.

Due to unavoidable positioning errors during subsequent travels along the same path, the line-of-sight of the robot sensor changes among different shoots of the same place taken at different instant of time, so the images are not registered among them. This fact and the huge amount of memory that must be used to store an entire set of reference images, prevent the use of change detection schemes based on the subtraction of previously stored reference frames from the current one [1]. Moreover, lightning conditions must be accurately compensated to use subtraction-based schemes and this can be quite a difficult task. The problem is complicated by the fact that different kind of processing tools and different fusion/integration strategies must be employed at different locations of the same scene. In other words the processing algorithms which are employed must generally change from place to place in the scene according to the kind of inspection and to the type of objects to be inspected.

In the system we propose the knowledge about the various entities to be inspected is introduced by means of both procedural and declarative representations. Such an aim is accomplished by means of a knowledge map (KM) associated to each scene [2]. In this way a synthetic description of the scenes is obtained (declarative representation of knowledge) along with adaptivity of the inspecting tools, which are chosen according to the characteristics of the entity to be inspected (procedural representation of knowledge). Finally the occurrence of a relevant change is not considered as an on/off event. On the contrary, when a change is detected a degree of reliability is associated to it. Such an aim is accomplished by resorting to the fuzzy theory [3].

2. SYSTEM DESCRIPTION

Fig.1 shows the architecture of the change detection chain we have implemented. In a first phase (supervised training) a KM is built for each scene to be inspected. The Kms give a description of all the objects that must be controlled and of the different processing algorithms that must be applied to each of them. Such maps are composed by logic objects which describe real world objects or homogeneous areas of interest. In the actual system two different kinds of logic object are allowed, namely *structures* and *Regions Of Interest* (ROIs). The structures are collections of line segments and their mutual spatial relations. They are used to define *skeletons* relative to some parts of the scene. For example, windows, doors, fixed architectural elements, are suitably described in terms of their more pronounced linear edges which can be logically grouped together to form a structure. If these structures are properly chosen, they can be detected even if the robot line-of-sight changes (of course not in a dramatic way). The structures are also used to evaluate image shifts caused by the robot repositioning errors.

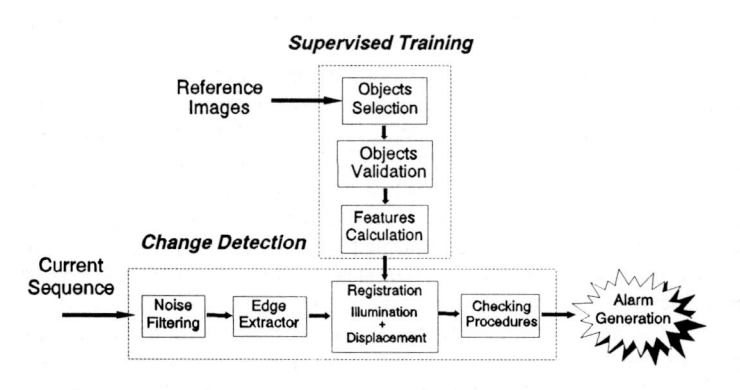

Fig.1 Overall system architecture.

The ROIs are defined by means of polygonal areas that enclose the objects to be monitored. In fact, objects are better analyzed by means of spatial features instead of by means of edges only. These features are extracted from the RGB levels of the images taken by the robot during its inspection travel, and are evaluated only inside the polygonal area that is associated to the ROI (see below in the text). A computation saving is obtained because only the regions of interest are analyzed and not the whole image.

To define the KM for a scene, semantic understanding of the scene itself is needed. Such an understanding, in the general case, can be obtained only with the aid of a human operator. It is such an operator, in fact, which decides what kind of modifications are of interest at the different places along each inspection path, and which are the objects that should be monitored in order to properly perform the inspection. During the supervised training phase, the robot takes some shoots of all the places that must be inspected. At this point the operator manually selects the objects of interest for each scene by means of a suitable graphic editor. The operator starts by selecting the appropriate logic object to be defined (i.e. a structure or a ROI). In order to define a structure the operator selects some linear edges, that will form the structure, by superimposing to the image some line segments drawn by hand. Once all the line segments have been drawn, the graphic editor validates the proposed structure. This is done by matching the segments given by hand with the edges extracted from the scene. Low confidence matches are signaled by highlighting the corresponding line segments. In the case of a ROI the operator draws polygonal closed contours which approximate the boundaries of the objects or areas to be analyzed and monitored. Once the contours have been defined, some feature-extraction tools are associated to them. The choice of the features is highly dependent on the kind of object. In our system a pool of 72 different low-level feature extraction algorithms has been developed, which warrants good change detection capabilities as well as an high noise immunity in different environments and operative conditions. Such a set of features includes: mean and variance of luminance, saturation and hue; texture classification based on the Laws matrices; uniformity, contrast, entropy, periodicity and energy parameters based on run lengths; texture and uniformity parameters based on cooccorrence matrices. Among them the most effective ones for the class of images we have used in our tests have proved to be the mean and variance of luminance and hue, since they couple high sensitivity and robustness against lightning changes and noise.

Once the supervised training has been performed the change detection chain is activated. Every time the robot takes a new shoot of a scene at one of the inspection sites the new image is filtered, then the edges contained in it are extracted and finally the image is registered from the point of view of both spatial displacement and lightning conditions. In order to perform the spatial registration the KM associated to the current scene is recovered. Then a particular structure, which has been edited for registration purposes (for example, the skeleton of architectural elements which are supposed not to move between two subsequent shoots) is matched against the edge map obtained in the previous step. Once the corresponding segments have been found, the displacement parameters for registration are computed and the coordinates of all the entities in the scene, are displaced in their proper positions according to the parameters found during the structure matching phase. The registration of the lightning conditions is performed by making the mean and the variance of the histogram relative to the luminance of the current image equal to those of the corresponding reference scene. Such an aim is achieved by means of the formula

$$V' = \frac{\sigma_c}{\sigma_r}(V - \mu_c) + \mu_r$$

where V' is the current grey level, V the transformed grey level, σ_c, σ_r, μ_c, μ_r the variance and the mean of the current and reference image respectively. At this point for each entity of the KM the corresponding checking procedure is activated. These procedures check if all the objects are in their safe conditions and if some irregularities are detected an alarm signal is generated. Finally it must be mentioned that the alarm generations is fuzzy in nature, so sophisticated decisions can be taken also in presence of uncertainties.

3. EXPERIMENTAL RESULTS

The whole change detection system has been tested on a great range of images. For all of them the whole process has been simulated. For all the images we have considered an alarm signal near or equal to the maximum has been generated when a change was present and no false alarm was given. Besides the position and lighting registration have always been correctly performed when the positioning errors were such that the structures used in such a phase did not disappear from the field of view of the robot sensor, and the lighting change was inferior to a camera stop (a camera stop halves the luminance of the scene). For higher values the registration have not always been achieved correctly. To achieve a good degree of robustness structures with only few short segments should be avoided as well as very little regions. From the point of view of computation time the system has proved to be very efficient. The time needed to rise an alarm is indeed very low. Of course, the speed of the algorithm depends on the number of objects in the scene, on their dimensions (regions) and on the number of segments contained in the structures. By implementing the whole chain on a DecStation 5000/240 (40 Mips), in normal cases (less then 6 ÷ 7 objects for each image) the computation time is less than 20 seconds.

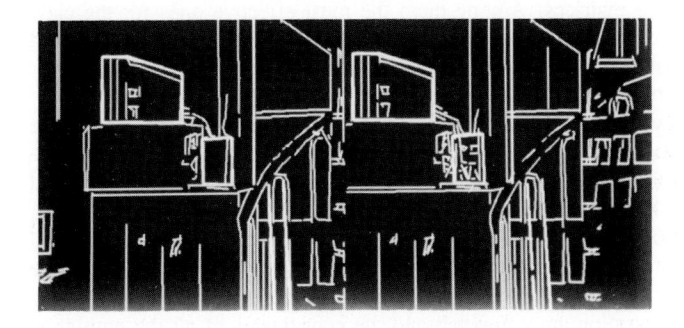

Fig.2 Example of change detection. The image on the left is the reference one. In the current image (on the right) an object is missing (small rectangle). The change is revealed since inside the rectangle the feature values are significantly changed.

4. REFERENCES

1   H.H. Nagel, Overview on Image Sequence Analysis, in Image Sequence Processing and Dynamic Scene Analysis, TS. Huang, Ed., NATO ASI Ser., Vol. F2. Berlin Springer, 1983.
2   Y. Shirai, Tridimensional Computer Vision, Springer Verlag, 1988 Berlin.
3   S.K. Pal, P.K. Dutta Majumder, Fuzzy Mathematical Approach to Pattern Recognition, Indian Statistical Institute, 1986 Calcutta.

A STRUCTURAL APPROACH IN STEREO VISION

C.Di Ruberto, N.Di Ruocco, S.Vitulano

Dipartimento di Informatica ed Applicazioni - Facoltà di Scienze
Università degli Studi di Salerno - Salerno (Italy)
tel. 089/822356, Fax 089/822272, E-mail SV@UDSAB.DIA.UNISA.IT

Abstract - From the geometric point of view, when certain conditions occour, it is possible to obtain the tridimensional reconstruction of a scene starting from two bidimensional images (left and right) of the same scene. In literature the methods used in stereo vision are based on local approaches (area-based). In these methods it is easy to find some mistakes chiefly in the matching phase where corresponding points are searched in left and right images. On the basis of such observations, we have developed a structural method according to it the points of the image are classified in objects. Our structural method adopts as similarity rule the relations among all the points in order to obtain a matching more and more careful and at the same time to obtain a whole interpretation of the scene. Our algorithm is divided in three phases: 1) segmentation; 2) features extraction; 3) matching phase.

1. STEREO GEOMETRY

A tipical system for the stereo vision is composed by two cameras t1 and t2 with the same focal lenght f placed at points O1 and O2. The distance between O1 and O2 is d and the angle between the two optic axes is 2θ. We introduce for the t1 and t2 cameras respectively two coordinate systems (X1,Y1,Z1) and (X2,Y2,Z2). We define a third coordinate system (X,Y,Z) so that the Z axis bisects the angle between the Z1 and Z2 axes.
The relations between these coordinate systems are the following:

$$\begin{vmatrix} X1 \\ Y1 \\ Z1 \end{vmatrix} = \begin{vmatrix} \cos\theta & 0 & \sen\theta \\ 0 & 1 & 0 \\ -\sen\theta & 0 & \cos\theta \end{vmatrix} \cdot \begin{vmatrix} x+d/2-f*\sen\theta \\ Y \\ Z \end{vmatrix}$$

$$\begin{vmatrix} X2 \\ Y2 \\ Z2 \end{vmatrix} = \begin{vmatrix} \cos\theta & 0 & -\sen\theta \\ 0 & 1 & 0 \\ \sen\theta & 0 & \cos\theta \end{vmatrix} \cdot \begin{vmatrix} x-d/2+f*\sen\theta \\ Y \\ Z \end{vmatrix}$$

Let us suppose that a point P(x,y,z) is observed by two cameras respectively at (x1,y1) on the image plane of t1 and at (x2,y2) on the image plane of t2. The transformation equations are the following:

$$x_i = f*X_i/(f-Z_i), \qquad y_i = f*Y_i/(f-Z_i) \qquad \text{per } i=1,2. \quad (*)$$

If P1(x1,y1) is a point on the image plane of t1 and P2(x2,y2) is the point corresponding on the image plane of t2, then the point P(x,y,z) in the coordinate system (X,Y,Z) is obtained from the former equations. Our resolvent system has ten equations and nine variables. This reflects the fact that the point corresponding to the point P1(x1,y1) on the image plane of t2 can lie on a straight line (epipolar line). If the angle θ is 0, that is the image planes are coplanar (Z1=Z2=Z), the epipolar straight line on the image plane of t2 is an horizontal line at same height of the point p1(x1,y1) on the image plane of t1. The depth (f-z) is given as :

$$f-z = f*d/(x1-x2).$$

Once z is determined the coordinates x and y are easily obtained from the equations (*).

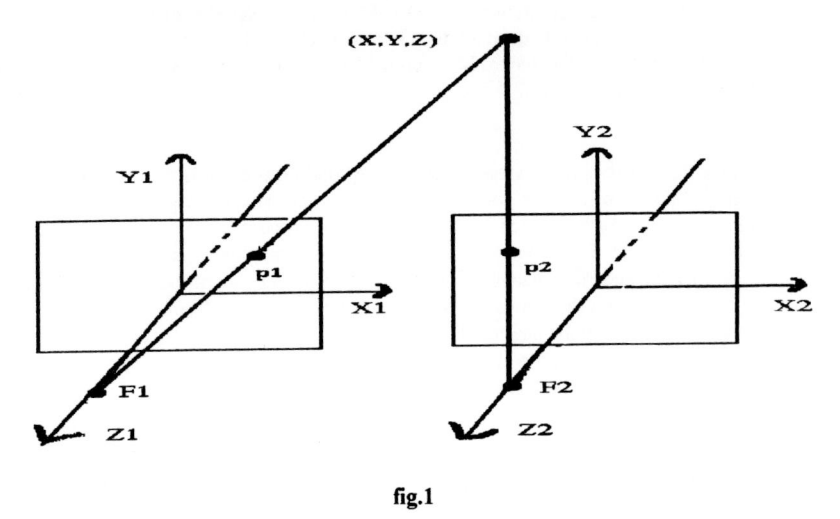

fig.1

From the former equations, we remark that the depth of a point is inversely proportional to the quantity (x1-x2), called stereoscopic disparity.

2. SEGMENTATION

This module is able to recognize the different structures in the picture. This module has been realized taking some theoretical considerations as a starting point [4-7]. The first is the theory of "Gestalt" that provides us with a set of rules, that are suitable to understand the perceptive act more deeply. The second theoretical consideration is given by V. Braitenberg in his book "The intelligent tissues" in which he asserts that in every perceptive act the eye can transmit only about ten bits among the million bits contained in a television image. We start from these considerations so our module tends to simulate the perceptive process. In the simulation process we resort to a hierarchic representation, the quadtree, in which each perceptive act is associated with every level .

3. FEATURES EXTRACTION

The features are represented by the contours of the regions determined in the segmentation phase. The contours of the regions are extracted from the left and the right image. The contours are classified on the basis of the area contained in them and on the basis of the minimum rectangle limiting the region. Moreover the contours are approximated by poligonals in order to solve the problems connected to the noise that producing the extraction of very irregular contours, and in order to obtain a very easy representation of the information (the shape), that is able to reduce the complexity of the matching algorithm.

4. MATCHING PHASE

Our method is not an area-based approach but it is based on a structural analysis in the feature detection phase and on a matching technique where the primitives consider exclusively the contours shape. For each feature of the left image we search for the correspondent feature in the right image, (two feature are correspondent if they are projections on the image planes of the same physical characteristic), among the ones verifing the epipolar constraint. Let us be S the left image and D the right image we denote with :

$$FS^l = \{(xs_i^l, ys_i^l)/0 \leq i < NS^l\} \quad \text{and} \quad FD^k = \{(xd_j^k, yd_j^k)/0 \leq j < ND^k\}$$

respectively the polygonal approximations of a feature of the left image and of a feature of the right image. We compare a feature FS^l of the left image with the features, of the set IS^l, that verify the epipolar constraint :

$$\left| \max_j yd_j^r - \max_i ys_i^l \right| < \varepsilon \; , \; \left| \min_j yd_j^r - \min_i ys_i^l \right| < \varepsilon \qquad r \in IS^l$$

and that verify the similitude predicates like the area and the size of the circumscribed rectangle. Defined $ps_i^l = (xs_i^l, ys_i^l)$ a vertex of a feature of the left image, ld_j^k the segment j of the feature k of the right image and $d(ps_i^l, ld_j^k)$ the euclidean distance between the point ps_i^l and the segment ld_j^k we denote as :

$$Eu(l,k) = \sum_{i=1}^{NS^l} \min_{j=1..ND^{k-1}} d(ps_i^l, ld_j^k)$$

an euristic function that is minimum when the two features have more or less the same shape. The feature corresponding to the FS^l will be the feature such that :

$$Eu(l,M) = \min_{FD^k \in IS^l} Eu(l,k)$$

Let us be FS^l and FD^M two corresponding features then the stereoscopic disparity of a point of FS^l is given by the sum of its distance from the projection of points of FD^M along the epipolar line and of the distance of the centroids of the two features.

5. EXPERIMENTAL RESULTS

We present in this section a result obtained using the proposed technique.

1. Left Image

2. Right Image

3. Segmented Left Image

4. Segmented Right Image

5. Features Matching

REFERENCES

[1] Yoshiaki Shirai, "Three Dimensional Computer Vision", Springer Verlag.
[2] G.Medioni, R.Nevatia, "Segment-based Stereo Matching", Computer Vision, Graphics and Image Processing, 1982.
[3] R.Mohan, G.Medioni, R.Nevatia, "Stereo Error Detection, Correction Evaluation", IEEE Transaction on Pattern Analysis and Machine Intelligence.
[4] V.Braitenberg, "I tessuti intelligenti", Boringhieri, Torino 1980.
[5] D.kats, "la psicologia della Gestalt", Feltrinelli, Milano 1979.
[6] R.Arnheim, "Art and Visual Perception", University of California 1974.
[7] B.De Gemmis, N.Di Ruocco, A.Vitale, S.Vitulano, "A Model of Visual Perception", Iasted Modelling, Identification and Control, Innsbruck, February 1992.

Classification of objects and patterns by affine moment invariants

J. Flusser and T. Suk

Institute of Information Theory and Automation, Czech Academy of Sciences, Pod vodárenskou věží 4, 182 08 Prague 8, Czech Republic; *E-mail:* flusser@cspgas11.bitnet

Abstract

The paper deals with moment invariants, which are invariant under general affine transformation and may be used for recognition of affine-deformed objects. Our approach is based on the theory of algebraic invariants. The invariants from 2nd and 3rd order moments are derived. The use of the affine moment invariants as the features for recognition of deformed characters is presented.

1. INTRODUCTION

A feature-based recognition of objects or patterns independent of their position, size, orientation and other variations has been the goal of ongoing research. Finding efficient invariant features is the key to solving this problem. There have been several kinds of features used for recognition . These may be divided into five groups as follows:

- Visual features (edges, texture and shape);

- Transform coefficient features (Fourier descriptors or Hadamard coefficients);

- Algebraic features (based on matrix decomposition of the image);

- Statistical features (moment invariants);

- Differential invariants (used especially for curved objects).

In this paper, attention is paid to the *statistical features.* Moment invariants are very useful tools for pattern recognition. They were derived by Hu [1] and they were successfully used in aircraft identification [2], remotely sensed data matching [3] and character recognition [4]. Further studies were made by Maitra [5] and Hsia [6] in order to reach higher reliability.

All above mentioned features are invariant only under translation, rotation and scaling of the object. In this paper, our aim is to find features which are invariant under *general affine transformation* and which may be used for recognition of affine-deformed objects. Our approach is based on the theory of algebraic invariants [7]. The first attempt to find affine invariants in this way was made by Hu [1], but his affine moment invariants were derived incorrectly.

Several correct *affine moment invariants* are presented in Section 2 and their use for recognition of deformed capital letters is shown in Section 3. For full derivation of the invariants and comprehensive discussion of their properties see the paper by Flusser and Suk [8].

2. AFFINE MOMENT INVARIANTS

The affine moment invariants (AMIs) are invariant under *general affine transformation*

$$u = a_0 + a_1 x + a_2 y$$
$$v = b_0 + b_1 x + b_2 y. \tag{1}$$

The central moment μ_{pq} of order $(p+q)$ of binary 2-D object G is defined as

$$\mu_{pq} = \int\int_G (x - x_t)^p (y - y_t)^q dx dy,$$

where (x_t, y_t) are the coordinates of the centre of gravity of object G.

The affine transformation (1) can be decomposed into six one-parameter transformations:

1) $u = x + \alpha, \quad v = y$
2) $u = x, \quad v = y + \beta$
3) $u = \omega \cdot x, \quad v = \omega \cdot y$
4) $u = \delta \cdot x, \quad v = y$
5) $u = x + t \cdot y, \quad v = y$
6) $u = x, \quad v = t' \cdot x + y$

Any function F of moments which is invariant under these six transformations will be invariant under the general affine transformation (1).

From the requirement of invariantness under these transformations, the type and parameters of the function F can be found.

Four simplest AMIs are listed below in the explicit form:

$$I_1 = \frac{1}{\mu_{00}^4}(\mu_{20}\mu_{02} - \mu_{11}^2),$$

$$I_2 = \frac{1}{\mu_{00}^{10}}(\mu_{30}^2\mu_{03}^2 - 6\mu_{30}\mu_{21}\mu_{12}\mu_{03} + 4\mu_{30}\mu_{12}^3 + 4\mu_{03}\mu_{21}^3 - 3\mu_{21}^2\mu_{12}^2),$$

$$I_3 = \frac{1}{\mu_{00}^7}(\mu_{20}(\mu_{21}\mu_{03} - \mu_{12}^2) - \mu_{11}(\mu_{30}\mu_{03} - \mu_{21}\mu_{12}) + \mu_{02}(\mu_{30}\mu_{12} - \mu_{21}^2)),$$

$$I_4 = \frac{1}{\mu_{00}^{11}}(\mu_{20}^3\mu_{03}^2 - 6\mu_{20}^2\mu_{11}\mu_{12}\mu_{03} - 6\mu_{20}^2\mu_{02}\mu_{21}\mu_{03} + 9\mu_{20}^2\mu_{02}\mu_{12}^2 + 12\mu_{20}\mu_{11}^2\mu_{21}\mu_{03} +$$
$$+ 6\mu_{20}\mu_{11}\mu_{02}\mu_{30}\mu_{03} - 18\mu_{20}\mu_{11}\mu_{02}\mu_{21}\mu_{12} - 8\mu_{11}^3\mu_{30}\mu_{03} - 6\mu_{20}\mu_{02}^2\mu_{30}\mu_{12} +$$
$$+ 9\mu_{20}\mu_{02}^2\mu_{21}^2 + 12\mu_{11}^2\mu_{02}\mu_{30}\mu_{12} - 6\mu_{11}\mu_{02}^2\mu_{30}\mu_{21} + \mu_{02}^3\mu_{30}^2).$$

The number of the invariants is dealt with in *Cayley - Sylvester theorem*. It is proved in the paper [8] that there is one invariant of weight 2 (I_1), one invariant of weight 4 (I_3) and two invariants of weight 6 (I_2 and I_4).

3. NUMERICAL EXPERIMENT

In order to show the performance of the affine moment invariants as the features for classification of deformed objects, the following experiment was carried out.

The experiment deals with the recognition of printed capital letters. Fig. 1a shows the original letters – templates. Fig 1b shows the same set of letters deformed by affine transformations with various parameters. The task was to classify the deformed letters.

Four features for each letter were computed – the invariants I_1, I_2, I_3 and I_4. Their values are given in Table 1. It can be seen clearly that I_1, I_2, I_3 and I_4 really are invariant under affine transform. However, the values of invariants *before* and *after* the transformation are not exactly equal but are slightly different. These differences are caused by digital character of the images.

The classification was performed by minimum-distance rule in 4-D Euclidean space. All letters were recognized correctly.

Many other experiments on test and real images were performed by the authors (see [8]). It was proved that the AMIs can be applied even in the case of projective deformation of the objects. This is very important for recognition of 3-D objects from their 2-D projections.

4. CONCLUSION

In this paper, a set of new affine moment invariants is presented. Results of the theory of algebraic invariants were used to derive them. It was shown by experiment that the AMIs can serve as the features for recognition of deformed objects.

From the theoretical point of view, our paper is a significant extension and generalisation of the recent works [1] - [6].

From the practical point of view, this approach may have numerous applications in many practical tasks in remote sensing, character recognition and robot vision.

5. REFERENCES

1 M.K. Hu, *IRE Trans. Inf. Theory* 8 (1962) 179-187.

2 S.A. Dudani, K.J. Breeding and R.B. McGhee, *IEEE Tr. Comput.* 26 (1977) 39-46.

3 R.Y. Wong and E.L. Hall, *Computer Graphics and Image Proc.* 8 (1978) 16-24.

4 S.O. Belkasim, M. Shridhar and M. Ahmadi, *Pattern Recognition* 24 (1991) 1117-1138.

5 S. Maitra, *Proc. IEEE* 67 (1979) 697-699.

6 T.C. Hsia, *IEEE Trans. System, Man, Cyb.* 11 (1981) 831-834.

7 I. Schur, Vorlesungen uber Invariantentheorie, Springer Verlag, Berlin (1968).

8 J. Flusser and T. Suk, *Pattern Recognition* 26 (1993) No.1

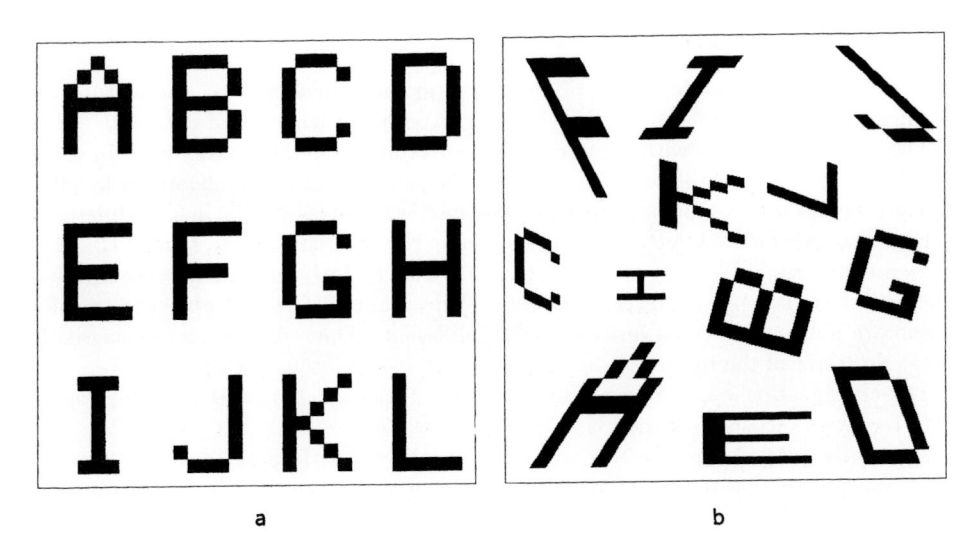

Figure 1: Recognition of deformed letters: (a) - templates, (b) - the letters to be recognized

Table 1
Affine moment invariants of the letters in Fig. 1a (left) and 1b (right)

Letter	$I_1[10^{-4}]$	$I_2[10^{-8}]$	$I_3[10^{-6}]$	$I_4[10^{-6}]$	$I_1[10^{-4}]$	$I_2[10^{-8}]$	$I_3[10^{-6}]$	$I_4[10^{-6}]$
A	330	6	-133	40	327	-3	-135	40
B	292	0	-15	4	292	0	-15	4
C	776	1286	192	75	719	1388	164	80
D	468	-8	-64	25	468	-8	-64	25
E	367	120	62	4	384	153	75	5
F	315	-245	-318	86	303	-210	-291	76
G	521	-2	-84	47	504	-4	-81	43
H	408	0	0	0	415	0	-4	2
I	218	0	0	0	230	0	0	0
J	601	3423	-1241	796	601	3458	-1234	792
K	512	550	187	10	499	465	184	7
L	523	-2295	-1950	916	522	-2338	-1943	902

This work was supported by the Czech Academy of Sciences, grant No. 27 529.

On substantial improvements in an image-processing-based high-resolution imaging scheme using multiple imagers

Takahiro Saito[a], Takashi Komatsu[a], Kiyoharu Aizawa[b]

a Department of Electrical Engineering, Kanagawa University, 3-27-1, Rokkakubashi, Kanagawa-ku, Yokohama, 221, Japan

b Department of Electrical Engineering, The University of Tokyo, 7-3-1 Hongo, Bunkyo-ku, Tokyo, 113, Japan

Abstract

Towards the development of a VHD image acquisition system, the work presents a new image-processing-based approach using multiple imagers. The image acquisition approach, integrating multiple images taken simultaneously with multiple imagers, produces an improved spatial resolution image with sufficiently high SNR. This method, however, has some disadvantage points. To remove these disadvantage points, we incorporate new techniques into the imaging scheme.

1. INTRODUCTION

Some research institutes are elaborating plans to develop very high definition (VHD) visual media beyond the HDTV[1]. The VHD visual media require high spatial and temporal resolution at least equivalent to that of 35mm films and/or motion pictures.

The construction of VHD visual media, however, involves overcoming many difficulties in different technologies, viz. imaging, processing, transmission, and display. Among them, image acquisition provides one of the most serious problems. CCD camera technology is considered promising as a high resolution imaging device. Although a CCD camera with two million pixels has been developed for HDTV, spatial resolution should be increased further to acquire VHD images. The most straightforward way to increase the spatial resolution is reducing the pixel size, viz. the area of each photo-detector, which causes decrease in the amount of light available for each pixel, and hence SNR is degraded because of the existence of shot noise. To keep shot noise invisible on a monitor, there needs to be a limitation in the pixel size reduction, and the limitation is estimated at about 50 μm^2[2]. The current CCD technology has almost reached this limit. Therefore, a new approach is required to enhance the spatial resolution further beyond this limit.

One promising approach towards improving spatial resolution further is to incorporate image processing techniques into the imaging process. From this point of view, recently we have presented a new method for acquiring an improved resolution image with sufficiently high SNR by integrating multiple images taken simultaneously with multiple imagers having the same pixel aperture[3],[4]. This method, however, has some disadvantage points. One of the most significant disadvantage points is that the improvements of spatial resolution depend on the arrangement of multiple imagers. Another disadvantage point is that the aliasing artifacts included in images taken with low-resolution imagers disturb the image processing for integrating multiple low-resolution images into an improved resolution image. To overcome these difficulties, this paper incorporates new techniques into the imaging scheme.

2. ORIGINAL CONCEPT[3],[4]

The concept of our image acquisition method is to unscramble the within-passband and aliased frequency components, which are weighted differently in undersampled images, by integrating multiple images taken simultaneously with multiple imagers, and then to restore the frequency components up to high frequencies so as to obtain an improved resolution image with sufficiently high SNR. Our image acquisition method consists of the two stages;
(1) Estimation of discrepancies and integration of sampling points (Registration):
Firstly, we estimate the relative discrepancies of pixels between two different images with fractional pixel accuracy, and then we combine together sampling points in multiple images, according to the estimated discrepancy, to produce a single image plane where sampling points are spaced nonuniformly. As for estimation of discrepancies in fractional pixel accuracy, the block-matching technique and the least square gradient technique may be used, but the work presented here employs the block-matching technique to measure relative shifts.
(2) Reconstruction of a high resolution image (Reconstruction):
We reconstruct a high resolution image from the integrated higher density sampling points spaced nonuniformly. This stage involves a reconstruction technique from nonuniformly-spaced sampling points. As for the reconstruction technique, we might use various methods. The methods suggested so far have individual limitations in the 2-D case, but we believe that the iterative method using the Landweber algorithm[5] is fairly flexible, comparatively suited to the reconstruction problem.

3. IMPROVEMENTS

3.1 Utilization of Multiple Imagers with Different Pixel Apertures

When we use multiple imagers having the same pixel aperture, the improvements of spatial resolution depend on the arrangement of multiple imagers. We assume that a two-dimensional test plate is shot with two imagers having the same pixel aperture and that the imagers are located with convergence. In this case, in the portion of the two-dimensional test plate where the optical axes of the two imagers intersect, a pixel of one image taken with one imager almost coincides with some pixel of another image taken with another imager. Hence, in this portion spatial resolution is not improved satisfactorily. That means the resolution improvements are not spatially uniform. On the other hand, however, the utilization of multiple imagers with different pixel apertures leads us to overcome this difficulty. Using different apertures, a pixel of one image would never coincide with any pixel of another image, irrespective of the arrangement of multiple imagers. Therefore, the utilization of multiple imagers with different pixel apertures seems advantageous.

3.2 Alternately Iterative Algorithm for Registration and Reconstruction

Undersampled images include aliased frequency components, which might cause errors in the registration process. Hence, the aliasing artifacts make it extremely difficult to estimate the relative discrepancies between two observed low-resolution images with subpixel accuracy. We should render the estimation method robust against the aliasing artifacts. The work presented here introduces an alternative algorithm. In the initial step of the algorithm, the relative discrepancies are estimated with comparatively low accuracy,e.g. a half pixel, by comparing the two observed low-resolution images with the variant of the block-matching technique, and then in the reconstruction stage according to the estimated discrepancies the observed low-resolution images are integrated into an improved resolution image with uniformly-spaced samples. In the second step of the algorithm, the relative discrepancies are updated with higher accuracy by comparing the reconstruction image and the observed low-resolution image with the variant of the block-matching technique, which somewhat alleviates errors in the registration process, because the aliasing artifacts are eliminated from the improved resolution image to some extent. After that, according to the updated discrepancies, an improved resolution image is anew reconstructed from the observed low-resolution images. The above-mentioned procedure for the registration and the reconstruction is repeated. With the advance of the repetition, estimation accuracy of the registration process are gradually improved, and thus the aliasing artifacts will be almost fully eliminated from a reconstructed improved resolution image.

4. SIMULATIONS

The whole process of the high resolution image acquisition scheme is examined. We used a printing digital image data with 2048 pixel x 2560 lines as a two-dimensional test plate. To make projected low resolution images whose size is 256 pixel x 640 lines and which are used in the simulation of the whole process, we modelled the imaging process as follows;
(1)The projection is perspective,
(2)The two imagers are located horizontally in parallel; their optical axes are parallel,
(3)The two imagers have different pixel apertures as shown in Fig.1, and the aperture ratio of each imager is assumed to be 100%.

Figure 2(a) shows a high resolution digital image used as a two-dimensional test plate. Figure 2(b) shows a part of the image captured by a single imager with a horizontally wide rectangular aperture shown in Fig.1(a), but in this figure for ease of comparison the image is magnified twice with the sinc function in the horizontal direction. Figure 3 shows a part of the improved resolution image reconstructed at each iteration step of the alternately iterative algorithm. In Fig 3, after a few iterations the algorithm provides an improved resolution image which is almost free of the aliasing artifacts. In Fig.3(d), compared to Fig.2(b), the aliasing artifacts are almost fully eliminated, and the details are well acquired.

5. CONCLUSIONS

Although the image-processing-based approach is still in its infancy and there remain a lot of unsolved problems towards the development of a VHD image acquisition system based on this approach, the experimental simulations demonstrate that the approach is promising.

6.ACKNOWLEDGMENT

This work has been supported in part by a grant of the TEPCO Research Foundation (Tokyo, Japan).

7. REFERENCES

1. S.Ono,T.Saito and K.Aizawa,Towards Super High Definition Image Communication, Proc. of Picture Coding Symposium of Japan, (1990),145-146, (in Japanese).
2. T.Ando, Trends of High-Resolution and High-Performance Solid State Imaging Technology, Journal of ITE Japan,44, (1990), 105-109, (in Japanese).
3. T.Komatsu,K.Aizawa and T.Saito,A Proposal of a New Method for Acquiring Super High Definition Pictures with Two Cameras, Journal of ITE Japan, 45,(1990),1256-1262,

(in Japanese).
4. T.Komatsu, T.Igarashi, K.Aizawa and T.Saito, A Signal-Processing Based Method for Acquiring Very High Resolution Images with Multiple Cameras and Its Theoretical Analysis, IEE Proceedings Part I ,(1993), (in the press).
5. H.Stark (ed.),Image Recovery : Theory and Application,Academic Press, (1977).
6. M.Irani and S.Peleg,Improving Resolution by Image Registration",CVGIP:Graphical Models and Image Processing,53,(1991),231-239.

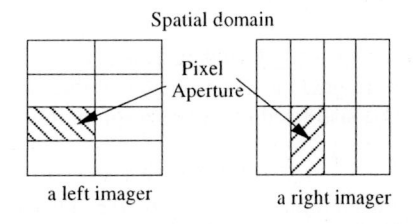

Figure 1- Two different pixel apertures for two imagers

(a)The two-dimensional test plate

(b)A part of the image captured by a single imager

Figure 2-The test plate and the image captured by a single imager.

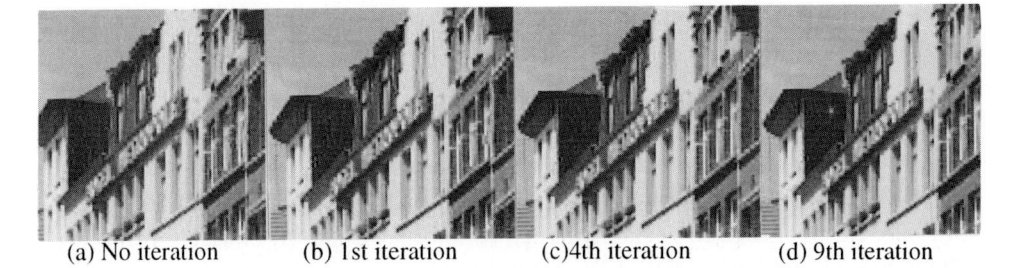

(a) No iteration   (b) 1st iteration   (c)4th iteration   (d) 9th iteration

Figure 3-Improved resolution images reconstructed at each iteration step of the alternately iterative algorithm.

Neural Defect Detection in Magnetic Particle Inspection Test's Images

Luigi Raffo[a], Daniele D. Caviglia[a] and Luisa De Vena[b]

[a]Department of Biophysical and Electronic Engineering
University of Genova-Via all'Opera Pia 11/A-16145 Genova-ITALY
ph: +39 10 3532163, fax: +39 10 3532795, e_mail: luigi@cpsi7.dibe.unige.it

[b]Automa-Sistemi di automazione industriale s.c.r.l
Via al Molo Vecchio, Calata Gadda-16126 Genova-ITALY
ph: +39 10 2092592, fax: +39 10 203987, e_mail: luisa@automa.it

Abstract
In this paper we present a method to detect defects in non destructive testing images using a modified neural principal component analyzer to extract features that characterize each discontinuity. Features so obtained are classified by back-propagation algorithm. Testing is realized using, as training set, two images that evidence defect, and other four images as test set. The results show good performances and this algorithm evidences high efficiency resulting suitable for this problem.
Keywords: *Neural Networks, Non Destructive Testing, Principal Component Analysis*

1. INTRODUCTION

Non destructive evaluation of defects or flaws in ferromagnetic parts is an important operation to verify the quality of the production and to test objects subjected to mechanical stress for particular purposes. One non-destructive evaluation methods in widespread use is *fluorescent magnetic particle inspection* (FMPI)[1].

The FMPI method is very simple in principle. Magnetic particles distributed onto a uniform bar of ferromagnetic material appropriately magnetized cover evenly its surfaces. On the contrary, if there are discontinuities at, or immediately below the surfaces, the density of the particles is higher next to the discontinuities. The final distribution of the powder on the surfaces is an indicator of the presence of discontinuities. This can be evidenced using fluorescent powders, which often give a clearer indication of small flaws when viewed under ultraviolet light.

FMPI consists of two operations: the first is to process the part in a magnetic particle solution; the second is for an operator to inspect the part visually. The first operation has been automated since a long time, but the second operation remains manual. Automation of the visual inspection would improve inspection quality and increase productivity. However ferromagnetic parts before machining usually have rough surfaces with granular texture and sharp edges, then FMPI images are characterized by low-contrast crack indications resulting from excessive visual noise in the defect-free background, large variations in crack-indication intensity, and uncertainty in distinguishing crack from edge indications.

Cheu [2], attempted to automatize FMPI method, using automated scanning of steel components for crack indications followed by digital image-processing techniques to enhance the results. This method, however, needs a complex mechanical pre-processing that could be considered too complex and expensive for testing in production lines.

FMPI defect detection problem is also considered in [3] and resolved with Hough transform. In [4] the similar problem of glass defect detection is considered.

2. THE NEURAL NETWORK

If we consider a neural network with N inputs and only one output linear processing unit (output neuron) and if this unit receives a set of zero-mean scalar-valued inputs $x_1,\ldots,x_N$ through N junctions with weights $w_1,\ldots,w_N$, the output of the unit (and thus of the network) can be written as: $y = \sum_{i=1}^{N} w_i x_i$. The junction can be assumed variable in time according to Hebbian hypothesis: the connections become stronger when both the pre- and post-synaptic signals are strong. In this model this means that the weights change with the product of the input and the output. However, as the basic Hebbian scheme would lead to unrealistic growth of strengths of connections, a saturation or normalization is to be assumed.

Oja [5] showed that the normalized Hebb rule can be realized with the following update rule:

$$w_i(t+1) = w_i(t) + \gamma y(t)\left[x_i(t) - y(t) w_i(t)\right] \tag{1}$$

Oja [5] proves that in a network so defined, the weights converge to the eigenvector with largest eigenvalue of the covariance matrix of the input set, and the output converges to the coefficient that represents the current input in the space generated by this eigenvector (also called principal component). The eigenvector of the covariance matrix with largest eigenvalue, corresponds to the direction of maximum variance (this could be expected from the use of Hebb rule), so if we consider a set of points in a bidimensional space distributed along a line the principal component finds the direction of this line. This property of the Oja's rule evidences that it is suitable for our problem, but our input data is not zero-mean: so it is necessary to evaluate the average of the data.

An iterative process that yields the average of data can be written as:

$$m_i(t+1) = m_i(t) + \eta(t)(x_i(t) - m_i(t)) \tag{2}$$

using the average definition, $\eta(t)$ will be: $\eta(t) = 1/(t+1)$ but a wide set of other functions decreasing to zero, can be used [6].

The main structure of the neural network is a standard one-layer with N inputs and one output, other N non- standard units are added to evaluate the average of each component of the input. These units can be considered as accumulators that at each step increment their output of a quantity proportional to the input. An intermediate layer was inserted to evaluate the zero-mean data set. The weights p_i of the connections between the input of the network and each of these units decrease with time. The weights w_i of the connections between the output and the intermediate layer units change with Oja's rule previously described.

3. DEFECT DETECTION

An image of a mechanical piece obtained with magnetic particle inspection technique is binarized activating the 2000 brightest points and setting all the remaining points to an inactive state (our test images are 256 × 256 pixels). This operation makes the network not sensitive to background luminosity. We subdivided the input image in partially overlapped square parts. We choose the size small enough to assume that only one defect is present in each square. Sub-images of 32 × 32 pixels are considered, each one is selected moving a window on the image by steps of 16 pixels. Sub-images with less than 15 active points are ignored; the others are used as input for the network presented, ignoring those that have averages near the border of the sub-images (if the points belong to a defect they will appear at least in one other sub-image).

The neural network (with N=2) extracts the principal component of the distribution of points in the sub-images (the weights) and gives the variance in this direction (the output). The same information can be extracted on the orthogonal direction, evaluating the variance also in this direction. We assume that these two variances (and especially the ratio) with information about the position (averages) and the directions (the weights) constitute a set of features sufficient to distinguish defects from sparse noisy point.

(a) (b) (c) (d)

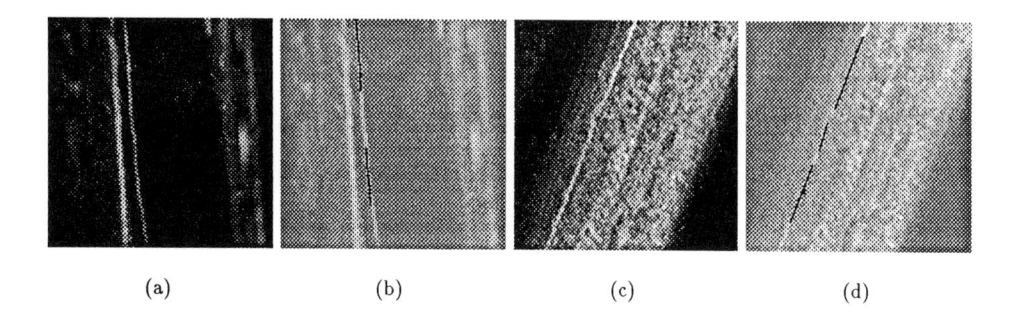

Figure 1: Defect detection in training images (a)(c). The network find the defects evidenced with a black line in (b)(d). The network is correctly trained (defect are identified).

(a) (b) (c) (d)

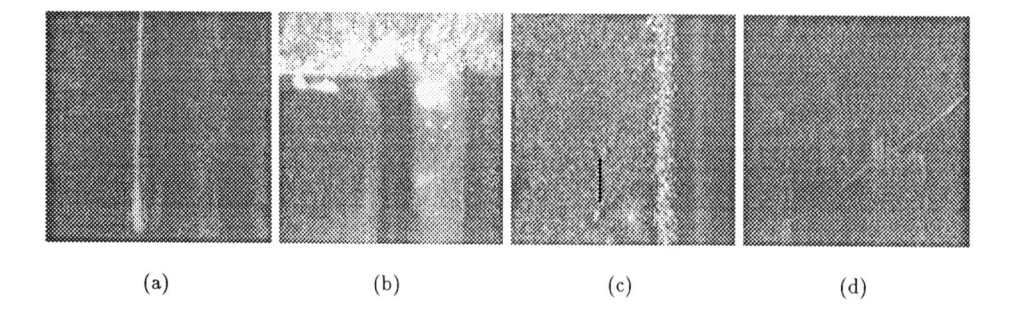

Figure 2: Defect detection in images not considered in training phase. The only defect can be found in (c), and is correctly identified (black line). The other lines represent discontinuities, not defects. The line in (d) is a superficial scratch

A simple discrimination based on the ratio of the two variances (that gives the aspect ratio) or on the variance on the direction orthogonal to the principal component (thinness)

gives good results, but an improvement can be achieved considering all the features (a defect along a diagonal is longer than one parallel to an axis). Thus, we have trained with back-propagation algorithm a one-hidden-layer perceptron [7] with these features, extracted from two test images and we have used for test set other four images. The images used for training and relative results are shown in Fig. 1. The images of Fig. 2 are the results with different tests. It can be noted that the defects are always detected in at least one segment in which the characteristic of linearity and brightness are more evident. Better results could be obtained with an appropriate choice of the number of points considered in the images.

4. CONCLUSION

We have considered the problem of processing images obtained by magnetic particle inspection in order to distinguish defects from various kinds of noise due to imperfect surfaces or the shape of the object. We have shown that the problem can be separated in smaller problems considering part of the original image, in which we suppose that at least only one defect is present. Each of these sub-images is processed with a modified Oja's neural network (with a sub-network that yields the averages of the input distribution) in order to obtain features of the distribution of the fluorescent points. The method is adequate only if at least one sub-image exists in which only the points belonging to a defect are evidenced. In our test images this is always true, so the above condition is not too restrictive. It is interesting to note the efficiency in distinguishing defects from superficial scratches, this is not the case for very short defects in which it is difficult to distinguish the characteristic shape (high ratio between the two dimension). The results are very interesting because the problem is not easy to solve, even for an expert human operator and taking into account the simplicity of the algorithm.

5. REFERENCES

1 D.C. Jiles. Review of magnetic methods for non destructive evaluation year-part 2. *NDT International*, 23(2):83–92, (1990).

2 Y.F. Cheu. Automatic crack detection with computer vision and pattern recognition of magnetic particle indications. *Material Evaluation*, 42:1506–1514, (1984).

3 L. Raffo, A. Blumenkrans, and D.D. Caviglia. Defect Detection in Non Destructive Testing Images by Hough Transform. In *Proc. Automation 1992 (Annual ANIPLA Congress)*, Genova ITALY, (1992).

4 K. Cios, R.E. Tjia, N. Liu, and R.A. Langenderfer. Study of Continous ID3 and Radial Basis Function Algorithms for the Recognition of Glass Defects. In *Proc. IJCNN*, pages I-49–54, (1991).

5 E. Oja. A simplified neuron model as a principal component analyzer. *Journal Math. Biol.*, 15:267–273, (1982).

6 L. Lijung. Analysis of Recursive Stochastic Algorithms. *IEEE Transaction on Automatic Control*, 22(4):551–575, (1977).

7 J. Hertz, A. Krogh, and R.G. Palmer. *Introduction to the theory of neural computation*. Addison Wesley, (1991).

Deinterlacing of HDTV images using a 2-layer perceptron

S. Carrato

D.E.E.I., University of Trieste, via A. Valerio, 10, 34127 Trieste, ITALY

Abstract

We present a novel approach for the deinterlacing of HDTV image sequences based on the use of a 2-layer perceptron. Its nonlinear behaviour permits an accurate edge preservation, even in presence of motion, while its simplicity makes it appropriate for use in low-constitutional environments. The results of various simulations confirm the validity of this approach.

1 INTRODUCTION

Multimedia applications have recently been receiving an increasing attention, due to the growing number and capabilities of electronic equipments in both commercial and consumer environments, together with the availability of various connections between different systems. In multimedia applications, different forms of information (text, sound, images...) have to be available on different platforms. Some data manipulations must obviously be performed when the received information has a different format with respect to the one used by the media on which it has to be presented. Within this group of problems, an interesting case is given by the visualization of sequences of HDTV images on workstation monitors. Here, three main issues are involved (i.e., deinterlacing, frame rate conversion, and format conversion [1]); the first one is addressed in this paper.

A trivial solution is given by linear interpolation, but some drawbacks are present. If the pixels of the adjacent lines of the same field are used, a severe smearing of the image occurs, which is particularly annoying in the stationary parts of a sequence, especially when considering the high quality of HDTV images; in turn, artifacts (as "image flow" [1]) are generated if temporal interpolation is considered. Various more sophisticated algorithms have already been presented in order to solve this problem. In general, they are based on the use of some motion information; in this way, the filtering is adapted in order not to blur the images. Sometimes, motion estimation data are available to the receiver, being they transmitted together with the images themselves. If this is not the case, motion can be estimated locally by means of sophisticated algorithms as block matching or phase correlation. In low-constitutional environments, however, a less complicated approach has to be followed in order to keep the complexity (and consequently the price) of the system within reasonable limits. In [2] and [3], motion is detected and estimated locally using simple algorithms and a reduced support (i.e., a filter window having small size in both space and time); some heuristics is used in the design of the algorithms, and extensive tests are needed to trim various parameters (e.g., gains, thresholds, etc).

In this paper, we too restrict the problem of image deinterlacing to applications in small-business environments, so that the constraint of low complexity still applies. However, our approach is based on the use of a neural network (NN), which solves the problem of motion-adaptive interpolation without an explicit evaluation of the motion itself, simply by searching the best solution for the sequence used for its training. Thanks to the so-called "generalization property" of the NN's, good results are obtained also if the test sequence is different from the training one.

In the following section, the proposed structure is described; in Sec. 3, some simulation results are presented and discussed. It is shown that good performances are obtained with this approach, notwithstanding the algorithm simplicity and the low hardware requirements; in fact, according to the above mentioned constraints, our scheme uses only three field memories and a filter window containing 12 pixels.

2 A 2-LAYER PERCEPTRON FOR THE DEINTERLACING OF IMAGE SEQUENCES

As already mentioned, our approach is based on the use of a neural network. In particular, a 2-layer perceptron (2LP) [4] is used, in which the input nodes are fed with the values of the known pixels which are adjacent to the one which has to be interpolated, while the output node computes the value of the interpolated pixel itself. It is well known [5] that a 2-layer perceptron is able to perform any partitioning of the input signal space, so that in principle any input-output function can be obtained, provided the number of hidden nodes is sufficiently high. This solution is therefore expected to give the best solution for the training sequence, given a certain support (i.e., number of pixels used for the interpolation) determined by the above mentioned contraints and supposing the learning process adequate.

The block diagram of the algorithm is presented in Fig. 1. Twelve pixels are used for the interpolation of the pixel labelled "0" in the figure: three on the same line of the pixel "0" but in the preceding field (pixels "1", "2", and "3"), six on the adjacent lines of the same field (pixels "4", "5", and "6" on the upper line, "7", "8", and "9" on the lower one) and three on the same line of the successive field (pixels "10", "11", and "12"). The number h of hidden nodes of the 2LP depends on the desired accuracy in the interpolation; results presented in the next section have been obtained using $h = 4$. The NN has one output, its value being assigned to the pixel being interpolated. The neurons of the 2LP are of the conventional nonlinear type, with threshold and nonlinear, sigmoid-shaped activation function $f(\cdot)$; more precisely, we chose a function $f(\cdot)$ of the type

$$f(x) = 1/(1 + e^{-x-\alpha x^3})$$

with $\alpha = .083$, in order to have a wider linear range with respect to the usual sigmoid function (i.e., with $\alpha = .0$). The total number of weights is $(12+1)*h + (h+1)$, i.e. 57 for $h = 4$.

An interesting feature of this approach is the ease of the 2LP design, as its parameters (i.e., the weights) have not to be choosen according to heuristic reasoning or repeated simulations, being they evaluated by the structure itself during training. The training phase, in fact, can be performed using a progressive sequence in which odd and even lines are alternatively supposed to be unknown to the 2LP input (i.e., the sequence is reduced to an interlaced one), while the (actually known) missing point is used as the desired output for the 2LP. The training can be performed via the usual backpropagation algorithm [4]—only slightly modified by the presence of α in the activation function—and is concluded when the mean error between the values of the interpolated and the desired pixel is sufficiently small.

3 RESULTS AND CONCLUSIONS

As already mentioned, the proposed deinterlacing scheme has been tested on a deinterlaced HDTV sequence obtained from a progressive one ("Renata", in our simulations), by

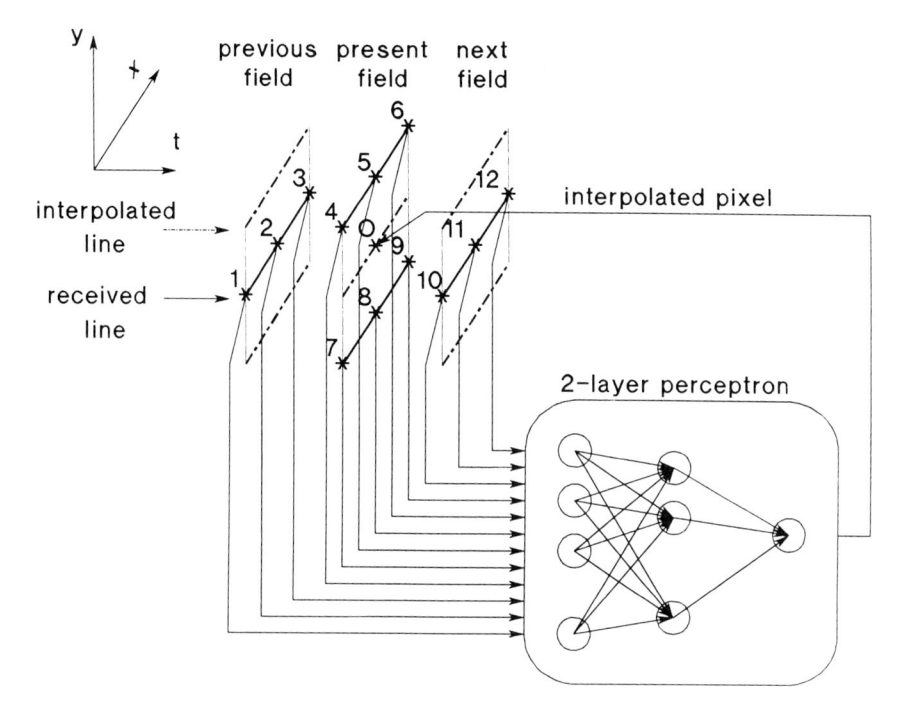

Figure 1: Block diagram of the proposed deinterlacing scheme.

considering only one line out of two, the other one being used for training and performance evaluation. Training has been performed using fields 11 to 13 of the same sequence; this part of the sequence has been chosen because it presents a reasonable mix of panning and medium speed motion.

Qualitative and quantitative analysis of the results confirm the validity of the approach. Visual inspection of the deinterlaced sequence shows that the algorithm is accurate and preserves a good image contrast in static or slow-moving areas. In the case of fast movement, no artifacts are generated; due to the reduced size of the filter support, some smearing inevitably occurs, but this is not particularly annoying due to the large movement itself. Quantitative measurements can be made by evaluating the mean square error with respect to the original progressive sequence. In Fig. 2, the results are shown for three different algorithms: a simple holder, a vertical FIR filter with a 2 × 1 window, and our 2LP-based structure. It can be seen that the best performance is given by the proposed solution, with the exception of the fields 72 to 88, where in fact motion is almost absent so that the holder, which produces large errors elsewhere, obviously outperforms the other schemes. We also tested the algorithm proposed in [3], and we found that its performance is very close to the one given by the vertical FIR filter. Similar results are obtained also if the maximum or mean absolute error are considered instead of the mean square error.

Other sequences ("salesman", "mobile and calendar", and "miss America") have also been used to test the algorithm. In any case, good results have been obtained thanks to the generalization property of the 2LP.

An interesting characteristic of this approach is that it is rather general. We are presently considering its use in order to solve the other two problems related to the

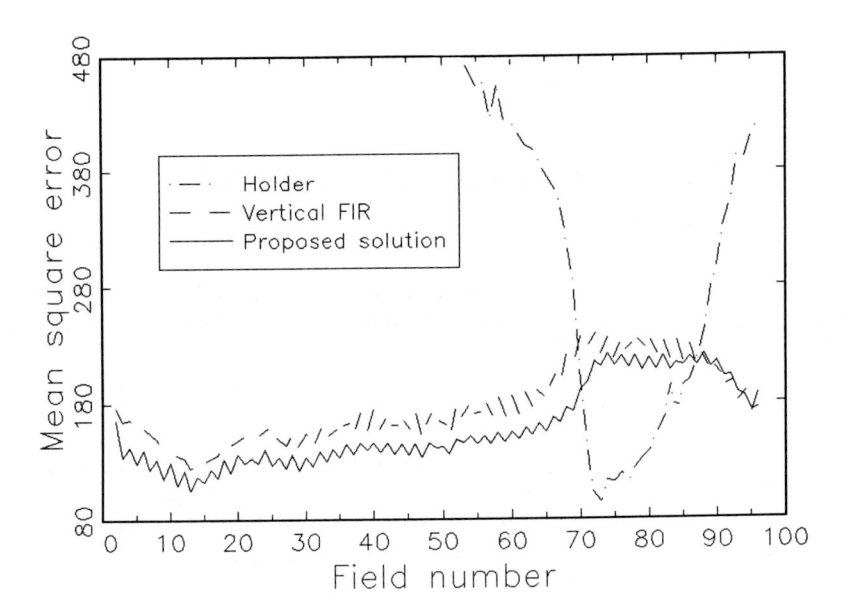

Figure 2: Mean square error in the deinterlaced sequence versus field number for the test sequence "Renata". Three different algorithms are used: a simple holder, a vertical FIR filter with a 2 × 1 window, and the proposed 2LP-based structure.

visualization of image sequences in a multimedia environment (see Sec. 1). In fact, by suitably decimating in time or in space a sequence and using the appropriate eliminated pixels as desired output for the training of the perceptron, both frame rate conversion and format conversion can be performed.

References

[1] C. P. Sandbank, *Digital Television*. John Wiley & Sons, 1990.

[2] P. Haavisto, J. Juhola, and Y. Neuvo, "Scan rate up-conversion using adaptive weighted median filtering," in *Proc. 3rd Int. Workshop on HDTV*, (Torino, Italy), Aug. 1989.

[3] R. Simonetti, S. Carrato, G. Ramponi, and A. Polo Filisan, "Deinterlacing of HDTV images for multimedia applications," in *Proc. International Workshop on HDTV'92*, (Kawasaki, Kanagawa, Japan), pp. 95.1–95.8, Nov. 1992.

[4] R. P. Lippmann, "An introduction to computing with neural nets," *IEEE ASSP Magazine*, pp. 4–21, Apr. 1987.

[5] M. A. Sartori and P. J. Antsaklis, "A simple method to derive bounds on the size and to train multilayer neural networks," *IEEE Trans. Neural Networks*, vol. 2, pp. 467–471, July 1991.

System of active coded targets for motion analysis

P.COSQUER[a] and M.CATTOEN[a]

[a]Groupe de Télévision et de Traitement du Signal Image (G.T.T.S.I.), Ecole Nationale Supérieure d'Electrotechnique, d'Electronique, d'Informatique et d'Hydraulique de Toulouse, 2, rue C.Camichel 31071 Toulouse Cedex France

Abstract

A motion analysis system is a device which allows the observation of the movement in space of a body. Amongst the existing apparatus, those which work with the aid of video sensors do not automatically identify the passive markers used. In order to eliminate this inconvenience, we propose a system of active coded targets, so as to automatically identify each active marker with its position within the acquired images.

1. INTRODUCTION

Motion analysis systems, known as "Motion Analysers" amongst other names, are available on the market. They permit the study without contact of the movement of a pliable body, by analysis of the images of a certain number of targets, or markers, placed on this body. These images are acquired by one or several sensors and the targetted field of view constitutes the calibrated experimental working volume [1]. The motion analysers which provide monitored images, employ video sensors. However, these systems posses the major disadvantage of not being able to automatically identify the detected markers. Consequently, these apparatus are very sensitive to noisy luminous environments. In fact, light sources or reflections are perceived as useful targets.

The manufacturers of motion analysers use two kinds of markers [1]. The first are passive targets which reflect in the same direction, the light stemming from a light source situated near the optical axis of each camera. In the acquired images, it is not possible to differentiate these markers from each other without the help of an operator. His decisions can be difficult in the case of complex motions. ELITE [2], Expert Vision, PRIMAS [3], VICON VX are motion analysers which employ passive markers.

The second type of markers are active targets which send out light. They either emit continuously and so are seen as passive markers, or they send out a luminous pulse one after another (time multiplexing). In the latter case, the target is automatically identified by the position in time of the luminous emission. However, the sensors employed in these systems (Position Sensitive Device or C.C.D. linear sensor coupled to a cylindrical lens) do not supply monitored images [1]. Moreover, they are sensitive to reflections and luminous environmental variations. SELSPOT II and OPTOTRAK [4] are the motion analysers with active targets which are the most widely used.

The motion analyser systems' applications are very extensive (e.g. domains of car industry, neurophysiology, sport, etc...). For the first time, in 1988, a motion analyser was used in space [5]. This spatial experiment highlighted several problems, which are the basis of new research on this kind of apparatus.

In this context, we have begun studies which aim at eliminating the detection and recognition problems of the targets held by the subject and we propose a *concept of active coded targets* for 3-D vision. Thus, it is possible to *identify automatically and without ambiguity, each active marker* held by the subject and its position, from images of the scene acquired by video sensor cameras which, furthermore, provide a monitored image. This concept is illustrated by the *temporal coding of active targets*.

2. TEMPORAL CODING OF ACTIVE TARGETS

For the setting up of this encoding (explained here for only one camera), two hypothesis are put forward. The active markers are more luminous than the majority of the other elements of the scene and an operator supplies various parameters useful for data treatment. The images are acquired by a Black and White camera with a C.C.D. sensor and the active targets are Infrared LEDs (light emitting diodes). Each marker emits light according to a *binary code* based on time. A high luminous emission from the active target provides the level "1" and a low luminous emission, level "0". To each scene image stemming from a frame sequence, is associated a bit from the binary code of the marker identification. The number of this bit is given by the acquired image position in the sequence. It is, therefore, necessary to acquire ($Log_2(N+2)$) images in order to code N markers (for example, 30 targets for 5 images). The sending-out of the markers being controlled during the experiment, a specific treatment of the acquired images allows their identification. The advantage of this two level encoding is that the active targets supply a luminous track in the images, for each bit of the code.

Firstly, each acquired image of the sequence is binarized with a threshold, computed according to the image statistical distribution. Then, after transformation to a run length coding [6], each resultant binary image is treated in order to separate each object. The surface and the barycentre position parameters of all the objects of the targetted scene are then computed and the only ones which are kept, are those whose objects satisfy the operator-introduced surface conditions. The datum which allows the definition of the code bit is the brightness information. In order to obtain stable values, the retained brightness is the average of the pixel brightnesses of the considered object, in the grey multilevel image.

The parameters of the objects acquired during a sequence are then grouped together in a file. The recognition and the identification of the markers are established by analysis of this file and the objects, whose sending-out encoding is not recognized, are eliminated. The different phases of this analysis are the following ones. At first, the brightness data stored in the sequence file permit the settlement of a decoding array for the active targets. The bits "0" or "1" being expressed by different brightnesses, the emission code of a marker is reconstituted by calculating the different luminous levels between two successive images of the sequence, for the same object. In fact, a "01" succession is coded by a "low-high" brightness transition and a "10" succession by a "high-low" brightness transition. After complete analysis of the file, the codes of all the objects are written down in the decoding array. By reading the file, the active marker codes are thus determined and the uncoded light sources, as well as parasite objects, are eliminated. The targets whose code is truncated (appearance or disappearance from the scene during a sequence) and those whose code does not correspond to the markers employed list, associated to each experiment, are not retained. When several objects posses the same code (case of reflections), only the one with the highest brightness is kept.

A visualization programme allows the operator to compare the results with the monitored images. By this encoding form, it is possible to determine if an active marker is hidden during the experiment because, in the opposite case, its luminous track is visible in every image, whatever the code bit sent out.

3. TEMPORAL CODING IMPLEMENTATION AND RESULTS

An *active targets lighting programmable device* has been developed in order to control the luminous sending-out of the markers (maximum 30). For the markers lighting, performed in synchronization with the camera, a value is atributed to each target, for each frame. The software computerized on a PC-AT micro-computer, controls the intensity level of the active targets and produces the temporal coding treatment on the acquired sequences. The active coded marker concept has been ratified by a certain number of experiments, performed on fixed subjects or in movements, with Infrared LEDs. Measurements on the markers, during analysis of the acquired images, have allowed the definition of the system's parameters, as well as the experimental working volume.

At the moment, this automatic perusal step of the sequences is not performed in real-time and a video-tape recorder is inserted between the processing unit and the camera (system configuration : 1 C.C.D. video camera, 1 S-VHS video-tape recorder, 1 digital image memory, 1 monitor, 1 computer, 1 active targets lighting programmable device, 30 active markers). Images stemming from experiments carried out with moving targets have been acquired, stored on the video-tape recorder and perused with the aid of the previous system. At the time of the analysis, all the parasite objects had been eliminated, the useful targets identified and their parameters stored in a file. The experiment results concerning an up and down movement (maximal speed of about 0.4 m/s), acquired at the rate of 50 Hz, are presented in figure I. It must be noted that, in spite of the fact that marker N°6 goes momentarily out of the camera field of view, it is always automatically identified when it is visible within the images.

4. CONCLUSION

A concept for a motion analysis system with automatic identification, using active coded targets synchronized with video cameras, has been proposed. In order to illustrate this concept, the active marker temporal coding has been studied. The treatment is performed on image sequences acquired in black and white and identifies the active targets, with the help of their brightness information. This encoding has been implemented, thanks to the development of the active targets lighting device and ratified by various experiments.

However, numerous problems remain. To date, the active markers are Infrared LEDs connected to the lighting device by wires. It would be desirable to study the target in a detailed way (optics and electric parameters), as well as the means to make it autonomous. Moreover, in order to render the markers independent from the acquisition system, two possibilities could be evaluated. Either a remote control can be developed, or the processing unit ought to be able to work in an asynchronous way. In the future, these diverse studies should provide a prototype of automatic active targets perusal for the motion analysis systems.

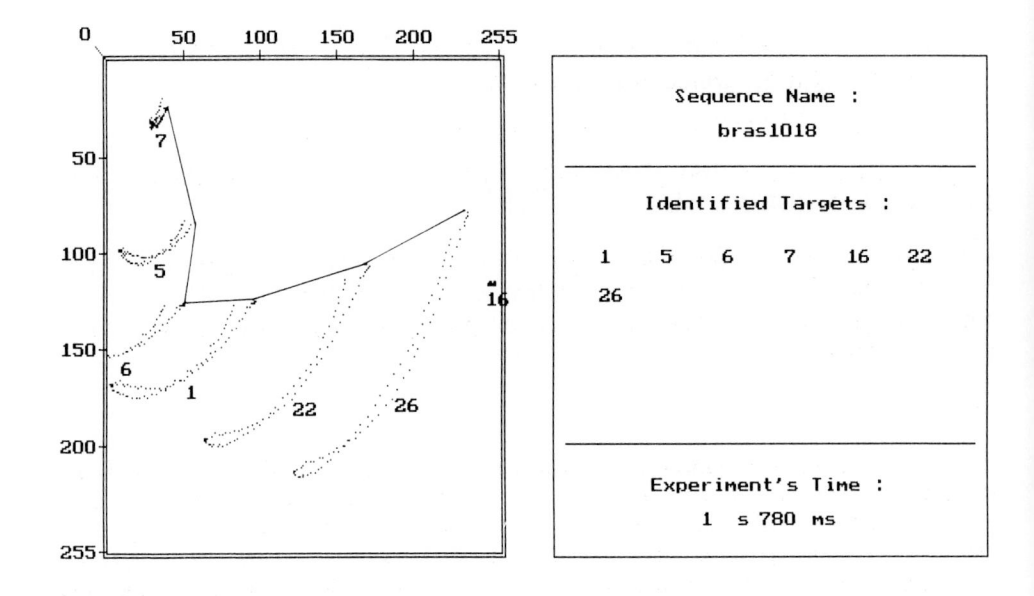

Figure I.

5. REFERENCES

1 E. H. FURNEE "TV/Computer motion analysis systems, the first two decades", TU Delft, 1989.
2 G. FERRIGNO, A. PEDOTTI "ELITE : a digital dedicated hardware system for movement analysis via real-time TV signal processing", IEEE Transactions on Biomedical Engineering, vol BME-32, N°11, November 1985.
3 E. H. FURNEE "High-resolution real-time movement analysis at 100 Hz with stroboscopic TV-camera and video-digital coordinate converter" Delft University of Technology, North American Congress on Biomechanics, Montréal, Août 1986.
4 "OPTOTRAK : 3-D motion measurement system" Technical product description, Northern Digital Inc., Octobre 1990.
5 L. DEUR "Suivi vectoriel d'un cosmonaute" Projet de fin d'études ENSEEIHT, Toulouse, Juin 1986.
6 P. BERE "Contribution à l'étude de méthodes de codage et de caractérisation d'images achromes ou trichromes. Mise en oeuvre sur un système multiprocesseur." Thèse de Docteur, INP Toulouse, 1991.
7 P. COSQUER, M. CATTOEN "Concept of active coded targets for 3-D vision." 2ème colloque Image'Com, Bordeaux, Mars 1993.

3-D Imaging of the Residual Limb of an Above-Knee Amputee using Morphological Edge Detection Algorithms

G. K. Matsopoulos[a], S. Marshall[a], J. C. H. Mackie[b] and S. E. Solomonidis[c]

[a] Signal Processing Division, University of Strathclyde, 204 George Street, Glasgow, Scotland, U.K., G1 1XW

[b] National Centre for Training and Education in Prosthetics and Orthotics, Curran Building, 131 St. James' Road, Glasgow, Scotland, U.K., G4 0LS

[c] Bioengineering Unit, Wolfson Centre, 106 Rottenrow, Glasgow, Scotland, U.K., G4 0NW

Abstract

In this paper, a new 3-D modelling scheme for the residual limb of an Above-Knee (AK) amputee is developed in order to establish a design process for manufacturing high quality prosthetic sockets. Firstly, transverse MRI scans of the residual limb are obtained. Morphological edge detection algorithms are then applied on each MRI slice to extract crucial information from the contours of the external and internal shapes of the limb. Finally, advanced high performance CAD software is employed to reconstruct 3-D solid graphics and to enhance visibility of the basic anatomical structures of the limb (skin, bone, muscles).

1. INTRODUCTION

In an attempt to develop a Computer Aided Socket Design (CASD) system, recently advanced imaging modalities, such as ultrasound imaging, computed tomography (CT), and nuclear magnetic resonance (NMR) imaging, have been employed as a preliminary stage in order to provide information below the skin reaching muscle and bone surfaces. Also, Torres-Moreno [1] developed a new version of the FE model with realistic geometrical characterisation of the inner residual tissues from MRI cross-sectional images of the residual limb. In the above techniques, the objective was to obtain external contours of tissues, such as skin, muscles and bone. For every MRI slice, the edges of the tissue were extracted interactively with the use of a mouse and a cursor. This manual method for edge detection used in existing technique is time consuming and subjective. The following proposed scheme employs morphological edge detection techniques as a more accurate, efficient and robust method in order to identify the basic external and internal anatomical structures of the residual limb.

2. MORPHOLOGICAL EDGE DETECTION SCHEME

2.1. Preprocessing stage

a) <u>Data Acquisition:</u> Transverse MRI scans of an AK residual limb are obtained. Two spin-echo sequences were required to complete the entire length of the residual limb. The first sequence covers the proximal half of the stump (distal from the femoral head) containing 18 transverse cross-sections while the second covers the second half containing 16 transverse slices. Each slice has a thickness of 10mm, separated by a gap of 10mm, 256 phase encoding steps and a 400mm field-of-view. The slices are transferred to a SUN-workstation creating a pixel matrix of 256x256 for each slice, with 256 gray-levels and a pixel size of 1.56mm.

b) <u>Background removal:</u> Since each of the slices contains noise, a background removal technique based on noise statistics is applied [2] by calculating the background noise.

2.2. Morphological edge detection techniques

The edge detection procedure consists of three filtering steps:

a) Morphological edge detection techniques are applied in order to identify the edges of basic anatomical structures of the residual limb (e.g. skin, muscle, femoral bone). Lee et al. [3] proposed an edge strength detector based on gray-scale morphology. Its sensitivity to noise is much less than conventional gradient operators. Extending the previous idea for morphological edge detection, a new morphological edge detector, based on grey-scale morphology, is developed as follows:

$$I_{edge} = max\{ (I_1)_B - (I_1 \ominus B), (I_1 \oplus B) - (I_1)^B \} \tag{1}$$

where $I_1 = ((I_{input})_B)^B$, thus I_1 is obtained by firstly opening and then closing the input image by a flat structuring element B. In (1), $\ominus$ symbol denotes grey-scale erosion, $\oplus$ denotes a grey-scale dilation, $(\ )_B$ denotes an opening operation by B and $(\ )^B$ denotes a closing operation by B as a structuring element.

The combination of an opening followed by a closing is applied as a preliminary edge detection stage in order to smooth the image and it has been quantitatively examined in [4]. The smoothing is essential for the new operator in order to assign only one pixel as an edge element when it has a grey-value between two extremes. Furthermore, the new edge detection operator is very robust to noise, removes isolated noise samples in the input image (due to the erosion operation of the smoothing operator) and produces edges with greater grey-value than the operator in [3]. This is a desirable property as edge points can be easily distinguished from the background points.

b) <u>Thresholding:</u> After the edge points are extracted, the threshold procedure is applied for each MRI slice. Because of the different resolution of the two sequences, an initial threshold level based on the grey-value histogram is estimated for each slice and this is then adjusted by visual inspection.

c) <u>Conditional Erosion:</u> Edges that correspond to other anatomical structures (muscles or ligaments) created by thresholding can be removed by using a new approach, the conditional binary erosion. The condition is to test if the structuring element completely covered an object. If the object under test is completely covered, the erosion process ensured and the object is eroded, otherwise it remains untouched.

3. RESULTS

The above scheme is applied to the first 22 MRI slices, starting from the most distal to the femoral head slice. as it is shown in Fig. 1. The algorithm is applied for the residual limb, and because of the contrast (high-to-low resolution) between skin-muscle and muscle-bone, it gave quite satisfactory results which allows the three contours corresponding to skin, muscle, and bone respectively to be obtained.

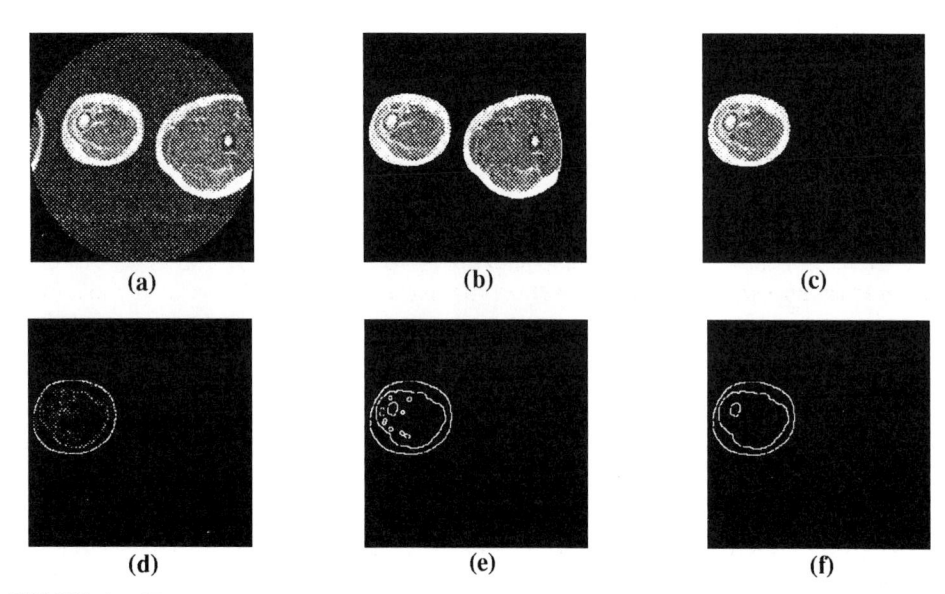

(a) (b) (c)

(d) (e) (f)

FIGURE 1. The morphological edge detection scheme. Fig. 1a shows a transverse cross-sectional MR image of the limbs of an AK amputee (on the left side: the residual limb). Fig. 1b shows the cleared MR image from noise, while Fig. 1c shows the clear residual limb. Fig. 1d shows the result of applying the new morphological edge detection operator and Fig. 1e shows the limb after thresholding. Finally, Fig. 1f shows the basic contours of the limb (skin, muscle, bone), after the conditional erosion operation by a 7x7 window was applied.

The above morphological edge detection scheme creates edges that can not be removed and unconnected contours when it has been applied to the last ten MR slices. As a result of that, another morphological filtering scheme is applied consisting of the following steps: 1) Morphological grey-scale opening: by a 3x3 flat structuring element in order to remove small regions. 2) Threshold: up to an appropriate level to create a binary image. 3) Conditional erosion: to remove small edges (noisy edges). 4) Binary closing: by a 3x3 structuring element to fill small gaps within the image. 5) Filling algorithm: to fill large regions of the image. 6) Morphological contour: by using the following expression $\partial X = X \cap [X \ominus B]^c$. Full details of these procedures will be given in the presentation.

4. 3-D RECONSTRUCTION ALGORITHM

After the edge detection stage, the data is transferred to an Apollo 590-T workstation where the reconstruction software resides. For each image frame, three independent data files of the coordinates are created representing the edges of the external contour of the skin, the muscle, and the femoral bone. From each file, 3-D surfaces are developed by successively adding adjacent slice data. The software is also capable of rotating, translating, hiding and removing any of the displayed structures. Furthermore, by using a feature of the projection editor, provided by the software, it is possible to arrange for the near and far clip planes to cut portions of the 3-D model and to enable visibility of the internal structures.

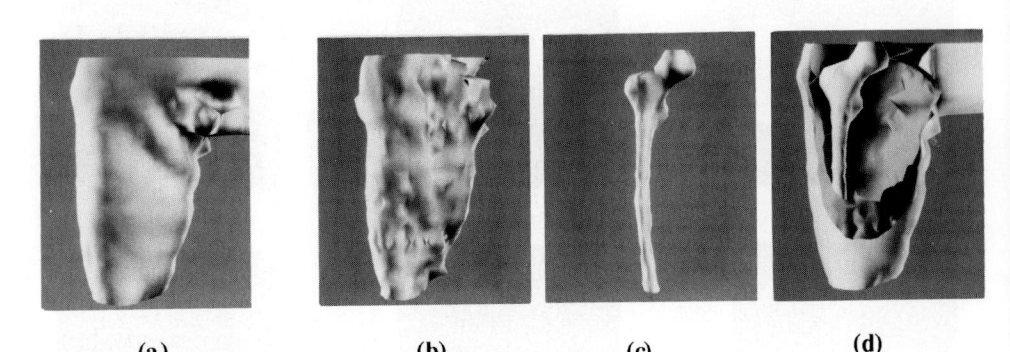

(a)   (b)   (c)   (d)

FIGURE 2.   3-D surfaces of the skin (a), muscle (b), femoral bone (c) and a combination of the above (d), for the residual limb of the AK amputee.

5. CONCLUSIONS

A new 3-D modelling scheme for the residual limb of an AK amputee is developed in this paper based on morphological edge detection and morphological filtering operators in order to enhance visibility of the basic anatomical structures of the limb (skin, muscle, bone). The 3-D reconstruction may be used to create a new version for the Finite Element (FE) model of the AK residual limb. The FE model is a method used to analyse the load bearing tissue and it consists of three phases: creation of a model, solution of a system of equations, and translating the numerical results into an understandable form.

6. REFERENCES

1. R.Torres-Moreno, "Biomechanical analysis of the interaction between the Above-Knee residual limb and the prosthetic socket", Ph.D thesis, University of Strathclyde, 1991.
2. S. Marshall and G. Matsopoulos, "Morphological Data Fusion in Medical Imaging", IEEE Winter Workshop on Nonlinear Digital Signal Processing, pp. 5.1-5.6, Tampere, Finland, January 1993.
3. J.S.J.Lee, R.M.Haralick and L.G.Shapiro, "Morphological edge detection", IEEE Journal of Robotics and Automation, vol.RA-3, No.2, pp.142-156, April 1987.
4. G.K.Matsopoulos and S.Marshall, "A new morphological scale space operator", IEE, 4th Inter. Conference on Image Process. and its applications, Maastricht, Netherlands, pp.246-249, April 1992.

ACKNOWLEDGEMENT

The authors are grateful to Mr. G. Whitefield, orthopaedic surgeon for arranging the MRI scanning at the Institute of Neurological Sciences, Southern General Hospital in Glasgow.

A Spatial-Frequency Selective Approach for Modelling the Phenomenon of Perceptual 'Filling In' in Human Vision

André Kaup and Til Aach

Institute for Communication Engineering
RWTH Aachen, Melatener Str. 23, W-5100 Aachen, Germany

Abstract

This paper is concerned with the phenomenon of 'filling in' in human visual perception. Proceeding from the assumption that localized spectral image decomposition plays an essential role in the human brain's analysis of pictures, a frequency selective approach is derived as a suitable model for describing how the visual system compensates for gaps in perception. Extrapolation is obtained by successive local series expansion of the observed image with respect to globally defined spatial basis functions. Simulation results of the proposed model based on blind spot experiments are given which are in accordance with the observed human visual perception.

1 Introduction

In the 17th century, the French scientist Edme Mariotte was the first to notice the optic disk, a small area of the retina where the optic nerves attach to the eyeball. From the fact that unlike other parts of the retina the optic disk is insensitive to light, he concluded that every eye should be blind in a small portion of its visual field. In human vision, however, the blind spot of the retina has only minor influence on visual perception, for which there are several reasons. First of all the blind spot is usually situated at the periphery of the eye's field of view. This makes it very difficult to observe any noticable effects caused by the spot. Secondly, the brain can mostly compensate for the incomplete visual perception from one eye by extracting the missing information from the slightly different image perceived by the other eye.

But even if only one perceptual representation of an object forms in the visual areas of the brain, we will usually not notice that object parts which may have fallen onto the blind spot are missing. This fact is due to a visual process known as filling in and is thought to be a manifestation of a more general perceptual mechanism called surface interpolation [1]. The leftmost pattern in Fig. 1 elucidates this process. When a person aimes his blind spot so that it covers the black disk, the line appears continuous and filled with transitional colours. The human visual system thus interpolates the missing observation using the given pictorial information of the surrounding area. From this it becomes clear that examining how the visual system compensates for gaps in perception will give valuable insight in how the brain processes images.

The aim of this paper is to introduce a spatial-frequency selective image model which allows to simulate the phenomenon of surface interpolation in human vision. The model we propose is based on linear approximation theory and has already led to substantial progress in region-oriented image coding [2]. The approach complies with the idea of linear system analysis being a suitable model for describing a variety of physical phenomena in the human visual system [3].

2 Series Expansion and Signal Extrapolation

Applying linear system analysis for the description of human vision corresponds to modelling the visual perception of an observed image based on its spatial-frequency representation. Transformed into the spatial domain this means that the perceived image signal is to be viewed as series expansion in terms of a number of two-dimensional basis functions with each basis function representing a specific two-dimensional spatial frequency.

Let $f(m,n)$ denote a physical world projection on the retina which is to be expanded into a number of N two-dimensional Fourier basis functions $\varphi_k(m,n)$ in the visual cortex. The expanded image can then be written as

$$g(m,n) = \sum_{k=1}^{N} c_k \varphi_k(m,n). \qquad (1)$$

The unknown weighting coefficients c_k are usually determined by solving the so-called Gaussian normal equations, a set of linear algebraic equations which is obtained by minimizing the error energy between the signal $f(m,n)$ and the sought expansion $g(m,n)$. The resulting expansion coefficients directly correspond to the Fourier spectral representation of the regarded image and thus contain visually important information about the presence of specific spatial frequencies in the observed scene.

If some part of the real scene cannot be observed because of it being projected onto the blind spot, the problem of finding an appropriate signal expansion gets somewhat more difficult. If we denote the brain's subjectively perceived image by $t(m,n)$, we may link $t(m,n)$ to the incomplete physical projection $f(m,n)$ on the retina by

$$f(m,n) = w(m,n) \cdot t(m,n) \qquad (2)$$

with $w(m,n)$ denoting a window function which is zero for all $\{(m,n)|(m,n) \in S\}$ covered by the area S of the blind spot and unequal to zero for all other values of (m,n).

Equ. (2) imposes consequences on the definition of an appropriate error function assessing the quality of the signal approximation $g(m,n)$. Whereas reliable information about physical objects is only accessible for light sensitive locations on the retina, the basis functions in (1) are defined such that they span the complete spatial range, i.e. the visible image area as well as those image parts which do not reveal real world information. Since the error between $t(m,n)$ and its approximation $g(m,n)$ can only be evaluated at those points where a physical object stimulus on the retina is present, we define a weighted error energy function with help of the above introduced window function $w(m,n)$:

$$E = \sum_{m,n} w(m,n) \cdot (t(m,n) - g(m,n))^2. \qquad (3)$$

The minimum of E is obtained by setting the partial derivatives of E with respect to c_k to zero. This results in a set of linear equations

$$\sum_{m,n} w(m,n) t(m,n) \varphi_k(m,n) = \sum_{m,n} w(m,n) \varphi_k(m,n) \sum_{j=1}^{N} c_j \varphi_j(m,n) \quad \forall \ k, \qquad (4)$$

the solution of which in principle yields the unknown expansion coefficients c_k. However, although (4) looks very similar to the Gaussian normal equations, there is a severe difference concerning the existence of a unique solution. Due to the definition of the window function there are in fact less equations than unknowns. Hence the system of equations is under-determined and cannot be solved uniquely without the help of additional constraints [4] (cf. also [2]).

3 Selective Approximation

Experiments with the blind spot reveal that the human visual system tries to fill in the missing area with the surrounding texture, whereby it gives preference to certain dominating pattern orientations of the observable image $f(m,n)$. This observation gives reason to believe that the process of filling in can be modelled by some kind of frequency selective extrapolation of $f(m,n)$ in those areas which due to the blind spot or scotomas contain only incomplete or no information about the real physical world.

With respect to the set of linear equations to be solved in (4) this strongly suggests that the human visual system tries to expand $t(m,n)$ such that as few basis functions $\varphi_k(m,n)$ as possible are needed for closely representing $t(m,n)$. This complies with the general assumption of parsimony in the human visual system implying that only very few key features are used by the visual cortex for the analysis of pictures. This behaviour in human perception can be modelled by iteratively solving (4) based on a steepest gradient method. Starting with all coefficients c_k initialized to zero, we successively approximate $t(m,n)$ by recalculating that coefficient which currently corresponds to the steepest slope of the error energy E. If the coefficient selected in this way is denoted by c_k and a_k is the weighted autocorrelation of basis function $\varphi_k(m,n)$, the update is performed according to

$$c_k^{(m+1)} = c_k^{(m)} - \frac{1}{2a_k}\frac{\partial}{\partial c_k}E. \qquad (5)$$

Thus the selected coefficient is adjusted in the downhill direction of the cost surface E by an amount proportional to the gradient of E with respect to the regarded coefficient. Since the partial derivative of E with respect to the selected coefficient c_k evaluates to

$$-\frac{1}{2}\frac{\partial}{\partial c_k}E = \sum_{m,n} w(m,n)t(m,n)\varphi_k(m,n) - \sum_{m,n} w(m,n)\varphi_k(m,n)\sum_{j=1}^{N} c_j^{(m)}\varphi_j(m,n), \qquad (6)$$

we choose that basis function $\varphi_k(m,n)$ in the m-th iteration the gradient of which maximizes the absolute value of (6).

Although the basis functions must have global spatial extent for an expansion in order to yield a signal extrapolation, early visual representation in human vision is known to rely on locally bounded image analysis [5]. In fact experiments with the blind spot reveal that for filling in the visual system gives more emphasis to visual perceptions close to the blind spot than to those having a greater distance. This local behaviour can be simulated by specifying the window function $w(m,n)$ appropriately. Apart from the fact that the samples covered by the blind spot must be zero, there are no restrictions for those samples of the window function corresponding to the observable part of $t(m,n)$. Thus we may well use a Gaussian shaped window function

$$w(m,n) = \frac{1}{2\pi\sigma_m\sigma_n}\exp\left\{-1/2\left[((m-m_s)/\sigma_m)^2 + ((n-n_s)/\sigma_n)^2\right]\right\} \quad \forall \ (m,n) \notin S \qquad (7)$$

centered at the location (m_s,n_s) of the blind spot with variable variances σ_m^2 and σ_n^2. In connection with Fourier basis functions this actually leads to a Gabor-like signal expansion where local image analysis is accomplished by explicitly using Gaussian weighted Fourier basis functions. There is large evidence that this type of image decomposition plays a significant role in human visual perception since at least simple cortical cells have receptive-field profiles which can well be modelled by Gabor elementary functions [5, 6].

Figure 1: Left: Test pattern for investigating the visual process of filling in, redrawn from [1]. Right: Simulation result obtained for the described selective extrapolation.

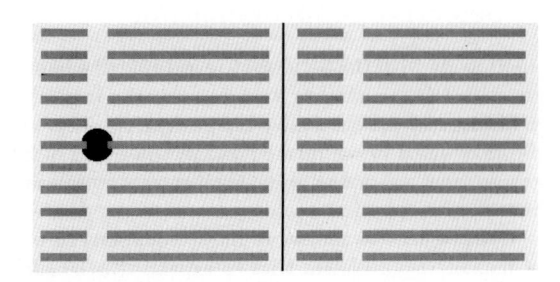

Figure 2: Left: A more complicated test pattern involving subjective contours, also based on [1]. Right: Applying the proposed approach yields the depicted vertical bar extrapolation which is identical to the observed human perception.

4 Results

The left image in Fig. 1 shows a test pattern used in [1] for investigating the visual process of filling in. If the black disk falls within a person's blind spot, a continuous line is perceived the colour of which seems to change rather gradually than abruptly from dark to bright. Application of the described frequency-selective extrapolation to this test pattern yields the result depicted in the right image in Fig. 1 which is in complete agreement with the subjectively perceived image.

A more complicated test pattern is shown in Fig. 2. If the blind spot is aimed at the black disk, the observer usually perceives a continuous *vertical* bar. This surprising fact is due to the extraordinary characteristic of the visual system to produce illusory or subjective contours when the eye records certain sets of incomplete shapes or broken lines [7]. The right image in Fig. 2 shows the extrapolated pattern obtained by the described model which is also identical to the observed human perception. In both experiments a symmetrical Gaussian window function was used the standard deviation of which was set to be half the diameter of the black disk.

References

[1] V. S. Ramachandran, "Blind spots," *Scientific American*, vol. 266, pp. 44–49, May 1992.

[2] A. Kaup and T. Aach, "A new approach towards description of arbitrarily shaped image segments," in *Proc. Int. Conf. on Intell. Sig. Process. and Com. Syst.*, pp. 543–553, 1992.

[3] M. D. Levine, *Vision in Man and Machine*. New York: McGraw-Hill Book Company, 1985.

[4] A. Kaup and T. Aach, "An approximation-theoretical approach for efficient texture analysis," in *Proc. 14. DAGM-Symp. Patt. Recogn.*, pp. 206–213, Berlin: Springer, 1992, (in German).

[5] D. A. Pollen and S. F. Ronner, "Visual cortical neurons as localized spatial frequency filters," *IEEE Trans. SMC*, vol. 13, pp. 907–916, 1983.

[6] J. G. Daugman, "Complete discrete 2-D Gabor transforms by neural networks for image analysis and compression," *IEEE Trans. ASSP*, vol. 36, pp. 1169–1179, 1988.

[7] D. Marr, *Vision*. New York: W. H. Freeman and Company, 1982.

Object Identification Using Shape and Color Information

R. Schettini

Institute of Cosmic Physics and Related Technologies
National Research Council
Via Ampere 56, 20131, Milano, Italy.

Abstract
A noise-robust and general method is presented for detecting in an image the occurrences of a sample object exploiting both shape and color information.

1. INTRODUCTION

Most early works on object recognition and location assumed the objects were monochromatic and that they could be discriminated using information on shape alone. However, everyday human experience and recent computer vision research have shown that the color distribution of multicolored objects also provides an efficient cue for their identification [1]. A novel computational strategy for recognizing and locating objects on a known background which exploits both shape and color information is presented here. It has been designed to provide for the following basic situations: objects in the scene having the same shape and different color distributions (or viceversa), objects in the scene having different scales and which may touch or overlap, giving rise to partial occlusions.
The recognized objects are classified as "overlapping", i.e. partially or totally covering other objects (isolated objects are included in this category) or "underlapping", i.e. partially covered by other unknown objects.

2. MODEL AND CANDIDATE OBJECT REPRESENTATION

Given any discrete color space, the 3D color histogram can be used to represent the color distribution of a multicolored object. Both the color space adopted and the number of bins of the histogram used to describe color distribution can influence the recognition rate. A simple opponent color space (RG, BY, WB) derived from the RGB values is employed here. This is defined as:

$$RG = R - G, \; BY = (2B - R - G)/2, \; WB = (R + G + B)/3 \qquad (1)$$

Four bins are used for each color axis as it has been shown that this resolution permits both good color image quality and a high recognition rate.

Color distribution is fairly insensitive to any translation, rotation or distortion of the objects, while it is influenced by changes of scale and occlusions. Appropriate scaling must be performed, exploiting shape information, before using it to identify a model in a scene.

As regards shape information both model and candidate objects are processed in the same way. In each case, the external boundary is first traced to obtain an ordered set of boundary points. A polygonal approximation is then introduced to decrease the number of boundary points and smooth out quantization noise and segmentation errors introduced into the objects' boundaries [2]. The approximation is found by first locating a few "true" breakpoints according to an iterative algorithm derived from the collinearity principle, and then applying to each boundary sector a modified version of Lowe's algorithm. This algorithm produces a high quality, but not optimal, approximately scale-invariant polygonal approximation, with the interesting property that no thresholds are needed to check the accuracy of the result.

A new polygonal approximation is then adopted to capture the essence of the shape with a lower number of segments. This is obtained by linking the vertices of local maxima, i.e. those vertices of the polygonal approximation with a local curvature greater than the local curvatures of adjacent vertices. The following function is used to estimate vertex curvature [3] :

$$k(V) = \frac{\pi - \phi}{\Delta l} \qquad (2)$$

where $k(V)$ represents the curvature of the vertex V, ϕ the inangle, and Δl the change in arc length.

A generic object shape is therefore represented at two different levels. The rough polygonal approximation (obtained linking local maxima vertices) is used in generating hypotheses, while the finer polygonal approximation is used in the evaluation phase.

The shape of an object is described using only simple-to-compute local properties of the vertices valid for the rough polygonal approximation. Each vertex is numbered progressively clockwise around the boundary and described by its inangle ϕ and by the length ratio R of the segments which form the vertex itself. The description is consequently invariant with translation, rotation, and scaling [2].

3. HYPOTHESES GENERATION

To generate a hypothesis about the presence of a model in the scene the algorithm exploits the rough approximation of the objects' boundaries. We begin by comparing the description of each local maxima of the model $V_i^m = (\phi_i^m, R_i^m)$ $i = 1, ..., n$, with all the local maxima of the object examined $V_j^o = (\phi_j^o, R_j^o)$ $j = 1, ..., m$, . Two vertices are judged similar if the following conditions are satisfied [4] :

$$C1: \quad |\phi_i^m - \phi_j^o| \leq T_\phi \qquad (3)$$

$$C2: \quad T_R \leq \frac{R_i^m}{R_j^o} \leq \frac{1}{T_R} \tag{4}$$

in which T_ϕ and T_R are thresholds chosen by the user.

This last phase identifies the local maxima in the two objects that show similar characteristics. Since similarity is not univocal, wrong and multiple assignments may occur. Consequently, the algorithm must now remove local maxima sequences in the model which do not match ordered sequences in the candidate object. If correct sequences are not identified, the algorithm considers the model not recognized in the candidate, and the hypothesis of a match is rejected. If more than one sequence is located, the algorithm will test each sequence separately.

4. HYPOTHESES VERIFICATION

For each transformed model TM_f (a model to which the average coordinate transformation has been applied to be superimposed on the candidate object) and candidate object, the shape matching error is defined as the normalized difference between the area of the polygon enclosing the candidate and the transformed model and the area of the polygonal approximation of the candidate object:

$$E(O, TM_f) = \frac{AREA(O \cup TM_f) - AREA(O)}{AREA(TM_f)} \tag{5}$$

The algorithm retains only those hypotheses with a matching error that is below than a given threshold. Then, for each candidate object, only the best hypothesis for each occurrence of the model object is retained.

At this stage all the hypotheses still valid correspond to objects in the scene having the same shape as the model searched. Since the scale of the possible model occurrences in the scene is now known, color information can be effectively exploited for the final identification step. The following test is performed for each possible model occurrence.

Let $I_k(M)$ and $I_k(TM_f)$ $k = 1....K$ (where K is the number of buckets) be the 3D color histograms of the scaled model object and of the region enclosed in the boundary of the transformed model respectively. Their histogram intersection $H(I_k(M), I_k(TM_f))$ is defined as [1]:

$$H(I_k(M), I_k(TM_f)) = \frac{1}{\sum_{1}^{K} I_k(M)} \sum_{1}^{K} MIN(I_k(M), I_k(TM_f)) \tag{6}$$

The result of the histogram intersection is a number normalized to 1, representing the percentage of pixels of the model that have a corresponding pixel of the same color in the candidate object (or more precisely in that part of the candidate object within the boundary of the transformed model).

The higher the value of $H(I_k(M), I_k(TM_f))$, the better the fit between the model and the candidate object. Therefore a value of $H(I_k(M), I_k(TM_f))$ greater than a given threshold means that an occurrence of the model object is actually present in the candidate object, possibly occluding other objects. In this case, the model is recognized and classified as overlapping.

A low histogram intersection value can be interpreted to mean either that the algorithm has identified an object having the same shape of the model but a different color distribution (and therefore the hypothesis must be rejected), or that the model has been correctly detected but it is partially covered by other objects. The verification of this last possibility is based on the fact that if the model object is actually present but is partially occluded by other objects, it is however likely that the part P included in the matched boundary part that has generated the hypothesis is not occluded (this may not be true when, for example, an occluding object protrudes into P). The algorithm now considers the model part P as the model object, and performs the color consistency test. If the histogram intersection value is high (i.e. greater than a given threshold) the model is considered recognised and classified as "underlapping"; otherwise the hypothesis is discarded.

5. FINAL REMARKS

According to the proposed strategy some threshold values must be established. These should be estimated on a training set, however, in our experience their heuristic definition is not troublesome. The method has been successfully tested using a large object database (90 objects).

The proposed method is effectively used to implement the search phase of a "search-and-replace" function for raster images [5] used for pattern editing in the field of textile design (in which the function provides an editing facility for images analogous to the well known "search and replace" of a text editor).

6. REFERENCES

[1] M.J. Swain, D.H. Ballard "Indexing Via Color Histograms" Proc. Image Understanding Workshop, Pittsburgh, Pennsylvania, pp. 623-630, 1990.

[2] R. Schettini "Recognition and Location of Two-dimensional Partially Occluded Objects" Signal Processing VI: Theories and Applications (J. Vanderwalle, R. Boite, M. Moonen, A. Oosterlinck eds.) Elsevier Science Pub., Vol. I, pp. 599-602, 1992.

[3] M.W. Koch R.L. Kashyap "Using Polygons to Recognize and Locate Partially Occluded Objects" IEEE Trans. on Pattern Analysis and Machine Intelligence Vol. PAMI-9, pp. 483-494, 1987.

[4] H-C. Liu, M.D. Srinath "Partial Shape Classification Using Contour Matching in Distance Transformation" IEEE Trans. on Pattern Analysis and Machine Intelligence, Vol. PAMI-12, pp. 1072-1079, 1990.

[5] A. Della Ventura, P. Ongaro, R. Schettini "Search and Replace of 2D-Objects in Digital Images" Visual Form: Analysis and Recognition, World Scientific press, pp. 205-212, 1992.

GCV-aided linear image regularizations in sparse-data computed tomography and their applications to plasma imaging

N. Iwama and M. Teranishi

Department of Electronics and Informatics, Toyama Kenritsu University, Kosugi, Toyama 939-03, Japan

Abstract

To regularize the ill-posed images in sparse-data computed tomography, two linear algebra methods of Phillips-Tikhonov and of pseudoinverse are studied with emphasis on the optimization by means of the generalized cross validation (GCV). The methods are examined and compared by simulations and experiments of the plasma imaging for nuclear fusion research, which is currently a typical example of image reconstruction based on poor projection measurement. The superiority of the former method and the usefulness the minimum GCV criterion are shown.

1. INTRODUCTION

The purpose of this paper is to study numerical techniques of image reconstruction for imaging smooth objects like plasmas under the condition that the projection data are sparse. The interest is given to the linear regularization and its neg-entropy aided optimization of the ill-posed images with the following formulation.

Let us suppose a two-dimensional intensity distribution $E(x,y)$ to be reconstructed from M line integration values, which are few. Regarding the sparseness of projection data and the expected smoothness of distribution which may allow a fairly rough partition of the object domain into K pixels, we take a fully discrete scheme of reconstruction using the linear regression model $S=LE+e$, where S and E are M and K dimensional vectors composed by the projection values and by the intensities lexicographically ordered, respectively, and L is the projection matrix composed with the geometrical weight L_{mk} of the k'th pixel in producing the m'th projection; e denotes the residual vector. With a fixed observation system L and from data S, we examine two linear algebra methods and the use of the GCV to get a reasonable estimate of E [1].

2. RECONSTRUCTION PROCEDURES

2.1 Phillips-Tikhonov Method

The solution obtained with strategy of minimizing the mean squares $\varepsilon^2 = |S-LE|^2/M$ may be well regularized, even under the sparse-data condition of M<K, when we take the

objective functions of Phillips-Tikhonov type. Taking the functional $Q = \int [\nabla^2 E(x,y)]^2 dxdy$ to be minimized with constraint of constant ε^2, we pursue the Lagrangian procedure of minimizing the quantity $\Phi=\gamma|CE|^2+|S-LE|^2/M$ with a square matrix C which approximates the Laplacian operator, and with a multiplier γ, which plays the role of regularization parameter. Then, the solution is given by $\hat{E}=(L^TL+M\gamma C^TC)^{-1}L^TS$, which is linear on S.

2.2 Pseudoinverse Method

The pseudoinverse solution with the minimized norm $|E|^2$ is also linear on S and may be well regularized by tapering the series expansion with use of the window function $w_i(\gamma)$, that is, a function taking nonnegative-values smaller for the terms of smaller singular values:

$$\hat{E} = \sum_{i=1}^{N} w_i(\gamma) a_i v_i, \quad a_i = \sigma_i^{-1} u_i^T S, \tag{1}$$

where u_i and v_i are the i-th column vectors of the unitary matrices U and V, respectively, and generated by the singular value decomposition (SVD) of L, that is, $L=U\Sigma V^T$, $\Sigma=\mathrm{diag}(\sigma_1,\sigma_2, \ldots ,\sigma_N)$ $(\sigma_1 \geq \sigma_2 \geq \ldots \geq \sigma_N; N=\min(M,K))$. When we use the window function of the form $w_i(\gamma)=(1+\gamma\sigma_i^{-4})^{-1}$ $(i \leq R)$, $=0$ $(i>R)$, we have γ and R as regularization parameters.

2.3 Expressions of GCV and PRESS

In the Phillips-Tikhonov method, the SVD of the matrix LC^{-1} as $LC^{-1}=XDY^T$ leads to the expressions of the GCV, $V(\gamma)$, and its primitive form, namely, the prediction sum of squares (PRESS), $P(\gamma)$, which are to be minimized at the optimum value of γ:

$$V(\gamma) = M\sum_{i=1}^{M}(\rho_i Z_i)^2 / (\sum_{j=1}^{M}\rho_j)^2, \quad P(\gamma) = M^{-1}\sum_{i=1}^{M}[(\sum_{k=1}^{M}\rho_k x_{ik} Z_k)/(\sum_{k=1}^{M}\rho_k x_{ik}^2)]^2, \tag{2}$$

where Z_i and x_{ik} are the components of X^TS and X, respectively, and we have $\rho_i=M\gamma(\sigma_i^2+M\gamma)^{-1}$, putting σ_i to zero for $i=K+1,K+2,\ldots,M$ in the case of $M>K$. Additionally, the SVD of LC^{-1} leads to a useful reduction of $\hat{E}$ to a computationally convenient form for which any large scale matrix inversion is unnecessary; that is, the k'th component $\hat{E}_k$ is rewritten as

$$\hat{E}_k = M^{-1}\sum_{i=1}^{M}[\sigma_i(\sigma_i^2+M\gamma)^{-1}](C^{-1}Y)_{ki} Z_i \quad (k=1, 2, \ldots, K) \tag{3}$$

with the ki component $(C^{-1}Y)_{ki}$ of $C^{-1}Y$.

Meanwhile, the SVD of L in Eq. (1) gives the GCV and PRESS expressions in the pseudoinverse method:

$$V(\gamma,R) = M\sum_{i=1}^{M}(\beta_i \zeta_i)^2 / (\sum_{i=1}^{M}\beta_i)^2, \quad P(\gamma) = M^{-1}\sum_{k=1}^{M}[b_{kk}(\gamma,R)(\sum_{i=1}^{M}\beta_i u_{ki}\zeta_i)], \tag{4}$$

where ζ_i and u_{ik} are the components of U^TS and U, respectively; and we have $\beta_i=\gamma\sigma_i^{-4}\alpha_i$ and $b_{kk}(\gamma,R)=[1-(UFU^T)_{kk}]^{-1}$ with $\alpha_i=(1+\gamma\sigma_i^{-4})^{-1}$ and $F=\mathrm{diag}(\alpha_1,\alpha_2,\ldots,\alpha_M)$, putting σ_i to zero for $R+1 \leq i \leq M$.

The GCV and the PRESS in each of two regularization methods are expected to take minimum as functions of the regularization parameters, and thereby give the optimally regularized image $\hat{E}$ for which the contradiction between the spacial resolution and the reliability is compromised in an appropriate manner. The behaviors of these quantities and $\hat{E}$ are examined below by simulations and experiments.

3. SIMULATIONS AND EXPERIMENTS

The above reconstruction procedures were examined on a visible light emission CT system mounted in a poloidal plane of torus as seen in Fig. 1. The results of numerical simulation and experimental data analysis are summarized in Figs. 2 and 3, where $\hat{\epsilon}^2 = M^{-1}|S-L\hat{E}|^2$, $\delta_E^2 = K^{-1}|E_o-\hat{E}|^2$ and $\delta_S^2 = M^{-1}|S_o-L\hat{E}|^2$ are the attained least mean squares, the mean squared reconstruction errors evaluated on the assumed image E_o and its projection S_o, respectively. With the pixel number up to K=900 for M≤600 in simulations of both methods, the GCV V(γ) suitably worked, taking minima at the γ values large for smooth shapes of E_o and small for steep shapes, and thereby yielding good approximates of E_o. It is noted in Fig. 2 that the γ values minimizing V(γ) agree well rather with those minimizing δ_S^2 as expected from the final prediction error concept in defining the GCV on projection data. Also, the behavior of P(γ) on point sources might suggest the breakdown of PRESS related to the nearly diagonalized matrix L. The window function regularization of the pseudoinverse proved too weak, as seen in Fig. 3(b), for the expected smooth shape of plasma under the sparse-data condition. The comparative excellence of the strongly regulated Phillips-Tikhonov image of plasma in Fig.3(a) is given a guarantee by a similar value of $\hat{\epsilon}^2$ obtained at the minimum of GCV, and surely by the lowered value of the minimum.

4. CONCLUSIONS

The image reconstructions from projections with linear algebra methods are well improved, under sparse-data conditions, by optimizing regularization parameters with the criterion of minimum GCV. The Phillips-Tihkonov method is powerful for the tomographic imaging of objects smooth like plasma.

The authors would like to thank Prof. S. Takamura and Dr. Y. Shen of the Nagoya University for providing experimental data and also for useful discussions.

Reference
1 N. Iwama, H. Yoshida, H. Takimoto et al., Applied Phys. Lett. 54 (1989) 502.

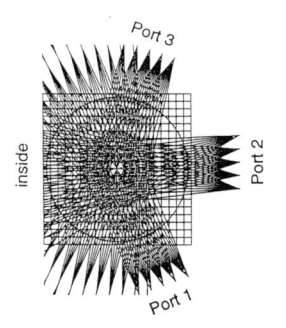

Figure 1. Layout of Hα line emission CT on the tokamak CSNT-III (Nagoya University) with three observation ports; line of sights of M≤600 in a poloidal plane of torus with optical fibers (plasma radius a_p=8cm).

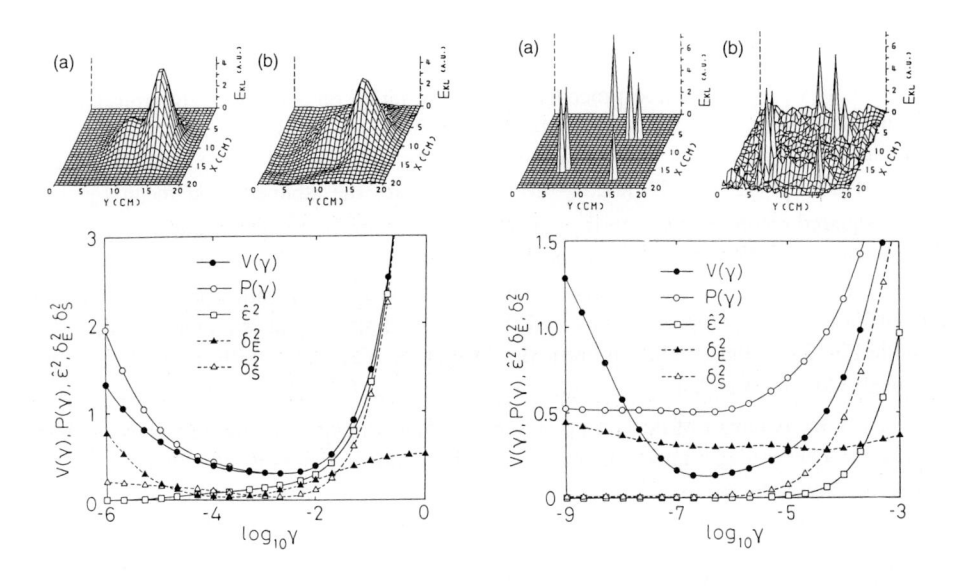

Figure 2. Simulation of Phillips-Tikhonov reconstruction. Left: (a) the assumed image, (b) the min. GCV image ($\gamma=2.0\times10^{-3}$), and γ-dependences of GCV etc. (M=600, K=900). Right: those on point sources; the min. GCV image was obtained for $\gamma=2.0\times10^{-7}$.

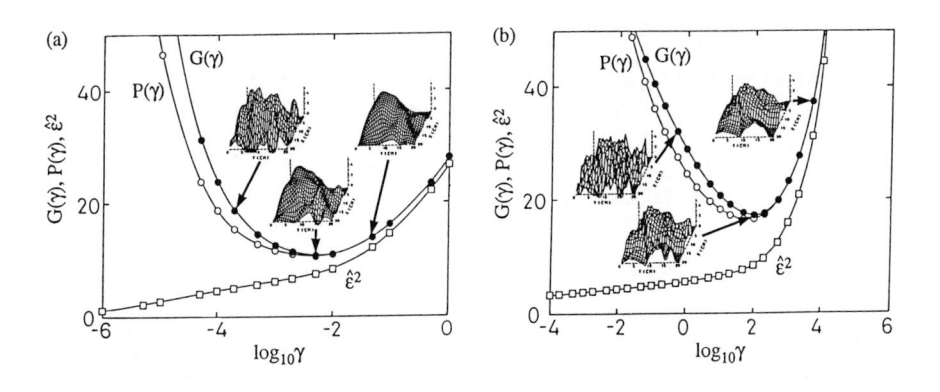

Figure 3. γ-dependences of $V(\gamma)$, $P(\gamma)$ and $\hat{\varepsilon}^2$ in plasma experiment with (a) Phillips-Tikhonov and (b) pseudoinverse methods (M=600, K=900; R=567). Obtained images are illustrated by insets.

> # Recovering 3–D Motion from Optical Flow Field

J. Heikkonen

Lappeenranta University of Technology, Department of Information Technology,
P.O. Box 20, SF–53850 Lappeenranta, Finland

Abstract
A new algorithm is proposed for the interpretation of optical flow fields obtained from consecutive pairs of images. This algorithm is based on on the ideas of Randomized Hough Transform (RHT) i.e. random sampling of velocity vectors and accumulation of 3–D motion parameters. The motion parameters are solved in a least–square sense from linear equations which relate the 2–D motion measurements and 3–D motion and structure together. Some experiments based on simulated data are presented and according to these results the algorithm is quite robust.

1. INTRODUCTION

The problem of recovering three–dimensional motion parameters of a moving observer relative to its stationary environment from a sequence of images is important for many applications. For instance in industrial robotics, the ability to estimate 3–D motion has become a basic task of robot vision system to determine and control motion of a robot in order to follow desired paths or in order to avoid collisions with obstacles.

A large number of algorithms have been proposed to compute motion parameters. These algorithms can be categorized to either optical flow approaches or correspondence approaches [2]. The optical flow approaches are based on two phases: computation of the optical flow field and interpretation of this field. Unfortunately the optical flow fields recovered by the existing techniques [2] are noisy and partially incorrect and thus this will lead into difficulties when interpreting of such fields.

In the literature the are many techniques for recovering 3–D motion parameters from optical flow field. Several researches have presented sets of nonlinear equations with motion parameters as unknowns [2]. These equations are usually solved by using iteration methods and they require good initial guesses of the unknowns. Longuet–Higgins and Prazdny [6] and Waxman and Ullman [8] have proposed methods for solving the motion parameters of local surface patch using the first and second spatial derivatives of the optical flow field. Adiv [1] has proposed an approach that searches in a least–square sense 3–D motion parameters which are compatible with optical flow field. Heeger and Jepson [3] have also described an algorithm which uses same least–squares residual function as Adiv.

Longuet–Higgins [5] and Tsai and Huang [7] have developed for motion estimation computationally simple techniques based on a set of linear equations and correspondence approach. However, experiments have shown that these linear algorithms cannot successfully deal with a realistic level of noise. Zhuang et. al. [9] have also given the differential counterpart of this 8–point discrete–time algorithm for use with optical flow field.

In this paper, an algorithm for the estimation of the 3–D motion parameters of a moving observer from the optical flow field is proposed. This algorithm is based on on the ideas of Randomized Hough Transform (RHT) [4] i.e. random sampling of velocity vectors and accumulation of 3–D motion parameters. The motion parameters are solved in a least–square sense from linear equations which relate the 2–D motion measurements and 3–D motion and structure together.

2. 3–D MOTION AND OPTICAL FLOW FIELD

In this section the relations between image velocity and 3–D structure and motion is reviewed according to Longuet–Higgins and Prazdny [6] and Waxman and Ullman [8]. From these basic formulas a linear model close to the model of Zhuang et al. [9] is derived for calculation of the 3–D motion parameters.

Let the moving observer's coordinate system be fixed to the camera with the Z–axis pointing along the optical axis as show in figure 1. It is well known that any rigid motion in space can be decomposed into a combination of translational and rotation. Here a vector $T = (U, V, W)$ is used to denote the translational velocity of a moving camera and vector $\Omega = (A, B, C)$ is used to denote its angular velocity.

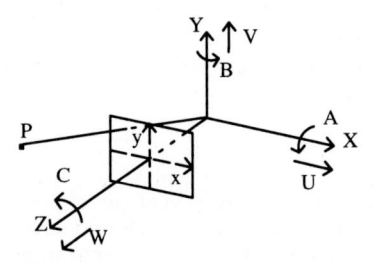

Figure 1. Observer coordinate system

Due to observer's motion, the velocity of a point P, located by position vector R, with respect to to the observer's coordinate system is given by $V_P = -(T + \Omega \times R)$. Assuming the focal length of the camera to be unity, the perspective image coordinates of a point P is given by $(x, y) = (X/Z, Y/Z)$. The corresponding image velocities (u, v) are the projections of spatial velocities and they are obtained by differentiating of the coordinates (x, y) with respect to time and utilizing the components V_P for the time derivatives of the spatial coordinates:

$$u = \dot{x} = \frac{\dot{X}}{Z} - \frac{X\dot{Z}}{Z^2} = \frac{-U + xW}{Z} + Axy - B(1 + x^2) + Cy \tag{1a}$$

$$v = \dot{y} = \frac{\dot{Y}}{Z} - \frac{Y\dot{Z}}{Z^2} = \frac{-V + yW}{Z} + A(1 + y^2) - Bxy - Cx \tag{1b}$$

From equations (1a–b) can be seen that the image flow reflects neither the absolute distance to points on scene nor the absolute translational velocities through space. They can be determined up to a scale factor i.e. only the direction of translation and the relative depth can be solved.

By solving Zs from (1a–b), equating them and rearranging terms, following equation is obtained:

$$-vx_F + uy_F - x(A + Cx_F) - y(B + Cy_F) - xy(Bx_F + Ay_F) +$$
$$(x^2 + y^2)C + (1 + y^2)Ax_F + (1 + x^2)By_F = uy - vx, \tag{2}$$

where $(x_F, y_F) = (\frac{U}{W}, \frac{V}{W})$ denotes the image coordinates of the focus of expansion (FOE). If $W = 0$, FOE will correspond to the infinity point on the image plane. For now on, it is assumed that $W \neq 0$. Finally, Given at least 8 image points (x_i, y_i) with corresponding velocities $(u_i, v_i), i \geq 8$ we have from (2)

$$\left(-v_i, u_i, -x_i, -x_iy_i, x_i^2 + y_i^2, 1 + y_i^2, 1 + x_i^2\right) H = \left(u_iy_i - v_ix_i\right), \tag{3}$$

where

$$H = \left(h_1, h_2, h_3, h_4, h_5, h_6, h_7, h_8\right)^T \equiv \left(x_F, y_F, A + Cx_F, B + Cy_F, Bx_F + Ay_F, C, Ax_F, By_F\right)^T \tag{4}$$

Obviously, by solving a system of linear equations expressed in (3), the motion parameters (A, B, C, x_F, y_F) can be solved uniquely from given $h_j, j = 1,...,8$. The conditions when these h_j's are unique i.e. when the the coefficient matrix in (3) is nonsingular, can be checked from a determinant of that matrix. If it is nonzero then the matrix is nonsingular and h_j-parameters can be solved from (3) with a least–square approach. It should be noted that the linear algorithm proposed by Zhuang et al. [9] is close to equation (3), however, the coefficient matrix and vector H differs from that of the Zhuang et al.'s [9] algorithm.

3. ALGORITHM FOR RECOVERING 3–D MOTION FROM OPTICAL FLOW

The Randomized Hough Transform (RHT) [4] first introduced a random sampling and a convergence mapping mechanics into the conventional Hough Transform (HT) method. For detecting curves which can be expressed by a N–parameter equation $f(a_1,...,a_N,x,y)$ the basic idea of the RHT is following: First the set of P is formed from edge points of a binary image. Then N point pairs (x_i, y_i) is picked randomly from set P and the parameter space point $(a_1,...,a_N)$ is solved from the curve equation. The cell $C(a_1,...,a_N)$ in the accumulator space using following matching criterion: If there exists a cell $C_i(a_{i1},...,a_{iN})$ in the accumulator, such that $|a_{ij} - a_j| \leq a_{tj}$, $j = 1,...,N$ then the score of C_i is incremented by one, otherwise a new cell $C(a_1,...,a_N)$ is created in the accumulator and its score is set to 1. Parameter a_{tj} controls the tightness of the matching criterion. This random sampling procedure is repeated until a predefined global maximum n_t in the accumulation space is detected. The cell which correspondes to this maximum gives the parameters of a detected curve.

The ideas of RHT are applicable for 3–D motion analysis. In the case of 3–D motion interpretation the set of P is formed from image points and their 2–D velocity information. The accumulated motion parameters should be solved using least–square approach from equations which relate the 2–D motion measurements and 3–D motion and structure together i.e. one should use the linear equation (3) given in section 2. The benefits of motion parameter accumulation is clear: It reduces the effect of noisy and partially incorrect optical flow field.

4. EXPERIMENTAL RESULTS

In this section some simulation tests are given. In all the experiments, values to appear in translation vectors are given in focal units and rotation parameters are given in degrees. Only rotation parameters (A, B, C) are used during the accumulation of motion parameters i.e. a cell $C_i(A_i, B_i, C_i)$ is search from the accumulator, such that $|A_i - A| \leq A_t, |B_i - B| \leq B_t$ and $|C_i - C| \leq C_t$. The translation direction and the rotation parameters were obtained by calculating the average value of (A, B, C, x_F, y_F)–parameters hitting to a cell $C_i(A_i, B_i, C_i)$. A flow field was synthesized from a random depth map (depth varying randomly between 10 and 200 in focal units) using equations (1a–b). Uniform noise of various amount was added to each component of each velocity vector according to formulas

$$(u_{noise}, v_{noise}) = \left(u(1 - \frac{noise}{100}(1 - 2rand())), v(1 - \frac{noise}{100}(1 - 2rand())) \right), \quad (5)$$

where *noise* is given in percents and *rand()* is a random number generator in the range from 0 to 1. Entire image was 1×1 in focal units, thus the field of view was 53 degrees. All these simulation results were run on a Convex 3420 computer.

The first experiment simulates a translational motion of the camera, represented by the vectors $T = (1,1,1)$ and $\Omega=(0°,0°,0°)$. Table 1 shows results in the estimates of translation direction (x_F, y_F) as a function of the noise level and number of points used in equation (3). The threshold value n_t was set to 10 and the parameters A_t, B_t and C_t were set to 0.10 degrees. In order to remove the effect of random variation of the results, all results are averaged over 10 trials at each noise level. Parameter N gives the number of points used in equation (3). Table 1 shows that if only 8 points are used then the error of translation is significant, but using 15 points or more then results are good even if there is 15 % noise.

Table 1. The translational motion parameters (x_F, y_F) obtained with input data $T = (1, 1, 1)$ and $\Omega=(0°,0°,0°)$. The correct value of (x_F, y_F) should be $(1, 1)$. For more information, see text.

Number of points used in equation (3)

Noise (%)	N=8	N=15	N=20	N=50	N=100
0	(1.00,1.00)	(1.00,1.00)	(1.00,1.00)	(1.00,1.00)	(1.00,1.00)
5	(0.92,0.91)	(0.98,0.98)	(0.98,0.99)	(0.99,0.99)	(0.99,0.99)
10	(0.66,0.66)	(0.88,0.87)	(0.90,0.89)	(0.91,0.90)	(0.93,0.93)
15	(0.63,0.57)	(0.81,0.80)	(0.81,0.82)	(0.83,0.82)	(0.84,0.85)
20	(0.45,0.42)	(0.58,0.57)	(0.66,0.68)	(0.71,0.71)	(0.75,0.75)
30	(0.42,0.40)	(0.55,0.55)	(0.61,0.62)	(0.64,0.65)	(0.69,0.68)

Next, a more complicated experiment which simulates translational and rotational motion of the camera. For this example the motion parameters are $T = (0.5, 0.5, 1)$ and $\Omega=(1.40°, 0.50°, 1.20°)$. The results are summarized in Table 2. The threshold value n_t was set to 10, the parameters A_t, B_t and C_t were set to 0.05 degrees and all results are averaged over 10 trials at each noise level. In equation (3) 20 flow vectors were used. Also the averaged count of random sampling and parameter accumulations N_c and averaged computation times t_c in seconds are presented. From table 2 one can clearly see that the results and computation times are favorable.

Table 2. The 3-D motion parameters (A, B, C, x_F, y_F) obtained with input data $T = (0.5, 0.5, 1)$ and $\Omega=(1.40°, 0.50°, 1.20°)$. For more information, see text.

Noise (%)	Motion parameters (A, B, C, x_F, y_F)	Sampling count N_c	Computation time t_c (in seconds)
0	(1.40°,0.50°,1.20°,0.50,0.50)	10	<0.1
5	(1.39°,0.53°,1.19°,0.47,0.47)	77	0.28
10	(1.33°,0.57°,1.18°,0.41,0.40)	290	1.93
15	(1.31°,0.58°,1.20°,0.40,0.40)	561	2.01
20	(1.26°,0.63°,1.19°,0.34,0.35)	818	3.36
25	(1.21°,0.70°,1.19°,0.27,0.29)	1004	4.27
30	(1.16°,0.72°,1.21°,0.24,0.24)	1202	5.52

5. CONCLUSION

In this paper, an algorithm is proposed for estimation of 3-D motion parameters of a moving observer (e.g. camera) from optical flow field. This algorithm can determine motion parameters given at least eight image points with velocities and these parameters can be obtained by solving only a set of linear equations. The final 3-D motion parameters are obtained by accumulating the motion parameters. The experimental results have demonstrate the potential of this new algorithm by showing that it is not too sensitive to errors in the image velocity measurements. The results in this paper should be of interest to numerous areas of research.

REFERENCES

1. Adiv, G., 1985, Determining Three-Dimensional Motion and Structure from Optical Flow Generated by Several Moving Objects, IEEE Transactions on Pattern Analysis and Machine Intelligence, Volume 7, Number 4, 384-401.
2. Aggarwal, J. K., Nandhakumar, N., 1988, On the Computations of Motion from Sequence of Images: A Review, Proceedings of the IEEE, Volume 76, Number 8, 917-935.
3. Heeger, D. J., Jepson, A. D., 1992, Subspace Methods for Recovering Rigid Motion I: Algorithm and Implementation, International Journal of Computer Vision, Volume 7, Number 2, 95-117.
4. Kultanen, P., Xu, L., Oja, E., 1990, Randomized Hough Tranform (RHT), In proc. 10th International Conference on Pattern Recognition, Atlantic City, USA, June, 631-635.
5. Longuet-Higgins, H. C., 1981, A Computer Algorithm for Reconstructing a Scene from Two Projections, Volume 293, September, 385-397.
6. Longuet-Higgins, H. C., Prazdny, K., 1980, The Interpretation of a Moving Retinal Image, Proc. R. Soc. Lond., B 208, 385-397.
7. Tsai, R. Y., Huang, T. S., 1984, Uniqueness and estimation of three-dimensional motion parameters of rigid objects with curved surfaces, IEEE Transactions on Pattern Analysis and Machine Intelligence, Volume 6, Number 1., 13-27.
8. Waxman, A. M., Ullman, S., 1985, Surface Structure and Three-Dimensional Motion from Image Flow Kinematics, The International Journal of Robotics Research, Volume 4, Number 3, 72-93.
9. Zhuang, X., Huang, T. S., Ahuja, N., Haralick, R. M., 1988, A Simplified Linear Optic Flow-motion Algorithm, Computer Vision, Graphics and Image Processing, 42, 334-344.

Analysis of semivariograms in satellite images for territorial applications

Pietro A. Brivio, Ilaria Doria and Eugenio Zilioli

Remote Sensing Dept., IRRS - National Research Council
56 Via Ampère - 20131 Milan, Italy

Abstract
This paper describes the contribution of the structure function, also called semivariogram, in the qualitative and quantitative analysis of Landsat Thematic Mapper images to explore spatial characteristics of different natural scenes over a range of spatial scale.

1. INTRODUCTION

Remotely-sensed images are spatial arrangements of set of radiometric measurements that capture radiances from natural scenes and then they reflect the spatial structure of the real scene represented. Quantitative measures of texture were proposed firstly in the field of computer vision and then extended also to remote sensing applications. Semivariogram, developed in geostatistics [1], more recently was also found to be valid in the area of remote sensing. We refer here to the application in the field of ground-based radiometry [2-3] and of image-processing [2-4-5-6-7-8-9].

This study deals with the application of semivariogram analysis to explore spatial pattern in Landsat Thematic Mapper (TM) images where different representations of Italian territory were considered. Analysis of semivariogram shapes is here presented at a regional scale, over a 200 km^2 window, for mountainous, rural and metropolitan areas. Same analysis was made at a local scale, over a 0.5 km^2 window, for industrial and urban centres. Parametrization of experimental semivariograms is proposed to give a quantitative measure of texture of different territorial samples.

2. SEMIVARIOGRAM IN DIGITAL IMAGES

Digital image, derived from scene scanning by a remote-sensing device, is a set of physical measurements each associated with a spatial position. In mathematical terms, the radiance or brightness B(x) as function of two-dimensional spatial position x can be considered as realisation of a "regionalized variable" [10].

Spatial information can be analysed by means of the semivariogram function defined by the relationship:

$$\gamma(h) = 1/2m \sum [B(x) - B(x+h)]^2 \quad (1)$$

in which B(x) is the radiance value measured at pixel (x), h is the lag or distance in the image, expressed as a number of pixels, defining the different locations (x + h) at which the regionalized variable B is observed.

Main characteristic features of the semivariogram are the *sill*, indicating the semivariogram value in which the function remains constant as separation distances increase and equals the general data variance and the *range* distance, indicating the lag-distance value at which the semivariogram reaches the sill. The range is the critical h at which the correlation structure ceases to exist and data vary randomly [2-10].

3. DATA USED AND SCENE SAMPLE SELECTION

This study was accomplished by means of Landsat-TM data whose ground cell resolution is about 30 m x 30 m and providing six optical spectral bands ranging from visible to middle infrared wavelengths. The semivariogram analysis was applied to digital image samples properly selected in order to represent different aspects of the Italian territory and landscape.

For the *regional scale* of analysis four large windows of 512 x 512 pixels (15 km x 15 km) were taken in the northern part of Italy: alpine range land (AL), Palmanova rural country (RU), South Milan Agricultural Park (AG), metropolitan area of Milan (ME).

For the *local scale* of analysis further five small sized windows of 24 x 24 pixels (720 m x 720 m) were taken in southern Italy (Sicily): large (IN/L) and small (IN/S) industrial settlements, nearby Priolo; suburban agglomerate of Siracusa (SU/S); urban centres of Siracusa (UR/S) and Floridia (UR/F).

4. QUALITATIVE ANALYSIS OF SEMIVARIOGRAMS

The $\gamma(h)$ function previously described was used in the analysis of the selected landscape samples. At regional scale, the most efficient band resulted to be TM4 (0.76 - 0.90 μm, near infrared); at local scale, dealing essentially with man-made features of industrial and urban agglomerates, the band TM1 (0.45 - 0.52 μm, blue) was preferable [9].

For large image windows, qualitative analysis of semivariogram shape shows two clear patterns separating the alpine site from the others. In fact the AL semivariogram appears to be unbounded, while semivariance in the other samples seems to reach the sill at a certain lag-distance around 600 meters. This fact suggests that in such landscape units the image autocorrelation stops for greater lag-distances. Distinction between the semivariograms also depends on different slope angle in the initial part of them. In this way, the metropolitan area of Milan shows the less steepened slope, in relation to an average low spatial contrast of its pixels in this band (Fig. 1).

At local scale, urban samples show flattened semivariograms, because of their variances being much lower respect to the industrial sites. The suburbs of Siracusa (SU/S) occupies a middle position, close to the small element industrial curve (IN/S). That should be due to the highest variance inferred by the presence of suburban vegetation features in the window SU/S. Finally, the highest semivariance occurs for large-element industrial site (IN/L) and it levels off at lag-distance between 60 and 90 meters which corresponds to the actual measure of petroleum refinery plants occurring in the scene (Fig. 1).

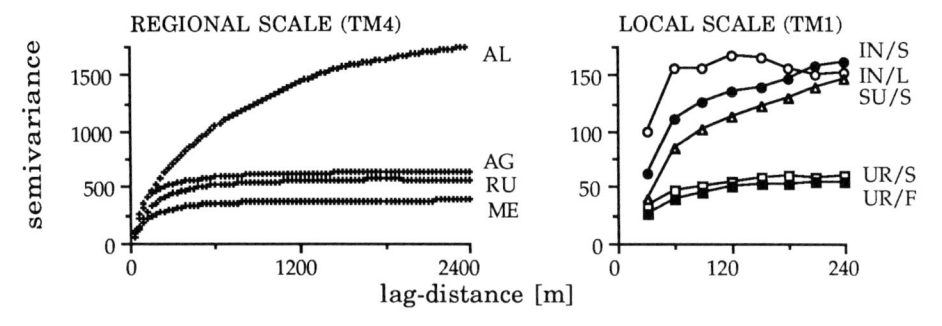

Figure 1. Experimental semivariograms of satellite image samples.

5. QUANTITATIVE ANALYSIS OF SEMIVARIOGRAMS

In order to standardize the results and to get a quantitative analysis, the sample semivariograms are fitted to the *authorised* models [1-11]. In this study the power model was adopted, for its simplicity and flexibility:

$$\gamma(h) = w\, h^a \tag{2}$$

where a must be positive and strictly lower than 2 [11].

Experimental semivariograms obtained were fitted by the least-squares method and the resulting w, a parameters are reported in Table 1. Correlation coefficient r is also indicated as evaluation of the fitness.

Table 1. Semivariogram parameters by fitting the power model.

	w	a	r		w	a	r
AL	274	0.410	0.949	IN/L	119	0.164	0.701
RU	358	0.129	0.746	IN/S	71	0.422	0.954
AG	265	0.169	0.723	SU/S	47	0.580	0.962
ME	184	0.174	0.842	UR/S	36	0.262	0.962
				UR/F	30	0.324	0.956

These spatial parameters were then plotted in a w-a space (Fig. 3). At regional scale, the mountainous area AL results well separated from the other three landscape types. The metropolitan, agricultural and rural samples show same low a values and quite different w values, with the highest distance between the rural and metropolitan areas. At local scale, it is possible to clearly distinguish two extreme situations: the urban centres of

Siracusa and Floridia from one part and the industrial agglomerate (IN/L) at the opposite site. The suburban zone (SU/S) and the industrial area (IN/S) are characterized by intermediate values.

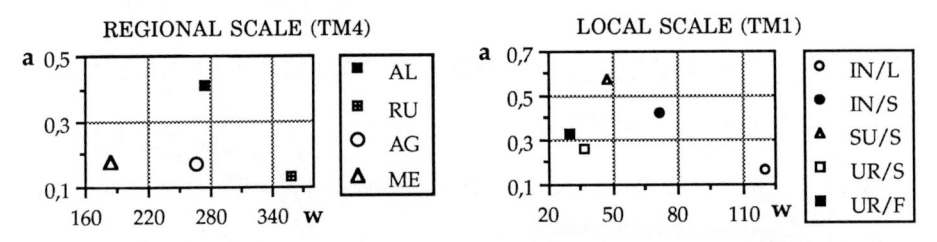

Figure 2. Diagrams of w, a parameters by fitting the power model.

6. CONCLUSION

The use of semivariograms provides insight to interpret the spatial characteristics of different landscape and territorial samples over a range of spatial scale. Regarding the results, the following points should be made:
• Experimental semivariogram shapes showed the band TM4 (infrared) and the visible band TM1 (blue) to be the most suitable for discriminating natural scenes and man-made features, respectively.
• The quantitative analysis was accomplished by fitting experimental semivariogram to the power model.
• Significant encouraging results came from the separability of landscape features here considered in a w-a parametrization space.

7. REFERENCES

1. I. Clark, Practical geostatistics. Applied Science Publishers, London,1979.
2. P.J. Curran, Remote Sensing of Environment, 24, 3 (1988) 493.
3. R. Webster, P.J. Curran and J.W. Munden, Remote Sensing of Environment, 29, 1 (1989) 67.
4. C.E. Woodcock, A.H. Strahler and L.B. Jupp, Remote Sensing of Environment, 25, 3 (1988) 323.
5. C.E. Woodcock, A.H. Strahler and L.B. Jupp, Remote Sensing of Environment, 25, 3 (1988) 349.
6. G. Ramstein and M. Raffy, Int. J. of Remote Sensing, 10, 6 (1989) 1049.
7. L. Wald, Photogrammetric Eng. and Remote Sensing, 55, 10 (1989) 1487.
8. W.B. Cohen, T.A. Spies and G.A. Bradshaw, Remote Sensing of Environment, 34, 3 (1990) 167.
9. P.A. Brivio, I. Doria and E. Zilioli, Photogrammetric Eng. and Remote Sensing, (in press).
10. D.L.B. Jupp, A.H. Strahler and C.E. Woodcock, IEEE Trans. on Geoscience and Remote Sensing, 27, 3 (1989) 247.
11. M. Oliver, R. Webster and J. Gerrard, Trans. Inst. Brit. Geogr., N.S. 14, 3 (1989) 259.

Radar image processing for automatic traffic control of sea-ports

G. Benelli[a], A. Garzelli[b] and A. Mecocci[a]

[a]Dipartimento di Elettronica, Università di Pavia, via Abbiategrasso 209, 27100 Pavia, Italy

[b]Dipartimento di Ingegneria Elettronica, Università di Firenze, via S. Marta 3, 50139 Firenze, Italy

Abstract
In this paper a system for the automatic control of sea-traffic in the access area of a port is proposed. The system employs digital processing techniques applied to sequences of radar images. Some new algorithms for the detection of drift-angles have been implemented. They provide useful information for collision avoidance systems.

1. INTRODUCTION

The main purpose of a tracking system for ship traffic control is the estimation of the target trajectory in the controlled area. An important aspect for the control of ship traffic is related to the estimation of the ship trajectory and the analysis of the prow position with respect to the motion trajectory.

A radar system can provide significant image sequences of the area near a sea-port. In this study, images formed by a radar simulator have been processed, in order to test precision and reliability of the ship location and tracking process.

The characteristics of the simulated radar system are: transmission frequency = 10 GHz; Pulse Repetition Frequency = 1600 Hz; quote of the radar site = 50 m; antenna rotation = 0.25 cps; antenna beam = 0.4°, 20° (azimuth, elevation).

The images are automatically processed to evaluate the motion direction and velocity of the ships for collision avoidance, in particular in presence of fog or low visibility.

2. SEGMENTATION AND CONTOUR EXTRACTION

Ships appear as bright regions in the radar image, while sea is represented as a dark area characterized by a higher value of local variance. The image regions that correspond to ships can be properly extracted by a thresholding operation. An adaptive threshold should be used, because of the changing conditions of the sea surface and the variable distances and positions of the ships with respect to the radar site. The solution chosen in this study for threshold computation uses a two-dimensional space (gray-level / mean-gray-level) [1]. Every pixel is mapped in a 2-D space whose axes are pixel gray-level and the local mean of gray-levels computed on a 5x5 neighbourhood of the given pixel.

Starting from this representation, points which belong to homogeneous regions are automatically selected, in order to generate a histogram which shows deeper valleys than the original one. Then the threshold value is chosen equal to the local minimum of the transformed histogram with the highest gray level. The value of this minimum must exceed a certain value (fixed at 120 in our experiments) in order to avoid false alarms in images without ships.

The binarization process also provides a method of classification of the image pixels, because only two classes, sea and ships, can be represented in these radar images. This fact results from the characteristics of the sensor and from the nature of the scene. Therefore, each region selected during the binarization phase is considered as an object of interest and its contour is extracted with a border following technique [2]. However some intolerable effects of the binarization step can occur and proper controls on the detected objects are thus requested. First, an object whose perimeter is lower than a fixed value is automatically discarded as it can represent neither a ship, nor an important sub-part of it. Moreover, it has been experimentally found that after the thresholding step a single ship can divide in two parts, because of its shape (particularly for oil tankers) and its position with respect to the radar site. This unsuitable effect is corrected by a simple automatic control. If the contour extraction detects two valid objects a and b and the distance d between their centroids is less than the mean length of a ship, a geometric operation that joins the two objects is activated. This operation, that iteratively works on couple of objects, corresponds to a morphological closing with a linear structuring element orientated along the line joining the centroids and whose length is ⅔ d.

3. SHIP-PROW DETECTION

For each detected object contour, a centroidal map is computed. This map gives the distance between a point of the contour and the object centroid versus the angle ϑ formed by these two points. Starting from this centroidal representation $\rho(\vartheta)$, it is possible to detect the prow of a ship, by using geometry and symmetry properties of the ship shape. This two characteristics are generally well preserved even if ship and radar site locations vary. However, it is important to study the precision of prow detection relating to the angle between radar beam and ship motion direction. The prow position is identified with that of the "sharpest" peak of $\rho(\vartheta)$. The point that satisfies the following two properties is selected:
1. high distance from the object centroid; 2. symmetrical high curvature.

The $\rho(\vartheta)$ profile is analyzed by means of an iterative process. For any ϑ, $\rho(\vartheta)$ is multiplied n times by the function

$$f(\rho, \text{Area}) = K_1 \, \rho(\vartheta) / \rho_{MAX} + K_2 \, \text{Area}(\vartheta) / \text{Area}_{MAX} \tag{1}$$

where $K_1 + K_2 = 1$ and $\text{Area}(\vartheta) = \min\{A_l(\vartheta), A_r(\vartheta)\}$. $A_l(\vartheta)$ and $A_r(\vartheta)$ are defined as

$$A_l(\vartheta) = \frac{W}{2} \rho(\vartheta) - \int_{\vartheta - W/2}^{\vartheta} \rho(x) \, dx \tag{2}$$

$$A_r(\vartheta) = \frac{W}{2}\rho(\vartheta) - \int_{\vartheta}^{\vartheta + W/2} \rho(x)\,dx \cdot \tag{3}$$

W is the dimension of the working window. The number n of iterations and the weights K_1 and K_2 are initially set to n = 10 and $K_1 = K_2 = 0.5$, respectively. If $A_l < 0$ or $A_r < 0$, then Area is forced to 0 and only local maxima of the centroidal map are considered.

If the ratio R between the ship area and the ship centroidal moment is greater than a given value (R > 2.5), the function f(ρ, Area) uses a different set of weights, namely $K_1 = 1$ and $K_2 = 0$. In this way the iterative process does not search for the sharp regions of the contour. This approach is effective when the ship-prow points towards the radar antenna, in fact in such cases the perimeter characteristics become useless and the distance from the centroid is the only effective parameter. The value of R gives an estimation of the elongation of ship shapes that is used in turn as an index to switch from one set of K_i value to the other.

The value of ϑ which corresponds to the maximum of $\rho(\vartheta)$ at the end of the iterative process gives the prow orientation.

4. REFINEMENT OF POSITION ESTIMATION AND TRACKING

For all the images of the radar sequence, the coordinates of prows and ship-centroids are memorized together with the angle of orientation of every ship.

Working on a whole sequence of images, the positions of all ship are tracked. First a matching operation is performed in order to pair each point from the set of the ship centroids at a given time with a point from the same set at the following time. The main requirement of this operation is to minimize the sum of the distances between the points in these pairs. This process is iterative and it is effective even if some ships appear in the scene or leave it [3].

The subsequent phase refines the ship-orientation estimates and provides a technique for motion prediction. The α-ß filter, which is derived from the Kalman filter theory has been used. In [4] a strategy to adaptively change the parameters α and ß in order to minimize the error variance is presented. This adaptation method belongs to the stochastic gradient family, because it updates α and ß proportionally to the negative gradient of the instantaneous quadratic innovation. This technique has been applied to solve the ship tracking problem.

5. EXPERIMENTAL RESULTS

Table 1 shows the results obtained for a ship moving on a straight line at an angle of 340 degrees with respect to the North and with a velocity of 10 knots (about 5.14 m/s). The actual position of the ship centroid has been computed from initial position, motion direction, velocity, and antenna rotation. The reference plane is centered on the antenna site.

Fig. 1.a shows a radar image which represents an oil-tanker at a distance of 3000 meters from the antenna site; the corresponding detected contour (obtained after morphological processing) is represented in Fig. 1.b. Ship-prow is accurately extracted from this map $\rho(\vartheta)$ as previously described.

Table 1
Example of computation of ship-centroid positions (x, y) (m)

	Actual coordinates	Estimated coordinates
Scan 1	(2890, 100)	(2890.36, 100.54)
Scan 2	(2918.15, 177.35)	(2917.51, 172.02)
Scan 3	(2946.30, 254.69)	(2944.58, 248.09)
Scan 4	(2974.46, 332.04)	(2974.44, 331.63)

The experimental data show that the proposed centroid procedure does not introduce relevant errors if compared with the sensor uncertainty (6.25 m in range and 0.4° in azimuth). Moreover, prow positions and ship orientations are computed with small errors (1-2 pixels in the image plane). The uncertainity in the computation of the ship direction is constantly less than 5 degrees.

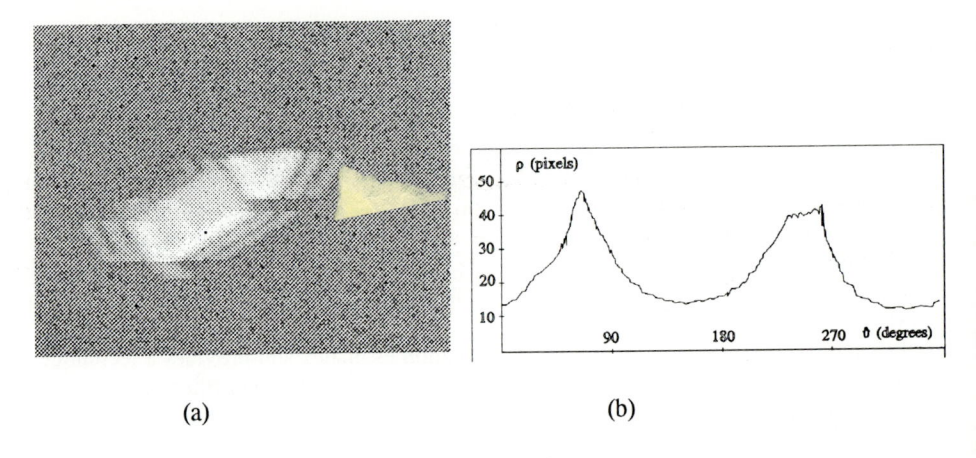

(a)   (b)

Figure 1. Radar image of an oil tanker (a) and centroidal map of its contour after morphological preprocessing (b). The left peak of $\rho(\vartheta)$ locates the ship-prow position.

6. REFERENCES

1 R.L. Kirby and A. Rosenfeld, IEEE Trans. on Systems, Man, and Cybernetics, 9 (1979) 860.
2 T. Pavlidis, Algorithms for Graphics and Image Processing, Springer Verlag, New York, 1982.
3 J. Wiklund and G. Granlund, Proc. of the Int. Work. of Time-Varying Image Processing and Moving Object Recognition, Firenze (1986) 241.
4 M.G. Otero and J.M. Páez Borrallo, Proc. Int. Conf. 'Digital Signal Processing-91', Elsevier Science Publisher (1991) 535.

Integration of Neural Network Techniques in Fuzzy Logic-Based Classification of Multisource Remote Sensing Data

E. Binaghi(*), A. Mazzetti(+), A. Rampini(*)

(*) Istituto di Fisica Cosmica e Tecnologie Relative - C.N.R.
Via Ampere 56, 20131 Milano, Italy

(+) CAP - Gemini Industria, Via A. Costa 31, 20131 Milano, Italy

Abstract

The paper empirically investigates the use of connectionist classifiers in classification of remote sensing images.
The novelty of the approach lies in applying neural network techniques to a classification task which includes multimembership patterns and in integrating results in a fuzzy rule-based system which combines other sources of data and incorporates techniques of approximate reasoning to reproduce the expert's decision behaviour. A specific example of a neural network classifier applied in the domain of fire hazard surveillance is presented, providing numerical results and experimental verification of the approach.

1. INTRODUCTION

In classification of remotely sensed images, many kinds of data collected from different sources regarding the same scene are used to extract more information and achieve greater accuracy in assigning image structures to classes. These data are not only spectral data, but include topographic information such as elevation and esposition, ground cover maps and forest maps.
In this context classes are imprecise in nature. Many relevant geographical concepts related to ground cover classes may not be defined precisely, and we do not have well-specified criteria for distiguishing between them. In talking about land cover, a piece of land with sparse grass, for example, can be classified into either grassland or soil. The impreciseness also results from natural variation. The case of multimembership patterns must be treated by interpreting pattern indeterminacy in the light of partial belonging to several categories at the same time [1].
The integration of spectral data with other ancillary data includes a multifactorial evaluation process in which data are not equally reliable and not necessarily expressed in common units but we find numerical, nominal, line-like, area-like data [2] . In most cases human experts easily perform this task by applying a cognitive process based on approximate modes of reasoning.

Factors involved in the decision process are described qualitatively, aggregated, and linked with classes according to the expert's subjective decision behaviour.

An approach that provides for all these aspects consists in dividing the analysis of multisource remote sensing data into two independent subtasks - classification of remote sensing images and integration of ancillary data by multifactorial evaluation - and in defining an unified strategy with specific solutions for the two subtasks concerned.

Proceeding from this assumption, the paper empirically investigates the use of connectionist classifiers [3] in classification of remote sensing images.

The novelty of the approach lies in applying neural network techniques to a classification task which includes multimembership patterns and in integrating results in a fuzzy rule-based system which combines other sources of data and incorporates techniques of approximate reasoning to reproduce the expert's decision behaviour [4] Expert-derived knowledge and data-derived knowledge are considered in a unified framework.

A specific example of a neural network classifier applied in the domain of fire hazard surveillance is presented, providing numerical results and experimental verification of the approach.

2. CLASSIFICATION OF MULTISPECTRAL DATA USING A FEED FORWARD NETWORK

Applications of neural networks have increased rapidly in the last years. New and improved models have been proposed which can be successfully trained to classify data of real complexity [5] . The question of how well neural network models perform as classifiers, compared with other methods, is much debated in the remote sensing community [6] . Neural networks have an advantage over the statistical methods in that they are distribution free and no a priori knowledge about the statistical distribution of the classes in the data sources is needed in order to apply these methods for classification. The performance of the neural network, however, is more dependent on a representative training sample.

The basic concept of our approach is to use clasifers which take into account the inherent variability and vagueness of real categories. In remote sensing images, the classes are frequently fuzzy, rather then sharply defined, collections of elements and boolean membership to classes is over restrictive. The problem has been addressed in a previous paper by defining a fuzzy supervised classification to explicitly represent partial membership in classes [7] . Bendiktsson et al. [6] have demonstrated that a three-layer back propagation network, in comparison with statistical methods, shows greater potentials for the processing of remote sensing data. We propose, on the bases of their results, the use of a feed forward one-hidden layer neural network with a error back propagation learning algorithm [8] . The neural network model is used to extract soft classification results from spectral data in the form of degrees of belonginess to the different classes and it is integrated with a fuzzy rule-based system which combines uncertain supports from ancillary data.

In our context, the neural network input nodes represent values for multispectral data. The output represent the gradual membership to ground cover classes. In this study the sigmoid function is used to determine the output state. The weighting factors are determined through a supervised learning algorithm. It is through this learning process that information is extracted from the data.

The basic idea of the learning procedure is to train the network with fuzzy classified spectral data. The back propagation method, which is also known as the generalized delta rule [8], is used to train the neural network.

3. INTEGRATION OF THE NEURAL NETWORK CLASSIFIER IN THE FUZZY RULE-BASED SYSTEM

The neural network classifier has been integrated with approximate reasoning techniques to model the multifactorial evaluation process in which classification results are interpreted with other data. The salient aspect of the approach is the use of fuzzy sets as the representation framework, providing knowledge structures and inference mechanisms that can model uncertain information and capture the intrinsic human variability in making rational decisions in an environment of uncertainty and imprecision. The approach used here employs fuzzy production rules as knowledge structures to represent all the ingredients of the decision-making process [9]. Fuzzy production rules are structured as evaluation-decision pairs defining a fuzzy relation between linguistic descriptions of observables and decision classes. An example of rule in the domain of fire hazard surveillance is the following: IF (Maquis is High) AND (Slope is High) THEN (Medium Fire Hazard is Low).

Terms, such as "Maquis is High", "Medium Fire Hazard is Low", are interpreted as fuzzy declarative propositions in which the linguistic qualifications "High" and "Low" are labels of fuzzy sets; the connective "and" is associated to a fuzzy logic aggregation operator.

The rule-based system interprets the results of the neural classifier as elements of the universe of discourse of fuzzy sets whose labels linguistically qualify the presence or the absence of a given ground cover class.

4. EXPERIMENTAL VERIFICATION OF THE APPROACH

The use of fuzzy reasoning techniques for the analysis of multisource remote sensing data has been validated by previous studies [10].

Here attention is focused on the operation of feed forward neural network under the conditions encountered in classifying spectral data.

For our study of neural network we have chosen a real image classification problem related to the identification of different land covers in an area of some $900 km^2$ on the north-eastern coast of Sardinia. The problem arises in the context of a project for the production of a fire hazard map of the region concerned, to be obtained by integrating image classification results with ancillary data derived from topography and from metereological aspects [10].

The data sources are three TM bands (3,4,5) of the satellite Landsat recorded in June 1989. Five land-cover classes have been pertinent as useful to the objective of the experiment.

Areas of known composition have been identified on the 512X512 pixel image, creating a training area is of 900 pixels. Each element in the training set is constituted by a vector containing the spectral values in the three bands and the classification result, expressed in terms of degree of membership in the five classes.

At the present stage of our study several experiments of Neural Networks implementations has been repeated with different coding mechanisms for input and output.

The best accuracy has been obtained using for input patterns a coding mechanism which ensures similar representations for close input values. The number of input neurons is 48. 50 outputs neurons have been implemented. The membership to one class is represented with 10 neurons.

The network, trained on the training set above described, learned with an accuracy of 95%. The learning rate was set 0.3, and 240.000 training epochs were performed. The cumulative error was approximately 10% and showed a decreasing trend. We are currently studying ways of improving these parameters, before evaluating the classification accuracy.

5. REFERENCES

[1] Wang F., Fuzzy Supervised Classification of Remote Sensing Images, IEEE Trans. on Geosci. Remote Sensing, 28, 2, 194-20, 1990.
[2] Hutchinson C.F., Techniques for Combining Landsat and Ancillary Data for Digital Classification Improvement, Photogrammetry Eng. Remote Sensing, 48, 1, 123130, 1982.
[3] Anderson J.A., Rosenfeld E. Eds., Neurocomputing, Cambridge, MA: MIT Press, 1988.
[4] L. A. Zadeh, "The Role of Fuzzy Logic in the Management of Uncertainty in Expert Systems", in Yager R.R., Ovchinnikov S., Tong R.M., Nguyen H.T. (Eds.), Fuzzy Sets and Applications, Jhon Wiley & Sons, 1987.
[5] Mazzetti A., Reti Neurali Artificiali, Apogeo, 1991.
[6] Benediktsson J.A. et al., Neural Network Approaches Versus Statistical Methods in Classification of Multisource Remote Sensing Data, IEEE Trans. Geosci Remote Sensing, 28, 4, 540-552, 1990.
[7] E. Binaghi, Anna Rampini, Enrico Zini, Multiple Uncertainty Management in Image Classification with Multisource data, in V. Cantoni, M. Ferretti, S. Levialdi, R. Negrini, R. Stefanelli Eds., Progress in Image Analysis and Processing II, pp. 443-447, World Scientific Press, 1992.
[8] Rumelhart, Hinton G.E., Williams R.J., Learning Internal Representation by Error Propagation, Parallel Distribute Processing, Rumelhart, McClelland, MIT Press, Cambridge MA, 1986.
[9] Binaghi E., A Fuzzy Logic Inference Model for a Rule-Based System in Medical Diagnosis, 7(3) Expert Systems, 7(3), 134-141, 1990.
[10] M. Antoninetti, E. Binaghi, M. D'Angelo, A. Rampini, The integrated use of satellite and topographic data for forest fire hazard mapping in NW Sardinia, 12th EARSeL Symposium on Remote Sensing for monitoring the changing geography of Europe, Eger, Hungary, 1992, (in press).

Accurate Calibration of a Binocular Stereoscopic System and its application to 3-D object measurements

F. Pedersini, S. Tubaro

Dipartimento di Elettronica e Informazione, Politecnico di Milano, P.zza Leonardo da Vinci 32, I-20133 Milano, Italy

Abstract

In this paper a technique to characterize and use a low-cost binocular stereoscopic system is presented. First an accurate camera model has been introduced, which takes into account lens distortion and displacement between optical axis and image centre. A calibrated binocular vision system has then been used for accurate measurements of rigid objects. Good performances have been achieved by considering a precise mathematical modeling of the whole stereoscopic process.

1. INTRODUCTION

In most part of binocular stereoscopic systems, a parallel optical axes geometry is used [1], because, in this case, algorithms for the detection of the stereo-correspondences are much simpler than in the general case of vergent axes. Generally these "simplified" algorithms work even with "roughly" parallel optical axes. On the other side, accurate 3-D back-projection operations require knowledge of position and orientation of the cameras with great precision. Parallelism between axes is a very hard goal to achieve with mechanical devices, so it is the main limitation to precision in all low-cost "a-priori" mechanically adjusted systems.

This problem can be overcome by means of *self-calibration*: the internal geometric and optical characteristics of the cameras (*intrinsic* parameters) and their position and orientation with respect to an external co-ordinate system (*extrinsic* parameters) are derived from a particular analysis of the images obtained framing an object (the *calibration target*) whose dimensions are known with great precision [2][3]. Moreover, a proper use of calibration results in the stereometric process allows to override the limitation of parallel optical axes and to choose the optimal framing geometry for any condition.

A mathematical model and a calibration technique for a stereoscopic system with "near" parallel optical axes has been carried out. Then optimized feature-matching and back-projection algorithms has been developed.

The system has been used to take measures on a vehicle suspension coil spring having variable radius and pitch (fig. 1).

2. SYSTEM CALIBRATION

Calibration of an image-acquisition system is a set of operations that allows to estimate all the parameters describing the acquisition process. An efficient and well tested calibration procedure is the one proposed by Tsai [2]. Its efficiency derives from having subdivided the global non-linear system in two independent subsystems, only one non-linear. This method is theoretically correct only under the following assumptions: *a)* the lens distortion has only a radial component, and *b)* the Optical Centre (OC), defined as the intersection between the image plane and the optical axis of the lens, lies in the image centre or, at least, its position is known.

While optical distortion of most part of the lenses can be considered as radial one, the assumption about the OC is generally contradicted in practice. Tsai's method gives anyway good performances, since errors in the estimation of the OC location change calibration results weakly. Nevertheless, knowledge of OC position is necessary to obtain, after calibration, good performances in the back-projection of correspondent points.

For this reason we have developed a calibration method capable to determine the position of the Optical Centre too. Estimation of its position is a critical operation because the sensitivity function of the available data (the target image co-ordinates) versus optical centre shifts is very low. So a very precise target and an accurate detection of the *fiducial points* in the images are necessary, since little data imprecision modify in a significant way the calibration results. The calibration set-up is presented in fig. 2.

Two different calibration algorithms have been implemented. The first is based on a sampling procedure where Tsai's method is used iteratively. At each iteration, Tsai's algorithm is run with different OC co-ordinates and the estimated parameters are introduced in calculations to predict the image position of the fiducial marks. Then an error function is calculated as the sum of the distances between real and predicted image points, and the OC position is found choosing the co-ordinates corresponding to the minimum error. The other algorithm is based on a global optimization of all calibration parameters, searching for a minimum of the error function described above. For this problem an algorithm for non-linear multivariable function minimization has been exploited.

The results given by the proposed algorithms have been compared with those obtained with Lenz's method [3] which provides a reliable optical way to estimate the OC position. The experiments have shown good accordance of our results with those derived optically.

3. THE STEREOMETRIC SYSTEM

A binocular stereometric system is able to derive a tridimensional description of the framed scene from an analysis of two 2-D scene projections (the images). The analysis of stereoscopic images consists of two main steps:

a) Matching: every scene element must be detected in both the images: this couple of element projections is said to be an *homologous couple* of features.

b) Back-projection: the tridimensional position of each scene element is calculated on the basis of the exact location, in the images, of its homologous.

It follows that, to make good measurements, some significant image features must be chosen: they must be easy and reliable (i.e. not ambiguous) to be recognized as homologous for a good matching and, moreover, they must be localizable with good precision, to achieve accurate measurements resulting from back-projection.

In an ideal stereo system model, epipolar lines are rectilinear, being defined as intersections between planes [4]. This is not true for real systems, because of lens distortion (fig. 3). So our modelization will take into account this effect and calculate a distorted epipolar line, simulating lens behaviour.

Distortion will be also corrected on every couple of image points to be passed to the back-projection algorithm.

4. EXPERIMENTAL RESULTS

The system has been calibrated and employed to measure a vehicle suspension coil spring, with variable radius and pitch (fig 1).

The choice of the feature to be matched is very critical in this case, because of the circular section of the steel rod. The matching feature chosen is the axis of the rod section all along the coil, which is localized as the medial line between the rod contours, revealed by edge detection algorithms. This is the only feature whose 3-D location is invariant with respect to the viewpoint.

During the experiments it's been realized that some data regularization is necessary to achieve good precision. So a polynomial curve fitting is applied to subsequent intervals of the curves formed by the two images of the helical rod axis.

The set of 3-D points provided by the back-projection is used to determine the diagrams of local radius and pitch (fig. 4) as functions of the angular abscissa. With our system the typical relative positioning error is 0.2% in the plane orthogonal to the optical axis and 1% along the axis. This anisotropy is due to the small parallax angle formed by the cameras.

5. CONCLUSIONS

The results of our experiments have shown the great importance of an accurate calibration to obtain good 3-D measures with a low-cost stereoscopic system. Precision depends much more on the accuracy in modeling the system than on the quality of instruments that build up the system itself.

For this experiment a near parallel optical axes geometry has been used, but the proposed stereometric algorithms are independent from this particular stereoscopic configuration. Thus, if the features used for matching are chosen "independent" from camera orientation, the proposed method can work with any stereoscopic geometry. It is then possible to choose the optimal system configuration depending on the particular scenic context, so that all the problems of too low measurement precision caused by a too small parallax can be eliminated.

Figure 1: Left image of the coil spring.

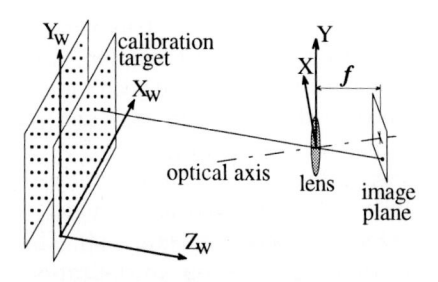

Figure 2: System calibration set-up.

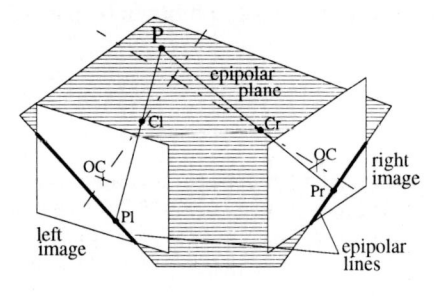

Figure 3: a) Definition of epipolar line (pin-hole model); b) ideal and real epipolar line.

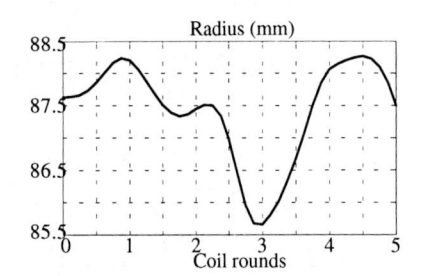

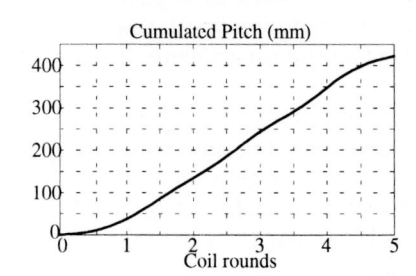

Figure 4: Final results: a) Spring coil radius (nominal:87.5 mm), b) Coil cumulated pitch.

6. REFERENCES

1. X. Tu, B. Dubuisson, *3D information derivation from a pair of binocular images*, Pattern Recognition, Vol. 23, No. 3/4, pp. 223-235, 1990.
2. R. Y. Tsai - *A versatile camera calibration technique for high-accuracy 3D machine vision metrology using off-the-shelf TV cameras and lenses* - IEEE Journal on Robotics and Automation, Vol. RA-3, No. 4, Aug 1987, pp. 323-344.
3. R. Lenz, *Lens distortion corrected CCD-camera calibration with co-planar calibration points for real-time 3D measurements*, Proc. Intercommission Conf. on Fast Processing of Photogrammetric Data, ETH Zurich, 1987, pp. 60-67.
4. N. Ayache - *Artificial vision for Mobile Robots* - MIT Press, 1990.

Extraction of Road Traffic Data from Image Sequences

M. Atiquzzaman[*] and M.G. Hartley[**]

[*]Department of Computer Science and Computer Engineering, La Trobe University, Melbourne 3083, Australia. Email: atiq@LATCS1.lat.oz.au

[**]Department of Electrical Enginering and Electronics, University of Manchester Institute of Science and Technology, Manchester M60 1QD, England.

Abstract

Image processing involves processing a vast amount of data, and hence requires a large amount of computing power. Multiprocessor systems have been suggested as sources of large amount of computing power. For efficient execution of an algorithm in a multiprocessor system, the structure of the algorithm should mirror the architecture of the multiprocessor system. A linear array multiprocessor architecture for low level image processing is described in this paper. The application under consideration relates to a road traffic situation where vehicles must be counted and velocities determined in order to permit online urban traffic control strategies and freeway incident detection. A parallel algorithm to perform the above task and its multiprocessor implementation are described in this paper.

1 Introduction

The processing of real-world images of road traffic scenes for real-time extraction of traffic flow parameters presents a number of problems. The most significant problem is the requirement of a large amount of computing power, arising due to the real-time constraint. Information regarding traffic flow consists of parameters like speed of vehicles, headway between vehicles, and classification of vehicles (e.g. cars, busses, etc.). The objective of the research work described in this paper is to propose a cost-effective and scalable multiprocessor architecture, supported by an efficient parallel algorithm to extract the traffic flow information in real-time.

This paper describes the architecture of a proposed linear-array multiprocessor system and the parallel algorithm to extract the traffic flow parameters. A prototype of the system can be built using off-the-shelf components to reduce the development cost and implementation efforts. Because of regularity in the structure of a linear array, it is scalable and suitable for implementation in VLSI. Parallel algorithms mirroring the linear array architecture are presented.

2 System Architecture

The architecture of the proposed linear array multiprocessor is shown in figure 1. It consists of a linear-array of P processor modules (PM), each of which (except the ones at the edges) is connected to its two neighbours by dedicated interprocessor links. Each module is also connected to two busses – the *video bus* (VB) and the *common bus* (CB). Images from a camera are loaded into the PMs through the video bus. The PMs pass the results of processing to the host computer through the CB which is also used for non-neighbour interprocessor communications, if such a need arises. The PMs are small computers in themselves with their own RAM, EPROM, and CPU. The PMs work asynchronously, but can also be made to work synchronously if SIMD operation is desired.

An $M \times M$ image is loaded into the system, each PM receiving a subimage of size $(\frac{M}{P}+2) \times M$. To reduce the interprocessor communication, two lines of pixels at the two boundaries of a subimage are replicated in the neighbouring PMs. Concurrent loading and processing of sub-images [1] is advocated in the design. This means that a PM starts processing as soon as its subimage from an image has been loaded, while other sub-images of the image are still being loaded into other PMs. Processed results from all PMs are passed through the CB to the host computer, which is responsible for accumulating the results and determining the required parameters. A chip-level design of the system can be found in [2].

3 Parallel Algorithm

The traffic parameters mentioned previously are determined from the area and centroid of the vehicle as shown below. A background image of the road scene is defined to be one with only stationary components, i.e., an image with no moving cars. In order to determine the area and the centroid, the moving vehicle in an image, $C = \{c_{i,j}\}$, is isolated from the image by subtracting the image from the background image, $B = \{b_{i,j}\}$, to produce a difference image, $D = \{d_{i,j}\} = \{c_{i,j} - b_{i,j}\}$. This is followed by construction of a binary edge image $\hat{D}$, by operating on each row of D using a first-order one-dimensional gradient edge-detector, and then applying a threshold τ to the output of the edge-detector. The vehicle is assumed to be of convex shape. The assumption has been found to be true for real-world images.

$$\hat{D}(i,j) = \begin{cases} 1 & \text{if } |d_{i,j-1} - d_{i,j+1}| > \tau \\ 0 & \text{otherwise} \end{cases} \quad (1)$$

where, τ is a threshold, whose value is selected by the operator depending on the type of image. Each PM contains a subimage from each of B and C as described in section 2. All PMs operate on their subimages to obtain subimages of $\hat{D}$. Because of the convex nature of the object, each PM will have two edges in a line of $\hat{D}$ which is overlapped by the vehicle. Each PM will, therefore, have between 0 to $2(M/P)$ edges. All PMs send coordinates of the edge pixels in its subimage to the host computer, which computes the area, centroid, and perimeter of the object using Simpson's rule. The area A is given by [3]

$$A = \frac{d}{3}(f_1 + f_n + 2\sum f_o + 4\sum f_e) \quad (2)$$

where, d is the width of a line, and f_1 and f_n are the lengths of the first and last strips of the object respectively as shown in figure 2. $\sum f_o$ and $\sum f_e$ are the sum of the lengths of the odd and even strips respectively. The host computer computes the length of a strip by subtracting the two pairs of edge coordinates corresponding to the strip.

$$f_i = |x_{1,i} - x_{2,i}| \quad (3)$$

where, $x_{1,i}$ and $x_{2,i}$ are the x-coordinates of the left and right edges respectively of the i-th strip of the image. The coordinates $(\bar{x}, \bar{y})$ of the centroid of the vehicle are determined by

$$\bar{x} = \frac{\sum_{i=1}^{n} x_{c,i} A_i}{\sum_{i=1}^{n} A_i} \quad (4)$$

$$\bar{y} = \frac{\sum_{i=1}^{n} y_{c,i} A_i}{\sum_{i=1}^{n} A_i} \tag{5}$$

where, A_i is the area of the vehicle in the i-th strip of the object, $(x_{c,i}, y_{c,i})$ are the coordinates of the center of the i-th strip, and n is the number of strips forming the object. $y_{c,i}$ is obtained from strip number i, and $x_{c,i}$ is obtained from $x_{1,i}$ and $x_{2,i}$ as follows.

$$x_{c,i} = x_{1,i} + \frac{x_{2,i} - x_{1,i}}{2} \tag{6}$$

If ΔP_i is the distance between two edge points, $(x_{1,i}, y_{1,i})$ and $(x_{1,i+1}, y_{1,i+1})$, of two consecutive strips, then the perimeter P is given by

$$P = \sum \Delta P_i \tag{7}$$

where, $\Delta P_i = \sqrt{(x_{1,i} - x_{1,i+1})^2 + (y_{1,i} - y_{1,i+1})^2}$. The area and perimeter of the vehicle is used to classify the vehicle. The speed S of the vehicle is given by

$$S = \frac{\sqrt{(\bar{x}(t) - \bar{x}(t + \Delta t))^2 + (\bar{y}(t) - \bar{y}(t + \Delta t))^2}}{\Delta t} \tag{8}$$

where, $(\bar{x}(t), \bar{y}(t))$ and $(\bar{x}(t + \Delta t), \bar{y}(t + \Delta t))$ are the coordinates of the centroid of a vehicle at time t and $(t + \Delta t)$ respectively.

Centroids of all the vehicles in a lane of the road are determined as described above. The algorithm is repetitively applied to all the lanes of the road. Distance between the centroids of two vehicles in a lane provides a measure of the distance between the vehicles.

4 Results

The above algorithm for collection of traffic flow parameters from real-world images were implemented in an 8086-based testbed system. The values of the parameters were found to be within 5% of the actual values. Results indicate that real-time operation (1 frame/sec) on 100 × 100 images is possible with a linear array multiprocessor system with ten 8086 processors. Each processor is then responsible for ten consecutive lines of the image. Simulation results show that the generation of a binary edge image on a linear array system using 10 processors can be carried out at a rate of 8 frames/sec. The algorithm described in section 3 was also implemented in a Transputer-based system. Detailed results regarding the speed of operation of some typical image processing algorithms, using Transputers in the proposed architecture are given in Table 1 [2]. The figures represent the maximum rate, in images per second, at which the operations can be carried out using Transputers in the PMs. The above architecture and algorithm would be very useful for implementation as a versatile and cheap traffic data collection equipment for use by traffic engineers.

5 Conclusion

A linear array multiprocessor system and a parallel algorithm to extract, in real-time, information regarding the traffic flow from images of road scenes have been described in this paper. The algorithm does not require any communication among processors, which is usually considered to be a bottleneck in multiprocessor systems.

Table 1: Performance figures for typical low level image processing algorithms on 100 × 100 images using the Transputer.

Operation	Maximum speed of operation (images per second)
3 × 3 averaging	1220
3 × 3 edge detection	1987
4-neighbor connectivity fill, shrink, & expand	600
Centroid determination	747

References

[1] M. Atiquzzaman, "Performance modeling of multiprocessor systems for different data loading schemes," Tech. Rep., Computer Science Dept., La Trobe University, Melbourne, Australia., 1993.

[2] M. Atiquzzaman, *Algorithms and architectures for automatic traffic analysis*. PhD thesis, University of Manchester Institute of Science and Technology, England, March 1987.

[3] P.S. Sarma, "A microcomputer-based system for area measurement," *IEEE Transactions on Instrumentation and Measurement*, vol. IM-33, no. 3, pp. 168–171, September 1984.

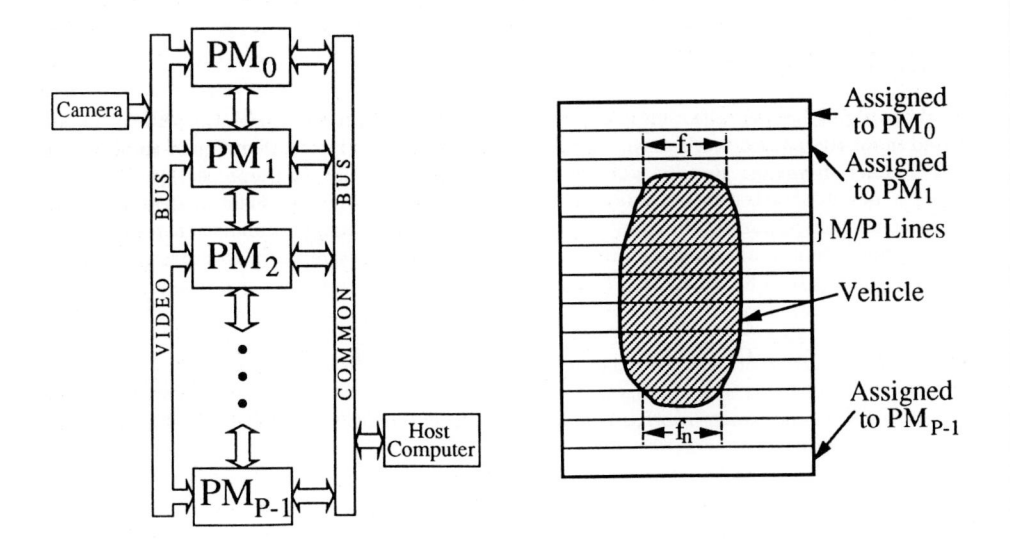

Figure 1: A linear array mulitprocessor system.

Figure 2: Partitioning an image among the processor modules.

Measuring tumor volume using pattern recognition in magnetic resonance images.
Velthuizen RP, Clarke LP, and Silbiger ML
Center for Engineering and Medical Image Analysis (CEMIA), 12901 Bruce B Downs Blvd, Box 17, Tampa, FL 33612, USA

Abstract
The use of image intensity based segmentation techniques are proposed to improve MRI contrast and provide greater confidence levels in 3D visualization of pathology. Pattern recognition methods are proposed using both supervised and unsupervised methods. In this paper some of the difficulties with volume measurements are addressed. Results are shown for a number of methods and compared with a commercial technique.

1. INTRODUCTION

The use of image intensity based methods for MRI image segmentation have been proposed to improve image contrast, or more recently for improved boundary definition between tumor, edema and normal tissues [1-3]. These methods are based on analysis of several images of the same anatomical location (i.e. a multispectral data set). Pattern recognition methods should permit objective criteria for removal of overlying normal tissues with 3D visualization of pathology as required for improved 3D display and diagnosis, 3D conformal radiation therapy treatment planning and electronic 3D surgery simulation. Contrary to MR parameter based segmentation methods that have not been particularly successful, intensity based methods are not restricted to specifically tailored RF pulse sequences. Hence, a wide selection of pulse sequences such as k space trajectory techniques, rapid acquisition with relaxation enhancement, fast magnetization preparation techniques or proton perfusion and diffusion imaging can be used that potentially provide either additional features for segmentation or improve the image signal/noise ratio. The application of pattern recognition methods to the tissue clusters generated by the above image pixel intensities result in higher order dimensional feature space that inherently should improve image segmentation. For example, these techniques potentially improve contrast between pathology, edema and normal tissues that have similar MR intrinsic relaxation. Hence, pattern recognition methods should provide a greater confidence level in boundary definition of pathology as compared to simpler gray scale approaches, using seed growing or region growing techniques, applied to a *single* image, as previously reported.

In this work a supervised segmentation technique (k Nearest neighbors) is compared with two clustering techniques: k-means and fuzzy c-means. A supervised method requires training data and hence the stability of segmentation for different training regions of interest (ROI's) within the same slice, or for segmentation based on interslice or interpatient training and classification, is therefore an important area of investigation. A comparison of supervised methods and an unsupervised method using fuzzy clustering techniques was also investigated to evaluate the feasibility of the latter method that does not require training data sets. This report will give a preliminary evaluation of the named methods for tumor volume definition based on MRI studies performed on brain tumor patients.

2. SEGMENTATION METHODS

The theoretical basis of the supervised and unsupervised segmentation methods have been reported in detail elsewhere and by these investigators for MRI segmentation [1-3], and will only be briefly reviewed.

2.1 k-Nearest Neighbors

k-NN calculates the distance from the pixel feature vector to all the training pixel feature vectors and calculates an estimate of the a posteriori probabilities from the frequency of the labels (classes) of the k-NN, where k is an odd integer. The labelled training data are obtained by operator interaction, by manually drawing regions on the raw data representing different tissue types. Even though the interface for the operator has been optimized, research has shown that results for MRI segmentation are strongly operator dependent.

2.2 k-Means

The clustering techniques used in our research greatly minimize the need for the expert's intervention. k-Means and FCM are clustering algorithms; they are unsupervised, that is, they do not need to be trained to perform the classification. They take a finite data set $X=\{x_1,x_2,...,x_n\}$ as an input, each $x_i \in X$ is a feature vector; $x_i=(x_{i1},x_{i2},...,x_{is})$, where x_{ij} is the jth feature of subject x_i, and s is the dimensionality of x_i. Let P_{km} denote partition m of X with k clusters. The k-means algorithm uses iterative optimization to approximate minima of an objective function J_m:

$$J_m(P_{km}, \mathbf{v}) = \sum_{i=1}^{n} d^2(\mathbf{x}_i, \mathbf{v}_{L_m(i)})$$

where $v=(v_1,v_2,...,v_c)$ with v_i being the cluster center of class i, $1 \leq i \leq c$, $L_m(i)$ the cluster to which x_i belongs in partition m, and $d^2_{ik} = \|x_k-v_i\|^2$. At each step, the k-means algorithm finds a partition whose error is a local minimum -- that is, J can not be decreased by moving just one case to another cluster. To obtain a partition with a selected number of clusters, the algorithm repeatedly splits of the case farthest from its cluster mean to form a new cluster and apply the k-means algorithm to the resulting partition.

2.3 Fuzzy c-means (FCM)

The third technique used in our research is as k-means, a clustering algorithm. FCM takes a finite data set $X=\{x_1,x_2,...,x_n\}$ as an input, each $x_i \in X$ is a feature vector; $x_i=(x_{i1},x_{i2},...,x_{is})$, where x_{ij} is the jth feature of subject x_i, and s is the dimensionality of x_i. A function $u: X \rightarrow [0,1]$ is defined; it assigns to each x_i in X its grade of membership in the fuzzy set u. The function u is called a fuzzy subset of X. The goal is to partition X by means of fuzzy sets. Let M_{fc} denote the fuzzy c-partitions of X, then $U \in M_{fc}$. The fuzzy c-means algorithm uses iterative optimization to approximate minima of an objective function J_m:

$$J_m(U, \mathbf{v}) = \sum_{k=1}^{n} \sum_{i=1}^{c} (u_{ik})^m (d_{ik})^2$$

where $v=(v_1,v_2,...,v_c)$ with v_i being the cluster center of class i, $1 \leq i \leq c$, and $d^2_{ik} = \|x_k-v_i\|^2$.

2.4 Brain extraction

The time complexity for the segmentation is such that it is important to segment only regions of interest, to reduce the size of the dataset X. An algorithm has been implemented to allow automatic extraction of the intracranial cavity based on a combination of adaptive thresholding and morphological filtering. It allowed to reduce the data set X for a 10-slice slab from 655360 input vectors to about 200,000.

3. RESULTS

Representative results for volume segmentations are shown in figure 1. The data was acquired in an MR scan of a patient with malignant meningioma, metastasized throughout the brain. There are several tumor sites, as can be seen in the display.

The size of the frontal tumor can be estimated from the number of pixels classified as tumor. The results are shown in table 1. The volumes found with the described pattern recognition methods are compared with the results of a seed-growing technique that is implemented on a commercially available workstation for display of volume data (ISG, Toronto, Canada). The two colums give the result for segmentation of the whole data set in one run of the segmentation algorithm, or on a slice-by-slice basis, in which case many runs of the algorithm have to be performed.

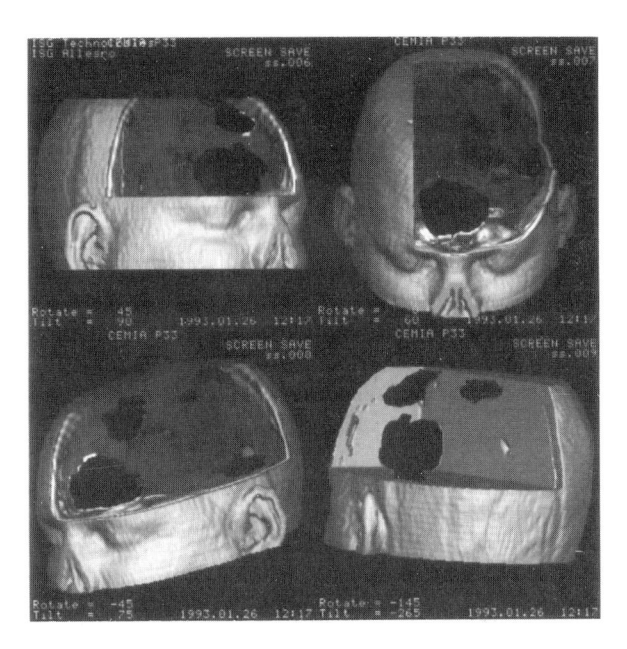

Fig 1 Four views of a segmentation result by fuzzy c-means with a clustering of the full volume data set.

4. DISCUSSION AND CONCLUSIONS

Image segmentation methods currently used in diagnostic radiology are generally supervised region growing and dilation techniques and are applied to a *single* image gray scale. Despite the recently proposed use of locally adaptive criteria, these methods do not provide either enhanced image contrast or adequate objective confidence levels in tissue segmentation. The use of pattern recognition methods as applied to multispectral image data sets should provide better image contrast and a quantitative confidence level in the segmentation achieved, particularly when unsupervised methods, such as FCM, are employed. We have recently demonstrated improved

TABLE 1
volume of the frontal lesion as in Fig 1 by different methods in mm^3.

Segmentation method	volume	slice-by-slice
k-Nearest Neighbors	7033	5726
k-Means	4588	n/a
fuzzy c-means	5203	6904
Seed-growing		
T2 weighted image	5957	n/a
enhanced T1 image	5386	n/a

contrast between tumor and edema for brain gliomas using FCM techniques with similar results to that observed with the use of MR contrast material [1].

The success of pattern recognition methods is currently limited by several factors that specifically apply to the acquisition of multispectral data sets: (a) partial volume effects related to slice thickness; (b) motion and flow artifacts, (c) image registration corrections and (d) image uniformity corrections relating to specific RF coil designs and RF field inhomogeneity. Future technology developments in MRI during the 1990's will address many of these problems. The variable motion and flow artifacts in multispectral data may be considerably reduced using 2DFT ultrafast echo planar or rapid gradient motion correction and imaging methods. The use of 3DFT techniques, in turn, will significantly reduce the effective slice thickness because of related improvements in signal/noise ratio. These technology developments will result in the acquisition of multiple 3D image data sets and will place greater demands for the development of objective and ideally unsupervised intelligent image fusion methods to increase the efficiency of 3D image display and diagnosis

Acknowledgments
This project is supported in part by NASA(NSTL), ACS - Florida Division, Sun Microsystems, Siemens Medical Systems and the H. Lee Moffitt Cancer Center and Research Institute.

REFERENCES
1   Hall LO, Bensaid A, Clarke LP, Velthuizen RP, Silbiger ML, Bezdek J. A Comparison of Neural Network and Fuzzy Clustering Techniques in Segmenting Magnetic Resonance Images of the Brain. *IEEE Trans on Neural Networks* 3: 672-682; 1992.
2   Clarke LP, Velthuizen RP, Phuphanich S, Silbiger ML, Schellenberg JD. MRI: Stability of Segmentation Techniques for Enhancing Gliomas. *Magnetic Resonance Imaging* 11 (1): 95-106; 1993.
3   Velthuizen RP, Hall LO, Clarke LP, Bensaid AM, Arrington JA and Silbiger ML. Unsupervised fuzzy segmentation of 3D magbnetic resonance brain images. *IS&T/SPIE international symposium on Electronic Imaging: Science and Technology*, Jan 31- Feb 4 1993, San Jose CA. Proceedings.

Nonlinear Filter and Neural Network for Nuclear Medicine Image Maximum Entropy Restoration

Laurance P.Clarke, Wei Qian and Huaidong Li

University of South Florida, Center for Engineering and Medical Image Analysis(CEMIA), College of Medicine, Engineering, Art and Science, and H.Lee Moffitt Cancer Center and Research Institute, MDC Box 17, Tampa, Florida,33612

Abstract

This paper proposes an order statistic-neural network hybrid filter for nuclear medicine image maximum entropy restoration based on our previouse work [1-2], This filter shares the advantages of both neural network for image maximum entropy deconvolution and advanced nonlinear filtering for noise removal and detail preservation. The filter performance is quantitatively evaluated and compared to that of the Wiener filter by investigating the relationship between the externally measured counts from cylinder sources containing phosphorous-32 (^{32}P) at various depths in water.

INTRODUCTION

A new class of filters, an Order Statistic and Neural Network hybrid filter (OSNNH), was proposed in [1] for the restoration of gamma camera images, based on the measured modulation transfer function. The use of neural network (NN) for the image restoration avoids ring effects caused by the ill-conditioned blur matrix and noise overriding caused by matrix inversion and has better performance than other restoration filters such as the Wiener, Metz, and Constrained-Least-Squares[1, 4]. As there is no optimal search for the parameter in the defined error function, maximum entropy(ME) was used in [2] as a constrained condition of the neural network and a single statistics to search for the parameter leading to better results.

Since nuclear medicine images, bremsstrahlung images in particular, are contained by various kinds of noise and the image restoration methods usually work poor in noise removing , these noise would degrade the restoring process. In our method, we use an order statistic filter to remove the noise before the neural network performs the restoration of the image.

1. ORDER STATISTIC(OS) FILTER

The class of smoothing OS filters is very efficient for noise suppression with adaptive flexibilities because (a) it offers a compromise in performance between linear filters and nonlinear filters and (b) it is possible to design an optimal (among OS filters) mean-square-error (MSE) filter for estimating the signal immersed in mixed noise. Smoothing OS filters operate in the following way. An arbitrary-extent L^2-dimensional signal $\{x_j\}$, $j=[j_1\ j_2\ \cdots\ j_m]^T$ $\in Z^L$ can be decomposed into overlapping vectors $\{x_j\}$ of length m, where x_j contains m elements from the set $\{x_j\}$ spanned by a fixed-extent m-dimensional window W_j, $x_j = \{x_j : j \in W_j\}$. The window W_j covers a fixed odd number (assume $m = 2l + 1$ for some integer $l > 1$) of connected signal samples and is symmetric with respect to the coordinate axes. In defining the OS filter, the spatial or temporal distribution of the elements of x_j within the window W_j is not of interest. Instead, define the vector of order statistics (algebraically ordered versions) $x_{(j)} = \text{order}[x_j] = [x_{(1)j}\ x_{(2)j}\ ...\ x_{(m)j}]^T$, such that $x_{(1)j} \leq x_{(2)j} \leq ... \leq x_{(m)j}$. Given a vector of real-valued coefficients $a = [a_1\ a_2\ ...\ a_m]^T$ with length m, the output of the OS filter is a matrix G. where m is the size of the selected window (number of pixels), $y_{(j)} = \text{order}[y_j]$. The coefficients a determine various OS filters, namely the median filter ($a_{l+1}=1$, $a_i=0$, i≠l+1), the more general k^{th} rank-order filter ($a_k=1$, $a_i=0$, i≠k), the averaging filter ($a_i=1/m$) which is the linear smoothing OS filter, and the α-trimmed mean filter ($a_j=1/(m-2[\alpha m])$, $[\alpha m]+1 \leq j \leq m-[\alpha m]$, $a_i=0$, i≠j) where $[\alpha m]$ corresponds to the greatest integer of αm and α is a control factor ($0 \leq \alpha \leq 0.5$) [1]. An optimal least-squares OS filter is used for estimating an arbitrary signal immersed in arbitrary noise.

The nature of the nuclear image and the contributing noise processes are such that the selection of an appropriate OS filter for a given noise smoothing application should balance the goals of both signal preservation and noise suppression. In the proposed filter, the α-trimmed mean filter is used for the OS filtering part providing robustness against a wide range of noise possibilities ranging from very shallow tailed to very heavy tailed.

2. NEURAL NETWORK FOR MAXIMUM ENTROPY (NN-ME) IMAGE RESTORATION

An artificial neural network (NN) system that can perform extremely rapid computations seems to be very attractive for image restoration. Zhou *et al* [4] formulated the restoration problem as one of minimizing an error function. Owing to the model's fault-tolerant nature and computation capability, better results can be obtained compared to the Wiener, inverse filter, and SVD filters. The error function used in [2] is defined as:

$$E = (1/2)\|Y-AX\|^2 + (1/2)\lambda\|DX\|^2 \qquad (1)$$

The parameter λ plays a significant role in the restoration process, especially in noisy cases. However, there is no optimal search for λ in Zhou's method and the result is a trade-off between resolution and noise suppression/ringing reduction. Since ME has demonstrated improved smoothness and super-resolution [7], we use it as a constraint to the NN forming a NN-ME algorithm. The following energy function is then defined:

$$E = -H(X) + \lambda_1 Q(X) + (\lambda_2/2)\|DX\|^2 . \qquad (2)$$

where $H(X)$ is the image entropy and $Q(X)=(1/2)\|Y-AX\|^2$ is a single statistics used to measure the misfit. Different choices for Q(x) are avaible depending on different circumstances. The spontaneous energy-minimization process of the NN is used to minimize the error function and the criterion of $Q(X)$ to search for c and stop the algorithm. The operation of smoothing operator, the third term of Eq. 2, is added in order to make the model more robust in restoring noisy images. Unlike Eq.(1), the experimental parameter λ_2 can be easily chosen according to the level of the noise; it can also be 0, if the noise level is low.

The parameters of the NN model, i.e. the interconnection strength between two neurons and the bias input, can be determined by comparing the error function to the energy function of the NN. Since the neurons representing the same image gray level have the same interconnection strengths and bias inputs, the NN described in [2] can be simplified. The new NN consists of only *LxL* mutually interconnected neurons (*LxL* being the image size). The neurons can be multi-state and the value of their states can be used to represent the image pixel's gray level. Because of the self-feedback, a deterministic rule should be designed to make the energy of the network converge to a minimum. It can be proved that under the condition of $x_i \gg 1$ (x_i being the state value of neuron i), the energy change ΔE due to the state change Δx_i of neuron i is given by

$$\Delta E \cong -\left(\sum_{j=1}^{L^2} T_{ij}x_j + I_i - 1\right)\Delta x_i - \frac{1}{2}T_{ii}(\Delta x_i)^2. \quad (3)$$

where T_{ij} is the interconnection strength of neuron i and j, I_i is the bias input of neuron i. The input of neuron i can be redefined as:

$$U_i = \sum_{j=1}^{L^2} T_{ij}x_j + I_i - 1 \quad (4)$$

The state of neuron i is updated by applying a decision rule:

$$\Delta x_i = g(U_i) = \begin{cases} 0, & U_i = 0 \\ 1, & U_i > 0 \\ -1, & U_i < 0 \end{cases} \quad \text{and} \quad x_i^{new} = \begin{cases} x_i^{old} + \Delta x_i, & \Delta E < 0 \\ x_i^{old}, & \text{else} \end{cases} \quad (5)$$

Using Eqs.(4) and (5), the energy function is warranted to converge to a minimum. The condition of $x \gg 1$ can be easily satisfied by adding a constant to the gray level of every image pixel; after restoration, the same amount can be deducted from the gray level of every image pixel. This procedure will not affect the restoration of the image.

Defining $W = Q(X)/(N/2)$ (N is the number of image pixels), the steps to be followed in the implementation of the algorithm are: (1) increase the degraded image gray level by a constant, e.g. 20, and consider it as the initial state of the neurons; (2) choose a suitable constant λ_1 and the initial λ_2 empirically; (3) sequentially visit all neurons to complete *one iteration*. For each neuron, use Eqs(4) and (5) to update it repeatedly until there is no further change; (4) check the energy function. If the energy does not change, stop the iteration; otherwise return to step (3) for the next iteration; (5) If $W \leq \varepsilon$ (ε being a prescribed small positive number), then a restored image is obtained after deducting the same constant from the image gray level. If $W > \varepsilon$, rechoose λ_1 and return to step (3).

3. EXPERIMENTAL RESULTS

First, different kinds of noise are added to a simulated noise-free image to generate the noisy images. These noisy images are used to test the noise removing performance of OS filter. The results show better performance of OS filter in noise removing and details perserving with comparison to standard median filter and nonlinear mean filter. The restoration performance and stability of our method is evaluated by applying it to clinical studies acquired with a gamma camera. The results are compared to that obtained from the Wiener filter [1].

Figure 1 shows the restoration results from a representative clinical study with β-emitting radionuclides; such images exhibit significantly greater degradation than those of single-photon emitters. The raw data image is degraded by a lot of noise. The image is restored with the Wiener and the proposing method. The Wiener filter (Fig. 2(d)) overcorrects the data resulting in a significantly higher number of total counts in the image while the OS filter and NN-ME method (Fig. 2(c) shows improved noise suppression and stable restoration, preserving the image's total counts.

4. REFERENCES

[1] W. Qian, M. Kallergi, L. P. Clarke, "Order-Statistic Neural-Network Hybrid Filters for Gamma Camera Bremsstrahlung Image Restoration," to be published in *IEEE Trans. Med. Imag.*, March 1993.

[2] H.D. Li, W. Qian, L. P. Clarke, and M. Kallergi, "Neural Network for Maximum Entropy Restoration of Nuclear medicine Images", Proceedings of ICASSP-93.

[3] L.P.Clarke, S.J.Cullom, R.S.Shaw, C.Reece, B.C.Penney, M.A.King, and M.Silbiger, "Bremsstrahlung Imaging Using the Gamma Camera:Factors Affecting Attenuation," *J. Nucl. Med.*, vol.33, pp.161-166, 1992.

[4] Y-T. Zhou, R. Chellappa, A. Vaid, and B.K Jenkins, " Image Restoration Using a Neural Network," *IEEE Trans. ASSP*, vol.36, pp. 1141-1151, 1988.

[5] J. B. Abbiss, B. J. Brames, and M. A. Fiddy, "Superresolution Algorithms for a Modified Hopfield Neural Network," *IEEE Trans. Signal Proc.*, vol.39, pp. 1516-1523, 1991.

[6] S. F. Burch, S. F. Gull, and J. Skilling, "Image Restoration by a Powerful Maximum Method," *CVGIP*, vol. 23, pp. 113-128, 1983.

[7] X. Zhuang, E. Stevold, and R.M. Haralick, " A Differential Equation to Maximum Entropy Image Restoration," *IEEE Trans. ASSP*, vol. 35, pp. 208-218, 1987.

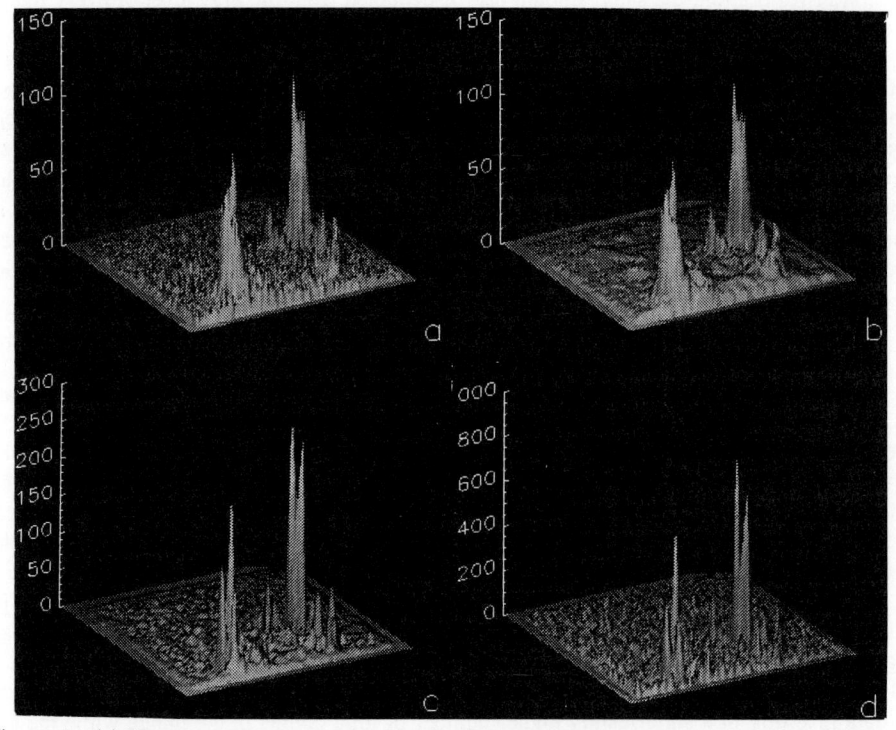

Figure 1. (a) Unprocessed planar image of the abdomen of a patient treated intraperitoneally with ^{32}P for ovarian cancer. (b) filtered image using OS filter. (c) Restored image using OS filter and NN-ME algorithm. (d) Restored image using the Wiener filter

AUTOMATIC HONEY QUALIFICATION VIA COMPUTERIZED MICROSCOPY

I. Erényi*   T. Holka,**   Z. Fazekas,*   and   A. Dékány*

* KFKI Research Institute for Measurement and Computing Techniques
P.O.Box 49, Budapest H-1525, Hungary
Phone: +36 1 169-6279   Fax: +36 1 169-5532   E-mail: h1715ere@huella.bitnet

** KFKI Raster Ltd., P.O.Box 49, Budapest H-1525, Hungary
Phone: + 36 1 169-9499   Fax: + 36 1 75-7054

Abstract

Automatic honey qualification methodology has been developed at KFKI Raster Ltd. and KFKI Research Institute for Measurement and Computing Techniques. Microscopical images of honey smears are sampled and processed via image analysis methods for identification and recognition of pollens of flowers typical in the Central European region. Based on pollen type identification and classification, the relative count of the pollens of different species is used for qualification.

Acknowledgements

The project is partially supported by Hungarian State Office for Technical Development under the contract number 91-97-16-0054. The honey smears were provided by Lukacs & Co. Biological and Chemical Laboratory, Budapest, (Hungary).

1. INTRODUCTION

Both the biological and trading value of various honey types depend on a great extend on the quality. Thus checking the type and quality of honey is of primary importance. Chemical and biological parameters for various types of honey are defined by a set of standards [STAND1987]. Quality criteria of honey-types are set in terms of pollens, ratio of pollens of various species (about 80 plants, typical for Central Europe).

The aim of the program is to take over the monotonous, subjective and tire-some work from skilled laboratory personnel, and to implement an objective, easy-to-use, low cost measurement tool. The usual analysis is carried out by observing microscopical images of stained honey-smears. Pollens of different plants are sought and counted separately. The total number of pollens for one qualification process usually is above 400.

2. POLLEN ANALYSIS PROBLEM, TASK CONSTRAINTS

Pollen analysis is based on the microscopical observation of stained honey-smears. Pollens are "floating" three-dimensional objects. There are over 80 pollen-types typical for the area, that should be classified. The number of pollens in a simple microscopical view differs from 0 to 5. Their size varies from 5 μm to 250 μm, however most of them are of 10-50 μm in diameter. Their shape has a wide variety of forms, similarly the texture of their surface varies significantly. Beside pollens, there are many other small objects: like flake of soot, parts of the body of insect, etc.

Although a fully automatic pollen recognition system would be desirable, it is not a realistic goal because of several reasons. These are the following:
- the recognition of pollens is a 3D problem,
- the pollens of one and the same species may differ radically from each other, not only in size, but also in shape,
- in some cases the discriminating features can hardly be formulated algorithmically,
- adultery of honey cannot be precluded and so not expected scenes may come up.

Therefore we confined ourselves to less ambitious project objectives. Our aim is to design and realize a pollen recognition system that takes over not all, but much of this monotonous, partially subjective and eye-probing work. The services of automatic honey qualification have been confined as follows:
- the lack of pollens (e.g. in the case of adulterate honey) can be easily detected,
- counting pollens (without their identification) can be also carried out efficiently,
- identifying pollens in special honey types using a priori information on pollen-content,
- counting selected pollens in special honey (e.g. acacia pollens in acacia honey).

It should be emphasized that these tasks use up a great part of the laboratory personnel's time if the microscopic examination is carried out conventionally. These services are supported by a low-cost Image Processing Workstation (in our case: ARGUS-IPW [BANGO 1991], [AMBROZY 1991]), an optical microscope and a camera.

3. POLLEN-IMAGE DATA-BASE

For pollen-analysis and classification, a collection of pollen-images was created and stored. The pollens have been identified by an expert. The expert has been thoroughly cross-questioned about the features he -- more or less consciously -- used for the discrimination of pollens. The image data and also this verbal information have been recorded to be used later in constructing the classification scheme.

Since the pollens are 3D objects and their size may vary significantly, the sharpness of their image cannot be achieved simultaneously (if several pollens are shown on the image). Even in case of an image with only one pollen on it, the optimal focusing is not evident. For each pollen type exhibiting spherical symmetry usually only two pictures are taken: one showing the pollen's upper surface and one focused on its horizontal great circle. For each pollen type not exhibiting spherical symmetry, image pairs from several characteristic views are taken, each image pair consists one image showing the pollen's upper surface and one showing the plasma of the pollen.

4. POLLEN ANALYSIS

The analysis of microscopic images and pollen classification is done as follows: autofocusing; pollen search; fine autofocusing for each pollen area; pollen classification, via extracting geometrical, morphological, and textural features; microscope positioning to shoot new images; repeated execution the above steps until the necessary pollen count is achieved.

Locating pollens

After an initial autofocusing procedure the slide is scanned and non-overlapping images of 512*512 pixels are taken. These square images are searched for dark regions (square subimages) having appropriate size to contain a pollen. These dark regions are obtained using automatic thresholding of images proposed by [REDDI 1984]. Then a finer autofocusing procedure is carried out for each region found. In order to obtain two images at different depth (presumably showing the plasma and the surface of the pollen), the sharpness measure is first calculated for the central part of the region and then for the peripheral part of the rectangle.

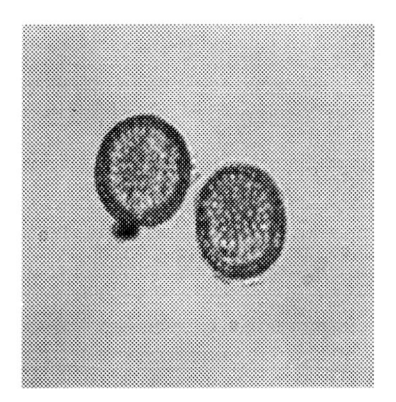

Figure 1. Pollens of *cruciferae* and regions after locating them

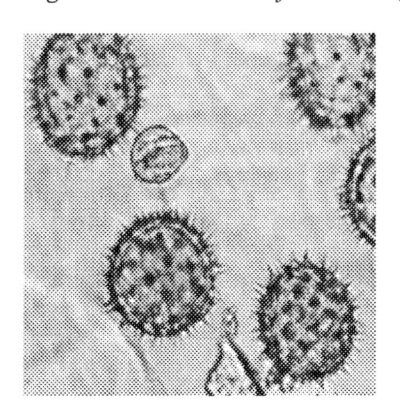

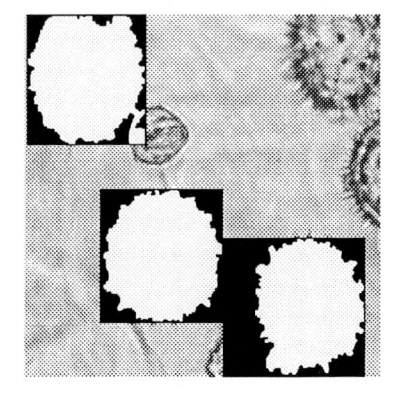

Figure 2. Pollens of *helianthus annuum L.* and regions after locating them

Classification of pollens

Consulting with our expert in honey qualification, we have chosen to set 6 pollen classes of primary interest. These are the basic pollen-constituents of special honey types collected in Hungary: white acacia, facelia, cruciferae, sunflower, sweet-chestnut and vetch. We have a class also for the anemophol pollens, that are usually extremely big or little with respect the pollens collected by the bees. The anemophiles are not "honey making" plants, they should be not counted into the sum of relevant pollens. An additional class comprises pollens not belonging to the first 7 classes. This class also comprises pollens which resemble to the pollens of primary interest. 10 subclasses are fixed for these pollens of secondary interest (e.g. apple-tree, raspberry, pear-tree, thistle, horse-chestnut-tree). Special investigation is needed to identify them. We have selected morphological (geometrical) and textural features for classification. Features, such as size, shape, pollen wall and plasma features are considered.

In spite of the importance of pollen size, this feature can only be used to a limited extent (as first filter for sizes of 10-40 µm). The pollens not in this range are taken as anemophiles. The next important feature is the pollen shape. We test whether the actual pollen's blob is shaped triangular, oval or circular. The pollen wall consists of two main layers (exine and intine). These layers are visible at the resolution we use. The wall of some pollen types is characteristic. Further features can be derived from the pollen plasma, for example the plasma of the facelia is typically ribbed, and the plasma of the acacia has low granularity.

A multi-stage, tree structured classifier using different features at different levels is used. The classification at the decision nodes is based upon the Mahalanobis distance. First, some geometrical features (e.g. the blob's greatest diameter, area, perimeter, shape factors [DANIELSON 1978] and shape descriptors for the blob's outer boundary's morphologic closure) are calculated and used for discrimination between pollen-types. At a further level of classification, the existence or the absence of some specially shaped structures within the blob at well-defined parts of the blob are adopted as feature. If these levels happen to be insufficient for unique classification (as often happens) then some sort of textural parameters should be calculated either for the blob in the binarized image, or for the same area in the original image.

5. CONCLUSION

The experiments are promising. Results of automatical honey qualification for a variety of honey types can be done with satisfactory reliability and in reasonable computing time. We expect further improvement after the real implementation work will fully be concluded.

6. REFERENCES

[AMBROZY1991] Gy. Ambrózy et al., *ARGUS - a PC based IPW, its software and application.* The EUROMICRO Journal, Vol. 34. (Feb 1992), No.1-5. pp. 135-138.
[BANGO1991] Gy. Bangó et al., *Design and realization of the hardware structure of ARGUS IPW.* The EUROMICRO Journal, Vol. 34 (Feb 1992), No.1-5. pp. 207-210.
[DANIELSSON1978] P.E. Danielsson, *A new shape factor.* CGIP Vol. 7, 1978, pp. 292-299.
[REDDI1984] S. S. Reddi et al., *An optimal multiple threshold scheme for image segmentation.* IEEE Trans. Systems, Man and Cybernetics, Vol. SMC-14, 1984, pp. 661-665.
[STAND1987] *Hungarian standard: The honey;* MSz 6950-87.

Feature Selection in Neural Networks for Image Classification

Sebastiano B.Serpico

Dept. of Biophysical and Electronic Eng. Univ. of Genoa, Via all' Opera Pia 11a, I-16145 Genova, ITALY, Phone: +39-10-3532752; Fax: +39-10-3532777

Abstract
This paper presents a technique for pruning the connections of artificial neural networks for classification, with neuron nonlinearities of the continuous type, and its application to feature selection. An upper bound of the influence of a connection on the outputs of a net is computed. If such an influence is below a threshold, then the connection can be removed so obtaining a smaller quasi-equivalent net. A similar concept can be applied to check if an input feature can be eliminated and all the connections starting from it removed from the net. Experimental results obtained in remote sensing image classification are reported.

1. INTRODUCTION

Artificial Neural Networks (ANNs) have proved to be a very effective tool for classification tasks [1]. Among the most important advantages they offer, we recall learning capability, fast classification time, and intrinsic parallelism. On the other hand, their use also gives rise to some problems, such as uncertain convergence, lack of precise rules for defining the architecture, and difficult interpretation of the network behaviour.
The purpose of this paper is to present a technique for pruning [2,3] the connections of neural networks for classification, with neuron nonlinearities of the continuous type, and its application to feature selection.
Pruning the connections of an ANN may be useful for two purposes, that is, for feature selection and to make architecture definition less critical.
 ANNs own some feature selection capabilities, in the sense that they are able to reduce the influence of some inputs on the outputs of the net. However, the result of this feature selection is not explicit [4]. We can make it explicit by pruning: if pruning allows to delete all the connections starting from some features, without affecting performances, then such features can be deleted, and the remaining ones are the selected features.
Two opposite requirements have to be considered when defining the architecture of an ANN. The network should be complex enough, so to guarantee a sufficient representation power. At the same time, the number of its parameters to be optimized (weights) should be small enough, with respect to the number of training samples, to guarantee generalization capabilities. A well known heuristic rule suggests to adopt the simplest architecture such that any increase of complexity does not improve performances. However, the application of this rule involves a time

consuming process, as it requires many architectures to be investigated. As an alternative, an iterative prune-and-retrain process allows to start with a rather large network and reduce progressively its size. The advantage is that we can make retraining start from the weights of the previous solution, so it is very fast.

The importance of the paper lies in that it addresses two important aspects of the use of ANNs for classification, namely, pruning and feature selection. The novelty of the paper lies in the originality of the proposed technique; one of its peculiarities is that it does not require the availability of a training set.

2. NETWORK PRUNING

Let us consider a multilayer feedforward ANN. Let c_j indicate a connection from the neuron $n_{i-1,k}$ to the neuron $n_{i,l}$ (Fig.1), and w_j indicate its weight (e.g., $w_j > 0$). We say that c_j can be deleted if there exists a value x such that, every output of the quasi-equivalent ANN (Fig.2), obtained by deleting c_j and adding x to the bias $b_{i,l}$ of the neuron $n_{i,l}$, differs from the output of the original ANN less than a threshold ε, for every possible value of the input vector ($\underline{F} = f_1,...,f_n$). (Intuitively, the role of x is to compensate for the average contribution of c_j). We can express this concept as follows:

Proposition 1: "A connection c_j can be deleted if

$$\min_{x \in R} \{ \max_{m} \{ \max_{\underline{F} \in D} \{ | out_m(\underline{F}) - out_m^*(\underline{F},x) | \} \} \} < \varepsilon \text{ "} \qquad (1)$$

where R is the set of real numbers, D is the domain of $\underline{F}$, $out_m(\underline{F})$ is the m-th output of the original net, and $out_m^*(\underline{F},x)$ is the m-th output of the quasi-equivalent net obtained by deleting c_j and changing the bias $b_{i,l}$ into $b_{i,l}^*$:

$$b_{i,l}^* = b_{i,l} + x . \qquad (2)$$

In general, it is difficult to compute the left part of eq.(1). Therefore, we define a sufficient condition for eliminating a connection that implies eq.1 (i.e., we compute an upper bound of the left part of eq.1 and compare it with ε).

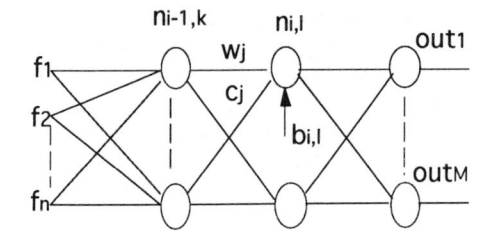

Fig.1. Original ANN. 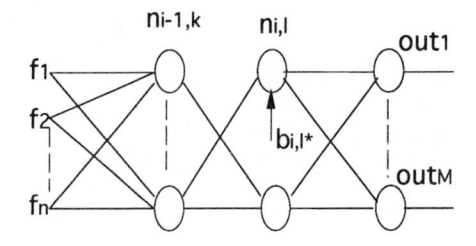   Fig.2. Quasi-equivalent ANN.

If $[\alpha,\beta]$ indicates the range of values exhibited by the output $o_{i-1,k}$ of the neuron $n_{i-1,k}$ for any value of the input vector $\underline{E}$, and w_j is greater than zero, then $o_{i-1,k} \cdot w_j$ belongs to the range $[\eta,\zeta] = [\alpha \cdot w_j, \beta \cdot w_j]$. A sufficient condition is the following one:

Proposition 2: "A connection c_j can be deleted if there exists $x_1 \in [\eta,\zeta]$ such that

$$\max_{x_2 \in [\eta,\zeta]} \{ \max_m \{ \max_{\underline{E} \in D} \{ | out_m^*(\underline{E},x_1) - out_m^*(\underline{E},x_2) | \} \} \} < \varepsilon \text{ "} \qquad (3)$$

In this case, we can use $x=x_1$ to generate a quasi equivalent ANN.
Proposition 2 expresses a sufficient condition as it can be proved that the left part of eq.(3) is always greater than the corresponding part of eq.(1), for all values of x_1. The advantage of the condition in eq.3 is that it can be more easily cecked. In addition, without loosing generality, we will consider, in the following, ANNs with just one output.

3. FEATURE SELECTION

To the purpose of feature selection, for simplicity, let us consider that the j-th input f_j corresponds to the j-th feature, after normalization in the range [0,1]. By applying a similar reasoning like above for connection pruning, we can adopt the following sufficient condition for eliminating a feature:

Proposition 3: "A feature f_j can be eliminated if there exists $f_j^1 \in [0,1]$ such that

$$\max_{f_j^2 \in [0,1]} \{ \max_{(f_1,...,f_{j-1},f_{j+1},...,f_n) \in D'} \{ | out(f_1,..f_j^1..,f_n) - out(f_1,..f_j^2..,f_n) | \} \} < \varepsilon \text{ "} \qquad (4)$$

where f_j^1 and f_j^2 are used as values of the input f_j, D' is equal to $[0,1]^{n-1}$.
The quasi-equivalent ANN is obtained by removing all the connections originating from the feature f_j, and modifying the biases of the neurons in which such

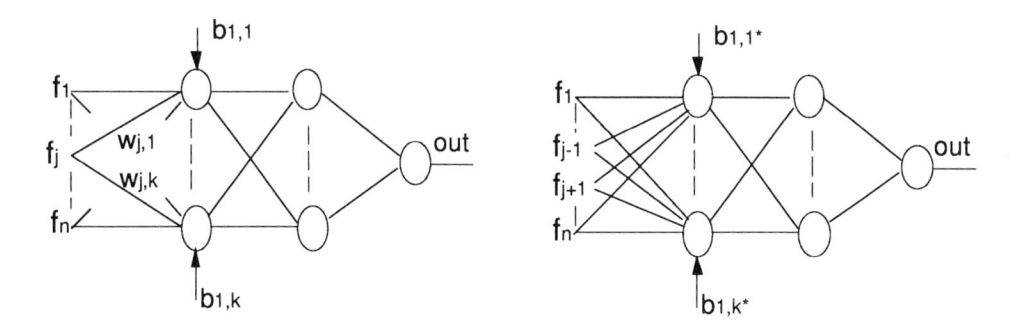

Fig.3. Original ANN. Fig.4. Quasi-equivalent ANN after removal of f_j.

connections arrives. If $c_{j,s}$ is a connection starting from the input f_j, $w_{j,s}$ is the related weight, and $b_{1,s}$ is the bias of the neuron in which $c_{j,s}$ arrives (Figs.3 and 4), then $b_{1,s}$ is substituted by $b_{1,s}^*$:

$$b_{1,s}^* = b_{1,s} + f_j^1 \cdot w_{j,s} \tag{5}$$

Different values of f_j^1 cause different overestimation of the importance of a feature.

4. CHEKING CONDITIONS FOR NETWORK PRUNING

Let us consider the case of connection pruning. We assume that the nonlinear tranfer function of neurons, S, is a continuous monotonic increasing function (e.g., the sigmoidal function). In addition, in order to make checking of eq.3 easier, we introduce a simplification: if t neurons are included in an ANN, and the output of a generic neuron n_i varies in a range D_i, then all t-uples of neuron outputs belonging to the cartesian product $D_1 \times D_2 \times ... D_t$ are considered possible.
The procedure to check the condition in eq.(3) consists of the following three steps (the procedure to check eq.(4), for feature selection, is very similar except for step 1, which is not needed).

1. Estimate the range $[\eta,\zeta]$ by estimating $[\alpha,\beta]$, which is the range of variations of the output $o_{i-1,k}$ of the neuron $n_{i-1,k}$ (Fig.5).
2. Point out all the paths from the connection c_j to the ANN output.
3. Estimate the difference in the left part of eq.(3).

Regarding step 1, an estimation of α and β can be performed by recursively using the following formula (and an analogous one for $\beta_{i,j}$):

$$\alpha_{i,j} = S \left(b_{i,j} + \sum_{k \in \Omega_{ij}^-} w_{i-1,k} \cdot \beta_{i-1,k} + \sum_{k \in \Omega_{ij}^+} w_{i-1,k} \cdot \alpha_{i-1,k} \right) \tag{6}$$

where $[\alpha_{i,j},\beta_{i,j}]$ is the range of values of the output $o_{i,j}$ of a generic neuron $n_{i,j}$; $[\alpha_{i-1,k},\beta_{i-1,k}]$ is the output range for a neuron of the preceding layer; Ω_{ij}^- (Ω_{ij}^+) is the set of the indexes "k" such that $w_{i-1,k}$ is a negative (positive) weight of a connection arriving in n_{ij}. The recursion stops at the input layer, whose range is known (e.g., [0,1]). In eq.(6) the above defined simplification has been used. In fact, the minimum and maximum output values, over the range of variation of the input vector, are computed for all neurons independently from one each other. This approximation causes an overestimation of the range of variation $[\eta,\zeta]$.
An example of the paths pointed out in step 2 is given is Fig.5.
Concerning step 3, we can note that the contributions of the set of connections on the paths found out in step 2 to the output depend on the values of x_1, x_2 and $\underline{F}$. On the contrary, the contributions of the set of all the other connections depend only on $\underline{F}$. In order to make the computation easier, we introduce a further simplification. We consider that a value of the input vector $\underline{F}$ exists such that the latter set of connections make all the neurons included in the paths singled out in step 2 work

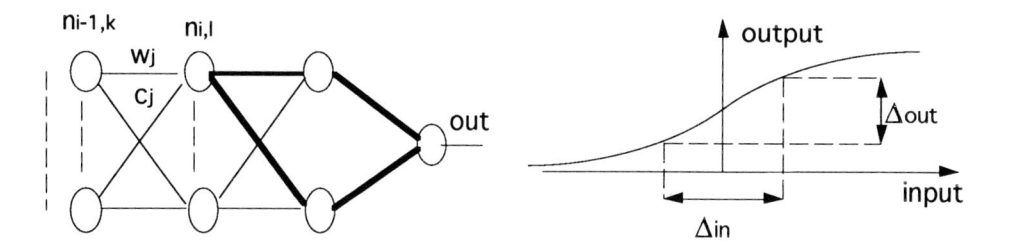

Fig.5. Thick lines indicate the paths from c_j to the output.

Fig.6. Sigmoidal function: situation of maximum Δout to Δin ratio.

in the part of the nonlinear function S with maximum slope. (The case of the sigmoidal function is depicted in Fig.6). Then we propagate the maximum difference between x_1 and x_2 (i.e., η-x_1 or ζ-x_1) along the paths found out in step 2, up to the output. Also in this computation, we disregard the interdependence among the various paths, that is, we propagate the maximum possible variations along all paths, contemporarily. This situation, if could take place, would obviously give rise to the maximum difference in the left part of eq.3.

By carrying out the above described three steps, we obtain an upper bound of the difference in eq.(3). In particular, both the simplifications introduced cause an overestimation of such a difference. If the computed upper bound is smaller than ε, then also eq.(1) holds (sufficient condition) and the connection c_j may be pruned.

5. RESULTS

Two experiments have been carried out on real image data to confirm the effectiveness of the proposed approach. In particular, the feature selection technique has been applied to multilayer feed-forward ANNs, with sigmoidal nonlinearity. Such ANNs have been trained by the backpropagation learning rule to classify remote sensing images.

In the first (multisensorial) experiment, we considered optical (ATM) and Synthetic Aperture Radar (SAR) images of the agricultural area of Feltwell (U.K.), taken from an aircraft. We filtered and segmented images into regions. Then, for each region, we extracted 11 features, that is, the average intensity in six optical bands, three radar channels, and two texture images derived from optical bands. We selected the four agricultural classes with the largest number of regions (carrots, sugar beet, potatoes and wheat) globally corresponding to 92 regions. Four ANNs (one for each class) were utilized, all based on the same architecture (Fig.7). Good classification results were obtained: 1.99% of average error rate on test sets independent from training sets (cross validation procedure). The proposed feature selection (with $\varepsilon=0.10$ and $f_j^1 = 0.5$) allowed four or five features to be eliminated from each ANN. After selection and retraining, the average error rate on test sets was slightly better: 1.59%.

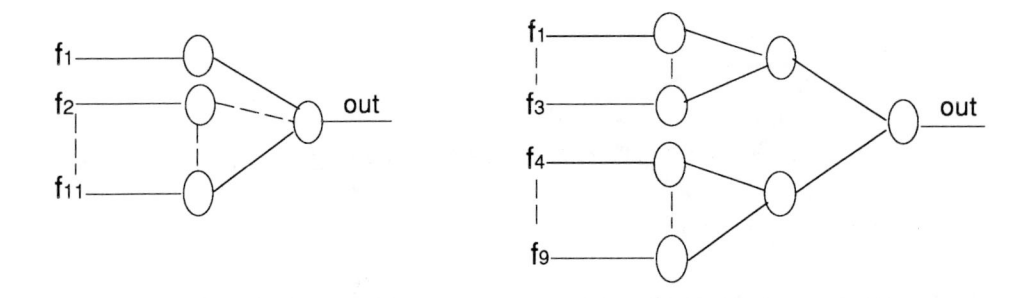

Fig.7. One-hidden layer tree-like architecture utilized for the first experiment.

Fig.8. Two-hidden layer tree-like architecture utilized for the second experiment.

Another experiment was carried out on optical images (ATM) of the Thetford forest (U.K.). In this case, segmentation provided 96 regions. Nine features were used: average intensity in six optical bands, and three geometrical attributes (elongatedness, compactness, and area). We considered the problem of distinguishing deciduous forest from all other cover types. An ANN with a slightly more complex architecture was necessary for this problem (Fig.8). Results were very good: 1.74% of error rate. Feature selection allowed two features to be eliminated, that is, compactness and area. Performances after selection and retraining were slightly worst: 2.08% of error rate.

In conclusion, both experiments can be considered successful, as we could reduce the number of input features without significant performance reduction. In addition, even though not considered here, interesting indications may derive from feature selection also from the application viewpoint. For example, it may suggest a different importance among different sensors and/or their channels.

REFERENCES

1 R.P.Lippmann, "Pattern Classification Using Neural Networks", IEEE Communications Magazine, vol.17, no.11, Nov.1989, pp.47-64.

2 S.J.Hanson, L.Y.Pratt, "Comparing Biases for Minimal Network Construction with Backpropagation", Advances in Neural Information Processing Systems vol. I, D.S.Touretzky, Ed., Morgan Kaufmann, 1989, pp.177-185.

3 M.C.Mozer, P.Smolensky, "Skeletonization: A Technique for Trimming the Fat from a Network via Relevance Assessment", Advances in Neural Information Processing Systems vol. I, D.S.Touretzky, Ed., Morgan Kaufmann, 1989, pp.107-115.

4 J.W.Shavlik, R.J.Mooney, G.G.Towell, "Symbolic and Neural Learning Algorithms: An Experimental Comparison", Machine Learning, 6, 1991, pp.120-122.

Isocontour detection from context-dependent region-growing

Silvana Dellepiane and Franco Fontana

Department of Biophysical and Electronic Engineering, University of Genoa
Via Opera Pia 11a, I-16145 Genova, ITALY

Abstract

Isoregions and isocontours are defined, which have been obtained by a new method for fuzzy image segmentation that exploits local and global contexts for the pixel labelling process.
They represent a multilevel segmentation result for an object pointed by a seed point. The result is insensitive to threshold and parameter values. A set of probable contours of the object can be derived from a single labelling session and changing only the visualization parameters.

1. INTRODUCTION

In this paper, we apply a novel method for fuzzy image labelling to obtain a multilevel segmentation of images acquired from the real world. Besides using the intensity component, the method of fuzzy labelling (described in detail in [1]) exploits contextual information, both local and global, to grow a fuzzy region, starting from a seed point.
Let $I = \{I_{i,j}\}$, defined on the integer lattice $Z_m = \{(i,j): 1 < i,j < m\}$, represent the random field of the original grey levels. Traditional region segmentation algorithms generate a label field $S = \{S_{i,j}\}$, where a scalar label is associated with each pixel with coordinates (i,j). This field may assume values ranging from 1 to a number of regions NR. The field S is therefore a new image with NR levels, and is referred to as the labelled version of the original image.
This result allows one to detect a region, Reg_a, as a set of connected pixels characterized by the same label a:

$$Reg_a = \{p_a\} \qquad (1)$$

Then a region consists of a crisp set of points.
The proposed method performs a fuzzy segmentation, i.e., it assigns a value (ranging from 0 to 1) to each pixel. Such a value denotes the degree of membership into the region Reg_a pointed by the reference seed a. As a result, a fuzzy field $F = \{E_{i,j}\}$ is obtained, where $\underline{F}$ is a vector of membership values with a number of components equal to the number of seed points used (corresponding to an equal number of objects searched for). For each seed used as a starting point, an image $M_a = \{\mu_{i,j,a}\} = \{\mu_{p_a}\}$ can be obtained, which represents a fuzzy labelled image.
The region Reg_a is therefore a fuzzy set R:

$$R = Reg_a = \{p, \mu_{pa}\} \tag{2}$$

made up of a seed point with a membership value equal to 1 and of surrounding pixels with constant or decreasing membership values.

On the basis of this result, the problem of extracting an object boundary, connected as expected by the user, is not so trivial as usual. Indeed, a fuzzy segmented image can be interpreted as a topological map from which one can derive a set of object areas or contours. The highest membership values correspond to points that are most likely to belong to a searched object.

A multilevel segmentation can thus be derived for each seed (and hence for each object): successive levels consider decreasing membership values, thus including new points in the probable object area. In this context, we define *isoregions* and *isocontours* at various levels for the fuzzy region Reg_a, which will be detected after the labelling process. Such definition makes it possible to obtain various contours of the same object, not by changing the parameters used during processing but by deriving a graduated result from only one labelling session.

2. PREVIOUS WORK ON FUZZY IMAGE SEGMENTATION

The fuzzy-set approach, often used to solve classification problems, seems very suitable for describing or segmenting data that represent ill-defined objects, (i.e., objects characterized by inaccurate and not well separated features, like real objects depicted in digital images).

The sensitivity of traditional segmentation algorithms to parameter and threshold values suggests that image segments should be described more appropriately by fuzzy image subsets, obtained independently of threshold and parameter values, as proposed by Rosenfeld [2]. The work described in [3,4], after providing the definitions of fuzzy image subsets and of their features, proposes the application of fuzzy membership functions and fuzzy operators to obtain segmented images. This method derives various fuzzy versions from the same original image and uses the geometry of fuzzy image subsets to favour the compactness of segmentation results, for the purpose of achieving the optimum thresholded image. The final result is then the most probable object silhouette, whereas object contours are not achievable in the usual way.

A similar concept, regarding the extension of fuzzy classification to images, was proposed in [5] and applied in [6,7], where a fuzzy clustering in the intensity space and an analysis of connected components in the spatial domain alternate. A fuzzy image subset is obtained for each found cluster, and a fuzzy region is obtained for each cluster. This ensures compactness only in a fuzzy way, that is, it does not allow the detection of traditional object boundaries.

In both cases, the fuzzy approach is applied at two levels: at the intensity level in a punctual fashion and at the spatial level by computing geometric features of intermediate fuzzy image subsets.

On the contrary, the proposed approach makes use of both the intensity and spatial domains during the one-shot processing step.

Intensity information is exploited through a comparison with the reference (seed) point intensity. Spatial information is taken into account by means of two mechanisms: selection of the growth direction and use of the neighbourhood in the assignment of a membership value to each pixel. This means that the local context is taken into account during the labelling process.

3. THE METHOD

The labelling process is sequential, as each pixel is examined and labelled only once.
The membership value μ_{pa} of a new pixel p is computed as a function of local and global contexts:

$$\mu_{pa} = \beta G_1(L_p) + \gamma G_2(SR_a) \tag{3}$$

where SR_a is the seed region obtained by aggregating a few pixels around the selected seed point sp_a.

The first term, G_1, propagates the membership values of the neighbouring pixels already labelled (L_p) on the basis of their similarities to the candidate pixel p.

The second term, G_2, represents the global context and is based on the seed region properties.

The weighting parameters β and γ are set so as to obtain a monotonical decreasing function as new pixels are added. Their sum is equal to 1, so the membership value in homogeneous areas is preserved. However, some pixels that have grey levels very similar to the seed grey level, but that are isolated from the seed area, should take on values higher than those of the neighbouring pixels, thus they do not fulfil the condition for monotonical decreasing behaviour. To reduce the risk of such a situation, the γ parameter is selected to be smaller than β, that is, the local context is privileged with respect to the global context. However, this situation may occur if the similarity of a new pixel to the seed is equal to 1 and the contrast in the local neighbourhood is higher than a fixed value. A typical case where β is equal to 0.6 and γ is equal to 0.4 gives rise to such a situation if the contrast value is higher than 84. In such rare cases, the new pixel is not assigned the μ value computed by equation (3), but will be constrained to assume a membership value that does not exceed the membership value of the neighbourhood.

At each step, the aggregation process selects as candidates only the neighbours of pixels characterized by the highest membership values. In other words, the growth is driven by contextual information, and the most probable pixels (i.e., the pixels with the highest membership values) are analyzed first.

If we define the path $P(q,SR_a)$ as an 8-connected sequence of points from a pixel q to any point belonging to the seed region, we can express the measure of connectedness (computed over a field $M_a=\{\mu_{pa}\}$) of q to the seed region as:

$$conn(\mu_{pa}, q, SR_a) = \max_{P(q,SR_a)} \min_{p \in P(q,SR_a)} \mu_{pa} \tag{4}$$

The mechanism of growing a region by means of successive additions of external layers, together with the decreasing behaviour of the membership value (as new pixels are added), ensures that the connectedness of q will always be equal to its membership value:

$$conn(\mu_{pa}, q, SR_a) = \mu_{qa} \tag{5}$$

In other words, the membership value obtained for each pixel is the measure of its connectedness with respect to the seed region.

As a result, in comparison with the methods outlined in section 2, the connectedness of the fuzzy set R is ensured and need not be verified.

Even though the process is sequential, the result is not sensitive to the analysis ordering, as the growth direction is decided on the basis of the local context, which does not change according to the selected direction. However, the result is partially biased on the position of the seed point inside a single object. The fuzzy labelling obtained by starting with a seed point sp_1 is exactly the same as that obtained by using the seed point sp_0 (with the same grey level), if the two points exhibit a connectedness equal to 1. In general, if their connectedness values are equal to any value c smaller than 1, points characterized by μ values smaller than c assume the same membership value, independently of the seed point. On the contrary, points labelled by μ values higher than c have different membership values, depending of the seed point used.

The labelling process expands independently for the two seeds until one seed is analyzed and labelled because of the other. Starting from this moment, the two growing processes are not independent anymore, since new pixels are added to the common external layer.

For more details on the method, the reader is referred to [1], where the aspects of connectedness and sensitivity are also discussed.

4. ISOREGIONS AND ISOCONTOURS

Every point characterized by a membership value higher than a fixed value α is defined as belonging to the "isoregion at level α of $Re g_a$". The related boundary is then named "isocontour at level α of $Re g_a$".

In accordance with the theory of fuzzy sets [2], the isoregion at level α is then the α-level set (or α-cut) R_α of R:

$$R_\alpha = \{p | \mu_{pa} \geq \alpha\} \tag{6}$$

The connectedness property described in (5) ensures that every isoregion will be a connected pixel set. More precisely, the isoregion at level α is connected to a degree equal to α, corresponding to the minimum connectedness value of its pixels. Moreover, every isocontour is a closed contour, except for possible internal holes.

By decreasing the α value, a set of isoregions (and related isocontours) can be extracted, for which:

$$R_{\alpha 2} \subset R_{\alpha 1} \quad \forall \alpha 2 < \alpha 1 \tag{7}$$

The set $\{R_{\alpha k}\}$ is then a set of crisp region entities representing a multilevel segmentation of the object pointed by the seed.

As a consequence of the connectedness of each isoregion, every segmentation level provides only one region for the object searched for. Each level is associated with a decreasing homogeneity, as α decreases; a large number of segmentation levels can be obtained by a dense sampling of the α parameter.

By means of a very simple and fast visualization process, various contours can be proposed to the user, who will select the best one according to his purpose.

To devise a method for automatic visualization, it is extremely important to reduce the number of levels to the significant ones, possibly defining an order for the most probable levels. To this end, a cost function is currently being studied to derive from the set $\{R_{\alpha k}\}$ a reduced ordered set $\{R_{\{\alpha h\})}\}$ containing only the h best regions.

5. RESULTS

An example is given to demonstrate the method performances and insensitivity to the position of the seed point inside the object area. Fig. 1a shows a Magnetic Resonance (MR) image, where two different seed points have been placed inside a pathological area. Series 1b,c,d and 1e,f,g reproduce isocontours (at different levels) obtained for each of the two seeds. It is worth noting that, after an initial evolution, the isocontours reach a common equilibrium point.

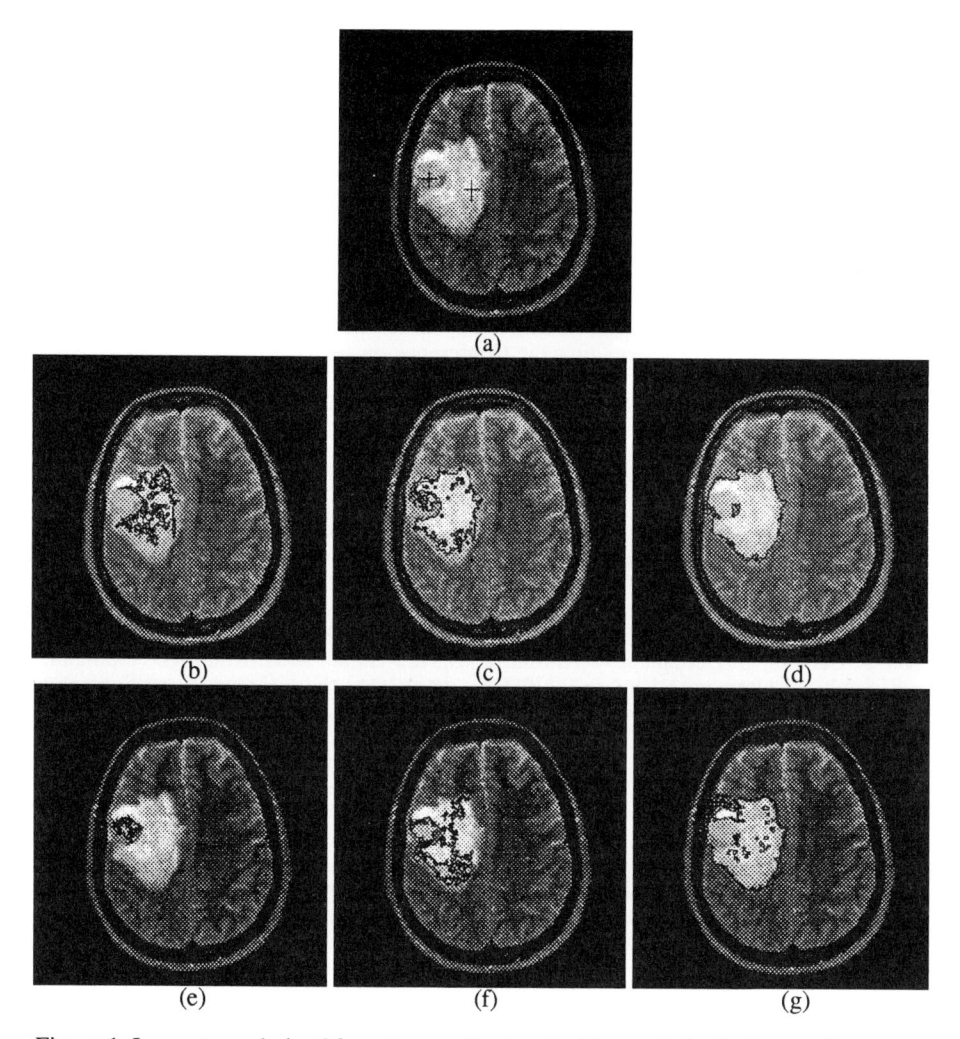

Figure 1. Isocontours derived from two seed points inside a pathological area of a Magnetic Resonance image.

6. CONCLUSIONS

In the paper, a method has been proposed, which is able to derive a fuzzy segmentation of an object pointed by a seed point. A single session of a fuzzy labelling algorithm allows the extraction of isoregions (and related isocontours), representing the segmentation results at various levels. The present approach overcomes the usual problem of sensitivity to algorithm parameters and thresholds by providing a multilevel result instead of a single-level one.

Isocontours turn out to be more accurate than contours extracted by single algorithms, based on either region extraction or edge extraction (as pointed out in [1]).

Up to now, the proposed segmentation process has been performed considering one seed at a time, related to only one object, but a multiseed version is currently under development.

ACKNOWLEDGMENTS

This work was partially supported by the CNR Target Project on Biotechnology and Bioinstrumentation and by the European Project COVIRA [9].

7. REFERENCES

1 F. Fontana and S. Dellepiane, "Context-Dependent Region Labelling by a Label-Driven Method", Proc. of the 8th Scandinavian Conference on Image Analysis (Norway, May 25-28, 1993) (in press).

2 A. Rosenfeld, "The fuzzy geometry of image subsets", Pattern Recognition Letters 2, pp.311-317, 1984.

3 S.K. Pal, and A. Rosenfeld, "Image enhancement and thresholding by optimization of fuzzy compactness", Pattern Recognition Letters, 8, pp.21-28, 1988

4 S. K. Pal and A.Ghosh, "Index of area coverage of fuzzy image subsets and object extraction" Pattern Recognition Letters,11, pp.831-841, 1990

5 Ortendahl and J.W. Carlson, "Segmentation of Magnetic Resonance Images Using Fuzzy Clustering", Proc. 10th IPMI, Utrecht, pp.91-106, 1987.

6 W. Menhardt, "Fuzzy Clustering and Component Analysis", Proc. 9th Ann. Sc. Meet. and Exhib. of the Society of Magnetic Resonance in Medicine, New York, p.557, 1990.

7 I.C. Carlsen, S. Dellepiane, P. Elliott, F. Galvez-Galan, M.H. Kuhn, H. Neumann, and H.S. Stiehl, "COmputer VIsion in RAdiology (COVIRA)- Knowledge Based Segmentation and Interpretation of Cranial Magnetic Resonance Images", Proc. of the AIM Euroforum, Sevilla 1990, Bouncopy, 1990, pp.169-180.

8 C. W. Therrien, "Decision, Estimation and Classification", John Wiley & Sons, 1989.

9 COVIRA - Project A2003 of the AIM programme of the European Community, Consortium Partners:
 - Philips Medical Systems, Best (NL) and Madrid (E); Corporate Research, Hamburg (D) (prime contractor) - Siemens AG, Erlangen (D) and Munich (D) - IBM UK Scientific Centre, Winchester (UK) - Gregorio Maranon General Hospital, Madrid (E) - University of Tuebingen, Neuroradiology and Theoretical Astrophysics (D) - German Cancer Research Centre, Heidelberg (D) - University of Leuven, Neurosurgery, Radiology and Electrical Engineering (B) - University of Utrecht, Neurosurgery and Computer Vision (NL) - Royal Marsden Hospital/Institute of Cancer Research, Sutton (UK) - National Hospital for Neurology and Neurosurgery, London (UK) - Foundation of Research and Technology, Crete (GR) - University of Sheffield (UK) - University of Genoa (I) - University of Aachen (D) - University of Hamburg (D) - Federal Institute of Technology, Zurich (CH).

Neural Implementation of Shape From Shading Algorithms

Chella A., Pirrone R., Sorbello F.

DIE - Dipartimento di Ingegneria Elettrica, Universita' di Palermo
Viale delle Scienze, 90128 Palermo, Italy
Email: chella@vlsipa.cres.it

Abstract

The tridimensional recognition of a complex scene is a fundamental problem in artificial vision which has been faced in different ways.
In this paper two original approaches are proposed both based on neural architectures: the first one consists of a cascaded architecture made up by a first stage named BWE (Boundary Webs Extractor) which is aimed to extract a brightness gradient map from the image, followed by a backpropagation network that estimates the geometric parameters of the object parts present in the acquired scene. The second approach is based on the extraction of the boundary webs map from the image and its comparison with boundary webs maps exhaustively generated from synthetic superquadrics. A purposely defined error figure has been used to find the best match between the two kinds of maps.
A functional comparison between the two systems is described and the quite satisfactory experimental results are presented.

1. INTRODUCTION

Classical approaches to tridimensional perception are generally based on several information processing stages, starting from features extracted from the 2-D image, e.g. to obtain the volumetric representation of the object [8, 9]. According to these approaches, functional and structural relations among objects may not be obtained by simple local analysis because heavy, global assumptions like regularity and isotropy would be necessary. Assumptions of this kind generally make the model not well suitable to describe the large variety of objects of the real world.
The general framework adopted by the authors about the 3-D description of a scene is based on the direct recognition of simple 3-D geometric primitives into which the objects in the scene may be decomposed [1]; the objects are therefore described as simple combinations of these "building blocks" [10].
The 3-D geometric primitives have to be simple and versatile; good candidates are the superquadrics [5]. Superquadrics are geometric forms that are derived from quadrics equation rising the trigonometric functions to two real exponents called shape factors. Modifying shape factors causes a surface deformation which can result in squared as welle as pinched or rounded shapes (fig. 1). Boolean combinations of superquadrics are able to represent with surprising realism scenes and objects of the real world that can therefore be described with extreme compactness by the list of their parameters.
In this work two different sytems developed by the authors are presented both based on neural network architectures that are able to directly extract 3-D shape caracterstichs from 2-D pictorial images following the idea just explained.
The first system is based on the cascade of the *BWE* (Boundary Webs Extractor) which

derives a brightness gradient map of the surfaces present in the scene and feeds its output to a backpropagation net that estimates the shape parameters of the superquadrics best fitting these surfaces. The second system is based on the minimisation of the error between the boundary webs map extracted by the BWE and the analytic maps exhausively generated from synthetic superquadrics by a stage called *SPS* (Superquadric Parameters Search).

The approach based on neural networks overcomes the shortcomings of the classical shape from shading algorithms: the operation of the neural network allows this kind of systems to be robust with respect to noise, change of illumination and environmental conditions; these features make them most suitable for real world application. Moreover, after the network is trained, the operation is very quickly and it not presents the heavy computational load typical of classic algorithms. The network operation is intrinsically parallel and it therefore well suited for implementation in parallel hardware architectures [4].

figure 1: superquadric examples: their shapes are obtained only acting on the shape factors.

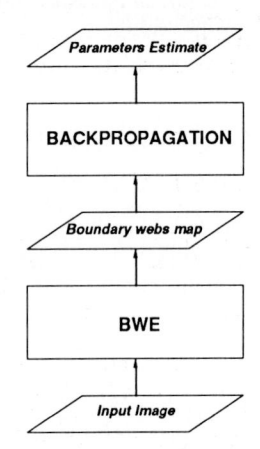

figure 2: an outline of the BWE+Backprop architecture.

2. FUNCTIONAL DESCRIPTION OF THE ARCHITECTURES

2.1 The BWE+Backprop architecture

The first architecture is based on the cascade of the BWE stage and a backpropagation network: an outline of this system is described in fig. 2.

The BWE is an architecture derived from Grossberg's *Boundary Contour System* (BCS) which is a neural architecture aimed to simulate some processes of early vision related to the perception of the boundaries and the local curvature of the objects in a natural scene [7].

The BWE is a multilayer network based on several competitive-cooperative processes between very simple cells. Units in the first two competitive stages detect local amount and direction of contrast in small windows of the image; then a cooperative stage performs an alignment of the local directions of contrast to extrct global iso-brightness contours. The process is tuned by a reaction loop.

The BWE is not an adaptive model and it does not learn from examples but the internal parameters are fixed from outside. The units of the BWE are sensitive to the brightness variations due to the surface shading: the BWE traces contour lines not only for the more sharp brightness variations near the external contours of the object, but also for the less sharp brightness variations growing up from the shading. The contours so generated are called *boundary webs* allowing to determine the shape information deriving from the image shading. The BWE filtering allows for

figure 3: an example of the BWE operation: on the right side is represented a scene made up by a superquadric and on the left one the boundary webs map is depicted.

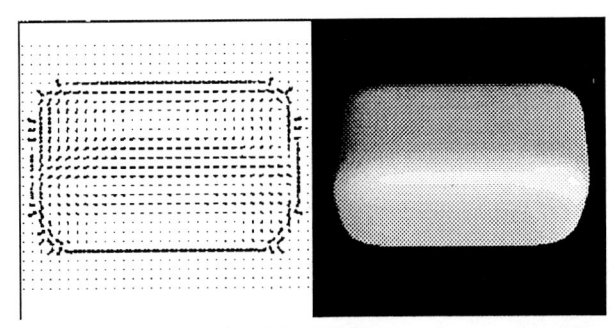

a substantial data compression without loss of information about the shapes of the objects; moreover its operation is invariant with respect to the direction of illumination and it shows a good behaviour with respect to the noise. An example of the BWE output is represented in fig. 3.

The BWE feeds its outputs to a neural network using the bakpropagation learning algorithm [11]; this architecture performs a local estimate of the shape factors of the superquadric best fitting the object surface over a small window of the image: the global estimate of the object shape derives from an average of the different local estimates.

2.2 The BWE+SPS architecture

The second system presented in this work is a hybrid architecture based on two stages: the first one is again the BWE while the second is a functional not neural block called *SPS* (Superquadric Parameters Search) which implements an algorithm of exhaustive search of the parameters of the superquadric whose boundary webs map best fits the object's boundary webs map. Fig. 4 reports a simple scheme of the presented architecture.

In particular the SPS analyses all the sets of parameters in a given range, moving by fixed steps. For each set it analytically derives the boundary webs map of the corresponding superquadric making use of an explicit form of its in-out equation [5]: the boundary webs are obtained as the projection on the image plane of the vectors point by point orthogonal to the brightness gradient vectors and laying in the tangent plane to the superquadric surface.

The SPS searchs exhaustively in the parameters range of variation and for each set of parameters derives the value of an error figure based on the minimisation of the differencies between the modules and the phases of the neural and analytic boundary webs. The minimum value of the error determines the parameters of the best fitting superquadric.

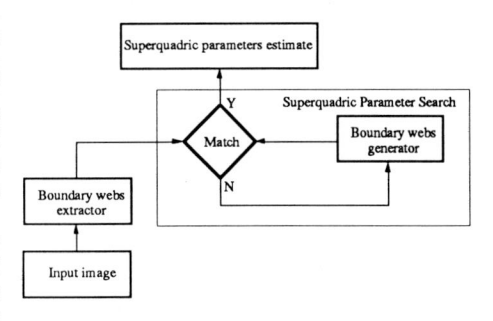

figure 4: an outline of the BWE+SPS architecture.

3. EXPERIMENTAL RESULTS

Fig. 5 reports some experimental recognitions of the BWE+Backprop architecture, while the results we obtained with the BWE+SPS architecture are reported in figg. 6 and 7.

Results are quite satisfactory in every case and in particular the BWE+SPS approach seems to exhibit a behaviour more likely than the other one: however it shares with tha classical Shape From Shading approaches some drawbacks related to its not neural nature. In facts the BWE+SPS system requires a considerable computational heavy because of both the need to generate the entire boundary webs map at every step and to perform an exhaustive search in the parameters range of variation.

On the contrary, the fully neural approach even if performing a little less tuned estimate of the shape parameters is able to obtain a suboptimal solution in every case generalising few meaningful training examples and, after the learning phase, its computational cost is very low being necessary only to calculate a rather simple in-out function.

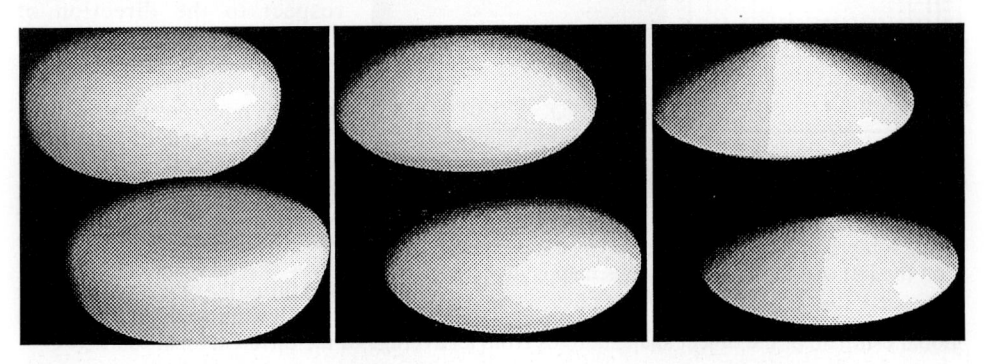

figure 5: examples of 3-D reconstructions of single superquadrics using the BWE+Backprop network;from left to right: input parameters: (0.5, 1) estimated: (0.8, 0.99); input parameters: (1, 1) estimated: (1.23, 0.92); input parameters: (1.5, 1) estimated: (1.72, 0.93).

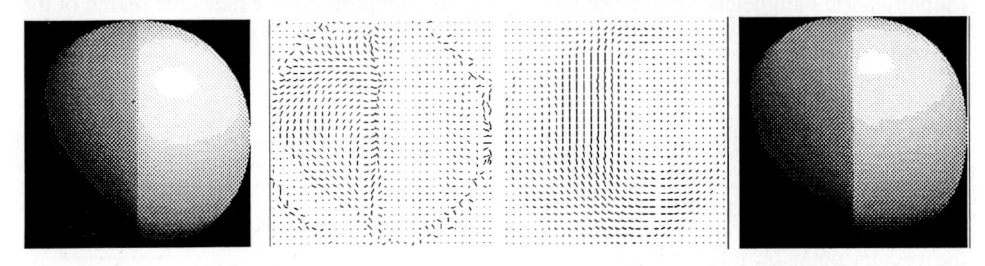

figure 6: the BWE+SPS process; from left to right: input image, its neural boundary webs map, the best fitting analytic map and the reconstructed superquadric.

figure 7: examlpes of reconstruction with the BWE+SPS architecture; from left to right: input parameters: (0.8,1.8) estimated: (0.6, 1.6); input parameters: (0.2, 1.4) estimated: (0.1, 1.4).

4. CONCLUSIONS

Both the approaches seem encouraging for different reasons: we are now trying to enhance the architectures.

As regards the BWE+Backprop system we are finding new topologies and learning paradigms to obtain a more tuned estimate, while relative to the BWE+SPS architecture we are implementing a version of the SPS with a different search criterion.

The possibility of merging the results of both the estimates togheter with other information sources (motion estimate and so on) is at present investigated.

REFERENCES

[1] Ardizzone, E., Gaglio, S., Sorbello, F. (1989). Geometric and Conceptual Knowledge Representation within a Generative Model of Visual Perception, *Journal of Intelligent and Robotic Systems*, **2**, 381-409.

[2] Ardizzone, E., Chella, A., Compagno, G., Pirrone, R. (1992). An Efficent Neural Architecture Implementing the Boundary Contour System, in: Aleksander, I., Taylor, J. (eds.) *Artificial Neural Networks, 2*. North-Holland, Amsterdam.

[3] Ardizzone, E., Chella, A., Gaglio, S., Pirrone, R., Sorbello, F. (1991), A Neural Architecture for the Estimate of 3-D Shape Parameters, in: Caianiello, E. (ed.): *Parallel Architectures and Neural Networks - Fourth Italian Workshop*, World Scientific Publishers, Singapore (in press).

[4] Ardizzone, E., Chella, A., Pirrone, R., Sorbello, F. (1991), A System Based on Neural Architectures for the Reconstruction of 3-D Shapes from Images, in: Ardizzone, E., Gaglio, S., Sorbello, F. (eds.): *Trends in Artificial Intelligence*, Springer Verlag, Berlin.

[5] Barr, A.H. (1981). Superquadrics and Angle-Preserving Transformations, IEEE *Computer Graphics and Applications*, **1**, 11-23.

[6] Callari, F., Chella, A., Gaglio, S., Pirrone, R. (1992). A New Hybrid Approach to Robot Vision, in: Aleksander, I., Taylor, J. (eds.) *Artificial Neural Networks, 2*. North-Holland, Amsterdam.

[7] Grossberg, S., Mingolla, E., (1987). Neural Dynamics of Surface Perception: Boundary Webs, Illuminants, and Shape-from-Shading. *Computer Vision, Graphics and Image Processing*, **37**, 116-165.

[8] Horn, B.K.P. (1986). *Robot Vision*. MIT Press, Cambridge, MA, USA.

[9] Marr, D.(1982) *Vision*, Freeman and Co., New York.

[10] Pentland, A.P., Perceptual Organization And The Representation Of Natural Form, *Artificial Intelligence*, **28**, 293-331, 1986.

[11] Rumelhart, D.E., Hinton, G.E. & Williams, R.J., Learning Internal Representations by Error Propagation, in: Rumelhart, D. E., McClelland, J. L. (ed.s) & PDP Research Group (1986), Parallel Distributed Processing, Vol 1, MIT Press, Cambridge, MA, USA.

A Multistage Synthesis of Modified NN Rule and Its Application for Remote-Sensing Images

Adam Jóźwik[a], Fabio Roli[b] and Carlo Dambra[b]

[a]Institute of Biocybernetics and Biomedical Engineering, Polish Academy of Sciences, 00-818 Warsaw, Twarda 55, Poland

[b]D.I.B.E. Dept. of Biophysical and Electronic Eng. University of Genova Via Opera Pia 11A - I-16145 Genova, Italy

Abstract

A new multistage NN classification rule is proposed. It is a modification of the simple 1-NN rule and the fuzzy k-NN rule. The way of synthesis of such a kind of classifier is presented on real problem.

1. PROBLEM STATEMENT

Most classification rules derived from training sets assign a class membership to a classified object without providing information in what degree we can believe the obtained decision. However, there appear to be some tasks which require sure decisions or indications that the decision is uncertain. Non-clearly clasified objects can be sent to be recognized by more sophisticated rule like the fuzzy k-NN classifier [1,2].

Another essential consideration might be the speed of classification. Thus, the classification rule should be computationally simple and eventually easily implemented in parallel computers.

Our requirements concerning the classification algorithm may be gathered:
1. It should be possible to state whether the recognized object is clearly assigned to one of the classes or not.
2. Classification algorithms and its synthesis ought to be fast and easily implementable in parallel computers.
3. Synthesis of the decision rule (i.e. learning) must comprise feature selection to increase classification quality.

4. The obtained decision-making algorithm should refuse to classify "strange" objects that differ too much from all objects in the reference set.

We propose the method of synthesis of the classification algoritm that satisfy all the above mentioned requirements.

2.THE PROPOSED METHOD

Any method of the synthesis of the classification rule should allow the quality of the obtained decision rule to be evaluated. Usually, as a quality criterion, a probability of correct classification is assumed. Such a kind of criterion is not satisfactory for our task. Still we would have no information whether the object classification is sure or not.

The percentage of objects recognized with sure decisions by the derived classification rule seems to be a suitable quality criterion for our aim. So, the area found as an overlapping one should be as small as possible.

The reference set consists of nc subsets and each of them contains points corresponding to only one class. Let them be denoted by $X_1, X_2, \ldots, X_{nc}$. With these sets we associate certain positive real numbers $e_1, e_2, \ldots, e_{nc}$ defined in the manner given below:

$$e_i = \max_j d(X_i - x_j, x_j),$$

where d denotes distance function and x_j is element of X_i. We define also areas $A_1, A_2, \ldots, A_{nc}$:

$$A_i = \{x: d(X_i, x) < e_i\}.$$

Now we can formulate the modified 1-NN rule. The object represented in the feature space by point x is classified to class i if and only if it belongs to the area A_i and does not belong to any other A_j, where j not equals to i. If x does not belong to any area A_i, $i=1, 2, \ldots, nc$, then the classification is refused. Let us denote by A the set of all points that belong exactly to one of the areas A_i, $i=1, 2, \ldots, nc$. It is strongly recommended to perform feature selection to maximise the size of the set A. Feature selection can be done by applying the forward, backward, or combined feature selection strategy.

The decision rule describeb below is proposed. If the point in the feature space that represents the classified object falls only in one area A_j, then the object is assigned to the class i. Otherwise, when it belongs to some of the areas A_j, then the object is classified by the fuzzy k-NN rule. It may happen that the classified point lies simultaneously in some A_j, where $i \in I$. Then such the object is recognized by the fuzzy k-NN rule operating with the reference set reduced to the classes i that appears in the set I. There are $ncI=2^{nc}-nc-1$ possible reduced reference sets. So, maximum ncI fuzzy k-NN classifiers may be required. A separate feature selection can be done for each of the fuzzy k-NN classifiers that most often leads to significant improvement of the classification quality. This time as a feature selection criterion minimum probability of misclassification can be assumed.

3. ANALYSED DATA SET DESCRIPTION

The above presented classifier was applied to remote-sensing images of the multisensorial type. In particular, we considered images acquired by two sensors installed on aircrafts: a Daedalus 1268 Airborne Thematic Mapper (ATM) scanner, and a PLC band, fully polarimetric, NASA/JPL airborne imaging radar system. The flights took place in July and August 1989, respectively. The geographical location was the Feltwell area (U.K.). The ATM image was scaled and registered on the radar image, with an average registration error of about 1 pixel.

The registered ATM image was filtered by a linear smoothing and a contex-sensitive enhancement filter, then it was segmented by a multiband region-growing technique. This procedure provided 192 regions. We selected 79 regions, belonging to the following classes: carrots, potatoes, stubble, sugar beat and wheat. For each region, a set of 16 features was computed, including original optical and radar channels, texture features calculated by using the above channels and combinations of optical bands.
Feature describtion:
features 1,2,3,4,5,6 - are the response of the optical sensor, i.e. the Daedalus sensor, for the band 2, band 3, etc., respectively, 7 - response of the radar sensor, i.e. the JPL one, for the band C with HH polarization, 8 - response of the radar sensor for the L band with HV polarization, 9 - responce of the radar sensor for the P band with VV polarization, 10 - Mandelbrot texture for the the C polarization HH, 11 - Mandelbrot texture for the band C polarization HV, 12 - Mandelbrot polarization for the band P polarization HV,

13 - synthetic feature computed as ratio between band7 and (band5+band7+band9), 14 - synthetic feature computed as ratio between band9 and (band5+band7+band9), 15 - synthetic feature computed as ratio between band7 and (band3+band5+band7), 16- synthetic feature computed as ratio between band5 and (band3+band5+band7).

4. OBTAINED RESULTS

By application the combined feature selection strategy a feature set {1,3,5,8,14} has been found that offers maximum number of points, from the reference set lying outside the overlapping areas. Seventeen points were simultaneuously in two or at least tree areas: 2 points in $A_1 \cap A_5$, 6 points in $A_1 \cap A_3$, 3 points in $A_2 \cap A_4$, 2 points in $A_3 \cap A_4$, and 4 points in $A_2 \cap A_3 \cap A_4$. Thus, four different fuzzy k-NN rules were applied and four additional feature sets were selected: {3}, {7,8}, {3} and {14}. Error rates (estimated by the leave-one-out method) were equal to zero for all these four classifiers. We can notice that the final classifier will use the set of six features: {1,3,5,7,8,14}. The numbers k of NN were found as it was shown in [2]. Only one point from the class 3 was misclassified to the class 4.

5. CONCLUDING REMARKS

The first stage contained the modified 1-NN rule since the second one consisted of four k-NN fuzzy rules. Of course, we could again use the modified 1-NN rule instead of fuzzy k-NN rules in the second stage. We have stopped at two stages to make our presentation more compact. The advantage of our method is the speed of classification as well as the large speed of classifier synthesis, i,e. very fast learning. The 1-NN rule, for instance , is considerably faster than the k-NN rule. More sophisticated methods are reserved for more difficult recognizable objects. The points that represents them in the feature space lie near class borders.

6. REFERENCES
1 A. Jóźwik, Pattern Recognition Letters 1 (1983) 287.
2 J.C. Bezdek, S.K. Chuah, D.Leep, Fuzzy Sets and Systems 18 (1986) 237.

Application of Moiré Fringes techniques to the inspection of aluminum bars [§]

A. Blumenkrans and G. Viano

Automa Scrl, Via al Molo Vecchio, Calata Gadda, I-16126 Genova, Italy

Abstract
This study is focused on the application of Moiré Fringes techniques to the inspection and 3D reconstruction of aluminum bars. The Moiré Fringes approach is realized by the Fourier-transform method. Several experiments were performed and their results are reported. Also practical acquisition and processing topics are addressed and the effectiveness of the technique is illustrated.

1. INTRODUCTION

The described application is being developed within a EEC BRITE project for the inspection of metal bars. X-ray Image Analysis is applied for the internal defects and vision techniques for the external defects. The project will lead to an on-line prototype for visual inspection in an industrial plant.

The need of a non contact inspection was evident due to the aggressive metallurgic environment, the speed of the object and the surface characteristics. Active range finders [1-2] such as laser devices were discarded in view of their low scanning speed. On the other hand, passive range finders such as stereo vision or depth from shading techniques did not seem suitable for our application. They require complex computation and do not take advantage of the possibility of projecting a structured light pattern that allows a higher depth resolution. The conventional lighting techniques were also discarded due to their inability to detect very slight and smooth variations on the surface shape.

Within the structured light methods an interferometric approach exploiting Moiré Fringes was considered. The surface analysis is based on a 3D reconstruction through the FFT and a detection phase enables to identify the differences respect to a reference surface model.

[§] *This work was partially supported by the EEC under contract BREU CT91 0478.*

2. MOIRE' TOPOGRAPHY BY THE FOURIER TRANSFORM METHOD

In Moiré topography [3-7] the shadow of a line grating back illuminated by an incoherent light projector is cast onto the scene. The strips shapes are modified by the surface elevations and depressions. Considering that the strips run parallel to the y axis on the (x,y) plane, the original algorithm is unidimensional along the x axis. The resulting image has a function of the following type:

$$g(x,y) = a(x,y) + b(x,y) \cdot \cos\{2\pi x f + \phi(x,y,z)\} \tag{1}$$

$a(x,y)$ and $b(x,y)$ are supposed to vary much slower than the grid of frequency f. On equation (1) the square wave is approximated to its first harmonic and $\phi(x,y,z)$ is the phase shift due to the scene geometry. $\phi(x,y,z)$ modulates in phase the periodic signal, the carrier, and depends on the image position (x,y), and the height $z(x,y)$ of the object at each point.

In Moiré Fringes Analysis the surface is recovered by demodulating the signal of equation (1) by observing the scene through a second grid. In the Fourier-transform method [6-10] the algorithm directly estimates the phase factor $\phi(x,y,z)$. Equation (1) may be written in the following way:

$$g(x,y) = a(x,y) + c(x,y) \cdot \exp(i2\pi xf) + c^*(x,y) \cdot \exp(-i2\pi xf) \tag{2.1}$$

$$c(x,y) = 1/2 \cdot b(x,y) \cdot \exp\{i\phi(x,y,z)\} \tag{2.2}$$

In the Fourier-transform method $c(x,y)$ is first extracted by filtering and shifting in the frequency domain the second term of equation (2.1). Then, considering that $b(x,y)$ is real, $\phi(x,y,z)$ may be directly extracted as the phase of $c(x,y)$ [equation(2.2)].

It is clear that the phase component is recovered as the inverse of the tangent function and therefore the result may differ on a $n.2\pi$ term from the real value. This issue is resolved assuming that the phase is a continuous function and then the difference between two neighboring points on the same row (or column) cannot have a phase step higher or equal to 2π. The basic unwrapping algorithm consists simply in adding the appropriate term $n.2\pi$ on each phase transition between adjacent pixels [6-8]. However, in some complex cases this unwrapping scheme might not work due to phase inconsistencies and therefore special unwrapping methods are applied [9-11].

3. ALUMINUM BARS INSPECTION

3.1. Generalities

The product to inspect is an aluminum bar produced by a continuous casting process. It appears as an infinite length bar of a 4 cm by 5 cm section moving at a fixed speed. The bar is used to produce a continuous rod by a rolling mill placed just after the casting. The defects to detect are the deformations and discontinuities of the external surface that would be amplified during rolling.

The external defects to detect can be classified according to three categories:
- Longitudinal segregation band (sweats) and convexity of the upper surface
- Lateral surface anomalies (cracks, sweats, and deformations)
- Lacking profile

All of them are shape defects, that is, they produce modifications onto the 3D structure of the external surface of the object. The characteristics of these shape changes differ from one class to another. Cracks involve rapid shape changes, sweats are smoother, and profile defects are very slow changes. In general the defects are present along longitudinal lines.

3.2. Acquisition Geometry

The main module, the acquisition system, comprises the structured light projector and the camera. The optical axis of the projector and the camera form an angle θ and both centers of pupil Y and Z are at the equal distance l from the reference plane R

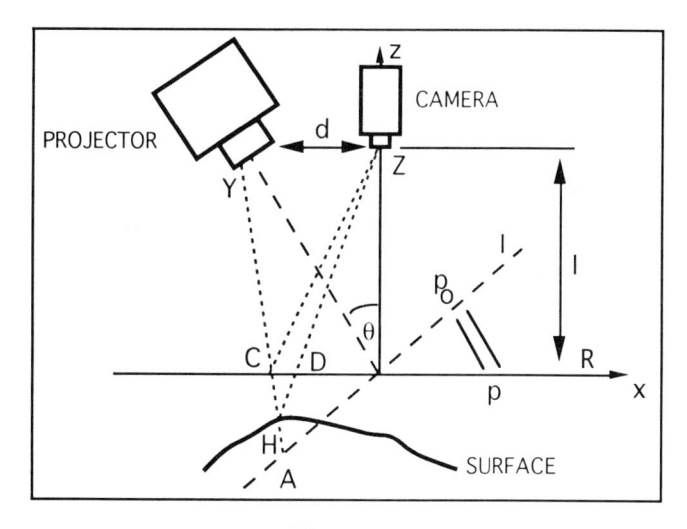

Figure 1

A regular grating of period p_0 is generated on the plane I, normal to the projector optical axis. The grating lines run parallel to the y axis that is normal to the figure plane. Then we have that the grid period p is equal to $p_0 / \cos(\theta)$. It follows then that the shift phase is:

$$\phi(x,y,z) = 2\pi \cdot CD / p \tag{3}$$

and considering that the triangles YZH and CDH are similar we have that:

$$CD = -d \cdot z(x,y) / [1 - z(x,y)] \tag{4}$$

and then from equations (3) and (4):

$$z(x, y) = 1 \cdot \phi(x,y,z) / [\phi(x,y,z) - 2\pi \cdot d / p] \tag{5}$$

One of the main technical problems encountered was that the projector is not perpendicular to the focusing plane thus requiring a large depth of field. For the acquisition a standard B&W CCD camera was used and the processing was performed on a Masscomp computer equipped with a DataCube Maxvideo vision system.

3.3. Preprocessing

The main deviation we found from the theoretical case was that the surfaces were far from smoothly colored and therefore the hypothesis of having only low frequency components is no loner valid. The solution that we applied was to subtract two identical images, one with the grid projected and the second one without it, in order to eliminate the surface coloration. Since this could not be done directly because the images had a different illumination one of them was normalized before subtracting. The model of illumination that we followed was a simplified approach were:

$$o(x,y) = A \cdot s(x,y) + B \tag{6}$$

where $o(x,y)$ is the function of the original image, $s(x,y)$ is the function of the surface with the grid projected on, and A and B are the normalizing constants.

The first step was to estimate the parameters A and B. This was accomplished by applying the least square estimation (LSE) that better fits equation (6) all over the image. After this normalization the images were subtracted on a pixel to pixel basis.

4. RESULTS

The experiments that evaluate the performance of the method were carried out using a set of samples provided by an aluminum producer.

Figure 2 shows the bottom of a bar with the defective irregular surface called *longitudinal segregation bands* (sweats). These defects are 2 mm high and have a diameter of 3 mm approximately. The same image with the grid projected on the surface is shown in figure 3. Its clear that this image is darker than the original one and that a direct pixel to pixel subtraction is not appropriate. In figure 4 the grid lines are shown as extracted following the scheme of section 3.3 and in figure 5 the depth map is displayed with the intensity at each point proportional to the corresponding surface height (the dark regions are higher).

Figure 6 shows the results for a second sample with the *longitudinal dark bands* defect and two depressions. Figure 7 shows the surface with the grid projected and figure 8 displays the grid lines as extracted with the scheme of section 3.3. The depth map in figure 9 clearly indicates the presence of the two depressions, and that the upper section of the sample is higher (darker).

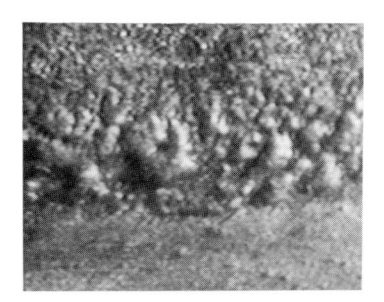

Figure 2

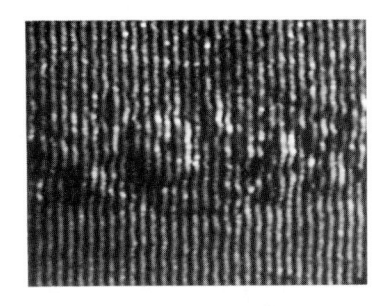

Figure 3

Figure 4

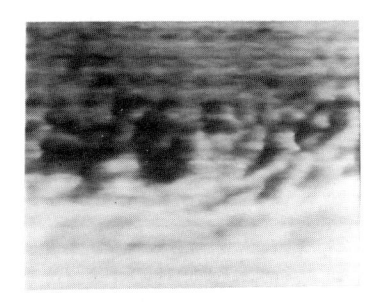

Figure 5

The theoretical limit of the system, that is, the steepest slope that the system allows, depends on the acquisition geometry [7] and is given by equation (7):

$$| \partial z(x,y) / \partial x |_{max} < 1/3d \qquad (7)$$

The system was specified for a maximum slope of 45 degrees because very steep slopes may be detected with conventional lighting. That means that l must be larger than $3d$ (see figure 1), a reasonable design criteria.

At present the system runs on a general purpose computer with non specially optimized algorithms. However, on a future prototype we may include dedicated hardware such as FFT processors that would enable us to have a real time system.

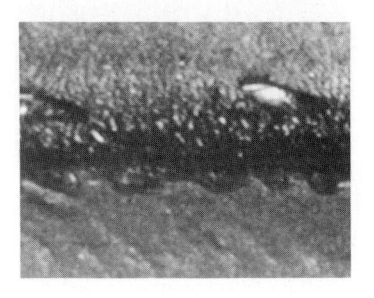

Figure 6

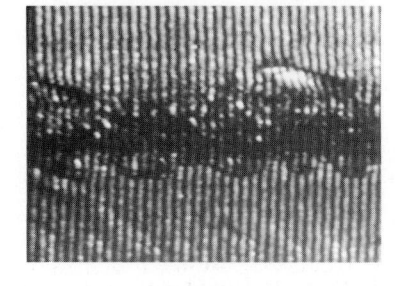

Figure 7

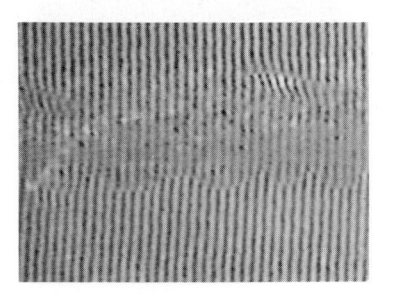

Figure 8

Figure 9

5. CONCLUSIONS

The experiments performed on the shown and further samples proved the effectiveness of the method.

On this first phase we were not so interested in verifying the precision of the method comparing it to an alternative technique such as laser range finder. The inspection task considered did not require to recover an absolute 3D map of the scene, but only to detect any change with respect to an expected surface. A model of the inspected object can be further exploited to specify thresholds on the slope of the observed surface and on the height of discontinuities.

Future work will be focused on the 3D reconstruction of different kind of objects and accurate surface measurements. We are also working on the defect detection methodologies as well as on the realization of the special projection and acquisition system able to cope with industrial requirements.

6. REFERENCES

1   R.A. Jarvis, "A perspective on range finding techniques for computer vision", IEEE Trans. Pattern Anal. Machine Intell., vol. PAMI-5, no. 2, March 1983.
2   P.J. Besl and R.C. Jain, "Three-dimensional object recognition", Computing Surveys, vol. 17, no. 1, March 1985.
3   D.M. Meadows, W.O. Johnson, and J.B. Allen, "Generation of surface contours by Moiré patterns", Applied Optics, vol. 9, no. 4, April 1970.
4   H. Takasaki, "Moiré topography", Applied Optics, vol. 9, no. 6, June 1970.
5   H. Takasaki, "Moiré topography", Applied Optics, vol. 12, no. 4, April 1973.
6   M. Takeda, H. Ina, and S. Kobayashi, "Fourier transform method of fringe-pattern analysis for computer-based topography and interferometry", J. Opt. Soc. Am, vol. 72, no. 1, January 1982.
7   M. Takeda and K. Mutoh, "Fourier transform profilometry for the automatic measurement of 3-D object shapes", Applied Optics, vol. 22, no. 24, December 1983.
8   L. Mertz, "Real-time fringe-pattern analysis", Applied Optics, vol. 22, no. 10, May 1983.
9   D.J. Bone, H.A. Bachor, and R.J. Sandeman, "Fringe-pattern analysis using a 2-D Fourier transform", Applied Optics, vol. 25, no. 10, May 1986.
10 D.J. Bone, "Fourier fringe analysis: the two-dimensional phase unwrapping problem", Applied Optics, vol. 30, no. 25, September 1991.
11 J.M. Huntley, "Noise-immune phase unwrapping algorithm", Applied Optics, vol. 28, no. 15, August 1989.

Image restoration using a standard Hopfield neural network

Mário A. T. Figueiredo   and   José M. N. Leitão

Centro de Análise e Processamento de Sinais, Departamento de Engenharia Electrotécnica e de Computadores, Instituto Superior Técnico, Lisboa, Portugal
Email: D2403@beta.ist.utl.pt

Abstract

In this paper, a fully standard Hopfield neural network (binary elements, threshold-based updating rule, symmetric interconnection weights, and zero auto-connections), which has well known convergence properties, is introduced to perform restoration of linearly blurred and white Gaussian noise contaminated images. The proposed scheme is a distributed structure exhibiting fault-tolerance. Its numerical precision can be adjusted simply by controling the number of elements in the network.

1. INTRODUCTION AND PROBLEM FORMULATION

Image restoration is an inverse *ill-posed* problem, i.e. its solution is not unique and/or does not depend continuously on the data [2]. After discretization (and some kind of ordering, e.g. lexicographical), the *linear blur – additive white Gaussian noise* problem, for $M \times N$ pixels images, can be stated as follows:

$$\text{Given } \mathbf{g} \in \mathbb{R}^{MN}, \quad \text{find } \mathbf{f} \in \mathbb{R}^{MN}, \quad \text{assuming that} \quad \mathbf{g} = \mathbf{A}\mathbf{f} + \mathbf{n}, \tag{1}$$

where $\mathbf{A}$ is a $MN \times MN$ matrix modelling the blur, $\mathbf{g}$ is the observed image MN-vector, $\mathbf{f}$ is the MN-vector (image) to be estimated, and $\mathbf{n}$ is the noise vector. In this discrete version, the problem's *ill-posedness* lies in the fact that $\mathbf{A}$ is not invertible and/or has very small singular values (ill-conditioned), its inversion (pseudo-inversion) being very noise sensitive [2]. Under a Bayesian framework, considering the original image as sample of a zero-mean Gauss-Markov random field, and adopting the *maximum a posteriori* (MAP) estimation criterion, one is led to the convex (i.e. having unique solution) optimization problem

$$\hat{\mathbf{f}} = \arg\min_{\mathbf{f}} \left\{ \mathbf{f}^T \mathbf{D} \mathbf{f} - \frac{1}{\sigma^2} \mathbf{f}^T \mathbf{A}^T \mathbf{g} \right\} \quad \text{with} \quad \mathbf{D} \equiv \frac{1}{2\sigma^2} \mathbf{A}^T \mathbf{A} + \mathbf{C}_\mathbf{f}^{-1}. \tag{2}$$

In (2), $\mathbf{C}_\mathbf{f}$ is the $MN \times MN$ covariance matrix of $\mathbf{f}$, σ^2 is the noise variance, and $\mathbf{D}$ is a symmetric positive definite matrix (SPDM) with positive diagonal elements [4]. Notice that solving (2) is equivalent to solving the system $\mathbf{D}\mathbf{f} = \mathbf{h}$, with $\mathbf{h} \equiv \mathbf{A}^T \mathbf{g}/(2\sigma^2)$. A Tikhonov regularization approach would lead to a similar problem, with a different interpretation of matrix $\mathbf{C}_\mathbf{f}$ [4].

2. PREVIOUS WORK

Hopfield-like neural networks have been recently used to solve optimization problems involved in signal and image restoration [1, 7, 8, 9, 10]. The commonly used strategy is the mapping of the function to be minimized into the energy of a Hopfield-type network in order to exploit its energy descent ability. However, none of the mentioned work was able to successfully map minimization problem (2) into a fully standard Hopfield network (binary elements, threshold-based updating rule, symmetric interconnection weights, and zero auto-connections) to invoke its well known convergence properties. The modifications proposed include graded elements [7], additional energy reduction checking [8, 10], and modified updating rules using several threshold levels [1, 9].

3. PROPOSED APPROACH

3.1 Introduction

The key feature of algorithms suitable for neural implementation is a local updating rule, i.e. each element only needing to access its neighbors in order to update its state. In [4], we proposed a sequential scheme and proved its equivalence to the Gauss-Seidel algorithm [3] to solve $\mathbf{Df} = \mathbf{h}$. Convergence is guaranteed since $\mathbf{D}$ is a SPDM with strictly positive diagonal elements [4]. Its neural implementation, introduced in [5], is as follows:

Network 1 *Define a network of MN real valued elements $\{f_i, i = 1, 2, \ldots MN\}$ each assigned to one image pixel. Let $W_{ij} = -D_{ij}/D_{ii}$ be the interconnection strength between elements i and j, and $I_i = h_i/D_{ii}$ be the bias input to element i. Let a cyclic sequential visiting schedule to the elements be adopted, and the network be initialized with any finite state. At time t, element i, chosen according to the visiting schedule, updates its state,*

$$f_i(t+1) = f_i(t) + u_i(t) \quad \text{with} \quad u_i(t) = \sum_{j=1}^{MN} W_{ij} f_j(t) + I_i. \tag{3}$$

The cyclic schedule can be replaced by any schedule that visits all elements (e.g. random) with convergence still being guaranteed [3].

3.2 An auxiliary subnetwork

To obtain a standard binary Hopfield network we adopt the BDC scheme in which each element of $\mathbf{f}$ is represented as the sum of a set of L binary elements

$$f_i(t) = \frac{1}{L} \sum_{j=1}^{L} x_{ij}(t), \quad \text{for } i = 1, 2, \ldots M, \tag{4}$$

where $x_{ij} \in \{0, 1\}$, as proposed in [10]. Each f_i takes values in a bounded and discrete set, $f_i \in \{0, \frac{1}{L}, \frac{2}{L}, \ldots 1\}$. The bounded nature of the set poses no convergence problems if we assume that all the solution vector coordinates belong to the interval $[0, 1]$; other bounds are simply a matter of renormalization. The discrete nature of the set has numerical precision implications that will be studied in Section 3.4.

In [6], we introduced an interconnected set of binary elements that "knows" how to evolve in order to converge to a BDC representation of its input:

Network 2 *Assume a standard Hopfield network of L binary (0 or 1) elements $\{x_i,\ i = 1, 2, ..., L\}$. Let the autoconnection weights be zero and all other interconnection weights be equal to $-1/L$. All elements have a bias input of $b - 1/(2L)$ (assume that $b \in [0; 1]$), and follow the standard Hopfield updating rule.*

It is easy to show that Network 2 evolves until exactly $\mathcal{R}(bL)$ elements are in state 1, where $\mathcal{R}(x)$ stands for the nearest integer to x.

3.3 The proposed network

The complete binary network is built by replacing each continuous element f_i of Network 1 by a subnet equal to Network 2, with each f_i represented as in (4). The following network is obtained:

Network 3 *Define a network of MNL binary valued elements $\{x_{ij},\ i = 1, 2, \ldots MN,\ j = 1, 2, \ldots L\}$. Let $T_{ij,kl}$, the interconnection weight between elements x_{ij} and x_{kl}, be given by*

$$T_{ij,kl} = \begin{cases} 0 & \Leftarrow i = k \text{ and } j = l, \\ -1/L & \Leftarrow i = k \text{ and } j \neq l, \\ W_{ik}/L & \Leftarrow i \neq k, \end{cases} \quad (5)$$

and the bias input to each element be $H_{ij} = I_i - 1/(2L)$, where I_i is as given in Network 1. The network follows the standard Hopfield updating rule.

Convergence is guaranteed by the fact that the autoconnections are zero and the connections are symmetric. Each time an elements x_{mn} changes state, the value $f_m = \frac{1}{L}\sum_{j=1}^{L} x_{mj}(t)$ moves one $1/L$ step in the direction of the BDC representation of $f_i(t+1)$, as would be given by *Network 1*. The steps of *Network 3* follow the same direction of those of *Network 1*, towards the solution, but with $1/L$ length. In conclusion, the binary network is a $1/L$ step size version of *Network 1*. Fault-tolerance is assured since the BDC scheme has many possible solutions, as is clear from (4). In other words, the energy of the binary network has many minima, all corresponding to the unique solution of (2).

3.4 The effect of finite numerical precision

Given the finite numerical precision implied by the quantization effect of the BDC scheme, the fixed points of *Network 3* may not coincide with those of *Network 1*, which are exactly $\mathbf{D}^{-1}\mathbf{h}$. Using known inequalities involving matrix and vector norms, we conclude that the root mean-squared error (RMSE) of the fixed points $\mathbf{f}^f$ of Network 3, with respect to the exact solution $\mathbf{D}^{-1}\mathbf{h}$ (not to the true original image) is upper bounded as

$$RMSE \equiv \frac{\|\mathbf{f}^f - \mathbf{D}^{-1}\mathbf{h}\|_2}{\sqrt{MN}} < \frac{\kappa(\mathbf{D})}{2L}. \quad (6)$$

The bound depends, in a natural way, on the number of quantization levels and on $\kappa(\mathbf{D})$, the condition number of $\mathbf{D}$ which measures the numerical difficulty in computing $\mathbf{D}^{-1}$ [3].

4. EXAMPLES

In Figure 1, the proposed scheme is applied to two degraded versions of the same image. In both cases a restored window in the center of the image is presented. In example (a), although not artificially added, noise with standard deviation $\sigma = 1.0$ was assumed to assure stability.

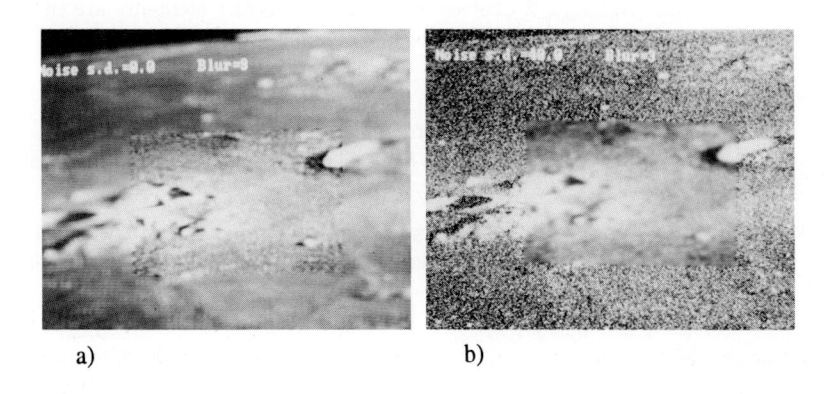

a) b)

Figure 1: (a) Blur: 9×9 uniform; noise $\sigma = 0.0$. (b) Blur: 3×3 Gaussian; noise $\sigma = 40.0$.

REFERENCES

1. J. Abbiss, B. Brames, and M. Fiddy. "Superresolution algorithms for a modified Hopfield neural network". *IEEE Trans. on Signal Proc.*, vol. 39, pp.1516–1523, 1991.
2. H. C. Andrews and B. R. Hunt. *Digital Image Restoration*. Prentice Hall, New Jersey, 1977.
3. D. Bertsekas and J. Tsitsiklis. *Parallel and Distributed Computation. Numerical Methods*. Prentice Hall, New Jersey, 1989.
4. M. Figueiredo and J. Leitão. "Sequential and parallel iterative image restoration". *Proc. of the 13th GRETSI Symposium*, pp. 789-792, Juan-les-Pins, 1991.
5. M. Figueiredo and J. Leitão. "Image restoration using neural networks". *Proc. of ICASSP'92*, pages II–409 – II–412, S. Francisco, 1992.
6. M. Figueiredo and J. Leitão. Signal restoration using a standard Hopfield neural network. In *NATO-ASI Acoust. Sign. Proc. for Ocean Exploration*, Kluwer, 1992.
7. J. Paik and A. Katsaggelos. "Image restoration using the Hopfield network with nonzero autoconnections". *Proc. of ICASSP'90*, pp.1909–1912, Albuquerque, 1990.
8. J. Paik and A. Katsaggelos. "Image Restoration Using A Modified Hopfield Network". *IEEE Trans. on Image Proc.*, vol. 1, pp.49-63, 1992.
9. S. Yeh, H. Stark, and M. Sezan. "Hopfield-type neural networks". In A.Katsaggelos, ed., *Digital Image Restoration*, pages 57–88. Springer Verlag, 1991.
10. Y. Zhou, R. Chellappa, A. Vaid, and B. Jenkins. "Image restoration using a neural network". *IEEE Trans. on ASSP*, vol. 36, pp. 1141–1151, 1988.

Associative reduction of dimensionality in image classification

D. Anguita, G.C. Parodi, and R.Zunino

DIBE - Dept. of Biophysical and Electronic Engineering - University of Genova
Via all'Opera Pia 11a, 16145 Genova, Italy

Abstract

The reduction of data dimensionality is a key issue in all applications involving image recongition. Associative techniques can contribute to this crucial task by providing robustness and generalization ability. After briefly outlining the basic models adopted, the paper presents several different architectures exploiting Associative Memories and Neural Networks, also evidencing the benefits deriving from hybrid integrated schemata.

1. INTRODUCTION

In classical approaches to image recognition, the reduction of data dimensionality involves some transformation in the representation of visual inputs; multiresolution techniques and feature extraction provide typical examples. In all cases, however, data representation must be decided a priori and it usually needs to be tuned to application domains. Although this might improve a system's effectiveness, it also affects the generality of the underlying method; in addition, specific representation techniques (e.g., transforms) might suffer from intrinsic noise-sensitivity, whcih should add to domain-originated noise sources - this lead to the search for robust techniques in visual recognition tasks.

In this context, the contribution of associative methodologies is interesting from a two-fold perspective: 1) the possibility of being trained by examples broadens the range of a system's applicability, still maintaining compatibility with other representation methods; 2) modelling associative behaviour results in a strong improvement of a systems' robustness. Not every associative model, however, can virtually fit visual domains: in the case of Neural Networks (NNs), for example, generalization constraints prevent direct applications to large-size images; on the other hand, Associative Memories (AMs) can handle high-dimensional patterns [1] but exhibit poor generalization ability and, ultimately, small storage capacity.

The research presented in this paper stems from such framework and aims at defining effective structures for visual tasks in real applications. Awareness of advantages and limitations of individual techniques lead to regard integration of different methodologies as a possible solution to the technical tradeoff involved. The overall result is a set of unconventional "fully associative" architectures to cope with image classification tasks. A special attention has been pai to the overall flexibility, which should ensure compatibility with other approaches.

Section 2 outlines the associative models adopted, whereas different experimental solutions are illustrated in Section 3. Section 4 draws concluding remarks.

2. THE ASSOCIATIVE MODELS

Two basic associative models characterize the presented research: the noise-like coding associative memory [2] and the feed-forward schema of neural network [3]. The former is entrusted with high-dimensional associative data processing, providing robustness by the memory's pattern-completion ability; the latter exploits generalization ability to achieve 'smart' performances in classification and invariant behaviour.

2.1. The Associative Memory

The noise-like coding model of associative memory [2] exploits convolution and correlation as complementary operators for memory writing and reading, respectively. Specific conditions characterize key patterns: zero-mean value, normalization, statistical independence among a key elements, and mutual uncorrelation among keys associated with different stored patterns. As a result of these conditions, keys appear as containing gaussian noise, whence the model's name. Some methods for key-generation are described in [4]. The memory is the sum of single key-image convolutions; it has been shown [3] that under the above conditions, correlation of a key with memory provides selectivity in information retrieval:

$$M = \sum_j K_j * I_j \quad \text{(memory writing)} \tag{1}$$

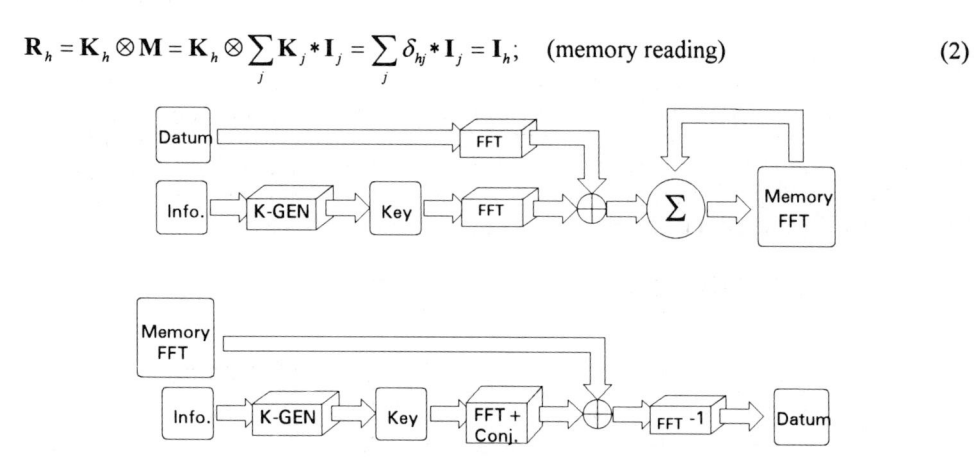

$$R_h = K_h \otimes M = K_h \otimes \sum_j K_j * I_j = \sum_j \delta_{hj} * I_j = I_h; \quad \text{(memory reading)} \tag{2}$$

Fig.1 - Schematic representation of the memory writing and reading processes

The model has been selected for the following reasons: 1) the simple mathematics involved can be easily computed using massive use of FFT-techniques, thus allowing implementations on parallel architectures for real-time applications; 2) the theoretical conditions underlying the memory's functioning seem more general than those imposed by other memory models [1].

2.2. The Neural Network

The neural structure adopted is the classical three-layer feedforward network; inputs to the neural net are general bit strings, whereas outputs can be either the same bit string (in which case the neural net operates as filter), or class representation (in which case the network operates as a classifier). The training algorithm for these networks is an improved version of standard back-prop named SuperSAB [3], providing notable convergence speed.

3. ASSOCIATIVE STRUCTURES

The presentation proceeds by increasing levels of integration and complexity, moving from simple classification schemata, whose primary goal was to evaluate the overall associative effectiveness, to fully integrated architectures, showing interesting invariant properties.

3.1. Associative Image Classification

The simplest way to approach image classification tasks by Associative Memories is to include them in the recognition module directly. After storing a set of "prototype" patterns in the memory using autoassociative information coding, classification involves: 1) addressing the memory with unrecognized images; 2) computing differences ("errors") between the recalled image and prototypes; 3) selecting the candidate issuing smallest error variance. The minimum-variance criterion is equivalent to a minimum-distance classifier: the AM operates as a sort of 'active filter', restoring incomplete information *before* computing pattern distances. This approach applies to classification of biomedical images [5], and, in general, to all domains requiring robust, "first-glance" selection of a few possible models. Comparative research [6] shows the advantages of associative classification over standard schemata (e.g., Hamming).

3.2. Message-based Reduction of Dimensionality

In the previous schema, all prototypes must be available at image-distance evaluation; this might be a significant drawback in terms of storage space. Explicit class representation can be avoided by a hetero-associative storage strategy [7] combined with a message-based reduction of dimensionality. The key idea is to store a set of couples (visual prototype, bit string) in the AM, thus exploiting the memory's robust contribution. Low-dimensional memory-recalled information can be supplied to the NN for intelligent, adaptive processing.

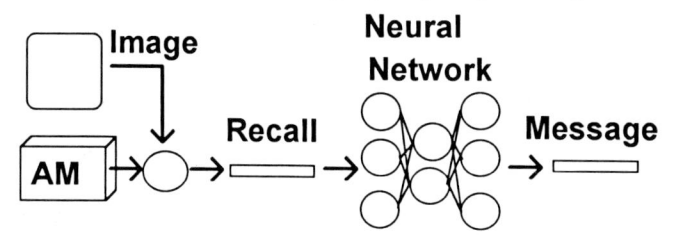

Fig.2 - The functional schema of message-based architectures

4. EXPERIMENTAL RESULTS

4.1. Robust Image Classification

The classifier's robustness has been evaluated by affecting reference images with increasing corrupting noise, and detecting the maximum tolerated damage percentage. To keep measures unbiased, binary images have been used. Results of over 6,000 runs exhibited a noise-insensitivity of about 80% on average [6], as opposed to the 60% robustness provided by standard 'passive' classifiers; the method's generality has been verified by using images drawn from different domains (e.g., geometrical shapes, human faces, cell nuslei photos, etc.).

4.2. Message-decoding Schema

As a first test for the message-based approach, the NN was trained to reproduce at the output the same message supplied in input; in such an autoassociative neural encoding, the NN operates as a filter that removes memory's inherent crosstalk noise. Massive experimentations to evaluate the efficacy of this nonlinear filtering involved the same image-corrupting procedure described previously; the test procedure stops when even one message bit is decoded incorrectly. Results proved that the neural contribution can increase the overall message-decoding performance up to 90% on average [7].

4.3. Scale-invariant Structure

The message-based schema can be easily generalized to an invariant classifier, by simply imposing that bit strings carry class information. The NN operates as a nonlinear, adaptive module, mapping memory-recalled messages into classes. In this experimentation, the system was trained, for each class, with images (shapes) drawn from a scale-graded sequence, and tested with images at scale factors not shown previously, checking classification outcome. On a total of 24 training images and 40 test images, we detected a 5% classification error percentage, which appears quite satisfactory, also considering that messages were randomly chosen and therefore the internal knowledge representation was unsupervised.

5. CONCLUSIONS

The role of associative techniques in image processing lies in providing robustness and generalization even in the crucial dimensionality-reduction phase. Future realted work includes the verification of other invariant properties, and applications to other domains (e.g., robotics).

6. REFERENCES

[1] G.E.Hinton, J.A.Anderson (eds.) Parallel Models of Associative Memory, Lawrence Erlbaum, New York, 1989
[2] S.Bottini "An algebraic model of an associative noise-like coding memory" Biological Cybernetics, No. 36, 1980, pp.221-228
[3] T. Tollenaere, SuperSAB: fast adaptive back propagation with good scaling properties, Neural Networks, vol.3, 1990, pp.561-573
[4] GC.Parodi, S.Ridella, R.Zunino, Distributed Key-generation Structures for Associative Image Classification, Proc.IEEE Symp.Circuits and Sys. ISCAS'92, SanDiego, 1549-1552
[5] A.Diaspro, GC.Parodi, R.Zunino, Classification of Optically-sectioned Images of Biopolymers by Means of Associative Noise-like Coding Memories, Studia Biophysica, vol.139, No.2, pp.69-76
[6] GC.Parodi, R.Zunino, Robust Associative Image-classification: a Comparative Analysis, in W.Forstner, S.Ruwiedel (eds.) Robust Computer Vision, Wichmann Press, Bonn, 1992, pp.204- 217
[7] A.Cerrato, GC.Parodi, R.Zunino, An Integrated Associative Structure for Vision, Proc. Int. Joint Conf. on Neural Networks IJCNN'92, Baltimore MD, June 1992, IEEE Press

AUTHOR INDEX

Aach T.	369	Carrai P.	201	Erenyi I.	413
Abrantes A.J.	91	Carrato S.	357	Ersoy O.K.	223
Aiazzi B.	119	Casini A.	119	Estola K.-P.	291
Aizawa K.	349	Cattoen M.	361	Fallside F.	295
Aldridge R.W.	309	Caviglia D.D.	353	Fazekas K.	127
Alparone L.	119	Chandrasekaran B.	29	Fazekas Z.	413
Alparone L.	271	Chella A.	429	Figueiredo M.A.T.	447
Anguita D.	451	Chen C.H.	25	Fini S.	321
Asselin de Beauville J.	143	Chen C.H.	213	Fioravanti S.	137
Astola J.	279	Chetverikov D.	159	Fitzgerald W.J.	239
Atiquzzaman M.	401	Chimienti A.	107	Flanders A.E.	325
Bangham J.A.	309	Chiuderi A.	321	Flusser J.	345
Barni M.	337	Clark R.	329	Fontana F.	423
Baronti S.	119	Clarke L.P.	329	Foresti G.L.	167
Baronti S.	271	Clarke L.P.	405	Frazer G.	325
Baseri R.	49	Clarke L.P.	409	Gabbouj M.	283
Bataouche S.	267	Clarke R.J.	99	Garbay C.	95
Bedini L.	263	Coble M.R.	31	Garzelli A.	389
Bellifemine F.	107	Cortelazzo G.M.	201	Giusto D.D.	137
Benelli G.	389	Corvi M.	103	Gonzalez C.F.	325
Berthod M.	79	Cosquer P.	361	Grattarola A.	131
Bertrand J.	333	Costamagna E.	227	Guertz A.M.	287
Bertrand P.	333	Csillag P.	127	Hall R.W.	243
Binaghi E.	393	Curinga S.	131	Hamdi M.	243
Blumenkrans A.	439	Dambra C.	435	Harris C.F.	111
Boroczky L.	127	De Vriendt J.	247	Hartley M.G.	401
Braccini C.	131	Dekany A.	413	Hatzinakos D.	189
Brelstaff G.J.	83	Dellepiane S.	423	Heikkonen J.	381
Brivio P.A.	385	Denasi S.	163	Higuchi T.	259
Broggi A.	147	DeNatale F.G.B.	137	Hinamoto T.	231
Bruyland I.	115	Desoli G.S.	137	Holka T.	413
Bundschuh B.	235	Destri G.	147	Hornegger J.	155
Bunke H.	303	DeVena L.	353	Huang Q.	173
Cappellini V.	321	DiRuberto C.	341	Huang T.S.	13
Cappellini V.	337	DiRuocco N.	341	Impey S.J.	309
Carlà R.	271	Doria I.	385	Iwama N.	377

Iwata Y.	259	Mertzios B.G.	219	Poggi G.	71
Jacovitti G.	313	Mertzios B.G.	251	Poggi G.	123
Jolion J.M.	205	Messmer B.T.	303	Pye C.J.	309
Jolion J.M.	267	Mian G.A.	201	Qian W.	329
Josephson J.	29	Mingyue D.	209	Qian W.	409
Jozwik A.	435	Minniti A.	313	Quaglia G.	163
Kalviainen H.	87	Mitchell D.G.	325	Raffo L.	353
Karasik Y.B.	317	Mitra S.K.	75	Rampini A.	393
Kaup A.	369	Mitzias D.	219	Redmill D.W.	67
Kawamata M.	259	Modestino J.W.	49	Regazzoni C.S.	167
Kilindris P.	193	Modestino J.W.	111	Roisin M.	333
Kim J.	19	Mohamed F.B.	325	Roli F.	435
King P.-S.	213	Monro D.M.	45	Saff E.	329
Kingsbury N.G.	67	Morris R.D.	239	Saito T.	63
Komatsu T.	63	Muneyasu M.	231	Saito T.	349
Komatsu T.	349	Munson D.C.	31	Salerno E.	263
Kunt M.	1	Murino V.	167	Salotti M.	95
Kunt M.	59	Najman L.	151	Samaria F.	295
Kunt M.	287	Najman L.	299	Santos-Victor J.	185
Lavagetto F.	131	Neri A.	313	Schettini R.	373
Leitao J.M.N.	447	Neuvo Y.	53	Schmitt M.	151
Leonardi R.	287	Neuvo Y.	279	Sentieiro J.	185
Li H.	329	Neuvo Y.	283	Serpico S.B.	417
Li H.	409	Ottaviani E.	103	Seshagiri S.	325
Li W.	59	Parodi G.C.	451	Silbiger M.L.	405
Lialios C.	177	Paulus D.W.R.	155	Simon S.	115
Lotti F.	119	Pedersini F.	397	Smucker A.J.M.	173
Lucier B.	329	Pendock N.	275	Solomonidis S.E.	365
Mackie J.C.H.	365	Pepe M.G.	263	Song X.	53
Marques J.S.	91	Pernot E.	299	Sorbello F.	429
Marshall S.	365	Petrou M.	197	Stockman G.C.	173
Matsopoulos G.K.	365	Picco R.	107	Suk T.	345
Mattioli M.	337	Pirrone R.	429	Sundaram R.	223
Mazzetti A.	393	Pitas I.	177	Suoranta R.	291
Mecocci A.	337	Pitas I.	193	Takemasa E.	63
Mecocci A.	389	Podda A.	227	Tang K.	279

Tasselli T.	313
Teranishi M.	377
Thami R.O.H.	143
Tonazzini A.	263
Trahanias P.E.	41
Tsirikolias K.	251
Tubaro S.	397
Turno A.	227
Urago S.	79
Vaillant R.	299
Vargiu A.	227
Velthuizen R.P.	405
Venetsanopoulos A.N.	41
Venetsanopoulos A.N.	189
Venturi G.	181
Viano G.	439
Vinitski S.	325
Vitulano S.	341
Wahl F.M.	209
Woods J.W.	19
Yang R.	283
Yao S.	99
Yin L.	283
You G.H.	213
Yu T.H.	75
Yusof S.	197
Zerubia J.	79
Zervakis M.E.	255
Zhang Y.	189
Zilioli E.	385
Zipoli S.	337
Zunino R.	451